Applied Survey Data Analysis

Highly recommended by the *Journal of Official Statistics, The American Statistician,* and other top statistical journals, *Applied Survey Data Analysis, Third Edition* provides an up-to-date overview of state-of-the-art approaches to the analysis of complex sample survey data. Building on the wealth of material on practical approaches to descriptive analysis and regression modeling from the first and second editions, this third edition further expands the topics covered and presents more step-by-step examples of modern approaches to the analysis of survey data using the newest statistical software procedures.

New to the Third Edition:

- Applied Bayesian methods for the analysis of complex sample survey data using available software implementing these methods
- State-of-the-art methods and software for the analysis of survey data collected from non-probability samples
- Software for modern applications of machine learning techniques to complex sample survey data
- A completely revamped website providing code for replicating all the analyses illustrated in the book using Stata, SAS, SPSS, R, Mplus, SUDAAN, WesVar, and IVEware
- New end-of-chapter exercises, allowing for practice implementing the methods, including Bayesian analysis exercises
- Updated summaries of the newest literature on the analysis of survey data collected from complex samples
- An updated review of software packages currently available for the analysis of complex sample survey data

Designed for readers working in a wide array of disciplines who conduct secondary analyses of survey data as part of their applied work, this book continues to provide a practical and accessible guide to the analysis of survey data. Continuing to use an example-driven approach to clearly illustrate analysis methods and software, the third edition contains many new examples and practical exercises based on recent versions of real-world survey data sets. Although the authors continue to use Stata for most examples in the text, they also offer the newest code for replicating the examples in other popular software packages on the book's revamped website.

Chapman & Hall/CRC
Statistics in the Social and Behavioral Sciences Series

Series Editors Jeff Gill, Steven G. Heeringa, Wim J. van der Linden, and Tom Snijders

Recently Published Titles

An Introduction to the Rasch Model with Examples in R
Rudolf Debelak, Carolin Stobl, and Matthew D. Zeigenfuse

Regression Analysis in R: A Comprehensive View for the Social Sciences
Jocelyn H. Bolin

Intensive Longitudinal Analysis of Human Processes
Kathleen M. Gates, Sy-Min Chow, and Peter C. M. Molenaar

Applied Regression Modeling: Bayesian and Frequentist Analysis of Categorical and Limited Response Variables with R and Stan
Jun Xu

The Psychometrics of Standard Setting: Connecting Policy and Test Scores
Mark Reckase

Crime Mapping and Spatial Data Analysis using R
Juanjo Medina and Reka Solymosi

Computational Aspects of Psychometric Methods: With R
Patricia Martinková and Adéla Hladká

Principles of Psychological Assessment
With Applied Examples in R
Isaac T. Petersen

Multilevel Modeling Using R, Third Edition
W. Holmes Finch, Jocelyn E. Bolin, and Ken Kelley

Polling, Prediction, and Testing, Second Edition
Ole J. Forsberg

Generalized Kernel Equating with Applications in R
Marie Wiberg, Jorge Gonzalez and Alina A. von Davier

Applied Survey Data Analysis, Third Edition
Brady T. West, Steven G. Heeringa, and Patricia A. Berglund

For more information about this series, please visit: https://www.routledge.com/Chapman--HallCRC-Statistics-in-the-Social-and-Behavioral-Sciences/book-series/CHSTSOBESCI

Applied Survey Data Analysis

Third Edition

Brady T. West, Steven G. Heeringa,
and Patricia A. Berglund

CRC Press
Taylor & Francis Group
Boca Raton London New York

CRC Press is an imprint of the
Taylor & Francis Group, an **informa** business
A CHAPMAN & HALL BOOK

Third edition published 2025
by CRC Press
2385 NW Executive Center Drive, Suite 320, Boca Raton FL 33431

and by CRC Press
4 Park Square, Milton Park, Abingdon, Oxon, OX14 4RN

CRC Press is an imprint of Taylor & Francis Group, LLC

ISBN: 9781032747286 (hbk)
ISBN: 9781032750408 (pbk)
ISBN: 9781003472094 (ebk)

DOI: 10.1201/9781003472094

Typeset in Palatino
by Apex CoVantage, LLC

***Brady T. West** dedicates this third edition to Laura, Carter, and Everleigh, in loving memory of a once-in-a-lifetime sabbatical adventure.*

***Patricia A. Berglund** dedicates this third edition to her ever-supportive UM-ISR professional family and to the memory of her late parents and sister, who always encouraged her to keep learning from others and growing intellectually.*

Contents

Bios

Brady T. West is a research professor in the Survey Methodology Program, located within the Survey Research Center at the Institute for Social Research on the University of Michigan–Ann Arbor (U-M) campus. He earned his PhD from the Michigan Program in Survey and Data Science in 2011. Before that, he received an MA in applied statistics from the U-M Statistics Department in 2002, being recognized as an outstanding first-year applied masters student, and a BS in statistics with highest honors and highest distinction from the U-M Statistics Department in 2001. His current research interests include the implications of measurement error in auxiliary variables and survey paradata for survey estimation, selection bias in surveys, responsive/adaptive survey design, interviewer effects, and multilevel regression models for clustered and longitudinal data. An author or co-author of more than 200 peer-reviewed publications in survey statistics, applied statistics, and public health, he is also the lead author of a book comparing different statistical software packages in terms of their mixed-effects modeling procedures (*Linear Mixed Models: A Practical Guide Using Statistical Software, Third Edition*, Chapman & Hall/CRC Press, 2022). He was elected as a fellow of the American Statistical Association in 2022. Brady lives in Dexter, Michigan, with his wife, Laura; his son, Carter; and his daughter, Everleigh.

Steven G. Heeringa is a research scientist emeritus at the University of Michigan Institute for Social Research (ISR) and the former associate director of the ISR Survey Research Center (SRC). He is a member of the faculty of the University of Michigan's Program in Survey and Data Science and the Joint Program in Survey Methodology. He is a fellow of the American Statistical Association and elected member of the International Statistical Institute. He is the author of many publications on statistical design and sampling methods for research in the fields of public health and the social sciences. Steve has over 48 years of statistical sampling experience in the development of the SRC national sample design, as well as research designs for ISR's major longitudinal and cross-sectional survey programs. Steve has collaborated extensively with scientific colleagues in the design and conduct of major studies in aging, psychiatric epidemiology, and physical and mental health. He has been a teacher of survey sampling and statistical methods to U.S. and international students and has served as a sample design consultant to a wide variety of international research programs based in countries such as Russia, Ukraine, Uzbekistan, Kazakhstan, India, Nepal, China, Egypt, Iran, the United Arab Emirates, Qatar, South Africa, and Chile.

Patricia A. Berglund is a semi-retired senior research associate in the Survey Methodology Program at the Institute for Social Research. She has extensive experience in the use of computing systems for data management and analysis of complex sample survey data. She works on research projects focused on pediatric health, youth substance abuse, adult mental health, and survey methodology using data from PediaTrac, Monitoring the Future, the National Comorbidity Surveys, and various other national and international surveys. In addition, she has been involved in the development and teaching of analysis courses and computer training programs at the U-M Survey Research Center–Institute for Social Research and previously lectured in the SAS Institute–Business Knowledge Series.

Introduction: Preface to the Third Edition

Applied Survey Data Analysis (ASDA), Third Edition, like the first two editions of this book, is written as a state-of-the-art guide to the applied statistical analysis and interpretation of survey data. When the editors initially raised the question of a third edition, we once again had to ask the important questions – "Is there sufficient new material?" and "Have there been important advances in theory, methods, and software?" – to justify an update to our 2017 second edition of *ASDA*. We agreed as a co-author team to keep a careful eye on new developments along all these dimensions after the publication of the second edition of *ASDA* and document these developments on the book's website. A quick look at the extensive lists of new publications and software developments on the website provided us with easy answers to these questions, and we proceeded with a proposal for this third edition in 2023. We thank the anonymous reviewers of this proposal for their constructive insights and suggestions, all of which have been incorporated into this third edition of *ASDA*.

Writing this third edition has not only provided us with an opportunity to update the previous material and communicate new advances in applied methods and software but also once again opened the door to covering new and emerging areas in the analysis of survey data. The third edition now includes:

- Applied Bayesian methods for the analysis of complex sample survey data (Chapter 3), including illustrations of alternative Bayesian approaches to existing analysis examples in all other chapters using available software
- Methods for the analysis of survey data collected from non-probability samples and applications of machine learning techniques to complex sample survey data sets, including worked examples using available software (Chapter 13)
- A completely revamped website for the book, containing updated code for each of the worked examples in the newest versions of each software package (https://websites.umich.edu/~surveymethod/asda/)
- New end-of-chapter exercises, including Bayesian examples, with the specific data sets required clearly indicated on the book's website, and solution sets prepared and available from the authors for any instructors using *ASDA* as a course textbook
- Updated summaries of important developments in the literature in all chapters
- An updated review of software procedures that are currently available for the analysis of complex sample survey data in all the major statistical software packages (Appendix A)

The chapters of this third edition continue to be organized in sequence, with each chapter building on material covered in the preceding chapters. Chapter 1 provides important context for the remaining chapters, briefly reviewing historical developments; discussing

DOI: 10.1201/9781003472094-1

the sensitive interplay in today's research environment between minimizing disclosure risk, differential privacy, and making survey data available to the public for research purposes; and laying out a step-by-step process for approaching a survey analysis problem. The updated and extended content of Chapters 2 through 4 of this third edition will introduce the reader to the fundamental features of **complex sample designs** and demonstrate how design characteristics such as stratification, cluster sampling, and weighting are easily incorporated into the statistical methods and software for survey estimation and inference. Chapter 3 provides a general overview of new developments in Bayesian approaches to the analysis of complex sample survey data that we will continue to refer to when considering alternative Bayesian approaches to the analysis examples.

Treatment of statistical methods for survey data analysis begins in Chapters 5 and 6 with coverage of univariate (i.e., single-variable) descriptive and simple bivariate (i.e., two-variable) analyses of continuous and categorical variables. Chapter 7 presents the linear regression model for continuous dependent variables. Generalized linear regression modeling methods for survey data are covered in Chapters 8 and 9. Chapter 10 pertains to methods for event history analysis of survey data, including models such as the Cox proportional hazards model and the discrete time hazard model, and provides new coverage of methods for interval censoring. Chapter 11 covers the analysis of longitudinal survey data, including new developments since the publication of the second edition. Chapter 12 introduces methods for handling missing data problems in survey data sets, including multiple imputation and design-based fractional imputation. Finally, the coverage of statistical methods for survey data analysis concludes in Chapter 13 with a discussion of new developments in advanced statistical techniques for survey data, including multilevel analysis, structural equation modeling, methods for the analysis of survey data from non-probability samples, machine learning, small-area estimation, and nonparametric methods. Each of these chapters presents alternative Bayesian approaches to the example analyses, when possible, with a focus on comparing the finite population inferences that arise when following the different approaches.

Survey data analysts continue to have access to many high-quality statistical software choices. We have once again emphasized Stata (Version 18) for most of the book examples, due to its ease of use and flexibility for survey data analysis, but this third edition integrates several new examples of Bayesian approaches using contributed packages in the freely available R software. We also continue to demonstrate the use of available software and code for SAS, SUDAAN, SPSS, and Mplus for specific types of analyses. All example analyses presented in the book chapters have once again been replicated to the greatest extent possible using the newest versions of the SAS, SPSS, SUDAAN, IVEware, R, Mplus, and WesVar software packages on the book's website (https://websites.umich.edu/~surveymethod/asda/). We have also provided information about available tools in Python for the analysis of survey data.

The *ASDA* website aims to provide survey analysts and instructors with additional resources in the following areas: links to new publications and an updated bibliography for the survey analysis topics covered in Chapters 5–13; links to sites for example survey data sets; replication of the code and output for the analysis examples in the Stata, SAS, SUDAAN, IVEWare, R, IBM SPSS, Mplus, and WesVar software; answers to frequently asked questions (FAQs); short technical reports related to special topics in applied survey data analysis; and reviews of statistical software updates and any resulting changes to the software code or output for the analysis examples. As was the case with previous editions of *ASDA*, we will strive to keep this website updated as new developments emerge.

We owe a debt to our many students from the Joint Program in Survey Methodology (JPSM), the Michigan Program in Survey and Data Science (PSDS), the Survey Research Center's Summer Institute for Survey Research Techniques, Coursera, Statistics.com, and Statistical Horizons who have studied with us over the years. We also owe a debt to our mentors and colleagues who, over the years, have instilled in us a passion for statistical teaching and consultation: Leslie Kish, Irene Hess, Graham Kalton, Morton Brown, Jim Lepkowski, Julian Faraway, Edward Rothman, Rod Little, Trivellore Raghunathan, Mike Elliott, Stas Kolenikov, and Yajuan Si. We wish to once again sincerely thank the staff at Chapman & Hall/CRC Press, especially Rob Calver, for their continued guidance and support over the years. The Survey Research Center at the University of Michigan's Institute for Social Research provided Dr. West with generous sabbatical time in early 2024 that was crucial for the development of this third edition, and for that he is very grateful. Finally, we express our sincere thanks and appreciation to our families for their love and support while work on this third edition was ongoing.

Brady T. West
Steven G. Heeringa
Patricia A. Berglund

Ann Arbor, Michigan
August 2024

1

Applied Survey Data Analysis: An Overview

1.1 Introduction

Our modern society—its governments, businesses, educational institutions, nonprofit organizations, and individuals—continues to apply the **survey method** as a principal tool for looking at itself—"a telescope on society," in the words of House et al. (2004). Scientific surveys designed to measure labor force participation and employment, earnings and expenditures, health and health care, commodity stocks and flows, business and agricultural production, environmental conditions, and many other topics are critically important to social scientists, health professionals, policymakers and administrators, and thus to society itself.

Since the first and second editions of this book were published in 2010 and 2017, respectively, the topics of "**big data**" and computationally intensive **data analytic methods**, along with the general field known as **data science**, have exploded onto the scene in the popular and business media as well as in scientific communications. This continues to leave many of us wondering what the future role will be for statistics in general and survey methodology and survey statistics in particular (Foster et al., 2016; Beatty, 2023). Indeed, since the publication of the second edition of *Applied Survey Data Analysis* (ASDA) in 2017, the Michigan Program in Survey Methodology (MPSM), where the authors of this book teach and consult regularly, changed its name to the Michigan Program in Survey and Data Science (PSDS), largely in response to the ubiquity of data science in all scientific fields that regularly work with data.

We now live in a world where scientific research often integrates survey data with multiple, novel sources of non-survey data to answer the pressing research questions facing society. Survey research that combines survey data, administrative records, and social media information is now commonplace, and the data science techniques taught in programs like PSDS are essential for modern survey researchers to master. While also continuing to focus on state-of-the-art analytic methods for **complex sample survey data**, this third edition of ASDA places more focus on analytic techniques needed by survey researchers for working with non-standard sources of data (e.g., data from non-probability samples), including model-based Bayesian approaches, and techniques frequently used by data scientists, such as machine learning.

We continue to adopt the viewpoints expressed by our colleagues: "[D]emand for survey statistics has never been higher," and "[s]urvey research is not dying, it is changing" (Groves, 2011; Couper, 2013; Groves et al., 2017; Kalton, 2019; Beaumont, 2020; Lyberg & Heeringa, 2021; Beatty, 2023). Survey research now relies on the blending of multiple sources of data, with some sources capturing the information directly collected from surveys, and other sources capturing administrative data or "found" data generated

DOI: 10.1201/9781003472094-2

from some process that is outside of the control of the survey researcher (such as social media data; Conrad et al., 2021). In fact, we can go further and say that since its origins in the early 20th century, the scientific survey method has undergone continual change: improvements in sampling design and methods, the emergence of new modes of data collection and new sources of data to enhance survey research, shifts in **essential survey conditions** (such as demands on survey respondents and declining response rates), and most importantly for readers of this book, an ongoing evolution of new statistical approaches and refinement of old approaches to the applied analysis of survey data. This third edition of ASDA continues to build on the foundational theory and methods of applied survey data analysis and, at the same time, introduces important new developments and tools (e.g., analytic methods for non-probability samples, Bayesian approaches to the analysis of survey data, machine learning techniques for surveys) that, in a constantly changing research environment, promise to continue updating and improving our statistical treatments of survey data.

The focus of this book will be on statistical methods—both old and new—for the analysis of **complex sample survey data** typically seen in large-scale scientific surveys, but the general approaches to survey data analysis and specific statistical methods described here should apply to all forms of survey data, including blended data sets that leverage multiple sources of information to enrich a survey data collection. To set the historical context for our present work, Section 1.2 briefly reviews the history of developments in theory and methods for applied survey data analysis, with a new focus on important ethical and data privacy considerations that are crucial for modern survey researchers conducting secondary analyses of survey data to understand. Section 1.3 provides some necessary background on the data sets that will be used for the analysis examples in Chapters 2–13. This short overview chapter then concludes in Section 1.4 with a general review of the sequence of steps required in any applied analysis of survey data.

1.2 A Brief History of Applied Survey Data Analysis

Today's survey data analysts will generally approach a problem armed with substantial background in the theory of survey statistics, at least some training in data science, a literature filled with empirical results, and high-quality software tools for the task at hand. However, before turning to the best methods currently available for the analysis of all forms of survey data, it is useful to look back at how we arrived at where we are today. The brief history described here is certainly a selected interpretation, chosen to emphasize the evolution of probability sampling design and related statistical analysis techniques that are most directly relevant to the material in this book. Readers interested in a comprehensive review of the history and development of survey research in the United States should see Converse (1987). Bulmer (2001) provides a more international perspective on the history of survey research in the social sciences. For the more statistically inclined, Skinner et al. (1989) provide an excellent review of the development of methods for descriptive and analytical treatment of survey data. A comprehensive history of the impacts of sampling theory on survey practice can be found in O'Muircheartaigh and Wong (1981).

1.2.1 Key Theoretical Developments

The science of survey sampling, survey data collection methodology, and the analysis of survey data dates back a little more than 100 years. By the end of the 19th century, an open and international debate established the **representative sampling method** as a statistically acceptable basis for the collection of observational data on populations (Kaier, 1895). Over the next 30 years, work by Bowley (1906), Fisher (1925), and other statisticians developed the role of randomization in sample selection and large-sample methods for estimation and statistical inference for simple random sample (SRS) designs.

The early work on the representative method and inference for simple random and stratified random samples culminated in a landmark paper by Jerzy Neyman (1934) which outlined a cohesive framework for estimation and inference based on estimated confidence intervals for population quantities that would be derived from the probability distribution for selected samples over repeated sampling. Following the publication of Neyman's paper, there was a major proliferation of new work on sample designs, estimation of population statistics, and variance estimation techniques required to develop confidence intervals for sample-based inference, or what in more recent times has been labeled **design-based inference** (Yates, 1949; Deming, 1950; Hansen et al., 1953; Sukatme, 1954; Kish, 1965; Cochran, 1977). House et al. (2004) credit J. Steven Stock (U.S. Department of Agriculture) and Lester Frankel (U.S. Bureau of the Census) with the first applications of area probability sampling methods for household survey data collections. Even today, the primary techniques for sample design and population estimation and inference developed by these pioneers and published during the period 1945–1975 remain the basis for almost all descriptive analysis of survey data.

The developments of the World War II years firmly established the probability sample survey as a tool for describing population characteristics, beliefs, and attitudes. Based on Neyman's theory of inference, survey sampling pioneers in the United States, Britain, India, and elsewhere developed optimal methods for sample design, estimators of survey population characteristics, and confidence intervals for population statistics. As early as the late 1940s, social scientists led by sociologist Paul Lazarsfeld of Columbia University began to move beyond using survey data to simply describe populations to using these data to explore relationships among the measured variables (see Kendall & Lazarsfeld, 1950 or Klein & Morgan, 1951). Skinner et al. (1989) and others before them labeled these two distinct uses of survey data as "descriptive" and "analytical." Hyman (1955) used the term "explanatory" to describe scientific surveys whose primary purpose was the analytical investigation of relationships among variables.

During the period 1950–1990, analytical treatments of survey data expanded as new developments in statistical theory and methods were introduced, empirically tested, and refined. Important classes of methods that were introduced during this period included log-linear models and related methods for contingency tables, generalized linear models (e.g., logistic regression), survival analysis models, general linear mixed models (e.g., hierarchical linear models), structural equation models, and latent variable models. Many of these new statistical techniques applied the method of maximum likelihood to estimate model parameters and standard errors of the estimates, assuming that the survey observations were *independent* observations from a known probability distribution (e.g., binomial, multinomial, Poisson, product multinomial, normal). As discussed in Chapter 2, data collected under most contemporary sample designs do not conform to the key assumptions of these methods.

As Skinner et al. (1989) point out, survey statisticians were aware that straightforward applications of these new methods to complex sample survey data could result in underestimates of variances and biased estimates of confidence intervals and test statistics. However, except in limited situations of relatively simple designs, exact determination of the size and nature of the bias (or a potential correction) was difficult to express analytically. Early investigations of such "design effects" were primarily empirical studies, comparing design-adjusted variances for estimates to the variances that would be obtained if the data were truly identically and independently distributed (equivalent to a simple random sample of equal size). Over time, survey statisticians developed special approaches to estimating these models that enabled the survey analyst to consider the complex characteristics of the sample design (e.g., Koch & Lemeshow, 1972; Kish & Frankel, 1974; Rao & Scott, 1981; Binder, 1983; Pfeffermann et al., 1998). These approaches (and related developments) are described in Chapters 5–13 of this text.

1.2.2 Key Software Developments

Development of the underlying statistical theory and empirical testing of new methods was obviously important, but the survey data analyst needed computational tools to apply these techniques. We can have nothing but respect for the pioneers who, in the 1950s, fitted multivariable regression models to survey data using only hand computations (sums, sums of squares, sums of cross products, matrix inversions, etc.), performed on a rotary calculator and possibly a tabulating machine (Klein & Morgan, 1951). The first Hewlett-Packard electronic calculator with square root, logarithm, and trigonometric functions was advertised in *Scientific American* in 1971. The origin of statistical software as we know it today dates to the 1960s, with the advent of the first mainframe computer systems. Software systems such as BMDP and OSIRIS, and later, SPSS, SAS, GLIM, S, and GAUSS, were developed for mainframe users; however, with limited exceptions, these major software packages did not include programs that were adapted for the analysis of complex sample survey data.

To fill this void during the 1970s and early 1980s, several stand-alone programs, often written in the Fortran language and distributed as compiled objects, were developed by survey statisticians (e.g., OSIRIS: PSALMS and REPERR, CLUSTERS, CARP, SUDAAN, and WesVar). By today's standards, these programs had a steep "learning curve," had limited data management flexibility, and typically supported only descriptive analysis (means, proportions, totals, ratios, and functions of descriptive statistics) and linear regression modeling of multivariable relationships. A review of the social science literature of this period shows that only a minority of researchers employed these special software programs when analyzing complex sample survey data, resorting instead to standard analysis programs with their default assumption that the data originated with a simple random sample of the survey population (see Chapter 3 for more on this type of "analytic error").

The appearance of microcomputers in the mid-1980s was quickly followed by a transition to personal computer versions of the major statistical software (BMDP, SAS, SPSS), as well as the advent of new statistical analysis software platforms (e.g., SYSTAT, Stata, S-Plus). However, except for specialized software (WesVar PC, PC CARP, PC SUDAAN, Micro-OSIRIS, CLUSTERS for PC, IVEware) that was often designed to read data sets stored in the formats of the larger commercial software packages, the microcomputing revolution still did not put tools for the analysis of complex sample survey data in the hands of most survey data analysts. Nevertheless, throughout the late 1980s and early 1990s, the scientific and commercial pressures to incorporate programs of this type into the major software packages were building.

Beginning with Version 6.12, SAS users had access to PROC SURVEYMEANS and PROC SURVEYREG, two new SAS procedures that permitted simple descriptive analysis and linear regression analysis for complex sample survey data. At about the same time, the Stata software for statistical analysis appeared on the scene, providing complex sample survey data analysts with the "svy" versions of the more important analysis commands. SPSS's entry into the world of complex sample survey data analysis came later with the introduction of the Complex Samples add-on module in Version 13. Procedures for the analysis of complex sample survey data are also now available in the contributed survey package for the R software (Lumley, 2010), along with many other contributed packages that work together with the survey package, and in the MPlus software (Muthén & Muthén, 1998–2017). Today's data scientist can even find a library of routines for analyses of complex sample survey data in the Python software, where the samplics library is now available (https://pypi.org/project/samplics/; see also Paczkowski, 2022). Appendix A of this third edition of ASDA presents updated coverage of the capabilities of all these different software packages in detail.

The survey researcher who sits down today on their personal computer, laptop, or tablet has access to powerful statistical software, high-speed processing, and high-density data storage capabilities that the analysts in the 1970s, 1980s, and even the 1990s could not have envisioned. All these recent advances have brought us to a point at which today's survey analyst can approach both simple and complex problems with the confidence gained through a fundamental understanding of the theory, empirically tested methods for design-based estimation and inference, and software tools that are sophisticated, accurate, and easy to use.

1.2.3 The Modern Era: Disclosure Risk, Differential Privacy, and Survey Data

The recent explosion of research into "big data" and "found" sources of organic data, which, as discussed earlier, are frequently blended with survey data, has also increased the concern of the general public about too much of their information being available for others to see and analyze. In some cases, this concern has resulted in new legislation and regulations designed to protect individuals from invasion of privacy (Paterson & McDonagh, 2018). The unfortunate result of this may be increased reluctance of the public to participate in honest, scientific survey data collections and share yet more of their information with unfamiliar outside parties and organizations: "You already have access to so much of my information, so why should I share even more of it with you?" Now more than ever, survey data collectors place the utmost importance on protecting respondent confidentiality and placing restrictions on exactly what data from a particular survey should be available for sharing in the public domain. Taking important steps to protect the integrity of survey research is crucial for all individuals who either collect or analyze survey data (Jamieson et al., 2023).

These societal trends have important implications for researchers who frequently perform secondary analyses of large survey data sets in the public domain, many of which arise from complex probability samples. A vast literature has emerged discussing **differential privacy** in the social sciences (Oberski & Kreuter, 2020), creation of **synthetic data** (Raghunathan, 2021), and other related techniques for perturbing collected (and possibly blended) survey data sets to limit **disclosure risk** (Shlomo, 2023; NASEM, 2024), preventing data snoopers from identifying survey respondents. Data users are frequently provided with data sets that have been manipulated in some way to maintain the essential distributional features of the data but prevent identification of survey respondents, and are responsible for being aware of the perturbation techniques used and the implications of them for analysis (e.g., generation of multiple synthetic data sets followed by subsequent applications of multiple imputation techniques; see Chapter 12). Or in other cases, data users may need to access survey data sets in restricted physical (PDE) or virtual (VDE) data enclaves. In these settings, it is important for the data user to ensure that appropriate statistical software is available in the secure access locations for analyzing the (restricted) survey data, including routines provided by the data producer for accounting for any perturbations introduced in the collected data.

The implementation of all these methods for limiting disclosure risk does have the unfortunate result of making survey data sets less "usable" than they have been in previous years. Hotz et al. (2022) discuss the importance of these ideas related to differential privacy and disclosure risk, and the sensitive trade-off with making survey data more widely available for advancing knowledge and contributing to the public good. In short, the research-related usability costs of these limiting procedures need to be carefully weighed against the benefits of these procedures (in terms of maintaining protection of respondents and limiting actual disclosure risk) and empirical evidence of the actual risk of disclosure associated with any given survey data collection (Shlomo & Skinner, 2022; NASEM, 2024). This is a delicate balance to consider and will likely be a source of debate for years to come. Readers of this third edition of ASDA just need to be aware that maintaining respondent confidentiality is crucial for high-quality survey data collection to continue as a research tool, and that the methods briefly discussed here are critical for allowing this to happen.

Now that we have had a glimpse at our history, let us begin our study of applied survey data analysis.

1.3 Example Data Sets and Exercises

Examples based on the analyses of real survey data sets are routinely used in this book to demonstrate statistical methods and software applications. To ensure diversity in sample design and substantive content, example exercises and illustrations in this third edition are largely drawn from four major U.S. and international survey data sets.

We continue to use many of the same data sets that were covered in the second edition of ASDA, primarily to maintain consistency in the examples and the end-of-chapter exercises. While newer versions of public-use data sets are clearly available for some of these studies at this point, the essential complex sampling features of these studies and the analytic approaches recommended by the producers of these data sets have changed very little since the publication of the second edition. As a result, the methods illustrated in this third edition can easily be applied to newer versions of each of these data sets once the data user has identified the appropriate variables containing the complex sampling features.

1. The National Comorbidity Survey Replication (NCS-R)

The NCS-R was a 2002 study of mental illness in the U.S. household population ages 18 and over funded by the U.S. National Institutes of Health (NIH). The core content of the NCS-R is based on a lay-administered interview using the World Health Organization (WHO) CIDI (Composite International Diagnostic Interview) diagnostic tool, which is designed to measure primary mental health diagnostic symptoms, symptom severity, and use of mental health services (Kessler et al., 2004). The NCS-R was based on interviews with randomly chosen adults in an equal-probability, multi-stage sample of households selected from the University of Michigan National Sample master frame. The survey response rate was 70.9%. The survey was administered in two parts: a Part I core diagnostic assessment of all respondents (n = 9,282), followed by a Part II in-depth interview with 5,692 of the 9,282 Part I respondents, including all Part I respondents who reported a lifetime mental health disorder and a probability subsample of the disorder-free respondents in the Part I screening.

Although somewhat dated, we continue to use the 2002 NCS-R as an example data set for this third edition for the following reasons: (a) the scientific content and, in particular, its binary measures of mental health status; (b) the multi-stage design with primary stage stratification and cluster sampling typical of many large-scale public-use survey data sets; and (c) the two-phase aspect of the data collection.

2. The Health and Retirement Study (HRS), 2012

The Health and Retirement Study (HRS) is a longitudinal study of the American population 50 years of age and older. Beginning in 1992, the HRS has collected data every two years on a longitudinal panel of sample respondents born between the years of 1931 and 1941. Originally, the HRS was designed to follow this probability sample of age-eligible individuals and their spouses or partners as they transitioned from active working status to retirement, measuring aging-related changes in labor force participation, financial status, physical and mental health, and retirement planning. The HRS observation units are age-eligible individuals and "financial units" or couples in which at least one spouse or partner is HRS-eligible. Beginning in 1993 and again in 1998, 2004, 2010, 2016, and 2022, the original HRS 1931–1941 birth cohort panel sample was augmented with probability samples of U.S. adults and spouses/partners from (a) pre-1924 (added in 1993); (b) 1924–1930 and 1942–1947 (added in 1998); (c) 1948–1953 (added in 2004); (d) 1954–1959 (added in 2010); (e) 1960–1965 (added in 2016–2017); and (f) 1966–1971 (added in 2022–2024). The baseline response rates for the original 1992 cohort sample exceeded 80%. The corresponding baseline survey response rates for the 1993–2016 new cohort updates ranged from 60% to 80%. In 2012, the HRS interviewed over 20,500 eligible sample adults age 50+ in the composite panel.

The HRS samples were primarily identified through in-person screening of large, multi-stage area probability samples of U.S. households. For the pre-1931 birth cohorts, the core area probability sample screening was supplemented through sampling of age-eligible individuals from the U.S. Medicare Enrollment Database. Sample inclusion probabilities for HRS respondents vary slightly across birth cohorts and are approximately two times higher for African Americans and Hispanics. Data from the 2012 wave of the HRS panel are used for many of the examples in this book, and we consider a longitudinal analysis of multiple waves of HRS data in Chapter 11.

3. The National Health and Nutrition Examination Survey (NHANES), 2011–2012

Sponsored by the National Center for Health Statistics (NCHS) of the Centers for Disease Control (CDC), the NHANES is a survey of the adult, non-institutionalized population of the United States. The NHANES is designed to study the prevalence of major disease in the U.S. population and monitor the change in prevalence over time, as well as trends in treatment and major disease risk factors, including personal behaviors, environmental exposure, diet, and nutrition. The NHANES survey includes both an in-home medical history interview with sample respondents and a detailed medical examination at a local mobile examination center (MEC).

The NHANES surveys were conducted on a periodic basis between 1971 and 1994 (NHANES I, II, III), but beginning in 1999, the study transitioned to a continuous interviewing design. Since 1999, yearly NHANES data collections have been performed in a multi-stage sample that includes 15 primary stage unit (PSU) locations, with new sample PSUs added in each data collection year. Approximately 7,000 probability sample respondents complete the NHANES in-home interview phase each year, and roughly 5,000 of these individuals also consent to the detailed MEC examination. To meet specific analysis objectives, the NHANES oversamples low-income persons, adolescents between the ages of 12 and 19, persons age 60 and older, African Americans, and Hispanics of Mexican ancestry. To ensure adequate precision for sample estimates, NCHS recommends pooling data for two or more consecutive years of NHANES data collection. The NHANES example analyses provided in this text are based on the combined data collected in 2011 and 2012. The unweighted response rate for the interview phase of the 2011–2012 NHANES was approximately 72%.

4. The European Social Survey (ESS): Russian Federation, Round 6, 2012

The European Social Survey (ESS) is a multi-national survey conducted in over 30 European nations. ESS is jointly funded by the European Commission (EC), the European Science Foundation, and collaborating academic institutions in the participating countries. The ESS follows a **repeated cross-sectional survey design** (see Chapter 11), with independent samples of persons from a given country measured in each survey year. Each country's ESS questionnaire includes a common core module of questions focusing on social, political, economic, cultural, and religious topics, as well as several special modules of questions that rotate across rounds of ESS data collection. The inclusion of the ESS data set for example exercises in this third edition of ASDA reflects not only the importance of this multi-national data resource to comparative social science research in Europe but also a deliberate decision to include analysis topics that are not so health research–focused, such as those based on the NHANES, HRS, and NCS-R data sets.

Beginning with round 1 in 2002, a new round of ESS sampling and interviewing has been conducted every two years. The examples in this book are based on data from the round 6 ESS: Russian Federation (ESS:RF) survey that was conducted in 2012. The sample design for round 6 of the ESS:RF survey can be described as a stratified, clustered, multi-stage probability sample of individual respondents 15 years of age and older. The round 6 ESS:RF response rate was just under 70%. Interviews were completed with $n = 2{,}484$ sample respondents.

Each record in the ESS:RF round 6 data file includes a STRATIFY variable containing codes for eight (8) primary stage strata geographically defined by "okrug" administrative units. A total of 184 primary stage sampling units (PSUs) were selected from the eight strata, with multiple PSUs selected per stratum. Codes for these 184 primary stage clusters are contained in the PSU variable. Each data record also contains a sample selection weight, DWEIGHT, and a post-stratified analysis weight (PSPWGHT). Unless otherwise stated, all ESS:RF analysis examples in this text will use the post-stratified weight (PSPWGHT). ESS public-use data files also include a population size weight (PWEIGHT) intended to be used when analysts wish to combine data from multiple ESS country samples in a single analysis on a proportionate-to-population-size basis. Since our analysis examples will focus only on the round 6 ESS:RF data, this "combining" weight will not be used in this text.

Public-use versions of each of these four major survey data sets are available online. The companion website for this book provides the most current links to the official public-use data and documentation archives for each of these example survey data sets. We also provide reduced versions of the public-use data sets containing only the variables of interest in our examples for the convenience of our readers and any instructors using the end-of-chapter exercises for coursework.

New to this third edition of ASDA, we also include publicly available data from a pre-election political poll conducted in Wisconsin by ABC and the *Washington Post* in the fall of 2020 (prior to the 2020 presidential election in the United States), primarily to demonstrate the ability of various state-of-the-art methods to compute improved estimates of finite population parameters in non-probability samples and low-response-rate probability samples (of which this poll is one example). These methods generally rely on data from a reference probability sample that includes measures on the same covariates collected in the selected sample, and for this purpose we use data from the 2020 American National Election Study (ANES). We describe each of these data sets in more detail in Chapter 13, where we present a new case study demonstrating the implementation of these various methods. Reduced versions of each of these data sets new to the third edition, without any potentially identifying information and restricted to only the variables of interest for our examples, are also available on the updated ASDA website.

1.4 Steps in Applied Survey Data Analysis

Applied survey data analysis—both in daily practice and here in this book—is a process that requires more of the analyst than simple familiarity and proficiency with statistical software tools. It requires a deeper understanding of the sample design, the survey data (and any other data sources integrated with it), and the interpretation of the results generated by applying any statistical procedures. Following a more general outline for applied statistical analysis presented by Cox (2007), Table 1.1 outlines a sequence of six steps that are fundamental to applied survey data analysis, and we describe these steps in more detail later.

1.4.1 Step 1: Definition of the Problem and Statement of the Objectives

The first step involves a clear specification of the problem to be addressed and the formulation of objectives for the analysis exercise. For example, the "problem" may be ambiguity among physicians over whether there should be a lower threshold for prostate biopsy following prostate-specific antigen (PSA) screening in African American men (Cooney et al., 2001). The corresponding objective would be to estimate the 95th percentile and a

TABLE 1.1

Steps in Applied Survey Data Analysis

Step	Activity
1	Definition of the problem and statement of the objectives.
2	Understanding the sample design.
3	Understanding design variables, underlying constructs, and missing data.
4	Analyzing the data.
5	Interpreting and evaluating the results of the analysis.
6	Reporting of estimates and inferences from the survey data.

95% confidence interval for this quantity (+/–.2 ng/ml PSA) in a population of African American men. The estimated 95% confidence limits can, in turn, be used by medical experts to determine if the biopsy threshold for African American men should be different than for men of other race and ethnicity groups.

As described earlier, the problems to which survey data analyses may be applied span many disciplines and real-world settings. Likewise, the statistical objectives may vary. Historically, the objectives of most survey data analyses were to describe characteristics of a target population: its average household income, the median blood pressure of women, or the proportion of eligible voters who favor candidate X. But survey data analyses can also be used for decision-making: For example, should a pharmaceutical company recall its current products from store shelves due to a perceived threat of contamination? In a population case-control study, does the presence of silicone breast implants significantly increase the odds that a woman will contract a connective tissue disease, such as scleroderma (Burns et al., 1996)? In recent decades, the objective of many sample survey data analyses has been to explore and extend the understanding of multivariable relationships among variables in the target population. Sometimes, multivariable modeling of survey data is seen simply as a descriptive tool, defining the form of a functional relationship as it exists in a finite population. But it is increasingly common for researchers to use observational data from complex sample surveys to probe causality in the relationships among variables.

1.4.2 Step 2: Understanding the Sample Design

The survey data analyst must understand the sample design that was used to collect the data that they are about to analyze. Without an understanding of key properties of the sample design, the analysis may be inefficient, be biased, or otherwise lead to incorrect inference. An experienced researcher who designs and conducts a randomized block experimental design to test the relative effectiveness of new instructional methods should not proceed to analyze the data as a simple factorial design, ignoring the blocking that was built into her experiment. Likewise, an economics graduate student who elects to work with the HRS data should understand that this national probability sample of older adults includes stratification, cluster sampling, and disproportionate sampling (i.e., compensatory population weighting), and that these design features may require special approaches to population estimation and inference.

At this point, we may have discouraged the reader into thinking that an in-depth knowledge of sample design is required to work with survey data or that they may need to re-learn what was learned in general courses on applied statistical methods. This is not the case. Chapters 2 through 4 will introduce the reader to the fundamental features of **complex sample designs** and demonstrate how design characteristics such as stratification, cluster sampling, and weighting are easily incorporated into the statistical methods and software for survey estimation and inference. Chapters 5–13 will then show the reader that relatively simple extensions of their current knowledge of applied statistical analysis methods provide the necessary foundation for efficient and accurate analysis of the data collected in sample surveys.

1.4.3 Step 3: Understanding Design Variables, Underlying Constructs, and Missing Data

The typical scientific survey data set is **multi-purpose**, with the final data sets often including hundreds of variables that span many domains of study—income, education, health, family. The sheer volume of available data and the ease by which it can be accessed can cause survey data analysts to become complacent in their attempts to fully understand the properties of the data that are important to their choice of statistical methods and the conclusions that they will ultimately draw from their analysis. Step 2 described the importance of understanding the sample design. In the survey data, the key features of the sample design will be encoded in a series of **design variables.** Before analysis begins, some simple questions need to be put to the candidate data set: What are the empirical distributions of these design variables, and do they conform to the design characteristics outlined in the technical reports and online study documentation? Does the original survey question that generated a variable of interest truly capture the underlying construct of interest? Are the response scales and empirical distributions of responses and independent variables suitable for the intended analysis? What is the distribution of missing data across the cases and variables, and is there a potential impact on the analysis and the conclusions that will be drawn? Chapter 4 discusses techniques for answering these and other questions before proceeding to statistical analysis of the survey data.

1.4.4 Step 4: Analyzing the Data

Finally, we arrive at the step to which many researchers rush to enter the process. We are all guilty of wanting to jump ahead. Identifying the problem and objectives seems intuitive. We tell ourselves that formalizing that step wastes time. Understanding the design and performing data management and exploratory analysis to better understand the data structure are boring. After all, the statistical analysis step is where we obtain the results that enable us to describe populations (through confidence intervals), to extend our understanding of relationships (through statistical modeling), and possibly even to test scientific hypotheses.

In fact, the statistical analysis step lies at the heart of the process. Analytic techniques must be carefully chosen to conform to the analysis objectives and the properties of the survey data. Specific methodology and software choices must accommodate the design features that influence estimation and inference. Treatment of statistical methods for survey data analysis begins in Chapters 5 and 6 with coverage of univariate (i.e., single-variable) descriptive and simple bivariate (i.e., two-variable) analyses of continuous and categorical variables. Chapter 7 presents the linear regression model for continuous dependent variables, and generalized linear regression modeling methods for survey data are treated in Chapters 8 and 9. Chapter 10 pertains to methods for event-history analysis of survey data, including models such as the Cox proportional hazards model and discrete time logistic models. Methods of longitudinal analysis for complex sample survey data are addressed in Chapter 11. Chapter 12 introduces methods for handling missing data problems in survey data sets. Finally, the coverage of statistical methods for survey data analysis concludes in Chapter 13 with a discussion of survey applications of specialized statistical models and techniques, including methods for non-probability samples and machine learning.

1.4.5 Step 5: Interpreting and Evaluating the Results of the Analysis

Knowledge of statistical methods and software tools is fundamental to success as an applied survey data analyst. However, setting up the data, executing code, and printing the results are not sufficient to constitute a thorough treatment of the analysis problem. Likewise, scanning a column of *p*-values in a table of regression model output does not inform us concerning the form of the "final model," or even the pure relationship of a single predictor with the outcome of interest. As described in Step 3 earlier, interpretation of the results from an analysis of survey data requires a consideration of the error properties of the data. Variability of sample estimates will be reflected in the **sampling errors** (i.e., the standard errors used to compute confidence intervals and test statistics) estimated in the course of the statistical analysis. **Nonsampling errors**, including potential bias due to survey nonresponse and item missing data, cannot generally be estimated from the survey data (Lessler & Kalsbeek, 1992). However, it may be possible to use ancillary data to explore the potential directions and magnitudes of such errors. For example, an analyst working for a survey organization may statistically compare survey respondents with nonrespondents in terms of known correlates of key survey variables that are readily available on the sampling frame to assess the possibility of nonresponse bias.

As survey data analysts have pushed further into the realm of multivariable modeling of survey data, care is required in interpreting fitted models. Is the model reasonably identified, and do the data meet the underlying assumptions of the model estimation technique? Are there alternative models that explain the observed data equally well? Is there scientific support for the relationship implied in the modeling results? Are interpretations that imply causality in the modeled relationships supported (Rothman, 1988)?

1.4.6 Step 6: Reporting of Estimates and Inferences from the Survey Data

The end products of applied survey data analyses are reports, papers, or presentations designed to clearly communicate the findings to fellow scientists, policy analysts, and administrators/decision-makers. This book includes discussions of standards and proven methods for effectively presenting the results of applied survey data analyses, including table formatting, statistical contents, and the use of statistical graphics.

With these six steps in mind, we can now begin our updated walk through the process of planning, formulating, and conducting analysis of survey data.

2

Getting to Know the Complex Sample Design

2.1 Introduction

The first step in any applied analysis of survey data involves defining the research questions that will be addressed. The next step is to study and understand the sampling design that generated the sample of elements (e.g., persons, businesses) from the target population of interest, given that the actual survey data with which the reader will be working were collected from the elements in this sample. This chapter aims to help readers understand the complex sample designs that they are likely to encounter in practice and identify the features of the designs that have important implications for correct analyses of the survey data.

Although a thorough knowledge of sampling theory and methods can benefit the survey data analyst, it is not a requirement. With a basic understanding of complex sample design features, including stratification, cluster sampling, and weighting, readers will be able to specify the key design parameters required by today's software procedures for survey data analysis. Readers who are interested in a more in-depth treatment of sampling theory and methods are encouraged to review work by Hansen et al. (1953), Kish (1965), or Cochran (1977). More recent texts that blend basic theory and applications include Levy and Lemeshow (2007), Valliant et al. (2018), and Lohr (2021). A short monograph by Kalton (1983) provides an excellent summary of common sample designs that researchers are likely to encounter in practice.

The sections in this chapter outline the key elements of complex sample designs that analysts need to understand to proceed knowledgeably and confidently to the next step in the analysis process.

2.1.1 Technical Documentation and Supplemental Literature Review

The path to understanding the complex sample design and its importance to the reader's approach to the analysis of the survey data should begin with a review of the technical documentation for the survey and the sample design. A review of the literature, including both supplemental methodological reports and papers that incorporate actual analyses of the survey data, will be quite beneficial for the reader's understanding of the data to be analyzed.

Technical documentation for the sample design, weighting, and analysis procedures should be part of the "metadata" that are distributed with a survey data set. In the real world of survey data, the quality of this technical documentation can be highly variable (Kolenikov et al., 2020). At a minimum, the reader should expect to find a summary description of the sample, a discussion of weighting and estimation procedures, and

DOI: 10.1201/9781003472094-3

ideally, specific guidance on how to perform analyses of the survey data that correctly account for the complex sample design features. Readers who plan to analyze a public-use survey data set may find that the documentation of the design is lacking or inadequate. In these cases, readers should contact the help desk for the data distribution website or inquire with the study team directly to obtain or clarify the basic information needed to correctly specify the sample design features when analyzing the survey data.

Before diving into the statistical analysis of a survey data set, time is well spent in reviewing supplemental methodological reports or published scientific papers that used the same data. This review can identify important new information or even guide the reader's choice of an analytic approach given the statistical objectives of their research.

2.2 Classification of Sample Designs

As illustrated in Figure 2.1, Hansen et al. (1983) define a sampling design to include two components: the sampling plan, and a method for drawing inferences from the data generated under the sampling plan. Most survey data sets will be based on sample designs that fall in cell A of Figure 2.1, that is, designs that include a sampling plan based on probability sampling methods and assume that statistical inferences concerning population characteristics and relationships will be derived using the "design-based" theory initially proposed by Neyman (1934). In the third edition of this book, we mainly focus on sample designs of this type, but we also place a larger emphasis on model-based methods for probability samples (cell B of Figure 2.1) than the first two editions.

2.2.1 Sampling Plans

Probability sampling plans assign each member of the population a known non-zero probability of inclusion in the sample. A probability sampling plan may include features such as stratification and clustering of the population prior to selection—it does not require that the selection probability of one population element be independent of that for another. Likewise, the sample inclusion probabilities for different population elements need not be equal. In a probability sample of students in a co-educational secondary school, it is perfectly appropriate to sample women at a higher rate than men with the requirement that weights will be needed to derive unbiased estimates for the combined population. Randomization of sample choice is always introduced in probability sampling plans.

	Method of Inference	
Sampling Plan	**Design-based**	**Model-based**
Probability Sample	A	B
Model-dependent Sample	C	D

FIGURE 2.1
Classification of sample designs for survey data.

Source: Hansen et al. (1983).

Model-dependent sampling plans assume that the variables of interest follow a known probability distribution and that the choice of sample elements maximizes the precision of estimation for statistics of interest. For example, a researcher interested in estimating total annual school expenditures for teacher salaries (y) may assume the following relationship of the expenditures to known school enrollments (x): $y_i = \beta x_i + \varepsilon_i, \varepsilon_i \sim N(0, \sigma^2 x_i)$, where i indexes a school. Model-dependent sampling plans may employ stratification and clustering, but strict adherence to randomized selection is not a requirement.

Model-dependent sampling plans have received considerable attention in the literature on sampling theory and methods (Valliant et al., 2000). In a recent example, Clark and Steel (2022) proposed a type of probability sampling where probabilities of selection are based on expected influence statistics (e.g., Cook's distance) from regression models of interest. However, these types of sampling plans are not common in survey practice due to several factors: the multi-purpose nature of most surveys; uncertainty regarding the most appropriate model, where inappropriate models can lead to biased estimates (Hansen et al., 1983); and lack of high-quality ancillary data (e.g., the x variable in the preceding example).

Though not included in Figure 2.1, **quota sampling**, **convenience sampling**, **snowball sampling**, and "**peer nomination**" are **non-probability methods** for selecting sample members (Kish, 1965; Baker et al., 2013). Practitioners who employ these methods base their choices on assumptions concerning the "representativeness" of the selection process and often analyze the survey data using inferential procedures that are appropriate for probability sampling plans. However, these sampling plans do not adhere to the fundamental principles of either probability sampling plans or a rigorous probability model–based approach, leaving the survey analyst with no theoretical basis to measure the variability and bias associated with their inferences.

Survey statisticians have long been aware of these important theoretical limitations of non-probability samples and have been working hard to address them. Indeed, recent years have seen a significant amount of new research on appropriate techniques for analyzing survey data collected from non-probability samples. We discuss these state-of-the-art methods in detail in Section 13.4 and present several examples of analyses employing them to make reasonable finite population inferences.

2.2.2 Other Types of Study Designs Involving Probability Sampling

Occasionally, **retrospective case-control designs** or **prospective cohort study designs** will be embedded in complex probability samples, where at the stage of sampling elements for eventual measurement, elements will be stratified based on whether they are "cases" (e.g., patients at a hospital with a particular type of illness or disease) or "controls" (e.g., patients at a hospital without the illness or disease). These studies are designed to enable the researcher to make population-level inferences about differences in exposures or outcomes between the cases and controls, and rely on the same principles of random selection as probability sampling. In the case that the sampled cases and controls are still different on available covariates or outcomes despite the randomized selection, various techniques may be employed to ensure balance among the two groups, such as **propensity score matching**.

There is a vast literature on the design and analysis of case-control studies, along with propensity score matching and related techniques—see, for example, Schlesselman (1982); Rosenbaum and Rubin (1983); and Schulz and Grimes (2002). Secondary analysts working with survey data from a complex sample with an embedded case-control design need to carefully consider the ways in which the complex sample design features should be incorporated into their analyses. Work by Hahs-Vaughn and Onwuegbuzie (2006), Landsman and Graubard (2013), and Lin et al. (2016) provides important practical guidance in this regard. Propensity score matching techniques are also crucial for efforts focused on making **causal inference** from survey data. Nattino and Lu (2018) present estimators of population-averaged causal effects based on a complex sample using the ideas of propensity score matching. See also Mukerjee et al. (2018) for more on the theory of making causal inferences based on complex probability samples.

Complex probability samples will often embed **randomized experiments** as well, often with a methodological focus—see the recent edited volume by Lavrakas et al. (2019) for an overview. When analyzing the results of these types of experiments, it is important that complex sampling features be considered. Van den Brakel (2019) provides important guidance in this regard, and Buchanan et al. (2018) discuss how to generalize results from randomized trials to larger finite populations when using weights that account for the inverse (estimated) probabilities of sampling. This approach is especially important in trials with volunteers where the probability of sampling is not known exactly.

Many research studies collect repeated measurements from individuals over time, in what are generally known as **prospective study designs**. These study designs often give rise to **longitudinal survey data** or **panel survey data**, where an initial probability sample of individuals or establishments is surveyed at some baseline time point and then measured repeatedly over time in follow-up waves of the survey. The Health and Retirement Study (HRS), which we analyze in this book, provides a good example of a prospective study design. Secondary analysts of survey data collected as part of a prospective study design need to carefully consider the design features of the initial probability sample and how these features (including time-varying survey weights for the respondents) should be accounted for in their analysis of the longitudinal data. We dedicate an entire chapter of this third edition of ASDA to current thinking on how one should perform these types of analyses (see Chapter 11).

In many cases, large probability samples selected from diverse target populations of interest will feature multiple stages of cluster sampling (discussed in more detail later in this chapter). For example, a sampling plan may feature random selection of counties, followed by smaller geographic areas, and finally, households within the sampled areas. In most design-based approaches to survey data analysis, the **primary sampling units** (PSUs, for example, counties, or groups of counties) are most critical for making robust inferences regarding the target population of interest, and the secondary analyst only needs to identify codes that indicate these PSUs (or use **replicate weights** reflecting the cluster sampling) in the analysis. However, some researchers may be interested in taking more of a **model-based approach** to the analysis and making inferences regarding the **components of variance** for a particular survey measure of interest across *all stages* of the multi-stage cluster sample design.

For example, what proportion of the variability in the survey measure of interest arises from counties? What proportion arises from the smaller areas within counties? If covariates are available describing features of these sampled "clusters," do they effectively explain variance at each stage of the sample design? To answer these types of questions, researchers can employ **multilevel models**, and it is important that relevant features of the probability sample design, including selection probabilities for the clusters at each stage of sample selection, be accounted for when estimating the components of variance that define these models. We discuss these types of analytic approaches in more detail in Chapters 11 and 13.

2.2.3 Inference from Survey Data

In the Hansen et al. (1983) framework for sample designs, a sampling plan is paired with an approach for deriving inferences from the data that are collected under the chosen plan. **Statistical inferences** from sample survey data may be "design-based" or "model-based." The natural design pairings of sampling plans and methods of inference are the diagonal cells of Figure 2.1 (A and D); however, the hybrid approach of combining probability sampling plans with model-based approaches to inference is not uncommon in modern survey research, as discussed in the earlier section.

Both approaches to statistical inference use probability models to establish the correct forms of confidence intervals or hypothesis tests for the intended statistical inference. Under the "design-based" or "randomization-based" approach formalized by Neyman (1934), the inferences are derived based on the **sampling distribution**, or the distribution of survey estimates from all possible samples that could have been chosen under the specified probability sampling design. This approach is sometimes labeled "nonparametric" or "distribution-free" because it relies only on the known probability that a given sample was chosen and not on the specific probability distribution for the underlying variables of interest, for example, $y \sim N(\beta_0 + \beta_1 x, \sigma^2_{y \cdot x})$ (see Chapter 3).

As the label implies, **model-based inference** is based on an assumed probability distribution for the random variable of interest—not the expected sampling distribution of a given estimate under the specified probability sampling design. Estimators, standard errors, confidence intervals, and hypothesis tests for the parameters of the assumed probability distribution (e.g., means, regression coefficients, and variances) are typically derived using the method of maximum likelihood, or possibly Bayesian models (see Little, 2003). We discuss Bayesian approaches in more detail in Section 3.8.

2.3 Target Populations and Survey Populations

The next step in becoming acquainted with the sample design and its implications for the reader's analysis is to verify the survey designers' intended study population—who, when, and where. Probability sample surveys are designed to describe a **target population**. The target populations for survey designs are **finite populations** that may range from as few as 100 **population elements** for a survey of special groups to millions and even billions of elements for national population surveys. Regardless of the actual population size, each discrete population element ($i = 1, \ldots, N$) could, in theory, be counted in a census or sampled for survey measurement.

In contrast to the target population, the **survey population** is defined as the population that is truly eligible for sampling under the survey design (Groves, 2004). In survey sampling practice, there are geographical, political, social, and temporal factors that restrict our ability to identify and access individual elements in the complete target population, and the de facto **coverage** of the survey is limited to the survey population. Examples of geographic restrictions on the survey population could include persons living in remote, sparsely populated areas, such as islands, deserts, or wilderness areas. Rebellions, civil strife, and governmental restrictions on travel can limit access to populations living in the affected areas. Homelessness, institutionalization, military service, nomadic occupations, physical and mental conditions, and language barriers are social and personal factors that can affect the coverage of households and individuals in the target population. The timing of the survey can also affect the coverage of the target population. The target population definition for a survey assumes that the data are collected as a "snapshot" in time when, in fact, the data collection may span several months (or even years).

We now consider an example of this distinction. The *target population* for the National Comorbidity Survey Replication (NCS-R) is defined to be adults aged 18 and over living in households in the United States as of July 1, 2002. Here is the exact definition of the *survey population* for the NCS-R:

> *The survey population for the NCS-R included all U.S. adults aged 18+ years residing in households located in the coterminous 48 states. Institutionalized persons including individuals in prisons, jails, nursing homes and long term medical or dependent care facilities were excluded from the survey population. Military personnel living in civilian housing were eligible for the study, but due to security restrictions residents of housing located on a military base or military reservation were excluded. Adults who were not able to conduct the NCS-R interview in English were not eligible for the survey.*
>
> **(Heeringa et al., 2004)**

Note that among the list of exclusions in this definition, the NCS-R survey population excludes residents of Alaska and Hawaii, institutionalized persons, and non-English speakers. Furthermore, the survey observations were collected over a window of time that spanned several years (February 2001–April 2003). For populations that remain stable and relatively unchanged during the survey period, the time lapse required to collect the data may not lead to bias for target population estimates. However, if the population is mobile or experiences seasonal effects in terms of the survey variables of interest, considerable change can occur during the window of time that the survey population is being observed.

As the survey data analyst, the reader will also be able to restrict their analysis to **subpopulations** of the survey population represented in the survey data set, but their analysis can only use the available data and cannot directly reconstruct coverage of the unrepresented segments of the target population. Therefore, it is important to carefully review the definition of the survey population and assess the implications of any exclusions for the inferences to be drawn from the analysis.

2.4 Simple Random Sampling: A Simple Model for Design-Based Inference

Simple random sampling with replacement (SRSWR) is the most basic of sampling plans, followed closely in theoretical simplicity by **simple random sampling without replacement (SRSWOR**, or the short form, SRS, used in this book). Survey data analysts are unlikely to encounter true SRS designs in survey practice. Occasionally, SRS may be used to select samples of small, localized populations or samples of records from databases or file systems, but this is rare. Even in cases where SRS is practicable (feasible), survey statisticians will aim to introduce simple stratification to improve the efficiency of sample estimates (see below). Furthermore, if an SRS is selected but weighting is required to compensate for nonresponse or to apply post-stratification adjustments (see Section 2.7), the survey data now include complex features that cannot be ignored in estimation and inference.

2.4.1 Relevance of SRS to Complex Sample Survey Data Analysis

So why is SRS even relevant for the material in this book? There are several reasons:

1. SRSWR designs produce samples that most closely approximate the assumptions (**i.i.d.**—observations are *independent* and *identical in distribution*) defining the theoretical basis for the standard estimation and inference procedures found in the major statistical software packages. Survey analysts who use standard analysis procedures in Stata®, SAS®, and SPSS® are essentially defaulting to the assumption that their survey data were collected under SRS. In general, the SRS assumption results in underestimation of variances of survey estimates of descriptive statistics and model parameters. Confidence intervals based on computed variances that assume independence of observations will be biased (generally too narrow) and design-based inferences will be affected accordingly. Likewise, test statistics (t, χ^2, F) computed in complex sample survey data analysis using standard procedures will tend to be biased upward and overstate the significance of tests of effects.
2. The theoretical simplicity of SRS designs provides a basic framework for design-based estimation and inference on which to build a bridge to the more complicated approaches used for complex samples.
3. SRS provides a comparative benchmark that can be used to evaluate the relative efficiency of the more complex sample designs that are common in survey practice.

Let's examine the second reason listed above more closely, using SRS as a theoretical framework for design-based estimation and inference. In Section 2.5, we will turn to SRS as a benchmark for the efficiency of complex sample designs.

2.4.2 SRS Fundamentals: A Framework for Design-Based Inference

Most students of statistics were introduced to SRS designs through the example of an urn containing a population of blue and red balls. To estimate the proportion of balls in the urn that were blue, the instructor described a sequence of random draws of $i = 1, \ldots, n$ balls from the N balls in the urn. If a drawn ball was returned to the urn before the next draw was made, the sampling was "with replacement" (SRSWR). If a selected ball was not returned to the urn until all n random selections were completed, the sampling was "without replacement" (SRSWOR).

In each case, the SRSWR or SRSWOR sampling procedure assigned each population element an equal probability of sample selection, $f = n / N$. Furthermore, the overall probability that a specific ball was selected to be in the sample was independent of the probability of selection for any of the remaining $N - 1$ balls in the urn. In survey practice, random sampling is not typically performed by drawing balls from an urn. Instead, survey sampling uses devices such as tables of random numbers or computerized random number generators to select sample elements from the population.

Let's assume that the objective of a sample design was to estimate the mean of a characteristic, *y*, in the population:

$$\bar{Y} = \frac{\sum_{i=1}^{N} Y_i}{N}. \tag{2.1}$$

Under simple random sampling, an unbiased estimate of the population mean is the sample mean:

$$\bar{y} = \frac{\sum_{i=1}^{n} y_i}{n}. \tag{2.2}$$

The important point to note here is that there is a true population parameter of interest, $\bar{Y}$, and the estimate of the parameter, $\bar{y}$, which can be derived from the sample data. The sample estimate $\bar{y}$ is subject to sampling variability, denoted as $Var(\bar{y})$, from one sample to the next. Another measure of the sampling variability in sample estimates is termed the **standard error**, or $SE(\bar{y}) = \sqrt{Var(\bar{y})}$. If one selected all possible samples of size *n* from the population of size *N* and computed the estimate of the population mean for each sample, the standard error would describe the standard deviation of the resulting distribution of the sample estimates. For simple random samples (SRS) of size *n* selected from populations of size *N*, the standard error for the estimated population mean is calculated as follows:

$$\begin{aligned} SE(\bar{y}) &= \sqrt{Var(\bar{y})} = \sqrt{(1-n/N)\cdot\frac{S^2}{n}} \\ &\approx \sqrt{\frac{S^2}{n}} \quad \text{if } N \text{ is large} \end{aligned} \tag{2.3}$$

where: $S^2 = \sum_{i=1}^{N}(Y_i - \bar{Y})^2/(N-1)$; *n = SRS sample size; and N = the population size.*

Note that the calculation of the standard error in (2.3) relies on the knowledge of values of *y* for all elements in the population. Since we observe only a single sample and not all possible samples of size *n* from the population of size *N*, the true (or theoretical) standard error, $SE(\bar{y})$, must be estimated from the data in our chosen sample:

$$\begin{aligned} se(\bar{y}) &= \sqrt{var(\bar{y})} = \sqrt{(1-n/N)\cdot\frac{s^2}{n}} \\ &\approx \sqrt{\frac{s^2}{n}} \quad \text{if } N \text{ is large} \end{aligned} \tag{2.4}$$

where: $s^2 = \sum_{i=1}^{n}(y_i - \bar{y})^2/(n-1)$, *n = SRS sample size; and N = the population size.*

The term $(1 - n / N)$ in the expressions for $SE(\bar{y})$ and $se(\bar{y})$ is the **finite population correction (fpc)**. It applies only where selection of population elements is without replacement (see Theory Box 2.1) and is generally assumed to be equal to 1 in practice if $f = n / N < 0.05$.

THEORY BOX 2.1: THE FINITE POPULATION CORRECTION (FPC)

The fpc reflects the expected reduction in the sampling variance of a survey statistic due to sampling without replacement (WOR). For an SRS sample design, the fpc factor arises from the algebraic derivation of the expected sampling variance of a survey statistic over all possible WOR samples of size n that could be selected from the population of N elements (see Cochran, 1977).

In most practical survey sampling situations, the population size, N, is very large, and the ratio n / N is so close to zero that the fpc ~ 1.0. As a result, the fpc can be safely ignored in the estimation of the standard error of the sample estimate. Since complex samples may also employ sampling without replacement at one or more stages of selection, in theory, variance estimation for these designs should also include finite population corrections. Where applicable, software packages such as Stata and SUDAAN provide users with the flexibility to input population size information and incorporate the fpc values in variance estimation for complex sample survey data. Again, in most survey designs, the size of the population at each stage of sampling is so large that the fpc factors can be safely ignored.

If the sample size, n, is large, then under Neyman's (1934) method of design-based inference, a 95% confidence interval for the true population mean, $\bar{Y}$, can be constructed as follows:

$$\bar{y} \pm t_{0.975,n-1} \cdot se(\bar{y}). \tag{2.5}$$

2.4.3 An Example of Design-Based Inference under SRS

To illustrate the simple steps in design-based inference from a simple random sample, Table 2.1 presents a hypothetical sample data set of $n = 32$ observations from a very large national adult population (because the sampling fraction, n / N, is small, the fpc will be ignored). Each subject was asked to rate his or her view of the strength of the national economy (y) on a 0–100 scale, with 0 representing the weakest possible rating and 100 the strongest possible rating. The sample observations are drawn from a large population with population mean $\bar{Y} = 40$ and population variance $S_y^2 \cong 12.80^2 = 164$. The individual case identifiers for the sample observations are provided in Column (1) of Table 2.1. For the time being, we can ignore the columns labeled Stratum, Cluster, and Case Weight.

If we assume that the sample of size $n = 32$ was selected by SRS, the sample estimate of the mean, the estimated standard error of the sample estimate, and the 95% confidence interval for the population mean would be calculated as follows:

$$\overline{y} = \sum_{i=1}^{n} y_i \,/\, n = \sum_{i=1}^{32} y_i \,/\, 32 = 40.77$$

$$se(\overline{y}) = \sqrt{\text{var}(\overline{y})} = \sqrt{\sum_{i=1}^{32} (y_i - \overline{y})^2 \,/\, [n \cdot (n-1)]} = 2.41$$

$$95\%\ CI = \overline{y} \pm t_{.975,n-1} \cdot se(\overline{y}) = (35.87, 45.68)$$

TABLE 2.1

Sample Data Set for Sampling Plan Comparisons

Case No. (1)	Stratum (2)	Cluster (3)	Economy rating score, y_i (4)	Case Weight, w_i (5)
1	1	1	52.8	1
2	1	1	32.5	2
3	1	1	56.6	2
4	1	1	47.0	1
5	1	2	37.3	1
6	1	2	57.0	1
7	1	2	54.2	2
8	1	2	71.5	2
9	2	3	27.7	1
10	2	3	42.3	2
11	2	3	32.2	2
12	2	3	35.4	1
13	2	4	48.8	1
14	2	4	66.8	1
15	2	4	55.8	2
16	2	4	37.5	2
17	3	5	49.4	2
18	3	5	14.9	1
19	3	5	37.3	1
20	3	5	41.0	2
21	3	6	45.9	2
22	3	6	39.9	2
23	3	6	33.5	1
24	3	6	54.9	1
25	4	7	26.4	2
26	4	7	31.6	2
27	4	7	32.9	1
28	4	7	11.1	1
29	4	8	30.7	2
30	4	8	33.9	1
31	4	8	37.7	1
32	4	8	28.1	2

The fundamental results that have been illustrated here for SRS are that for a sample of size n, unbiased estimates of the population value and the standard error of the sample estimate can be computed. For samples of reasonably large size, a confidence interval for the population parameter of interest can be derived. As our discussion transitions to more complex samples and more complex statistics, we will build on this basic framework for constructing confidence intervals for population parameters, adapting each step to the features of the sample design and the analysis procedure.

2.5 Complex Sample Design Effects

Most practical sampling plans employed in scientific surveys are not SRS designs. Stratification is introduced to increase the statistical and administrative efficiency of the sample. Sample elements are selected from naturally occurring clusters of elements in multi-stage designs to reduce travel costs and improve interviewing efficiency. Disproportionate sampling of population elements may be used to increase the sample sizes for subpopulations of special interest, resulting in the need to employ weighting in the descriptive estimation of population statistics. All these features of more commonly used sampling plans will have effects on the accuracy and precision of survey estimators, and we discuss those effects in this section.

2.5.1 Design Effect Ratio

Relative to SRS, the need to apply weights to complex sample survey data changes the approach to estimation of population statistics or model parameters. Also relative to SRS designs, stratification, cluster sampling, and weighting all influence the sizes of standard errors for survey estimates. Figure 2.2 illustrates the general effects of these design features on the standard errors of survey estimates. The curve plotted in this figure represents the SRS standard error of a sample estimate of a proportion P as a function of the sample size n. At any chosen sample size, the effect of sample stratification is generally a reduction in

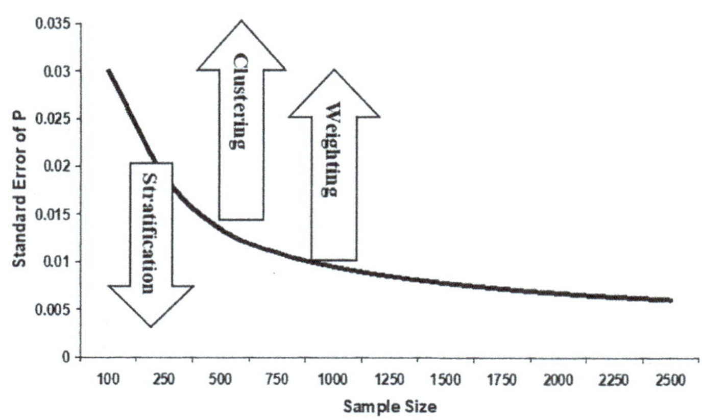

FIGURE 2.2
Complex sample design effects on standard errors of prevalence estimates.

standard errors relative to SRS. Cluster sampling and designs that require weighting for unbiased estimation generally tend to yield estimates with larger standard errors than an SRS sample of equal size.

Relative to an SRS of equal size, the complex effects of stratification, cluster sampling, and weighting on the standard errors of estimates are termed the **design effect**, and are measured by the following ratio (Kish, 1965):

$$D^2(\hat{\theta}) = \frac{SE(\hat{\theta})^2_{complex}}{SE(\hat{\theta})^2_{srs}} = \frac{var(\hat{\theta})_{complex}}{var(\hat{\theta})_{srs}} \tag{2.6}$$

where:

$D^2(\hat{\theta})$ = design effect for the sample estimate, $\hat{\theta}$;
$var(\hat{\theta})_{complex}$ = the complex sample design variance of $\hat{\theta}$; and
$var(\hat{\theta})_{srs}$ = the simple random sample variance of $\hat{\theta}$.

A somewhat simplistic but practically useful model of design effects that survey statisticians may use to plan a sample survey is:

$$D^2(\hat{\theta}) \approx 1 + f(G_{strat}, L_{cluster}, L_{weighting})$$

where:

G_{strat} = the relative gain in precision from stratified sampling compared to SRS;
$L_{cluster}$ = the relative loss of precision due to clustered selection of sample elements; and
$L_{weighting}$ = the relative loss of precision due to unequal weighting for sample elements.

The value of the design effect for a particular sample design will be the net effect of the combined influences of stratification, cluster sampling, and weighting. In Sections 2.5 to 2.7, we will introduce very simple models that describe the nature and rough magnitude of the effects attributable to these three common features of complex sample designs. The relative increase in variance measured by D^2 will be a complex and most likely nonlinear function of the influences of stratification, cluster sampling, and weighting, and their interactions. Over the years, there have been several attempts to analytically quantify the anticipated design effects for specific complex samples, estimates, and subpopulations (Skinner et al., 1989; Kish, 1995; Spencer, 2000; Park & Lee, 2004; Gabler et al., 2006; Lohr, 2014; Henry & Valliant, 2015; Chen & Rust, 2017).

While these more advanced models are instructive, the sheer diversity in real world sample designs and analysis objectives generally requires the empirical approach of estimating design effects directly from the available survey data:

$$d^2(\hat{\theta}) = \frac{se(\hat{\theta})^2_{complex}}{se(\hat{\theta})^2_{srs}} = \frac{var(\hat{\theta})_{complex}}{var(\hat{\theta})_{srs}} \tag{2.7}$$

where:

$d^2(\hat{\theta})$ = the estimated design effect for the sample estimate, $\hat{\theta}$;

$var(\hat{\theta})_{complex}$ = the estimated complex sample design variance of $\hat{\theta}$; and

$var(\hat{\theta})_{srs}$ = the estimated simple random sample variance of $\hat{\theta}$.

As a statistical tool, the concept of the complex sample design effect is more directly useful to the designer of a sample than to the analyst of the survey data. The sample designer can use the concept and its component models to optimize the cost and error properties of specific design alternatives or to adjust simple random sample size computations for the design effect anticipated under a specific sampling plan (Kish et al., 1976). Using the methods and software presented in this book, the survey data analyst will compute confidence intervals and test statistics that incorporate the estimates of standard errors adjusted for the complex sample design—generally bypassing the need to estimate the design effect ratio.

Nevertheless, knowledge of estimated design effects and their component factors does permit the analyst to gauge the extent to which the sampling plan for their data has produced efficiency losses relative to a simple random sampling standard, and to identify features such as extreme cluster sampling or weighting influences that might affect the stability of the finite population inferences that they will draw from the analysis of the data. In addition, there are several analytical statistics, such as modifications of the Pearson χ^2 or Likelihood Ratio χ^2 test statistics (Rao & Scott, 1984; Rao & Thomas, 1988), where estimated design effects are used directly in adapting conventional hypothesis test statistics for the effects of the complex sample design (see Chapter 6).

2.5.2 Generalized Design Effects and Effective Sample Sizes

The design effect statistic permits us to estimate the variance of complex sample estimates relative to the variance for an SRS of equal size:

$$\begin{aligned} var(\hat{\theta})_{complex} &= d^2(\hat{\theta}) \cdot var(\hat{\theta})_{srs}; \text{ or} \\ se(\hat{\theta})_{complex} &= \sqrt{d^2(\hat{\theta})} \cdot se(\hat{\theta})_{srs} \end{aligned} \tag{2.8}$$

Under the SRS assumption, the variances of many forms of sample estimates are approximately proportionate to the reciprocal of the sample size, that is, $var(\hat{\theta}) \propto 1/n$.

For example, if we ignore the fpc, the simple random sampling variances of estimates of a population proportion, mean, or simple linear regression coefficient are:

$$var(\hat{p}) \approx \frac{\hat{p}(1-\hat{p})}{(n-1)};$$

$$var(\bar{y}) \approx \frac{s^2}{n};$$

$$var(\hat{\beta}) \approx \frac{\hat{\sigma}^2_{y.x}}{\sum_{i=1}^{n}(x_i-\bar{x})^2} = \frac{\sum_{i=1}^{n}(y_i-\hat{y})^2/(n-2)}{\sum_{i=1}^{n}(x_i-\bar{x})^2}$$

Before today's statistical software was conveniently available to analysts, many public-use survey data sets were released without the detailed stratification and cluster variables that are required for complex sample variance estimation. Instead, users were provided with tables of generalized design effects for key survey estimates that had been computed and summarized by the data producer. Users were instructed to perform analyses of the survey data using standard SAS, Stata, or SPSS software procedures under simple random sampling assumptions, obtain SRS sampling variance estimates, and then apply the design effect factor as shown in (2.8) to approximate the correct complex sample variance estimate and corresponding confidence interval for the parameter being estimated. Even today, several major public use survey data sets, including the National Longitudinal Survey of Youth (NLSY) and the Monitoring the Future (MTF) Survey, require analysts to use this approach.

Survey designers make extensive use of design effects to translate between the simple analytical computations of sampling variance for SRS designs and the approximate variances expected from a complex sample design alternative. In working with clients, samplers may discuss design effect ratios, or they may choose a related measure of design efficiency termed the **effective sample size**:

$$n_{eff} = \frac{n_{complex}}{d^2(\hat{\theta})} \tag{2.9}$$

where:

n_{eff} = the effective sample size, or the number of SRS cases required to achieve the same sample precision as the complex sample design; and

$n_{complex}$ = the actual or "nominal" sample size selected under the complex sample design.

The design effect ratio and effective sample size are, therefore, two means of expressing the precision of a complex sample design relative to an SRS of equal size. For a fixed sample size, the statements "the design effect for the proposed complex sample is 1.5" and "the complex sample of size $n = 1000$ has an effective sample size of $n_{eff} = 667$" are equivalent statements of the precision loss expected from the complex sample design.

2.6 Complex Samples: Cluster Sampling and Stratification

We noted above that survey data collections are rarely based on simple random samples. Instead, sample designs for large survey programs often feature stratification, cluster sampling, and disproportionate sampling. Survey organizations use these "complex" design features to optimize the variance/cost ratio of the final design or meet precision targets for subpopulations of the survey population. The authors' mentor, Leslie Kish (1965, 1987), was fond of creating classification systems for various aspects of the sample design process. One such system was a taxonomy of complex sample designs. Under the original taxonomy, there were six binary keys to characterize all complex probability samples. Without loss of generality, we will focus on four of the six keys that are most relevant to the survey data analyst and aim to correctly identify the design features that are most important in applications:

Key 1: Is the sample selected in a single stage or multiple stages?

Key 2: Is clustering of elements used at one or more sample stages?

Key 3: Is stratification employed at one or more sample stages?

Key 4: Are elements selected with equal probabilities?

In the full realm of possible sampling approaches, this implies that there are at least 2^4 or 16 possible permutations of general choices for complex sample designs. In fact, the number of complex sample designs encountered in practice is far fewer, and one complex design—multi-stage stratified cluster sampling with unequal probabilities of selection for elements—is used in most in-person surveys of household populations. Because they are so important in major programs of household population survey research, we will cover these multi-stage probability sampling plans in detail in Section 2.8. Before we do that, let's take a more basic look at the common complex sample design features of cluster sampling, stratification, and weighting for unequal selection probabilities and nonresponse.

2.6.1 Cluster Sampling Plans

Cluster sampling of elements is a common feature of most complex sample survey data. In fact, to simplify our classification of sample designs, it is possible to view population elements as "clusters of size 1." By treating elements as single-unit clusters, the general formulas for estimating statistics and standard errors for clustered samples can be applied to correctly estimate standard errors for simpler stratified element samples (see Chapter 3).

Survey designers employ cluster sampling for several reasons:

- Geographic clustering of elements for household surveys reduces interviewing costs by amortizing travel and related expenditures over a group of observations. Multi-stage sample designs such as the area probability samples employed in the NCS-R, NHANES, HRS, and ESS incorporate clustering at one or more stages of the sample selection.
- Sample elements may not be individually identified on the available sampling frames but can be linked to aggregate cluster units (e.g., voters at precinct polling stations, students in colleges and universities). The available sampling frame often identifies only the clusters of elements. Identification of the sample elements requires an initial sampling of clusters and on-site work to select the elements for the survey interview.
- One or more stages of the sample are deliberately clustered to enable the estimation of multilevel models and components of variance in variables of interest (e.g., students in classes, classes within schools). This is not common but does occur in educational research and other areas of scientific investigation.

Therefore, while cluster sampling can reduce survey costs or simplify the logistics of the actual survey data collection, the survey data analyst must recognize that clustered selection of elements affects their approach to variance estimation and developing inferences from the sample data. In almost all cases, sampling plans that incorporate cluster sampling result in standard errors for survey estimates that are greater than those from an SRS of equal size; furthermore, special approaches are required to estimate the correct standard errors. The SRS variance estimation formulae and approaches incorporated in the standard procedures of most statistical software packages no longer apply, because they are based on assumptions of independence of the sample observations, and sample observations from within the same cluster generally tend to be correlated (e.g., students within a classroom, or households within a neighborhood).

The appropriate choice of a variance estimator required to correctly reflect the effects of cluster sampling on the standard errors of survey statistics depends on the answers to several questions:

1. Are all clusters equal in size?
2. Is the sample stratified?
3. Does the sample include multiple stages of selection?
4. Are units selected with unequal probability?

Fortunately, modern statistical software includes simple conventions that the analyst can use to specify the complex sample design features. Based on a simple set of user-supplied "design variables," the software selects the appropriate variance estimation formula and computes correct design-based estimates of standard errors.

The general increase in design effects due to either single-stage or multi-stage cluster sampling is caused by correlations (non-independence) of observations within sampled clusters. Many characteristics measured on sample elements within naturally occurring clusters, such as children in a school classroom or adults living in the same neighborhood, are correlated. Socio-economic status, access to health care, political attitudes, and even environmental factors such as the weather are all examples of characteristics that individuals in sample clusters may share to a greater or lesser degree. When such group similarity is present, the amount of "statistical information" contained in a cluster sample of n persons is less than in an independently selected SRS of the same size. Hence, cluster sampling increases the standard errors of estimates relative to an SRS of equivalent size. A statistic that is frequently used to quantify the amount of homogeneity that exists within sample clusters is the **intraclass correlation**, ρ (Kish, 1965). See Kish et al. (1976) for an in-depth discussion of intraclass correlations observed in the World Fertility Surveys.

A simple example is useful to explain why the intraclass correlation influences the amount of "statistical information" in clustered samples. A researcher has designed a study that will select a sample of students within a school district and collect survey measures from the students. The objective of the survey is to estimate characteristics of the full student body and the instructional environment. A probability sample of $n = 1{,}000$ is chosen by randomly selecting 40 classrooms of 25 students each. Two questions are asked of the students:

1) *What is your mother's highest level of completed education?*

   Given the degree of socio-economic clustering that can exist even among schools within a district, it is reasonable to expect that the intraclass correlation for this response variable is positive, possibly as high as $\rho = 0.2$.

2) *What is your teacher's highest level of completed education?*

   Assuming students in a sampled class have only one teacher, the intraclass correlation for this measure is 1.0! The researcher need not ask the question of all $n = 1{,}000$ students in the sample. An identical amount of information could be obtained by asking a single truthful student from each sample classroom. The effective sample size for this question would be 40, or the number of sample classrooms.

When the primary objective of a survey is to estimate proportions or means of population characteristics, the following model can be used to approximate the design effect that is attributable to the clustered sample selection (Kish, 1965):

$$D^2(\bar{y}) \;=1+L_{cluster} \approx 1+\rho\cdot(B-1) \tag{2.10}$$

where:

ρ = the intraclass correlation for the characteristic of interest; and

B = the number of respondents in each cluster or primary sampling unit (PSU).

The value of ρ is specific to the population characteristic (e.g., income, low-density cholesterol level, candidate choice) and the size of the clusters (e.g., counties, enumeration areas (EAs), schools, classrooms) over which it is measured. Generally, the value of ρ decreases as the geographical size and scope (i.e., the heterogeneity) of the clusters increases. Typical values of ρ observed for general population characteristics range from 0.00 to 0.20, with most between 0.005 and 0.100 (Kish et al., 1976).

In practice, it is sometimes valuable to estimate the value of ρ for a survey characteristic y. While it is theoretically possible to estimate ρ as a function of the sample data (y_i, $i = 1, \ldots, n$) and the sample cluster labels ($\alpha = 1, \ldots, a$), in most survey designs such direct estimates tend to be unstable. Kish (1965) suggested a synthetic estimator of ρ which he labeled the **rate of homogeneity (roh)**. Using the simple model for the estimated design effect for a sample mean, the expression is rearranged to solve for *roh* as:

$$roh(y) = \frac{d^2(\bar{y})-1}{\bar{b}-1} \tag{2.11}$$

where $\bar{b}$ is the average number of respondents per cluster.

Algebraically, this synthetic estimate of the intraclass correlation is restricted to a range of values from $-1/(\bar{b}-1)$ to 1.0. Since ρ is almost always positive for most survey variables of interest, this restricted range of the synthetic approximation *roh* is not a problem in practice. As the analyst (and not the designer) of the sample survey, the calculation of *roh* is not essential to the reader's work; however, when the reader begins to work with a new survey data set, it can be valuable to compute the *roh* approximation for a few key survey variables simply to get a feel for the level of intraclass correlation in the variables of interest.

We note that the above formulations do not reflect the possibility that the probability sample design also featured stratification (see Section 2.6.2). Chen and Rust (2017) derived expressions of the design effect in (2.10) that also account for stratification and two or three stages of cluster sampling. Their expressions also permit the estimation of *roh* within specific strata, which would be helpful for sampling statisticians looking to define probability samples in specific strata in the future.

Let's return to the hypothetical sample data set provided in Table 2.1. Previously, we assumed an SRS of size $n = 32$, and estimated the mean, the SRS standard error, and a 95% confidence interval for the mean of the economic conditions index. Now, instead of assuming SRS, let's assume that eight clusters, each of size four elements, were selected with equal probability. The cluster codes for the individual cases are provided in Column (3) of Table 2.1. Without digressing to the specific estimators of the mean and variance for this simple clustered sample (see Cochran, 1977), let's look at the correct numerical values for the estimated mean, the estimated standard error of the mean, and the 95% CI for $\bar{Y}$:

$$\bar{y}_{cl} = 40.77,\; se_{cl}(\bar{y}) = 3.65,\; CI(\bar{y}_{cl}) = (32.12, 49.49).$$

Comparing these results to those for the SRS example based on the same data (Section 2.4.3), the first thing we observe is that the estimate of $\bar{y}$ is identical. The change that occurs when clustering is introduced to the example is that the standard error of the mean is increased (due to the presence of the intraclass correlation that was used in developing the clusters for this example). The width of the 95% CI is also increased, due to the increase in standard error and the reduction in the degrees of freedom for the Student *t* statistic used to develop the CI limits (degrees of freedom determination for complex sample designs will be covered in Chapter 3). The relative increase in $se(\bar{y})$ for the simple cluster sample compared to an SRS of $n = 32$ and the synthetic estimate of the intraclass correlation are computed as follows:

$$d(\bar{y}) = \sqrt{d^2(\bar{y})} = \frac{se_{cl}(\bar{y})}{se_{srs}(\bar{y})} = \frac{3.65}{2.41} = 1.51;$$

$$roh(y) \approx \frac{1.51^2 - 1}{4 - 1} = 0.4267$$

2.6.2 Stratification

Strata are non-overlapping, homogeneous groupings of population elements or clusters of elements that are formed by the sample designer prior to the selection of the probability sample. In multi-stage sample designs (see Section 2.8), a different stratification of units can be employed at each separate stage of the sample selection. Stratification can be used to sample elements or clusters of elements. As noted above, if elements are viewed as "clusters of size $B = 1$," then the general formulas for estimating statistics and standard errors for stratified clustered samples can be applied to correctly compute these statistics for simpler stratified element samples (see Chapter 3). To take this "unification" one step further, if the sample design does not incorporate explicit stratification, it may be viewed as sampling from a single stratum (i.e., $H = 1$).

In survey practice, stratified sampling serves several purposes:

- Relative to an SRS of equal size, stratified samples that employ **proportional allocation** or **optimal allocation** of the sample size to the individual strata have smaller standard errors for sample estimates (Cochran, 1977);
- Stratification provides the survey statistician with a convenient framework to disproportionately allocate the sample to subpopulations, that is, to oversample specific subpopulations to ensure sufficient sample sizes for analysis;
- Stratification of the probability sample can facilitate the use of different survey methods or procedures in the separate strata (see Chapter 3).

Every stratified sample design involves the following four steps in sample selection and data analysis:

1. Strata ($h = 1, \ldots, H$) of N_h clusters/elements are formed;
2. Probability samples of a_h clusters or $a_h = n_h$ elements are independently selected from each stratum;
3. Separate estimates of the statistic of interest are computed for sample cases in each stratum and then weighted and combined to form the total population estimate;

4. Sampling variances of sample estimates are computed separately for each stratum and then weighted and combined to form the estimate of sampling variance for the total population estimate.

Because stratified sampling selects independent samples from each of the $h = 1, \ldots, H$ explicit strata of relative size $W_h = N_h / N$, and the same H strata would be used to select every hypothetical sample, any variance attributable to differences among strata is eliminated from the sampling variance of the estimate. Hence, the goal of any stratification designed to increase sample precision is to form strata that are "homogeneous within" and "heterogeneous between"—units assigned to a stratum are like one another and different from those in other strata, in terms of the measurements being collected.

To illustrate how stratification works to reduce sampling variance, let's consider the case of a simple stratified random sample of size $n = \sum_{h=1}^{H} n_h$, where the stratum sample sizes are proportional to the sizes of the individual strata: $n_h = n \cdot N_h / N$. This **proportionate allocation** of the sample ensures that each population element has an equal probability of being included in the sample, denoted by $f_h = n_h / N_h = n / N = f$. Let's compare $var(\bar{y}_{st,pr})$ for this stratified sample to $var(\bar{y}_{srs})$, ignoring the fpc:

$$\begin{aligned}\Delta &= var(\bar{y}_{srs}) - var(\bar{y}_{st,pr}) \\ &= \frac{S^2_{total}}{n} - \frac{S^2_{within}}{n} = \frac{\left[S^2_{within} + S^2_{between}\right]}{n} - \frac{S^2_{within}}{n} \\ &= \frac{S^2_{between}}{n} \\ &= \frac{\sum_{h=1}^{H} W_h(\bar{Y}_h - \bar{Y})^2}{n}\end{aligned} \tag{2.12}$$

The simple random sample of size n includes the variance of y both within the strata, S^2_{within}, and between the strata, $S^2_{between}$, but the stratified sample has eliminated the between-stratum variance component and consequently yields a sampling variance for the mean that is less than or equal to that for an SRS of equal size. The expected amount of the reduction in the variance of the mean is a weighted function of the squared differences between the stratum means, $\bar{Y}_h$, and the overall population mean, $\bar{Y}$. Hence the importance of forming strata that differ as much as possible in the $h = 1, \ldots, H$ values of the $\bar{Y}_h$.

Let's return to the example data set in Table 2.1 and assume that the $n = 32$ observations were selected from four population strata of equal size, such that $W_h = N_h / N = 0.25$. The stratum identifiers for each sample case are provided in Column (2) of Table 2.1. Under this proportionately allocated, stratified random sample design, we compute the following:

$$\begin{aligned}\bar{y}_{st,pr} &= \sum_{h=1}^{H} W_h \cdot \bar{y}_h \\ &= 0.25 \times 51.1 + 0.25 \times 43.32 + 0.25 \times 39.60 + 0.25 \times 29.04 \\ &= 40.77\end{aligned}$$

$$
\begin{aligned}
se(\bar{y}_{st,pr}) &= \sqrt{var(\bar{y}_{st,pr})} = \sqrt{\sum_{h=1}^{H} W_h^2 s_h^2 / n_h} \\
&= 2.04 \\
d(\bar{y}_{st,pr}) &= \frac{se(\bar{y}_{st,pr})}{se(\bar{y}_{srs})} = \frac{2.04}{2.41} = 0.85 \\
d^2(\bar{y}_{st,pr}) &= 0.85^2 = 0.72 \\
n_{eff} &= n / d^2(\bar{y}_{st,pr}) = 32 / 0.72 = 44.44
\end{aligned}
$$

Note that the unbiased sample estimate, $\bar{y}_{st,pr} = 40.77$, is identical to that obtained when we assumed that the 32 observations were sampled using SRS, and identical to that obtained when we assumed that the 32 observations arose from a cluster sample. The estimated standard error, $se(\bar{y}_{st,pr}) = 2.04$, is much smaller, resulting in an estimated design effect ratio of 0.72. For this example, it appears that the stratified sample of size $n = 32$ (8 cases from 4 strata) yields the sample precision of an SRS of slightly more than 44 cases.

The simple algebraic illustration and exercise calculation above demonstrate how stratification of the sample can lead to reductions in the variance of sample estimates relative to SRS. In contrast to this simple example, the precision gains achieved by stratification, G_{strat}, in complex samples are difficult to determine analytically. The gains will be a function of the stratum means, the variances of y within individual strata, and whether the sampling within strata includes additional stages, cluster sampling, unequal probabilities of selection, or other complex sample design features.

2.6.3 Joint Effects of Sample Stratification and Cluster Sampling

To see how the two design features of stratification and cluster sampling contribute to the design effect for a complex sample, let's revisit the sample data in Table 2.1, treating it as a stratified sample from $h = 1, \ldots, 4$ strata. Within each stratum, a proportionately allocated sample of two clusters ($a_h = 2$) of size 4 is selected with equal probability. This is a very simple example of a stratified **"two-per-stratum"** or "paired selection" cluster sample design that is very common is survey practice (see Chapter 4). Table 2.2 compares the results for

TABLE 2.2

Sample Estimates of the Population Mean and Related Measures of Sampling Variance under Four Alternative Sample Designs

Sample Design Scenario	Estimator	$\bar{y}$	$se(\bar{y})$	$d(\bar{y})$	$d^2(\bar{y})$	n_{eff}
SRS	$\bar{y}_{srs}$	40.77	2.41	1.00	1.00	32
Cluster Sample	$\bar{y}_{cl}$	40.77	3.66	1.51	2.31	13.9
Stratified	$\bar{y}_{st}$	40.77	2.04	0.85	0.72	44.7
Stratified Cluster Sample	$\bar{y}_{st,cl}$	40.77	2.76	1.15	1.31	24.4

Source: (Data Source: Table 2.1)

this stratified cluster sample design scenario to those for the simpler SRS, cluster sample, and stratified designs considered earlier.

Note that for the stratified cluster sample design scenario, the estimated design effect falls between the high value for the "cluster sample only" scenario and the low value for the scenario where only stratification effects are considered. As illustrated in Figure 2.2, real survey designs result in a "tug of war" between the variance inflation due to cluster sampling and the variance reduction due to stratification. From Table 2.2, the design effect for the stratified cluster sample is greater than 1, indicating a net loss of precision relative to an SRS of size n—sample clustering "won" the contest.

Estimates of design effects greater than 1 are common for most complex sample designs that combine stratification and cluster sample selection. In population surveys, large stratification gains are difficult to achieve unless the survey designer can select the sample from a frame that contains rich ancillary data, **x**, that are highly correlated with the survey variables, **y**. The survey data analyst needs to be aware of these design effects on the precision of sample estimates. Correctly recognizing the stratum and cluster codes for sample respondents provided by a data producer for use by data analysts is essential for correct and unbiased analyses of the survey data. We discuss the importance of identifying these codes in a survey data set in Chapter 4.

2.7 Weighting in Analysis of Survey Data

2.7.1 Introduction to Weighted Analysis of Survey Data

When probability sampling is used in survey practice, it is common to find that sample inclusion probabilities for individual population elements vary. Weighting of the survey data is thus required to "map" the sample back to an unbiased representation of the survey population. A simple but useful device for "visualizing" the role of case-specific weights in survey data analysis is to consider the weight as the number (or share) of the population elements that is represented by the sample observation. Observation i, sampled with probability $f_i = 1/10$, represents 10 individuals in the population (herself and nine others). Observation j, selected with probability $f_j = 1/20$, represents 20 population elements. In fact, if each sample case is assigned a weight equal to the reciprocal of its probability of selection, $w_i = 1/f_i$, the sum of the sample weights will, in expectation, equal the population size: $E\left(\sum_{i=1}^{n} w_i\right) = N$. As a survey data analyst, the reader can use this fact during their pre-analysis "checklist" to understand the distribution of the sampling weights that the data producer has provided (see Chapter 4).

Generally, the **final survey weights** in survey data sets (sometimes referred to as **estimation weights**) are the product of the sample selection weight (w_{sel}), a nonresponse adjustment factor (w_{nr}), and the post-stratification factor (w_{ps}):

$$w_{final,i} = w_{sel,i} \times w_{nr,i} \times w_{ps,i} \tag{2.13}$$

The sample selection weight factor (or **base weight**) is simply the reciprocal of the probability that a population element was selected in the sample, $w_{sel,i} = 1/f_i$. Under the theory of design-based inference for probability samples, weighted estimation using these "inverse probability" weight factors will yield unbiased (or nearly unbiased) estimates of population quantities. For example:

$$\bar{y}_w = \frac{\sum_{i=1}^{n} w_{sel,i} \cdot y_i}{\sum_{i=1}^{n} w_{sel,i}} \text{ is unbiased for } \bar{Y};$$
$$s_w^2 = \frac{\sum_{i=1}^{n} w_{sel,i} \cdot (y_i - \bar{y}_w)^2}{\sum_{i=1}^{n} w_{sel,i} - 1} \text{ is an unbiased estimate of } S^2. \quad (2.14)$$

Note how in each of these estimators, the selection weight "expands" each sample observation's contribution to reflect its share of the population representation. Throughout this book, weighted estimation of population parameters will follow a similar approach, even for procedures as complex as the pseudo maximum likelihood (PML) estimation of the coefficients in a multivariable logistic regression model.

If we were able to always complete an observation on each selected sample case, the development of the survey weights would be very straightforward. The computation of the weight would only require knowledge of the sample selection probability factors (see Section 2.7.2). In a carefully designed and well-managed survey, these probability factors are known and carefully recorded for each sample case. The computation of w_{sel} is therefore an accounting function, requiring only multiplication of the probabilities of selection at each stage of sampling and then taking the reciprocal of the product of the probabilities.

As we will see in Chapter 4, "**survey data producers**" who in most cases have the responsibility for developing the final survey weights have a more difficult task than the simple arithmetic required to compute w_{sel}. First, weighting by w_{sel} will only yield unbiased estimators of population parameters if all n elements in the original sample are observed. Unfortunately, due to **survey nonresponse**, observations are only collected for r cases of the original probability sample of n elements (where $r <= n$). Therefore, survey data producers must develop statistical models of the conditional probability that a sample element will be an observed case. In general terms, the nonresponse adjustment factor, w_{nr}, in the survey weight is the reciprocal of the estimated conditional probability that the sample case responds. The objective in applying nonresponse adjustment factors in survey weights is to attenuate bias due to differential nonresponse across sample elements. A price that may be paid for the bias reduction through nonresponse adjustment takes the form of increases in standard errors for the weighted estimates. The potential magnitude of the increases in standard errors is discussed below in the context of weighting for unequal selection probabilities.

Even after computing w_{sel} and adjusting for nonresponse through the factor w_{nr}, most survey data producers will also introduce the post-stratification factor, w_{ps}, into the final survey weight. As its label implies, the post-stratification factor is an attempt to apply stratification corrections to the observed sample "post" or after the survey data have been collected. While not as effective as estimation for samples that were stratified at the time that they were selected, the use of post-stratification weight factors can lead to reduced standard errors (variance) for sample estimates. Furthermore, post-stratification weighting can attenuate any sampling biases that may have entered the original sample selection due to sample frame noncoverage or omissions that occurred in implementing the sample plan. An example of the latter would be systematic under-reporting of young males in the creation of household rosters for the selection of the survey respondent.

In the sections that follow, we describe the development of these three important weighting factors in more detail.

2.7.2 Weighting for Probabilities of Selection (w_{sel})

As described above, the selection weight factor, w_{sel}, is introduced in the calculation of the final survey weight to account for the probability that a case was selected for the sample. Common reasons for varying probabilities of case selection in sample surveys include:

1. disproportionate sampling within strata to achieve an optimally allocated sample for a specific population estimate (Cochran, 1977);
2. disproportionate sampling within a stratum or group of strata to deliberately increase the sample size and precision of analysis for certain domains of the survey population;
3. the use of sample screening across the sample strata and clusters to identify and differentially sample subpopulations, for example, the NHANES oversampling of household members with disabilities;
4. unequal probabilities that arise from subsampling of observational units within sample clusters, for example, the common procedure of selecting a single random respondent from the eligible members of sample households; and
5. unequal probabilities that reflect information on the final sampling probability that can only be obtained in the process of the survey data collection, for example, in a **random digit dialing (RDD)** telephone sample, the number of distinct mobile telephone numbers that serve a respondent's household.

Due to the multi-purpose nature of modern population surveys (many variables, many statistical aims), the use of disproportionate sampling to optimize a sample design for purposes of estimating a single population parameter is rare in survey practice. Far more common are designs in which geographic domains or subpopulations of the survey population are oversampled to boost the sample size and precision for stand-alone or comparative analysis. For example, the HRS employs a roughly two-fold oversampling of age-eligible African Americans and Hispanic individuals to improve precision of cross-sectional and longitudinal analyses of these two population subgroups (Heeringa & Connor, 1995). The NCS-R employed a Part 1 screening interview to obtain profiles of respondents' mental health related symptoms, but disproportionately sampled persons with positive symptom counts for the more intensive Part 2 questionnaire (Kessler et al., 2004).

As these examples illustrate, the selection probability for individual observational units often includes multiple components—e.g., a base factor for housing unit selection, a factor for subsampling targeted groups within the full population, and a probability of selecting a single respondent at random within the household. For HRS sample respondents, the selection weight factor, w_{sel}, is computed as the product of the reciprocals of three probabilities: 1) $w_{sel,hh}$, the reciprocal of the multi-stage probability of selecting the respondent's

TABLE 2.3

Illustration of Example w_{sel} Computations

Description of Sample Case						
Sample ID	**Race/Ethnicity of Respondent**	**Eligible Rs/ Household**	$w_{sel,hh}$	$w_{sel,sub}$	$w_{sel,resp}$	w_{sel}
A	Black	2	2000	1	2	4000
B	Black	1	2000	1	1	2000
C	White	2	2000	2	2	8000
D	White	1	2000	2	1	4000

household (HH) from the area frame (see Section 2.8); 2) $w_{sel,sub}$, the reciprocal of the probability of sampling an individual under the design objective of a 2:1 oversampling of eligible African American and Hispanic adults; and 3) $w_{sel,resp}$, the reciprocal of the conditional probability of selecting a single financial unit within the eligible household. This calculation is illustrated below:

$$f_{sel} = f_{sel,hh} \times f_{sel,sub} \times f_{sel,resp}$$

$$\Rightarrow$$

$$w_{sel} = \frac{1}{f_{sel}} = w_{sel,hh} \times w_{sel,sub} \times w_{sel,resp}.$$

Drawing on the general features of the HRS sample selection, Table 2.3 illustrates the sample selection weight calculations for four hypothetical respondents from different race/ethnicity groups who have different numbers of eligible respondents (Rs) in their sample households.

2.7.3 Nonresponse Adjustments (w_{nr})

As described above, the most widely accepted approach to compensate for **unit nonresponse** in surveys is for the data producer to develop and apply a nonresponse adjustment factor to the sample selection weight that is used in analysis. Underlying the weighting adjustment for nonresponse is a model of the **response propensity**—conditional on sample selection, the probability that the unit will cooperate with the survey request. In a sense, the concept of response propensity treats response to the survey as another step in the "sample selection process." But unlike true sample selection in which the sampling statistician predetermines the sampling probability for each unit, an underlying propensity model—for the most part outside the control of a statistician—determines the probability that a sampled case will be observed. The multiplication of the original sample selection weight for each sample unit by the reciprocal of its modeled response propensity creates a new weight, which, if the model is correct, enables unbiased or nearly unbiased estimation of population quantities from the survey data.

Two related methods for estimating response propensity and computing the nonresponse adjustment are commonly used in survey practice: a simple weighting class adjustment method and a propensity score weighting approach.

2.7.3.1 Weighting Class Approach ($w_{nr,wc}$)

The "weighting class method" (Little & Rubin, 2019) assigns all eligible elements of the original sample—survey respondents and nonrespondents—to classes (or "cells") based on categorical variables (age categories, gender, region, sample stratum, etc.) that are predictive of response rates. Original sample cases that were found to be ineligible for the survey are excluded from the nonresponse adjustment calculation. The weighting class method makes the simple assumption that the response propensities for cases within a given weighting class are equal (i.e., nonrespondents are MAR, or missing at random, conditional on the categorical variables and interactions implicit in the defined cells). The common response propensity for cases in a cell is estimated by the empirical response rate for the cases assigned to that cell. The weighting class nonresponse adjustment is then computed as the reciprocal of the response rate for the cell $c = 1, \ldots, C$ to which the case was assigned:

$$w_{nr,wc,i} = \frac{1}{rrate_c} \tag{2.15}$$

where: $rrate_c$ = the response rate for weighting class $c = 1, \ldots, C$.

Little and Vartivarian (2005) showed that reductions in both the bias and variance of an estimate are possible if weighting classes are formed based on variables related to *both* response propensity and the survey variable of interest. This important result was explored in more detail by Kreuter et al. (2010) and Krueger and West (2014), who considered the utility of survey paradata and other auxiliary information in improving nonresponse adjustments.

Aside from identifying effective auxiliary variables for forming weighting classes, another important consideration is the role of the selection weights in forming the nonresponse adjustments. Should the estimate of the response rate within a given nonresponse adjustment cell be weighted by the selection weights? Kott (2012) addressed this question in theoretical detail, concluding that use of the selection weights in these adjustments will improve the mean squared error of the resulting estimates. West (2009) and Flores Cervantes and Brick (2016) provided further empirical support for this idea, finding evidence of substantial benefits of using the selection weights in estimating the response rates in each cell. Flores Cervantes and Brick (2016) show that this is especially true if the implicit model for response propensity is misspecified, and the weighting classes are not "ideal" classes that vary in terms of both response propensity and the survey measures of interest. In short, a failure to use the selection weights in estimation of the response rates in each weighting class has the potential to lead to estimates with substantial bias. We revisit this idea in Theory Box 2.2.

2.7.3.2 Propensity Cell Adjustment Approach ($w_{nr,prop}$)

The propensity cell adjustment approach also assigns an adjustment factor to each respondent's sample selection weight that is equal to the reciprocal of the estimated probability that they participated in the survey. However, in the propensity adjustment method, the assignment of cases to adjustment cells is based on individual response propensity values estimated (via logit transform) from a logistic regression model:

$$\hat{p}_{resp,i} = prob(respondent = yes \mid X_i) = \left(\frac{e^{X_i \hat{\beta}}}{1 + e^{X_i \hat{\beta}}} \right) \tag{2.16}$$

where:

X_i is a vector of values of response predictors for $i = 1, \ldots, n$; and

$\hat{\beta}$ is the corresponding vector of estimated logistic regression model coefficients.

Grau et al. (2006) and Wun et al. (2007) suggest that the selection weights should once again be used when estimating the coefficients in this response propensity model. Given the estimated coefficients, adjustment cells are then defined based on quantile ranges—often deciles, 0–0.099, 0.10–0.199, etc.—of the distribution of predicted response propensities (Little & Rubin, 2019). The propensity score nonresponse weighting adjustments are then computed as the reciprocal of the response rate in the quantile range cell $d = 1, \ldots, D$ to which the case was assigned:

$$w_{nr,prop,i} = \frac{1}{rrate_d} \tag{2.17}$$

where: $rrate_d$ *= the weighted response rate for propensity cell* $d = 1, \ldots, D$.

The use of individual estimated response propensities for these adjustments (i.e., $\hat{p}_{resp,i}$ as opposed to the response rates, $rrate_d$, within the quantile range cells, as shown in (2.17)) is often not recommended, as extremely low estimated response propensities (e.g., $\hat{p}_{resp,i} = 0.001$) could result in more variance in the adjusted weights. This increased variance in the weights would decrease the precision of weighted estimates based on the survey data (see Section 2.7.5).

THEORY BOX 2.2: ESTIMATING RESPONSE PROPENSITIES IN NONRESPONSE ADJUSTMENTS

In our discussion, both the weighting class and propensity cell adjustments for nonresponse require estimation of response rates within defined cells. Traditionally, many survey statisticians have employed the sample selection weight, w_{sel}, to compute weighted estimates of response rates in these cells:

$$rrate_{weighted,cell} = \sum_{i \in cell}^{n_{cell}} w_{sel,i} \cdot y_i \Big/ \sum_{i \in cell}^{n_{cell}} w_{sel,i} \tag{2.18}$$

where: $y_i = 1$ if sample person i is a respondent, 0 otherwise;

$w_{sel,i}$ = the sample selection weight for case i; and

n_{cell}= the total count of eligible sample cases in cell $c = 1, \ldots, C$.

Little and Vartivarian (2003) suggested that the unweighted response rate may be the preferred estimate of the response propensity for the cell, depending on what design variables are used to construct the cells:

$$rrate_{cell} = \sum_{i \in cell}^{n_{cell}} y_i / n_{cell} \tag{2.19}$$

As mentioned earlier, Kott (2012) presented theoretical rationale arguing for the use of the selection weights in estimating the within-cell response rates more generally (2.18). Flores Cervantes and Brick (2016) would later provide additional theoretical and empirical support for this practice. We advise readers responsible for developing nonresponse adjustments to selection weights to consider both approaches and examine the sensitivity of a variety of estimates to the approach used (along with any shifts in standard errors due to the different approaches). Given the literature in this area, we would recommend the use of selection weights when estimating the response rates within cells for nonresponse adjustment purposes, especially since the models underlying the cells may be misspecified in practice (Flores Cervantes & Brick, 2016).

Note that under the propensity modeling approach, the computed value of the empirical response rate for cases in a modeled propensity quantile need not always fall within the range of scores for the quantile. For example, the range of propensity scores used to define an adjustment cell might be 0.75–0.79, but the corresponding response rate for cases in the cell could be 0.73. This reflects a lack of fit of the response propensity model. Standard tests may be used to evaluate the goodness of fit of the logistic regression model of response propensity, and various machine learning techniques (e.g., classification trees) could be employed to improve the predictions of response propensity (e.g., Earp et al., 2018). We recommend including as many theoretical predictors of responding as possible in whatever model is used, and especially predictors that are also correlated with the survey variables of primary interest (Little & Vartivarian, 2005).

To employ either method for nonresponse adjustment, the characteristics used to define the cells or model the response propensities must be known for both respondents and nonrespondents. In cross-sectional sample surveys such as the NCS-R or NHANES, this limits the nonresponse adjustment to characteristics of sampled persons or households that are known from the sampling frame or are completely observed in the screening process. In the case of the 1992 HRS baseline sample, a simple weighting class adjustment approach was employed to develop the nonresponse adjustment (Heeringa & Connor, 1995). Region (Northeast, South, Midwest, West), urban/rural status of the PSU, and race of the respondent (Black, Hispanic, White, and Other) were used to define weighting class adjustment cells. See Table 2.4 for an illustration of typical values of w_{nr} for various cells and how the values of this adjustment factor influence the final composite survey weight.

2.7.3.3 Ignorable vs. Non-ignorable Unit Nonresponse

The weighting adjustments for unit nonresponse described in Sections 2.7.3.1 and 2.7.3.2 assume that conditional on the auxiliary variables used to form weighting classes or estimate response propensities, unit nonresponse is occurring at random. Put differently, unit nonresponse is **ignorable** conditional on these auxiliary variables, and the weighting

TABLE 2.4

Illustration of Example w_{final} Computations, Including w_{sel}, w_{nr}, and w_{ps}

	Description of Sample Case					
Sample ID	**Nonresponse Adjustment Cell**	**Post-stratum**	w_{sel} **(from Table 2.3)**	w_{nr}	w_{ps}	w_{final}
A	Black, Northeast, Urban	Age 50–54, Male, Married	4000	1.3	1.04	5408
B	Black, South, Rural	Age 55–61 Female, Single	2000	1.15	.96	2208
C	White, West, Urban	Age 50–54, Male, Married	8000	1.25	1.06	10600
D	White, Midwest, Rural	Age 55–61, Female, Single	4000	1.18	.97	4578

adjustments described above would be expected to reduce any bias arising from the differential unit nonresponse. In this setting, if one was to fix values on all the auxiliary variables and study the respondents and nonrespondents sharing these fixed values, the respondents would be a simple random sample of all sampled cases sharing these values.

If, even after conditioning on the values of these auxiliary variables, unit nonresponse is still a function of the variables of interest that could not be observed for the non-respondents, unit nonresponse is *not* occurring at random. We would say in this setting that unit nonresponse is **non-ignorable**, and specialized methods are needed to develop weighting adjustments for unit nonresponse in these settings (e.g., Andridge & Little, 2011, 2020; West & Little, 2013). Riddles et al. (2016) proposed an approach for estimating propensity scores based on maximum likelihood estimation in this setting of non-ignorable unit nonresponse, leveraging distributional features of the observed part of the sample, and via simulation demonstrated improved performance of this method relative to methods assuming ignorable unit nonresponse.

2.7.4 Post-Stratification and Calibration Adjustments (w_{ps})

The nonresponse adjustment procedures described above have the property that only data available for sampled respondent and nonrespondent cases are used to compute weighting adjustments. Another weighting technique used in practice to improve the quality of sample survey estimates is to incorporate known information for the full survey population—borrowing strength from data sources external to the sample. **Post-stratification** is one such method for using population data in survey estimation. **Raking ratio estimation, generalized regression (GREG) estimation**, and **calibration** are other forms of post-survey weight adjustment that may be employed to improve the precision and accuracy of survey estimates.

For example, Kott and Liao (2012) present a technique based on calibration weighting that can provide "double protection" against nonresponse bias, which means that as long as the variables used in the calibration procedure define either a well-specified substantive model for a variable of interest or a well-specified response propensity model, weighted estimates based on the technique (implemented in the WTADJUST procedure of the SUDAAN software) will still be unbiased asymptotically. Andersson (2019) presents another example of calibration weighting that makes use of estimated response propensities to further reduce the nonresponse bias in weighted estimates. To introduce these ideas, we will focus on post-stratification, which is the most common type of adjustment technique that leverages external information available for the population to further improve the weights in general social and health survey settings.

Simple post-stratification forms $l = 1, \ldots, L$ post-strata of respondent cases (just as the sample designer might form $h = 1, \ldots, H$ design strata prior to sample selection). The auxiliary variables used to form the L post-strata should satisfy the following three criteria: 1) they should be variables such as age, gender, and region for which accurate population control totals are available from external sources; 2) they should be highly correlated with key survey variables; and 3) they may be predictive of noncoverage in the sampling frame. Fuller (2022) proposes that post-strata be created based on the sample quantiles of available auxiliary variables so as to prevent the risk of empty or small post-strata. Cross-classifications of these variables are then used to define the $l = 1, \ldots, L$ post-strata. To ensure efficiency in the post-stratification, post-strata are generally required to include a minimum of n_l = 15 to 25 observations. If the cross-classification of post-stratification variables results in cells with fewer than this minimum number, similar post-strata are collapsed to form a larger grouping.

Post-stratification weighting involves further adjusting the survey weights for respondents so that weighted sample distributions conform to the known population distributions across the $l = 1, \ldots, L$ post-strata. The post-stratification factor applied to each respondent's adjusted survey weight is computed as follows:

$$w_{ps,l,i} = \frac{N_l}{\sum_{i=1}^{n_l} (w_{sel,i} \times w_{nr,i})} = \frac{N_l}{\hat{N}_l} \tag{2.20}$$

where:

$w_{ps,l,i}$ = the post-stratification weight for cases in post-stratum $l = 1, \ldots, L$; and

N_l = the population count in post-stratum l obtained from a recent Census, administrative records, or a large survey with small sampling variance.

In the case of the 1992 HRS baseline sample, post-strata of sample respondents were defined based on age category (50–54, 55–61) for the 1931–1941 birth cohorts, gender, and marital status (married, single). Population control totals, N_l, for each post-stratum were obtained as weighted population totals from the March 1992 demographic supplement to the U.S. Current Population Survey (Heeringa & Connor, 1995). See Table 2.4 for an illustration of typical values of w_{ps} for various post-strata and how the values of this adjustment factor influenced the final survey weight (w_{final}).

The benefits of post-stratification adjustments to working survey weights (which may have already been adjusted for nonresponse) in terms of the efficiency of weighted survey estimates will not be fully realized unless the post-strata and the known population sizes of the post-strata from the external source are explicitly specified to the software used to analyze the survey data. If software procedures only process the final adjusted survey weight after all of these steps, and the default variance estimation technique of Taylor Series Linearization (TSL) is used (see Chapter 3), the software will treat the processed weights as sampling weights *without recognizing the fact that the weights were calibrated to known population control totals*. In this sense, the gains in efficiency from calibrating the survey weights to known population controls (which in theory would be done for *every* hypothetical repeated sample, reducing the sampling variance in expectation) would not be realized.

The software procedures in Stata and R (see Valliant et al., 2018, for details regarding the approaches in R) for complex sample survey data analysis make it easy to do this, but specifying this information does require the data producer to make post-stratum identifiers and post-stratum population sizes available as variables in the survey data set (and not all data producers may wish to do this). For example, suppose that a complex sample survey data set has a variable named PSTRATUM available, containing unique codes that identify the post-strata of the respondents that were used for weighting adjustment, along with a variable named PSTRATUM_SIZE, which contains the known population sizes for each unique post-stratum (and takes a constant value for all respondents from the same post-stratum). When preparing for data analysis using TSL in Stata, we would specify these design variables using the `svyset` command below:

```
svyset PSUID [pweight = FINAL_WEIGHT], strata(STRATUMID) ///
poststrata(PSTRATUM) postweight(PSTRATUM_SIZE)
```

In this command, the PSUID variable contains the PSU codes, the STRATUMID variable contains the stratum codes (used for the *original* selection of the sample; note that these are *not* the post-stratum codes), and the FINAL_WEIGHT variable is the final adjusted survey weight for respondents. This setup would ensure correct variance estimation using TSL and would incorporate the efficiency introduced by post-stratification in all variance estimation, reducing the standard errors of the weighted estimates. This is in contrast to the following `svyset` command, which ignores the post-stratification information:

```
svyset PSUID [pweight = FINAL_WEIGHT], strata(STRATUMID)
```

Any design-based analyses conducted after using the `svyset` command above would result in linearized standard errors that are unnecessarily high and do not incorporate the efficiency gains from post-stratification. If data producers do not provide this information for whatever reason, data users unfortunately need to accept this loss in efficiency and acknowledge that their inferences will be conservative. We note that this issue primarily applies to the use of TSL for variance estimation; these types of calibration adjustments are typically applied to each replicate weight created for a replicate variance estimation procedure, increasing the efficiency of weighted estimates. See Chapter 3 for details.

We have covered the basics of weight construction thus far in this chapter, providing enough background to help survey analysts understand what information goes into the final survey weights that are typically provided in public-use survey data sets. For more detailed practical guidance on the construction of weights for complex sample surveys, we refer readers to Valliant et al. (2018).

2.7.5 Design Effects Due to Weighted Analysis

The weights that readers will be using in their survey data analysis are therefore compound products of several adjustments—unequal probabilities of selection, differential patterns of response, and post-stratification (or calibration more generally). Each of these adjustment factors can have a different effect on the precision and bias reduction for a design-based population estimate (Valliant, 2004). Along with stratification and cluster sampling, weighted estimation also contributes to the final design effect for a survey estimate. In Figure 2.2 shown earlier in this chapter, the large arrow representing the net effect of weighting could be split into three smaller arrows—the small arrows for selection weighting and nonresponse adjustment pointing up to higher variance, and the small post-stratification arrow pointing down to lower variance.

It is difficult if not impossible in most surveys to analytically partition out the size of the contribution of each factor (stratification, cluster sampling, selection weights, nonresponse adjustment, post-stratification adjustment) to the variance of sample estimates; see Park and Lee (2004) for discussion of this point. Henry and Valliant (2015) described how to compute design effects due to calibration weighting, and Chen and Rust (2017) present formulae for design effects that account for stratification, cluster sampling, and final survey weights (but not the components of the weights).

Changes in standard errors due to weighting are related to the variance of the weight values assigned to the individual cases and the correlations of the weights with the values and standard deviations of the variable(s) of interest (Spencer, 2000). We generally find in survey practice that the net effect of weighted estimation is inflation in the standard errors of estimates. This reflects the empirical fact that through the sequence of steps—selection weighting, nonresponse adjustment, etc.—the final survey weights can be quite variable. Furthermore, most survey weights are at best only moderately correlated with the survey variables of interest.

In recent years, the survey literature has used the terms "**weighting loss**" and "**unequal weighting effect**" to describe the inflation in variances of sample estimates that can be attributed to weighting. A simple approximation used by sampling statisticians to anticipate $L_{weighting}$, the proportional increase in variance of the sample mean due to weighted estimation, is:

$$L_{weighting}(\bar{y}) \approx cv^2(w) = \frac{s^2(w)}{\bar{w}^2} = \left\{ \frac{\sum_{i=1}^{n} w_i^2}{\left(\sum_{i=1}^{n} w_i \right)^2} \cdot n - 1 \right\} \tag{2.21}$$

where:

$cv^2(w)$ = the relative variance of the final survey weights;

$s(w)$ = the standard deviation of the final survey weights; and

$\bar{w}$ = the mean of the final survey weights.

This simple model of weighting loss was introduced by Kish (1965), and despite its widespread use, it was intended more as a design tool than as a strong model of true weighting effects on the variances of sample estimates. The Kish model of weighting loss in (2.21) was originally presented in the context of proportionate stratified sampling and represented the proportional increase in the variance of means (and proportions as means of binary variables) due to arbitrary disproportionate sampling of strata. It assumes that proportionate allocation is the optimal stratified design (i.e., variances of y are approximately equal in all strata) and that the weights are *uncorrelated* with the values of the random variable y. Spencer (2000) would later develop design effects due to weighting that relaxed the latter assumption. When using the Kish model in practice, analysts should consider the weighting effect produced for a given sample as an *upper bound* of what the true weighting effect might be (i.e., a "worst case scenario" for the variance inflation due to the use of weights in estimation).

Little and Vartivarian (2005) provide clear examples where this simple Kish model of weighting loss breaks down. As in any model, the quality of the predictions is tied to how closely the data scenario matches the model assumptions. Even within a survey data set, there may be variables for which the weighting improves the precision of estimates (due to intended or chance optimality for that variable) and many others for which the variability and randomness of the weights simply produces an increase in the variance of estimates. Mendelson and Huang (2016) demonstrated that design effects due to the use of weights in estimation (as captured by Kish's simple model) also tend to be much larger when sample sizes are smaller.

To illustrate the application of Kish's weighting loss model to a sample design problem, consider the following example. A survey is planned for a target population that includes both urban and rural populations of children. Eighty percent of the target population lives in the urban domain and 20% lives in the rural villages. The agency sponsoring the survey is interested in estimating the mean body mass index (BMI) of these children. The agency would like to have roughly equal precision for mean estimates for urban and rural children and accordingly decides to allocate the sample equally (50:50) to the two geographic domains (or strata). They recognize that this will require weighting to obtain unbiased estimates for the *combined* area. Urban cases will need to be weighted up by a factor proportional to 0.80 / 0.50 = 1.6, and rural cases will need to be down-weighted by a factor proportional to 0.20 / 0.50 = 0.4. To estimate $L_{weighting}$, they compute the relative variance of these weights for a sample that is 50% urban and 50% rural. They determine that $L_{weighting} \approx cv^2(w) = 0.36$. Ignoring any cluster sampling that may be included in the sampling plan, this implies that the final sample size for the survey must be $n_{complex} = n_{srs} \times 1.36$, or 36% larger than the SRS sample size required to meet a set precision level for estimating mean BMI for the combined population of urban and rural children.

The reader should note that $L_{weighting}$ is often non-trivial. Survey data sets that include disproportionate sampling of geographic areas or other subpopulations should be carefully evaluated: Despite what may appear to be large nominal sample sizes for the total survey, the effective sample sizes for pooled analysis may be much smaller than the simple case count suggests.

Fortunately, a detailed understanding of each of these contributions to the final design effect for a complex sample survey is more important to the survey designer than it is to the survey analyst. In the role of survey analyst, the reader will often not have access to the detailed information that the data producer's survey statisticians used to develop the final weights. As analysts, we must certainly be able to identify the correct weight variable to use in our analysis, be able to perform checks that the designated weight has been correctly carried forward to the data set that we will use for analysis, and be familiar with the syntax required to perform weighted estimation in our chosen statistical analysis software. Subsequent chapters in this book will provide the explanations and examples needed to become proficient in applying weights in survey data analysis.

2.8 Multi-Stage Area Probability Sample Designs

The applied statistical methods covered in Chapters 5–13 of this book are intended to cover a wide range of complex sample designs. However, because most large public-use survey data sets that are routinely used in the social and health sciences are based on multi-stage area probability sampling of households, it is important to describe this particular probability sampling technique in greater detail. The generic illustration presented here

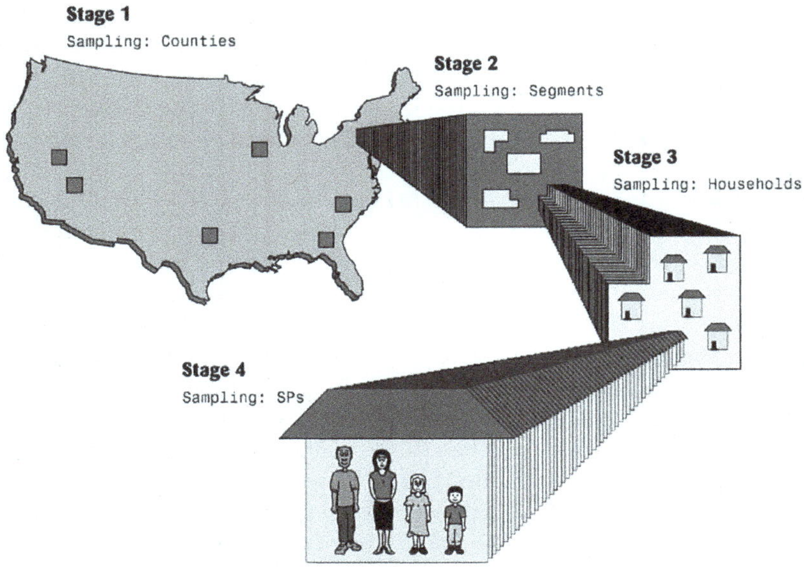

FIGURE 2.3
Schematic illustration of multi-stage area probability sampling.

Source: Reprinted by permission from: Mohadjer, L. (2002). The National Assessment of Adult Literacy (NAAL) Sample Design, Rockville, MD: Westat.

is most applicable to national household sampling in the United States and Canada. With minor changes in the number of stages and the choice of sampling units, similar household sample designs are used throughout Central and South America, Africa, Asia, and Europe (Heeringa & O'Muircheartaigh, 2010).

Figure 2.3 illustrates the selection of a multi-stage, area probability sample of individuals. The design illustrated in this figure conforms closely to the actual procedures used in the selection of the NCS-R, HRS, and NHANES samples of U.S. household populations. Under the general procedure illustrated by the figure, the selection of each study respondent requires a four-step sampling process: a primary stage sampling of U.S. counties or groups of adjacent counties, followed by a second stage sampling of area segments and a third stage sampling of housing units within the selected area segments, and concluding with the random selection of eligible respondents from the sampled housing units.

2.8.1 Primary Stage Sampling

The **primary stage units** (PSUs) of the multi-stage sample are single counties or groupings of geographically contiguous counties. The PSUs are therefore the highest-level groupings or "clusters" of sample observations (this will be important to remember when we discuss variance estimation in Chapters 3 and 4). All land area in the target population is divided into PSUs (e.g., the land area of the 50 United States is uniquely divided into roughly 3,100 county and parish (Louisiana) divisions). For comparison, the land area of Chile is uniquely divided into *communa* units, Russia into *raions* (regions), and South Africa and many other countries into Census *enumeration areas.*

Ideally, the populations within PSUs will be reasonably heterogeneous to minimize intraclass correlations for survey variables [and therefore cluster sample design effects, per (2.10)]. At the same time, the geographically defined PSUs must be small enough in size to facilitate cost-efficient travel to interview locations. Each designated PSU in the population is assigned to one of $h = 1, \ldots, H$ sampling strata based on region of the country, urban/rural status, PSU size, geographic location within regions, and population characteristics. Designs with approximately $H = 100$ primary stage strata are common in multi-stage samples of U.S. households, but the actual number of strata employed in the primary stage sample can range from less than ten to hundreds, depending on available stratification variables and the number of PSUs, a_h, that will be selected from each stratum.

Depending on the study sample design, some of the primary stage strata contain only a single **self-representing (SR)** metropolitan PSU. Each SR PSU is included in the sample with certainty in the primary stage of selection—for these strata, "true sampling" begins at the second stage of selection. The remaining **nonself-representing (NSR)** primary stage strata in each design contain more than one PSU. From each of these NSR strata, one or more PSUs is sampled with **probability proportionate to size (PPS)** as measured in occupied housing unit counts reported at the most recent Census. The number of PSUs selected from each primary stage stratum is decided by the sample allocation.

The University of Michigan Survey Research Center (SRC) National Sample Design (Heeringa & Connor, 1995) allocates one primary stage selection per stratum. More commonly, a minimum of two sample PSUs is selected from each primary stage stratum. A "two-per-stratum" design in which exactly two PSUs ($a_h = 2$) are selected with PPS from each stratum maximizes the number of strata that can be formed for selecting a primary stage sample of $a = \sum_{h=1}^{H} a_h$ PSUs. The two-per-stratum allocation is also common because a minimum of two PSUs per primary stage stratum is required to estimate sampling variances of estimates from complex samples (see Chapter 3). One-per-stratum primary stage samples, like that used by the University of Michigan's SRC for HRS and NCS-R, maximize the stratification potential but require a collapsing of design strata to create pseudo-strata for complex sample variance estimation (see Chapter 4 for details).

2.8.2 Secondary Stage Sampling

The designated second stage sampling units (SSUs) in multi-stage samples are commonly termed **area segments**. Area segments are formed by linking geographically contiguous Census blocks to form units with a minimum number of occupied housing units (typically 50–100 based on the needs of the study). Within sample PSUs, SSUs may be stratified at the county level by geographic location and race/ethnicity composition of residents' households. Within each sample PSU, the actual probability sampling of SSUs is performed with probabilities proportionate to Census counts of the occupied housing units for the Census blocks that comprise the area segment. The number of SSUs that are selected within each sample PSU is determined by survey statisticians to optimize the cost and error properties of the multi-stage design. The number of SSUs selected in an SR PSU varies in proportion to the total size of the PSU selected with certainty (e.g., HRS and NCS-R use approximately 48 area segments in the New York self-representing PSU). Depending on the total size of the survey and the cost structure for data collection operations, typically from 6 to 24 SSUs are selected from each NSR PSU.

2.8.3 Third and Fourth Stage Sampling of Housing Units and Eligible Respondents

Prior to the selection of the third stage sample of households, field staff from the survey organization might visit each sample SSU location and conduct an up-to-date enumeration or "listing" of all housing units located within the physical boundaries of the selected area segments. In modern survey research, lists of housing units in the sampled area segments are often purchased from commercial vendors at this stage. A third stage sample of housing units is then selected from the enumerative listing according to a pre-determined sampling rate. The third stage sampling rates for selecting households in the multi-stage area probability samples are computed using the following "selection equation" (Kish, 1965):

$$\begin{aligned} f &= f_1 \times f_2 \times f_3 \\ &= \frac{MOS_\alpha \times a_h}{MOS_h} \times \frac{b_\alpha \times MOS_{\beta(\alpha)}}{MOS_\alpha} \times \frac{C_h}{MOS_{\beta(\alpha)}} \end{aligned} \tag{2.22}$$

In the final selection equation derived in (2.22), we have the following notation: f = the overall multi-stage sampling rate for housing units; MOS_α = total population measure of size in the selected PSU α; MOS_h = total population measure of size in the design stratum h; a_h = number of PSUs to be selected from design stratum h; b_α = number of area segments selected in the PSU α; $MOS_{\beta(\alpha)}$ = total household measure of size for the SSU $\beta = 1, \ldots, b_\alpha$; and C_h = a stratum-specific constant = $(f \times MOS_h) / a_h b_\alpha$.

For example, the third stage sampling rate for selecting an equal probability national sample from the listed housing units for the selected PSUs and area segment SSUs is:

$$f_3 = \frac{f}{f_1 \times f_2} = \frac{f \times MOS_h}{a_h \times b_\alpha \times MOS_{\beta(\alpha)}} \tag{2.23}$$

The third stage sampling rate is computed for each selected SSU in the sample design. This rate is then used to select a random sample of actual housing units from the area segment listing.

Each sampled housing unit is then contacted in person by an interviewer. Within each cooperating sample household, the interviewer conducts a short screening interview with a knowledgeable adult to determine if household members meet the study eligibility criteria. If the informant reports that one or more eligible adults live at the sampled housing unit address, the interviewer prepares a complete listing of household members (or a "roster") and proceeds to randomly select an eligible respondent for the study interview. The random selection of the respondent is often performed using a special adaptation of the objective household roster/selection table method developed by Kish (1949).

Despite the obvious effort and complexity that goes into fielding a multi-stage area probability sample, relatively simple specifications of primary stage stratum, primary stage cluster (PSU), and final survey weight variables will be required for analysts desiring appropriate design-based analyses of survey data from this common household sampling design. A detailed discussion of a unified approach to variance estimation for multi-stage samples is given in Chapter 4.

2.9 Special Types of Sampling Plans Encountered in Surveys

As described above, most large-scale social, economic, demographic, and health-related surveys are designed to provide the capability to make descriptive inferences about specific survey populations or to analyze multivariable relationships in a population. Although the techniques for applied survey data analysis presented in the following chapters are generally applicable to all forms of probability sample designs suitable for population estimation and inference, some areas of population survey research (such as surveys of businesses, hospitals, and other non-household units that vary in size and "importance") have developed specialized methods that will not be covered in detail in this book. Researchers who are working with survey data for these populations are encouraged to use the survey literature to determine current best practices for these special population surveys.

Survey research on natural populations in environmental (e.g., forestry or fisheries), geological, and some human and animal epidemiological studies is increasingly turning to **adaptive sample designs** to optimize observation, estimation, and inference. If the reader's data are of these types, they will need to use special procedures for estimation and inference. An excellent reference on adaptive sample design can be found in Thompson and Seber (1996). In survey research on human populations, **adaptive survey designs** (Schouten et al., 2017) refer to specialized designs that allocate specific data collection protocols to certain pre-defined strata based on available auxiliary information at the onset of a data collection (e.g., using higher incentives for subgroups expected to have lower response propensity), in hopes of increasing response propensity. Any implications of such designs for appropriate analyses of the resulting data will be clearly communicated in analytic guidelines and data user documentation for these data sets.

Groves and Heeringa (2006) applied the term **responsive survey design** to surveys that adapt sampling, survey measurement, and nonresponse follow-ups to empirical information that is gathered *during* the survey process (as opposed to information available at the onset of a data collection, in the case of adaptive survey design). One sampling technique that is critical to the responsive design of surveys is the use of **multi-phase sampling**. Following this approach, active sample cases that have not yet responded to the survey request may be subsampled for further contact or implementation of a new data collection protocol at a time point, t, conditional on the prior disposition (e.g., number of calls, success with contact, resistance to interview, etc.) of the case. Presently, the stochastic nature of the sample disposition of each case at time t is ignored and the data are weighted for estimation as though the disposition of cases was a deterministic (fixed) outcome. Current and future research is expected to lead to improved procedures for estimation and inference in multi-phase sample designs (Brick & Tourangeau, 2017).

Occasionally, population-based survey methods are employed to perform research that is purely analytical. These include studies that fall in the category of epidemiological case-control designs; randomized population-based experiments including "group randomized trials"; or model-based designs for research on hierarchical or multilevel populations (e.g., research on student, classroom, and school effects). As discussed earlier in this chapter, the analysis of "survey" data from these types of analytical research designs requires special approaches. Chapter 13 will explore approaches to several of the more common analytical designs that use survey-like procedures to collect data; but again, with data of this type, the reader is encouraged to also turn to the statistical literature for an up-to-date and more in-depth description of best practices. See, for example, Burns et al. (1996), Heeringa et al. (2001), and Raudenbush (2000).

3

Foundations and Techniques for Estimation and Inference

3.1 Introduction

The fundamental theory and practical procedures for estimation and inference for complex sample survey data have been under development for almost a century. The foundations are present in the work on randomization and the "representative method" (Bowley, 1906; Fisher, 1925) and what is generally agreed to be the breakthrough paper by Neyman (1934) on the theory of design-based inference from probability sample designs. Even today, estimation and inference for survey data remains an evolving field, with important new developments in multilevel and latent variable modeling (Rabe-Hesketh & Skrondal, 2006; Silva & Moura, 2022; Burgard & Dörr, 2022), estimation of small area statistics (Rao & Molina, 2015), and Bayesian approaches to estimation and finite population inference using survey data (Little, 2003; León-Novelo & Savitsky, 2019; Williams & Savitsky, 2021).

The general concept of design-based inference and its application to survey data analysis was introduced in Section 2.2. The aim of this chapter is to expand the reader's understanding of the key components of:

1. Design-based approaches to estimation and finite population inference from survey data, including consistent estimation of population statistics, robust, distribution-free methods for estimating the sampling variance of estimates, construction and interpretation of confidence intervals, and design-adjusted test statistics for hypothesis testing; and
2. Bayesian approaches to estimation and finite population inference from survey data, including formation of relevant prior distributions, development of a posterior distribution for parameters of interest that are sensitive to the features of a complex sample design, construction of credible intervals for inference, and key differences between Bayesian and design-based approaches.

The chapter opens in Section 3.2 with a simple introduction to finite populations and a superpopulation model—two theoretical concepts that have only subtle implications for how survey data are analyzed but are important concepts for interpretation and reporting of the survey findings. Confidence intervals are a primary tool for presenting survey estimates and the corresponding degree of uncertainty due to population sampling. Section 3.3 introduces the confidence interval as the organizing framework for discussion of the three main components of design-based inference: weighted sample estimates of population parameters (Section 3.4); critical values from the sampling distribution of the

DOI: 10.1201/9781003472094-4

estimates (Section 3.5); and robust estimates of the standard error of the weighted sample estimate (Section 3.6).

Inference from survey data is not limited to design-based confidence interval construction and interpretation. Tests of specific hypotheses concerning the true population value of a parameter may also be performed based on survey data. Section 3.7 compares approaches to common hypothesis tests under simple random sampling assumptions to the corresponding test procedures for complex sample survey data. Section 3.8 then shifts the focus to Bayesian approaches for making finite population inferences, reviewing the basic theory of Bayesian inference and then discussing new developments in Bayesian methods for analyzing data from complex samples. The chapter concludes in Section 3.9 with a discussion of the potential impact of sources of total survey error on estimation and inference for survey data.

3.2 Finite Populations and Superpopulation Models

Chapters 1 and 2 have touched on the healthy tension in the historical statistical literature between design-based and model-based approaches to survey inference. There has generally been little debate over the application of design-based inference for **descriptive analysis** of probability samples of survey populations. The objectives of the analysis are clear—to describe with measurable uncertainty due to sampling a population characteristic or process as it exists in a defined **finite population**. More controversy arises when researchers go beyond descriptive analysis to **analytic uses** of survey data sets—uses that estimate models to explore more universal multivariate relationships or even to possibly attempt to understand causality in relationships among variables. To unify our thinking about finite population statistics versus a more universal model of processes or outcomes, sampling theoreticians have introduced the concept of a **superpopulation model**.

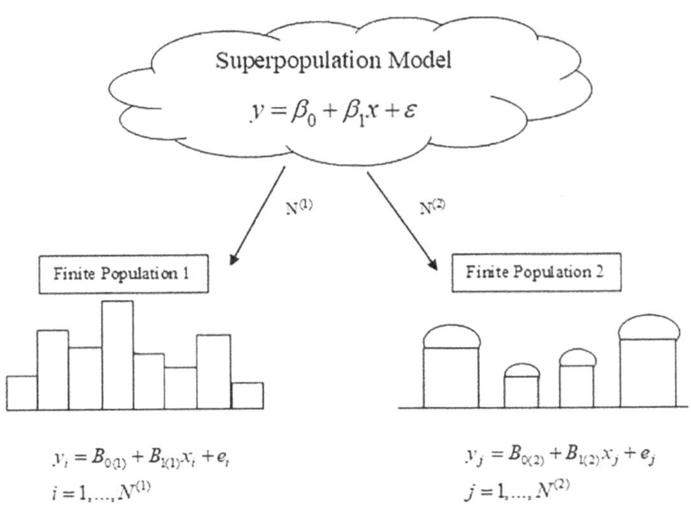

FIGURE 3.1
Superpopulation Model and Finite Populations.

Figure 3.1 illustrates this concept. The figure shows two finite survey populations, each somewhat different, yet each generated by the same over-arching superpopulation model. Research teams conducting probability sample surveys in the two respective finite populations would estimate their finite population regression parameter as:

$$b_{1(1)} = \hat{B}_{1(1)} = \frac{\sum_{i=1}^{n_{(1)}} w_i \cdot y_i \cdot x_i}{\sum_{i=1}^{n_{(1)}} w_i \cdot x_i^2}; \quad b_{1(2)} = \hat{B}_{1(2)} = \frac{\sum_{j=1}^{n_{(2)}} w_j \cdot y_j \cdot x_j}{\sum_{j=1}^{n_{(2)}} w_j \cdot x_j^2} \tag{3.1}$$

In each case, the design-based estimates, $b_{1(1)}$ or $b_{1(2)}$, would be design-consistent estimates of the corresponding finite population regression coefficients, $B_{1(1)}$ or $B_{1(2)}$. Note that sampling weights (denoted by w_i or w_j) are used to compute the estimates.

What if the statistical objective of each research team was not to simply estimate the best regression model for their survey population but to infer about the universal model that governs the relationship between y and x? The concept of a superpopulation model provides a bridge to this broader inference by postulating that the two finite populations are simply a size $N^{(1)}$ or $N^{(2)}$ "sample" of pairs of (x,y) observations generated from the superpopulation regression model: $y = \beta_0 + \beta_1 \cdot x + \varepsilon$. (See Theory Box 7.1 later in Chapter 7 for more details.)

The theoretical linkage of finite population samples to a superpopulation model is not foremost in the mind of applied survey data analysts as they sit down at their computer to begin their work with a survey data set. It will also have little bearing on how that analyst applies the statistical analysis techniques described in the second half of this book. Many survey researchers can become careless about reporting conclusions from their data analysis, suggesting universality of their findings when in fact the sample data may have been drawn from a small or unique survey population. It is important for research analysts reporting analysis results to be clear with their audiences concerning the nature of their inferential statements. Are the inferences targeted to the specific survey population, or is the aim to broaden the scope of the inference to imply more generalizable findings that extend beyond the boundaries of the finite population? If it is the latter, one should expect to provide scientific support for the inferential arguments. Is there a replication of one's findings in surveys of other distinct finite populations? Can one make a convincing theoretical argument for a single universal model for the relationships observed in the finite population?

3.3 Confidence Intervals for Population Parameters

> ". . . The form of this solution consists in determining certain intervals which I propose to call confidence intervals . . ."
>
> **[Jerzy Neyman, 1934]**

This quote from Neyman (1934) capped nearly three decades of research and experimentation by leading statisticians of the day (including R.A. Fisher and E.S. Pearson) to establish a theory of estimation for the representative method of sampling (i.e., simple random samples and other probability samples). Since the mid-1930s when the paper was published, Neyman's confidence intervals (CIs) have played a leading role in population estimation and inference for survey data. A generic form for an approximate 100(1 – α)% CI for a population parameter, where α represents a level of significance (e.g., 0.05, corresponding to a 95% confidence interval), is:

$$\hat{\theta} \pm t_{1-\alpha/2,df} \cdot se(\hat{\theta}) \tag{3.2}$$

where:

$\hat{\theta}$ is an unbiased or consistent estimate of the population parameter, θ;

$t_{1-\alpha/2,df}$ is the value of the Student *t* distribution with *df* degrees of freedom;

df is the degrees of the freedom for variance estimation under the sample design; and

se($\hat{\theta}$) is a robust estimate of the standard error of $\hat{\theta}$.

The confidence interval summarizes our uncertainty in inferring a true population value from a single sample. In simple terms, if the identical sample design was used to repeatedly draw samples from the survey population and the 95% CI was created for each sample, 95 in 100 of all such sample CIs would be expected to contain the true population value. Note that it is the theoretical distribution of sample estimates over repeated sampling from the population that provides "support" for this statement of uncertainty.

Construction of confidence intervals for population parameters therefore requires three components: an unbiased or consistent estimate, $\hat{\theta}$; the appropriate critical value from $t_{1-\alpha/2,df}$; and the standard error of the estimate, $se(\hat{\theta})$. The following three sections describe the underlying theory and derivation of each component.

3.4 Weighted Estimation of Population Parameters

Survey weights play a key role in consistent, design-based estimation of finite population parameters. **Consistent estimators** may be **unbiased estimators**. They may also be estimators such as the ratio mean for stratified, cluster sample data that have a small bias that decreases as the sample size, *n*, increases and disappears as the sample size approaches the population size, *N*. Some examples of consistent, weighted estimators of population parameters include:

$$\bar{y}_w = \frac{\sum_{i=1}^{n} w_i \cdot y_i}{\sum_{i=1}^{n} w_i} \text{ to estimate the population mean, } \bar{Y}; \tag{3.3}$$

$$b_{1,w} = \frac{\sum_{i=1}^{n} w_i \cdot y_i \cdot x_i}{\sum_{i=1}^{n} w_i \cdot x_i^2} \text{ to estimate the simple linear regression coefficient, } B_1. \tag{3.4}$$

Weighted estimation is also employed in iterative estimation of parameters for a large class of generalized linear models (GLM). Estimation of GLMs from survey data involves a "weighted" likelihood function, weighted first derivatives (the score or estimating equations) and then application of the **Newton-Raphson algorithm** to find the coefficient estimates that maximize the weighted **pseudo-likelihood**. To illustrate how weights enter GLM estimation, the weighted pseudo-log-likelihood function used in fitting a logistic regression model to survey data takes the following form:

$$\text{Pseudo ln(Likelihood):}$$
$$\sum_{h=1}^{H}\sum_{\alpha=1}^{a_h}\sum_{i=1}^{n_\alpha} w_{h\alpha i}y_{h\alpha i}\cdot\ln(\hat{\pi}(\boldsymbol{x}_{h\alpha i}))+\sum_{h=1}^{H}\sum_{\alpha=1}^{a_h}\sum_{i=1}^{n_\alpha} w_{h\alpha i}(1-y_{h\alpha i})\cdot\ln(1-\hat{\pi}(\boldsymbol{x}_{h\alpha i})) \tag{3.5}$$

where:

$y_{h\alpha i} \in (0,1)$ is the binary dependent variable in the logistic model;

$w_{h\alpha i}$ is the weight value for case i in cluster α and stratum h;

$\boldsymbol{x}_{h\alpha i}$ is the vector of predictor variables in the logistic regression model; and

$\hat{\pi}(\boldsymbol{x}_{h\alpha i}) = e^{\boldsymbol{x}_{h\alpha i}\boldsymbol{b}}/(1+e^{\boldsymbol{x}_{h\alpha i}\boldsymbol{b}})$ is the predicted probability that $y_{h\alpha i}=1$.

These and other weighted approaches to estimating population parameters and model coefficients will be introduced as we cover each of the major procedures for survey data analysis.

Since the early 1950s, survey statisticians, economists, and researchers in other disciplines have debated the role that weights should play in model-based inferences from survey data. Lohr (2021) provides an excellent review of this debate, including a discussion of contributions from Brewer and Mellor (1973), Hansen et al. (1983), and DuMouchel and Duncan (1983). A strong argument that emerges from papers by Pfeffermann and Holmes (1985) and Kott (1991) is that the survey weights protect the analyst against unknowingly misspecifying the model by omitting predictor variables or variable interactions that may be associated with the survey weight value (see Section 2.7). Leading statisticians who approach survey data from a model-based perspective have endorsed this view that survey weights can serve as a "proxy" for important features of the sample design that have a bearing on the response variable of interest (Little, 1991).

To illustrate this property of weighted estimation, consider the Health and Retirement Study (HRS) sample of Black adults. HRS employs a two-fold oversampling of households in U.S. Census blocks with 10% or greater Black households to increase the size of the Black subpopulation sample in a cost-effective manner. Black adults living in the geographic domain of higher density blocks have twice the selection probability of those who live on a block with fewer than 10% Black households. The HRS weights for Black respondents adjust for these differential selection probabilities. The HRS public use data set does contain a variable that is coded for the race/ethnicity of the respondent; however, the only source of information about whether the respondent lives in a low- or high-density Black neighborhood is encoded in the relative value of the survey weight. If the survey weights were ignored in a regression analysis, the potentially crucial influence of the disproportionate representation of Black adults from higher density (and often lower income) blocks would be omitted from the model.

Table 3.1 compares two estimated regression models where the dependent variable is the natural log of household income for Black respondents in the 2006 HRS. Each model

TABLE 3.1

Regression models of log-transformed household income for the 2006 HRS Black subpopulation ($n = 2,465$)

Independent Variable	Regression Parameter Estimate (Standard Error, p-value)	
	Unweighted	**Weighted (Design-based SE)**
Age (continuous)	0.0026 (0.0058, $p = 0.66$)	0.0056 (0.0093, $p = 0.54$)
Gender		
Female	−0.4629 (0.1246, $p < 0.001$)	−0.3034 (0.2199, $p = 0.17$)
Male	Reference category	Reference Category
Education		
Grade 0–11	−1.5585 (0.1991, $p < 0.0001$)	−1.9016 (0.2610, $p < 0.0001$)
Grade 12	−1.0304 (0.2011, $p < 0.0001$)	−1.5177 (0.2871, $p < 0.0001$)
Grade 13–15	−0.5145 (0.2152, $p < 0.0001$)	−0.7114 (0.1330, $p < 0.0001$)
Grade 16+	Reference category	Reference Category
Region		
Northeast	0.0804 (0.2743, $p = 0.77$)	0.1462 (0.1680, $p = 0.38$)
Midwest	−0.3331 (0.2635, $p = 0.21$)	−0.2423 (0.2614, $p = 0.36$)
South	−0.2525 (0.2476, $p = 0.31$)	−0.3519 (0.2405, $p = 0.15$)
West	Reference Category	Reference Category
Urbanicity		
Urban	−0.0697 (0.1690, $p = 0.68$)	−0.0553 (0.3751, $p = 0.88$)
Suburban	0.0588 (0.1965, $p = 0.76$)	−0.0262 (0.4764, $p = 0.58$)
Rural	Reference Category	Reference Category

includes demographic covariates of interest (age, gender of head of household, education level) as well as predictors that control for geographic region and urbanicity of residence. The first estimated model ignores the weights in the computation of model coefficients and their standard errors. The estimated coefficients in the second model correctly incorporate the HRS weights, and the estimated standard errors for the estimated coefficients correctly incorporate design effects for stratification, cluster sampling, and weighting.

Both estimated models point to the importance of education level in determining the income level of HRS Black households; however, the estimated relationship of a bachelor's degree or even some college training (13–15 years) is much larger when the survey weights are employed in the estimation. This can be attributed to the fact that a higher proportion of the more highly educated HRS Black respondents will reside in the more integrated neighborhoods of the low-density Black domain. Consequently, their contribution to the estimated model will be "up-weighted" when the survey weights are applied. (For a second illustration of how survey weights can influence the estimation of a regression model, see the Chapter 7 example of estimation of a model of diastolic blood pressure based on the NHANES data.)

There will be situations where the application of survey weights will have no real impact on a survey analysis and may even result in a loss of statistical precision relative to the unweighted alternative. Chambers and Skinner (2003) use a likelihood framework to demonstrate that the benefit of weighted analysis occurs when the sample and corresponding weights reflect an **informative sample design**. In simple terms, an informative design is one in which the design features—stratification, cluster sampling, disproportionate sampling—are associated with the response variable of interest. Korn and Graubard (1999, Sections 4.3 and 4.4) also provide examples of situations where the application of survey weights does not have a substantial impact on inferences and discuss methods for computing the inefficiency in survey estimates that can arise when survey weights are used unnecessarily.

The general approach that we adopt in this book is to begin with the assumption that the features of the complex sample design and the associated survey weights are in fact related to the survey variable of interest. Consequently, the design-based analyses illustrated in Chapters 5 through 13 will employ weighted estimation of population parameters. However, we agree with Lohr (2021) on the value of comparing models estimated both with and without the survey weights. If the estimated model parameters are very different, the analyst should be able to explain the difference based on documented knowledge about how the weights were constructed—e.g., oversampling of low-income households, persons with disabilities, older adults, etc. Bollen et al. (2016) provide a systematic evaluation of more formal procedures that have been proposed in the literature for testing the need to use weights when estimating regression models, and we revisit this idea in Chapter 7.

3.5 Probability Distributions and Design-Based Inference

> "The whole procedure consists really in solving the problems which Professor Bowley termed direct problems: given a hypothetical population, to find the distribution of certain characters in repeated samples. If this problem is solved, then the solution of the other problem, which takes the place of the problem of inverse probability, can be shown to follow."
>
> **[Jerzy Neyman, 1934]**

3.5.1 Sampling Distributions of Survey Estimates

Professor Neyman's "distribution of certain characters in repeated samples" is termed the **sampling distribution** of a sample estimate. The theoretical sampling distribution is based on all possible samples of size n that could be selected from a finite population of N elements. Using a single sampling plan, if all possible samples of size n from the N population elements were drawn in sequence, sample estimates were computed for each selected sample, and a histogram of the estimated values was plotted, the shape of the sampling distribution would emerge. Provided that the sample size, n, was sufficiently large, the distribution that would begin to appear as each new sample estimate was added to the histogram would be the familiar bell-shaped curve of a normal distribution.

Figure 3.2 illustrates a set of nine simulated sampling distributions for sample estimates of the population mean. Each individual graph in this figure represents the histogram of sample estimates, $\bar{y}$, computed from 5,000 independent samples from a large finite population with known mean $\bar{Y} = 25$. The nine simulated sampling distributions displayed in this figure represent nine different probability sampling plans—three levels of sample size (n = 500, n = 1,000, n = 5,000), and three levels of cluster sampling (no cluster sampling, and clusters of size B = 10 and B = 50). Since $\bar{y}$ is an unbiased estimator regardless of the sampling plan or the sample size, each sampling distribution is centered at the population mean, $\bar{Y} = 25$. As the sample size decreases or the number of elements sampled from

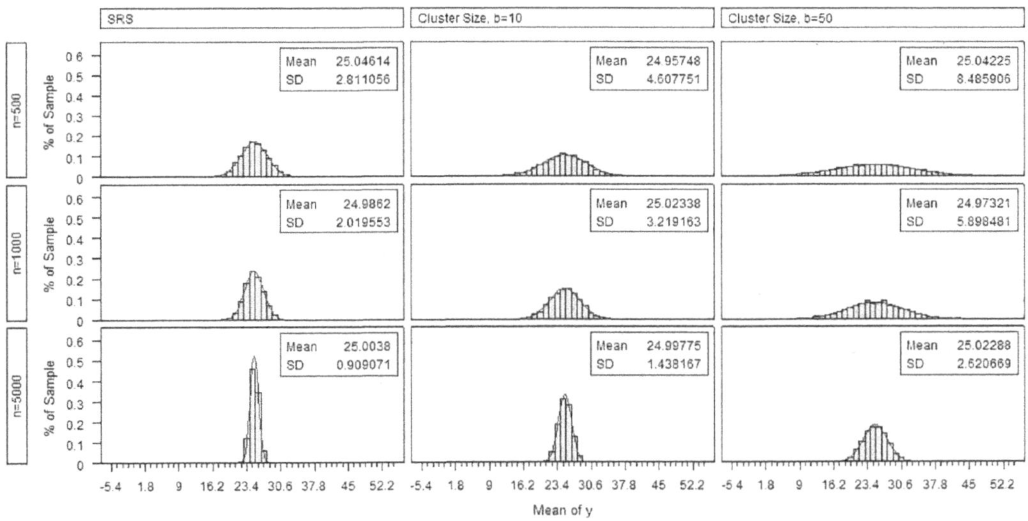

FIGURE 3.2
Sampling distributions for a survey estimate (5,000 simulated samples, $\overline{Y} = 25$).

each cluster increases, the dispersion of sample estimates about the population mean value increases. The degree of dispersion of the sample estimates about the mean of the sampling distribution is the **sampling variance** associated with the sample design, which can be written as:

$$var(\overline{y}) = \sum_{s=1}^{S} P(s) \cdot (\overline{y}_s - E(\overline{y}_s))^2 \tag{3.6}$$

where:

$s = 1,\ldots,S$ indexes all possible samples of size n under the design;

$P(s)$ = the probability that sample s was chosen from the set of S possibilities; and

$\overline{y}_s$ = the estimate for sample s.

The square root of the sampling variance is the **standard error** of a probability sample estimate, denoted by $SE(\overline{y})$. Or equivalently, the standard error of a design-based estimate is simply the **standard deviation** of the sampling distribution.

In real world survey samples, a single sample is observed. It is never practically feasible to observe the full sampling distribution of an estimate, its mean, its variance, or its distributional shape. So how is it possible to make inferential statements based on a sampling distribution that is never observed? Briefly, statistical theory shows that if the sample size n is sufficiently large (e.g., 100 cases), and $\hat{\theta}$ is an unbiased or otherwise consistent estimator of the population parameter θ, the sampling distribution converges to an approximately normal distribution: $f(\hat{\theta}) \sim N(\theta, Var(\hat{\theta}))$. Consequently, the test statistic $t = \dfrac{\hat{\theta} - \theta}{se(\hat{\theta})}$, where both $\hat{\theta}$ and $se(\hat{\theta})$ are *estimated* from the survey sample data, follows the Student t probability distribution with *df* degrees of freedom (to be defined in the following section). This test statistic can be pivoted to derive the following probability statement: $P_s\left\{\hat{\theta}-t_{1-\alpha/2,df}\cdot se(\hat{\theta})\leq\theta\leq\hat{\theta}+t_{1-\alpha/2,df}\cdot se(\hat{\theta})\right\}\cong 1-\alpha$. This re-expression of the test statistic as a range of values for θ is the basis for the 100(1 – α)% confidence interval presented in (3.3) above.

3.5.2 Degrees of Freedom for *t* under Complex Sample Designs

Probability distributions such as the Student *t*, χ^2, and *F* play a critical role in the construction of confidence intervals for population values or as the reference distributions for formal tests of hypotheses concerning population parameters. Included in the quantities that define the shape of these distributions are **degrees of freedom** (*df*) parameters. The degrees of freedom are indices of how precisely the true variance parameter(s) of the reference distribution have been estimated from the sample design. Sample designs with large numbers of degrees of freedom for variance estimation enable more precise estimation of the true variance parameters of the reference distribution. Conversely, the smaller the degrees of freedom afforded by the sample design, the less precisely these variance parameters are estimated. Consider the (1 – $\alpha/2$ = 0.975) critical values for the Student *t* distribution with varying degrees of freedom: $t_{.975,1}$ = 12.706; $t_{.975,20}$ = 2.0860; $t_{.975,40}$ = 2.0211; $t_{.975,\infty} = Z_{.975}$ = 1.9600. Whenever an analyst (or their statistical software) derives a confidence interval or test statistic from sample data, variance parameters that define the appropriate t, χ^2, or *F* reference distribution must be estimated from the sample data.

Precise determination of the degrees of freedom for variance estimation available under the complex sample designs that are often used in practice is difficult. Currently, statistical software for the analysis of complex sample survey data employs a **fixed degrees of freedom rule** to determine the degrees of freedom for the reference distribution used to construct a confidence interval (e.g., $t_{1-\alpha/2,df}$) or a p-value for a hypothesis test (e.g., $P(F > F_{k,d})$):

$$df_{des}=\sum_{h=1}^{H}(a_h-1)=\sum_{h=1}^{H}a_h-H=\#clusters-\#strata \tag{3.7}$$

Interested readers are referred to Theory Box 3.1 for a more in-depth (yet not strictly theoretical) explanation of the basis for this rule.

THEORY BOX 3.1: DEGREES OF FREEDOM FOR VARIANCE ESTIMATION

Consider the pivotal *t* statistic, $t=(\bar{y}-\bar{Y}_0)/se(\bar{y})$. Under simple random sampling:

$$t_{n-1,SRS}=\frac{(\bar{y}-\bar{Y}_0)}{se(\bar{y})}=\frac{(\bar{y}-\bar{Y}_0)}{\sqrt{s^2/n}}=\frac{(\bar{y}-\bar{Y}_0)}{\sqrt{\sum_{i=1}^{n}[(y_i-\bar{y})^2/(n-1)]/n}}.$$

A total of $n - 1$ independent contrasts (squared differences) contribute to the SRS estimate of S^2 (once the mean is known, there are only $n - 1$ unique pieces of information for estimating the variance). Consequently, for an SRS design, the *t* statistic is referred to a Student *t* distribution with $n - 1$ degrees of freedom.

Now, consider this same test statistic, only now under a more complex stratified cluster sample design:

$$t_{df,complex} = \frac{(\bar{y}_{st,cl} - \bar{Y}_0)}{\sqrt{var(\bar{y}_{st,cl})_{complex}}} = \frac{(\bar{y}_{st,cl} - \bar{Y}_0)}{se(\bar{y}_{st,cl})_{complex}}$$

$$var(\bar{y}_{st,cl}) = \sum_{h=1}^{H} W_h^2 \frac{1}{a_h}\left[\frac{1}{(a_h-1)} \cdot \frac{1}{b_h^2}\left\{\sum_{\alpha=1}^{a_h} y_{h\alpha}^2 - \frac{y_h^2}{a_h}\right\}\right]$$

$$= \sum_{h=1}^{H} \frac{W_h^2}{a_h b_h^2}\left[\frac{1}{(a_h-1)}\left\{\sum_{\alpha=1}^{a_h}(y_{h\alpha} - \frac{y_h}{a_h})^2\right\}\right]$$

Under a design with $h = 1, \ldots, H$ strata and $\alpha = 1, \ldots, a_h$ equal-sized PSUs per stratum (i.e., the sample size in each PSU in stratum h is b_h), each stratum contributes $a_h - 1$ independent contrasts to the estimated variance of the mean, var($\bar{y}_{st,cl}$). The t statistic for the complex sample design is no longer referred to a Student t distribution with $n - 1$ degrees of freedom. Instead, the correct degrees of freedom for variance estimation under this complex sample design are:

$$design\ df_{fixed} = \sum_{h=1}^{H}(a_h - 1) = \sum_{h=1}^{H} a_h - H = \#clusters - \#strata.$$

The variance estimation technique known as Taylor Series Linearization in particular (see Section 3.6.2) involves approximating complex non-linear statistics with simpler linear functions of linear statistics, and then estimating the variances of the linear statistics (enabling the use of known variance estimation formulas for complex designs, as above). This gives rise to the use of the fixed degrees of freedom rule when using this variance estimation procedure.

For additional theoretical details on the determination of degrees of freedom in variance estimation under complex sample designs, we refer readers to Valliant and Rust (2010), Korn and Graubard (1999, Section 5.2), and Rust and Rao (1996).

The fixed rule for determining the degrees of freedom is applied by software procedures designed for the analysis of complex sample survey data whenever the full sample is being analyzed. However, different software procedures may use different rules for determining the degrees of freedom for subpopulation analyses. For subpopulations, improved confidence interval coverage of the true population value is obtained using a "variable" degrees of freedom calculation method (Korn & Graubard, 1999):

$$df_{var} = \sum_{h=1}^{H} I_h \cdot (a_h - 1) \tag{3.8}$$

where:

$I_h = 1$ if stratum h has 1 or more subpopulation cases, and 0 otherwise; and

a_h = the number of cluster (PSUs) sampled in stratum $h = 1, \ldots, H$.

The variable degrees of freedom are determined as the total number of clusters in strata with 1+ subpopulation cases minus the number of strata with at least one subpopulation observation. Rust and Rao (1996) suggest the same rule for calculating degrees of freedom for test statistics when replicated variance estimation methods are used to develop standard errors for subpopulation estimates. (See Section 4.5 for a more in-depth discussion of survey estimation for subpopulations.)

In theory, both the "fixed" and "variable" rules only yield an approximation to the true degrees of freedom for a complex sample design. Valliant and Rust (2010) propose a simple estimator of degrees of freedom that leads to improved confidence interval coverage relative to the simple rule above that is currently used by most software packages. At the time of the writing of this third edition, the approach suggested by Valliant and Rust (2010) still has not made its way into statistical software packages including procedures for the analysis of survey data. Our hope is that there will be more work in this area soon, including implementation of improved methods for estimating degrees of freedom based on different types of complex sample designs.

3.6 Variance Estimation

Survey analysis that employs confidence intervals (or hypothesis test statistics and p-values) requires estimation of the sampling variability of the sample estimates. For simple statistics computed from data collected under equal-probability SRS designs or stratified random sample designs, exact expressions for the estimator of the sampling variance may be derived (Cochran, 1977). Survey analysis procedures in Stata or SUDAAN and many other software packages allow the user to choose the exact formula for these simple sample designs.

Consider estimating the population mean under a stratified (*st*), random sample (*ran*) design:

$$\bar{y}_{st,ran} = \sum_{h=1}^{H} W_h \cdot \bar{y}_h = \sum_{h=1}^{H} \frac{N_h}{N} \cdot \bar{y}_h \tag{3.9}$$

An unbiased estimate of the sampling variance of the estimated mean is computed using the following analytical formula:

$$var(\bar{y}_{st,ran}) = \sum_{h=1}^{H} W_h^2 \cdot var(\bar{y}_h) = \sum_{h=1}^{H} \left(1 - \frac{n_h}{N_h}\right) \cdot W_h^2 \cdot \frac{s_h^2}{n_h} \tag{3.10}$$

Note that this exact expression includes the finite population correction factor for each stratum ($1 - f_h$, where $f_h = n_h / N_h$). The variance expression for this most basic of stratified sample designs illustrates an important fact. For stratified samples, the sampling variance of an estimator is simply the sum of the variance contributions from each stratum. There is no between-stratum component of variance! This simple rule applies equally to more complex designs that utilize cluster sampling or unequal probability sampling within strata.

When survey data are collected using a complex sample design with unequal size clusters or weights are used in estimation, most statistics of interest will not be simple linear functions of the observed data, as in the case of (3.9) above, and alternative methods for estimating the variances of the more complex statistics are required. Over the past 50 years, advances in survey sampling theory have guided the development of several methods for correctly estimating variances from complex sample survey data sets. The two most common approaches to the estimation of sampling errors for estimates computed from complex sample survey data sets are: 1) Taylor Series Linearization (TSL) of the estimator (and the corresponding approximation to its variance), or 2) through the use of resampling variance estimation procedures, such as Balanced Repeated Replication (BRR), Jackknife Repeated Replication (JRR), or bootstrapping (Rust, 1985; Rao et al., 1992; Wolter, 2007). BRR, JRR, and the bootstrap technique comprise a second class of nonparametric methods for conducting estimation and inference from complex sample survey data. As suggested by the generic label for this class of methods, BRR, JRR, and the bootstrap utilize replicated subsampling of the sample data to develop sampling variance estimates for linear and non-linear statistics.

The following sections describe the basic features of the Taylor Series Linearization technique and the alternative replication methods for estimating the variance of sample estimates from complex sample survey data.

3.6.1 Simplifying Assumptions Employed in Complex Sample Variance Estimation

Before turning to the computational methods used for variance estimation, it is important to consider two simplifying assumptions that are routinely used in the TSL, JRR, BRR, and bootstrapping procedures available in today's software for survey data analysis:

1. Primary stage units (PSUs) in multi-stage sample designs are selected *with replacement* from the primary stage strata. Any finite population correction for the primary stage sample is ignored. The resulting estimates of sampling variance will be slight overestimates.
2. Multi-stage sampling within selected PSUs results in a single **ultimate cluster** of observations for that PSU. Variance estimation methods based on the ultimate clusters "roll up" the components of variance for the multiple stages into a single-stage formula that only requires knowledge of the primary stage stratum and PSU identifiers to perform the final calculations. All sources of variability nested within the PSU are captured in the composite variance estimate.

Figure 3.3 illustrates the concept of an ultimate cluster for a single sample PSU. Within the sample PSU, two additional stages of sampling identify a unique cluster of 4 elements (units 2 and 5 within sample secondary stage unit (SSU) 1, and units 1 and 4 within sample SSU 2). Under assumption 1 (above), PSUs are initially selected with replacement within first-stage sampling strata. Following this with-replacement selection of PSUs, all possible ultimate clusters that could arise from the multi-stage sampling procedure could be enumerated (in theory) within each selected PSU. (The multi-stage selection process emulates this process of enumerating and sampling a single ultimate cluster of elements to represent the PSU, but at substantially reduced cost and effort.)

Then, the survey statistician could (in theory) select a without-replacement sample of these ultimate clusters from all the PSUs. In practice, one ultimate cluster is generally

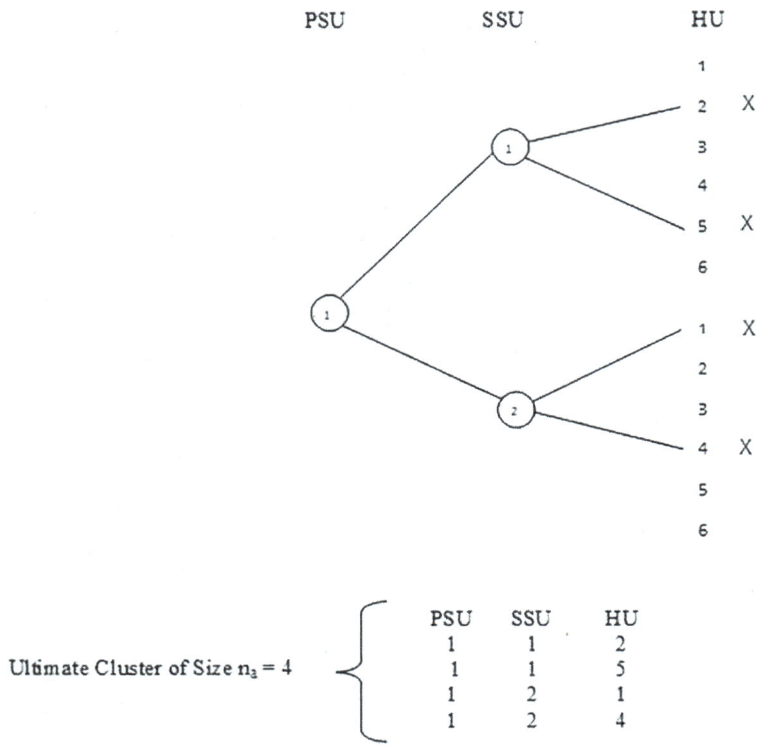

FIGURE 3.3
Schematic illustration of the Ultimate Cluster concept.

selected within each PSU, as shown in Figure 3.3. The resulting sample can be thought of as a single-stage, with-replacement selection of ultimate clusters, where all elements within each selected ultimate cluster are sampled. This greatly simplifies variance estimation via the use of formulae for stratified, with-replacement sampling of ultimate clusters of observations. Since PSUs are typically sampled *without* replacement from within the primary stage strata in practice, use of the simpler variance estimation formulae results in a slight overestimate of the true variance (Kish, 1965, Section 5.3), which is an acceptably conservative approach for making inferences.

Figure 3.4 provides a simple illustration of a survey data set that is ready for analysis. The data set includes $n = 16$ cases. From each of $h = 1, \ldots, 4$ strata, a total of $a_h = 2$ PSUs has been selected *with replacement*. An ultimate cluster of $b = 2$ observations has been selected from each sample PSU. The data set contains a unique stratum code, a PSU (or cluster) code within each stratum, and a case-specific value for the weight variable, w_i. As discussed in more detail in Section 4.3, the stratum, cluster, and weight variables represent the minimum set of **design variables** that are required to perform analyses of sampling variability for complex sample designs. (Note: As an alternative to releasing the detailed stratum and cluster codes, survey data sets may include "replicate weights" for use by software that supports replicated variance estimation—see Section 4.2.1.) This simple example data set will be used in the following sections to illustrate the computational steps in the TSL, JRR, and BRR variance estimation methods.

Stratum	PSU (Cluster)	Case	y_i	w_i
1	1	1	.58	1
	1	2	.48	2
1	2	1	.42	2
	2	2	.57	2
2	1	1	.39	1
	1	2	.46	2
2	2	1	.50	2
	2	2	.21	1
3	1	1	.39	1
	1	2	.47	2
3	2	1	.44	1
	2	2	.43	1
4	1	1	.64	1
	1	2	.55	1
4	2	1	.47	2
	2	2	.50	2

FIGURE 3.4
Data set for example TSL, JRR, and BRR sampling variance calculations: 4 Strata, 2 PSUs per Stratum, Ultimate Clusters of 2 Elements per PSU.

3.6.2 The Taylor Series Linearization (TSL) Method for Variance Estimation

Taylor series approximations of complex sample variances for weighted sample estimates of finite population means, proportions, and linear regression coefficients have been available since the 1950s (Hansen et al., 1953; Kish & Hess, 1959). Woodruff (1971) summarized the general application of the TSL methods to a broader class of survey statistics. Binder (1981) advanced the application of the TSL method to variance estimation for analysis techniques such as logistic regression and other generalized linear models that require iterative maximum likelihood estimation procedures. In general, the TSL approach to variance estimation involves a non-iterative process of five steps.

TSL Step 1: The estimator of interest is written as a function of weighted sample totals.

Consider a weighted, combined ratio estimator of the population mean of the variable y (Kish, 1965):

$$\bar{y}_w = \frac{\sum_h \sum_\alpha \sum_i w_{h\alpha i} y_{h\alpha i}}{\sum_h \sum_\alpha \sum_i w_{h\alpha i}} = \frac{\sum_h \sum_\alpha \sum_i u_{h\alpha i}}{\sum_h \sum_\alpha \sum_i v_{h\alpha i}} = \frac{u}{v} \tag{3.11}$$

Notice that the estimator of the ratio mean can be expressed as a ratio of two weighted totals, u and v, which are sums over the design strata ($h = 1, \ldots, H$), PSUs ($\alpha = 1, \ldots, a_h$), and individual ($i = 1, \ldots, n_{h\alpha}$) cases of the constructed variables $u_{h\alpha i} = w_{h\alpha i} \cdot y_{h\alpha i}$ and $v_{h\alpha i} = w_{h\alpha i}$.

The concept of a sample total is not limited to sums of single variables or sums of weighted values. For reasons that will be explained more fully in later sections, the individual case-level variates used to construct the sample totals for complex sample survey data may be more complex functions involving many variates and functional forms. For example, under a stratified, cluster sample design, the following estimated totals are employed in TSL variance estimation for simple linear and logistic regression model coefficients:

$$u=\sum_{h=1}^{H}\sum_{\alpha=1}^{a_h}\sum_{i=1}^{n_\alpha} w_{h\alpha i}\cdot y_{h\alpha i}\cdot x_{h\alpha i};$$

$$v=\sum_{h=1}^{H}\sum_{\alpha=1}^{a_h}\sum_{i=1}^{n_\alpha} w_{h\alpha i}\cdot x_{h\alpha i}^2;\ \text{or even}$$

$$\text{sc}=\sum_{h=1}^{H}\sum_{\alpha=1}^{a_h}\sum_{i=1}^{n_\alpha}\left[w_{h\alpha i}\cdot x_{h\alpha i}\cdot y_{h\alpha i}-w_{h\alpha i}\cdot x_{h\alpha i}\cdot\frac{e^{\hat{\beta}_0+\hat{\beta}_1\cdot x_{h\alpha i}}}{1+e^{\hat{\beta}_0+\hat{\beta}_1\cdot x_{h\alpha i}}}\right]$$

TSL Step 2: Like many other survey estimators for complex sample survey data, the weighted estimator of the population mean is a *non-linear function* of the two weighted sample totals. Consequently:

$$var(\bar{y}_w)=var\left(\frac{u}{v}\right)\neq\frac{var(u)}{var(v)}$$

To solve the problem of the non-linearity of the sample estimator, a standard mathematical tool, the Taylor Series expansion, is used to derive an approximation to the estimator of interest, rewriting it as a linear combination of weighted sample totals:

$$\bar{y}_{w,TSL}=\frac{u_0}{v_0}+(u-u_0)\left[\frac{\partial\bar{y}_{w,TSL}}{\partial u}\right]_{u=u_0,v=v_0}+(v-v_0)\left[\frac{\partial\bar{y}_{w,TSL}}{\partial v}\right]_{v=v_0,u=u_0}+\textit{remainder}$$

$$\bar{y}_{w,TSL}\cong\frac{u_0}{v_0}+(u-u_0)\left[\frac{\partial\bar{y}_{w,TSL}}{\partial u}\right]_{u=u_0,v=v_0}+(v-v_0)\left[\frac{\partial\bar{y}_{w,TSL}}{\partial v}\right]_{v=v_0,u=u_0}$$

$$\bar{y}_{w,TSL}\cong\text{constant}+(u-u_0)\cdot A+(v-v_0)\cdot B$$

where: A and B symbolically represent the deriatives with respect to u and v, evaluated at the expected values of the samples estimates u_0 *and* v_0.

The quadratic, cubic, and higher-order terms in the full Taylor series expansion of $\bar{y}_w$ are dropped (i.e., the *"remainder"* is assumed to be negligible). Further, consistent (and preferably unbiased) sample estimates are generally used in place of the expected values of the sample estimates (Rao & Wu, 1985).

TSL Step 3: A standard statistical result for the variance of a linear combination (sum) is applied to obtain the approximate variance of the "linearized" form of the estimator, $\bar{y}_{w,TSL}$:

$$\begin{aligned} var(\bar{y}_{w,TSL}) &\cong var[\text{constant} + (u-u_0)\cdot A + (v-v_0)\cdot B] \\ &\cong 0 + A^2 var(u-u_0) + B^2 var(v-v_0) + 2AB cov(u-u_0, v-v_0) \\ &\cong A^2 var(u) + B^2 var(v) + 2AB cov(u,v) \end{aligned}$$

where: $A = \left.\frac{\delta \bar{y}_{w,TSL}}{\delta u}\right|_{u=u_0, v=v_0} = \frac{1}{u_0}$; $B = \left.\frac{\delta \bar{y}_{w,TSL}}{\delta v}\right|_{u=u_0, v=v_0} = -\frac{u_0}{v_0^2}$; *and* u_0, v_0 *are the weighted sample totals computed from the survey data.*

$$\Rightarrow$$

$$var(\bar{y}_{w,TSL}) \cong \frac{var(u) + \bar{y}_{w,TSL}^2 \cdot var(v) - 2\cdot \bar{y}_{w,TSL} \cdot cov(u,v)}{v_0^2}$$

The sampling variance of the non-linear estimator, $\bar{y}_{w,TSL}$, is thus approximated by a simple algebraic function of quantities that can be readily computed from the complex sample survey data. The sample estimate of the ratio mean, $\bar{y}_{w,TSL}$, and the sample total of the analysis weights, v_0, are computed from the survey data. The estimates of var(u), var(v), and cov(u, v) are computed using the relatively simple computational formulas described in Step 4 (below).

TSL Step 4: Under the TSL method, the variance approximation in Step 3 has been derived for most survey estimators of interest, and statistical software such as Stata and SUDAAN provides procedures that permit TSL variance estimation for virtually all the analytical methods used by today's survey data analyst. The sampling variances and covariances of individual weighted totals, u or v, are easily estimated using simple formulae (under the assumption of with-replacement sampling of PSUs within strata at the first stage) that require only knowledge of the subtotals for the primary stage strata and clusters:

$$var(u) = \sum_{h=1}^{H} \frac{a_h}{(a_h - 1)} \cdot \left[\sum_{\alpha=1}^{a_h} u_{h\alpha}^2 - \frac{u_h^2}{a_h} \right];$$

$$var(v) = \sum_{h=1}^{H} \frac{a_h}{(a_h - 1)} \cdot \left[\sum_{\alpha=1}^{a_h} v_{h\alpha}^2 - \frac{v_h^2}{a_h} \right];$$

$$cov(u,v) = \sum_{h=1}^{H} \frac{a_h}{(a_h - 1)} \cdot \left[\sum_{\alpha=1}^{a_h} u_{h\alpha} \cdot v_{h\alpha} - \frac{u_h \cdot v_h}{a_h} \right];$$

where: $u_{h\alpha}$, u_h, $v_{h\alpha}$, v_h *are defined below.*

Provided that the sample stratum codes, cluster codes, and weights are included in the survey data set (see Chapter 4), the **weighted totals** for strata and clusters are easily calculated. These calculations are illustrated here for u: $u_{h\alpha} = \sum_{i=1}^{n_\alpha} u_{h\alpha i}$; $u_h = \sum_{\alpha=1}^{a_h} \sum_{i=1}^{n_\alpha} u_{h\alpha i}$.

Returning to the example data set in Figure 3.4 (see the ASDA website for calculation details):

$$\bar{y}_{w,TSL} = \frac{\sum_h \sum_\alpha \sum_i w_{h\alpha i} y_{h\alpha i}}{\sum_h \sum_\alpha \sum_i w_{h\alpha i}} = \frac{\sum_h \sum_\alpha \sum_i u_{h\alpha i}}{\sum_h \sum_\alpha \sum_i v_{h\alpha i}} = \frac{u}{v} = \frac{11.37}{24} = 0.47375$$

$$v_0 = \sum_h \sum_\alpha \sum_i w_{h\alpha i} = 24$$

$$var(u) = 0.9777;\ var(v) = 6.0000;\ cov(u,v) = 2.4000$$

$$\begin{aligned} var(\bar{y}_{w,TSL}) &\cong \frac{var(u) + \bar{y}_{w,TSL}^2 \cdot var(v) - 2 \cdot \bar{y}_{w,TSL} \cdot cov(u,v)}{v_0^2} \\ &= \frac{0.9777 + 0.4737^2 \cdot 6.0000 - 2 \cdot 0.4737 \cdot 2.4000}{24^2} \\ &= 0.00008731 \end{aligned}$$

$$se(\bar{y}_{w,TSL}) = 0.009344$$

TSL Step 5: Confidence intervals (or hypothesis tests) based on the estimated parameter(s), the linearized standard errors, and the correct degrees of freedom based on the complex sample design are then constructed and reported as output from the TSL variance estimation procedure. We show this calculation for a 95% confidence interval for the population mean based on the example data set in Figure 3.4 (note that the degrees of freedom are computed as df = 8 clusters – 4 strata = 4):

$$CI(\bar{y}_{w,TSL}) = \bar{y}_{w,TSL} \pm t_{1-\alpha/2,\, df} \cdot \sqrt{var_{TSL}(\bar{y}_{w,TSL})}$$

$$\begin{aligned} e.g.,\ CI(\bar{y}_{w,TSL}) &= 0.47375 \pm t_{1-\alpha/2,4} \cdot 0.009344 \\ &\doteq 0.47375 \pm 2.7764 \cdot (0.009344) = (0.44780, 0.49969) \end{aligned}$$

Most contemporary software procedures for analyzing complex sample survey data employ the TSL approach as the default method of computing sampling variances. For example, the following Stata commands illustrate an analysis of the prevalence of at least one lifetime episode of major depression in the NCS-R adult survey population.

```
. svyset seclustr [pweight=ncsrwtsh], strata(sestrat) vce(linearized)
singleunit(missing)

      pweight: ncsrwtsh
          VCE: linearized
  Single unit: missing
     Strata 1: sestrat
         SU 1: seclustr
        FPC 1: <zero>

. svy: mean mde
(running mean on estimation sample)
```

```
Survey: Mean estimation

Number of strata =        42          Number of obs      =      9282
Number of PSUs   =        84          Population size    =      9282
                                      Design df          =        42

--------------------------------------------------------------------
                  |             Linearized
                  |       Mean    Std. Err.     [95% Conf.  Interval]
------------------+-------------------------------------------------
              mde |   .1917112    .0048768      .1818694     .201553
--------------------------------------------------------------------
```

Note that Stata explicitly reports the linearized estimate of the standard error (0.0049) of the weighted estimate of the population mean (0.1917). These Stata commands will be explained in more detail in the upcoming chapters.

3.6.3 Replication Methods for Variance Estimation

Jackknife Repeated Replication (JRR), Balanced Repeated Replication (BRR), and the Bootstrap form a second class of nonparametric methods for computing sampling variances of survey estimates. As suggested by the generic label for this class of methods, JRR, BRR, and the Bootstrap use replicated subsampling of the database of sample observations to develop sampling variance estimates for linear and non-linear statistics (Rust, 1985; Rao et al., 1992; Shao & Tu, 1995; Wolter, 2007).

The precursor to today's JRR, BRR, and Bootstrap techniques for replicated variance estimation was a simple yet statistically elegant technique of interpenetrating samples or replicated sampling developed by Mahalanobis (1946) for agricultural surveys in Bengal, India. Mahalanobis's method requires selecting the complete probability sample of size n as a set of $r = 1, \ldots, R$ independent sample replicates from a common sampling plan. The replicated estimate of θ and the sampling variance of the simple replicated estimate are:

$$\hat{\theta}_{rep} = \sum_{r=1}^{R} \hat{\theta}_r / R, \text{ the mean of the replicate sample estimates;}$$
$$var(\hat{\theta}_{rep}) = \sum_{r=1}^{R} (\hat{\theta}_r - \hat{\theta}_{rep})^2 / R \cdot (R-1). \qquad (3.12)$$

Despite their simplicity, simple replicated sample designs are rarely used in population survey practice today. If many replicate samples are used to obtain adequate degrees of freedom for variance estimation, the requirement that each replicate sample be a "miniature" of the full sample design restricts the design efficiency by limiting the numbers of strata that can be employed at each stage of selection. If a highly efficient stratified design is employed, the number of independent replicates and corresponding degrees of freedom for variance estimation may be seriously limited.

During the late 1950s and 1960s, Mahalanobis' pioneering idea that replicated samples could be used to estimate sampling variances was extended to the Balanced Repeated

Replication (BRR), Jackknife Repeated Replication (JRR), and Bootstrap methods. These techniques draw on the principles of the simple replication method but yield a more efficient procedure for creating the replicates and provide greater efficiency for estimating sampling variances by increasing degrees of freedom for comparable sample sizes and survey costs.

Each of these replication-based methods employs a generic sequence of five steps:

1. Sample replicates ($r = 1, \ldots, R$) of the full sample are defined based on assignment rules tailored to the BRR, JRR, and Bootstrap techniques;
2. Full sample survey weights are revised for each replicate to create $r = 1, \ldots, R$ **replicate weights**;
3. Weighted estimates of a finite population parameter of interest are computed for the full sample and then separately for each replicate (using the replicate weights);
4. A replicated variance estimation formula tailored to the BRR, JRR, or Bootstrap method is used to compute standard errors;
5. Confidence intervals (or hypothesis tests) based on the estimated statistics, standard errors, and correct degrees of freedom are constructed.

JRR, BRR, and the Bootstrap employ different methods to form the replicates (Step 1), which in turn requires minor modifications to Steps 2–4 of the general replicated variance estimation algorithm. The following sections describe each of the basic steps for the most common JRR and BRR methods as well as the Resampling Bootstrap method of Rao et al. (1992).

3.6.3.1 Jackknife Repeated Replication (JRR)

The JRR method of replicated variance estimation is applicable to a wide range of complex sample designs. There are several approaches to constructing JRR replicates and estimating the variance of a survey estimate, including the JK1 and JKn approaches (Wolter, 2007; Valliant et al., 2018). We describe these approaches for JRR replicate formation and the corresponding variance estimation formulae below.

The **JK1 method** assumes that the sample of $\alpha = 1, \ldots, a$ PSUs (or elements) is selected under an unstratified design. Replicates (indexed by r) are formed by dropping a single unit (PSU or element) from the sample and rescaling the full sample weights for all observations in the undeleted units by a factor of $a \,/\, (a - 1)$. The JK1 estimate of variance for a statistic, q, is then computed using the formula:

$$var_{JRR,JK1}(\hat{q}) = \frac{a-1}{a}\sum_{r=1}^{a}(\hat{q}_r - \hat{q})^2$$
$$df = a - 1$$

The **JKn method**, which is the default method in most survey analysis software (e.g. SAS, Stata, SUDAAN, and R), applies to stratified sample designs with two or more PSUs selected per stratum and is the form of JRR estimation that will be used here for an example. This method employs the following five-step sequence:

JRR Step 1: Each JRR replicate is constructed by "deleting" a single PSU from a single primary stage stratum. Operationally, the software procedures that provide

JRR capability do not physically remove the cases for the deleted PSUs in creating each replicate sample. Instead, by assigning a zero or missing weight value to each case in the deleted PSU, the "replicate weight" sweeps the deleted cases out of the computation of the replicate estimate. Sample cases in all remaining PSUs (including those in all other strata) constitute the JRR replicate. See Figure 3.5 below, in which the first PSU in the first stratum is deleted to form a first JRR replicate.

For the JKn variance estimator, each stratum indexed by h will contribute a_h replicates, where for each replicate one of the clusters in the stratum is deleted. In our data example, $H = 4$ and $a_h = 2$ for $h = 1, \ldots, 4$, so the total number of required JRR replicates for the JKn variance estimation is:

$$R = \sum_{h=1}^{4}(a_h) = 8$$

Each stratum will contribute only $a_h - 1$ *unique* JRR replicates, yielding a total of $df = \sum_{h=1}^{H}(a_h - 1) = a - H = \#clusters - \#strata$ degrees of freedom for estimating standard errors of sample estimates. In our example, with two PSUs per stratum, once the total for the stratum and one PSU is known, the total for the second PSU is known (meaning that it is not unique information).

JRR Step 2: A new **replicate weight** is computed for each of the JRR replicates. Replicate weight values for cases in the deleted PSU are assigned a value of "0" or "missing." The replicate weight values for the retained PSU cases in the "deletion stratum" are formed by multiplying their survey weights by a factor of $a_h / (a_h - 1)$. This factor equals 2 in our example. Note that for all strata except the deletion stratum, the replicate weight value is the unaltered survey weight.

Stratum	PSU (Cluster)	Case	y_i	$w_{i,rep}$
1	1	1	.	.
	1	2	.	.
1	2	1	.42	2×2
	2	2	.57	2×2
2	1	1	.39	1
	1	2	.46	2
2	2	1	.50	2
	2	2	.21	1
3	1	1	.39	1
	1	2	.47	2
3	2	1	.44	1
	2	2	.43	1
4	1	1	.64	1
	1	2	.55	1
4	2	1	.47	2
	2	2	.50	2

FIGURE 3.5
Formation of a single JRR replicate for the example data set.

The second JRR replicate is formed by dropping the second PSU in Stratum 1 and weighting up the first PSU in Stratum 1, with the process continuing until all replicates have been formed.

JRR Step 3: Using the replicate weights developed for each JRR replicate sample, the weighted estimate of the population parameter of interest is computed for each replicate. The statistic of interest in our example is the weighted estimate of the population mean, so each replicate estimate takes the following form:

$$\hat{q}_r = \bar{y}_r = \frac{\sum_{i \in rep}^{n_{rep}} y_i \cdot w_{i,\,\text{rep}}}{\sum_{i \in rep}^{n_{rep}} w_{i,\,\text{rep}}} \quad \text{for each of } r = 1,\ldots,8 \text{ replicates.}$$

$$\hat{q}_{1\,(\text{del1\_1})} = 0.47240; \quad \hat{q}_{3(\,\text{del2\_1})} = 0.46958; \quad \hat{q}_{5(\text{del3\_1})} = 0.47434; \quad \hat{q}_{7(\text{del4\_1})} = 0.46615;$$

$$\hat{q}_{2\,(\text{del1\_2})} = 0.47521; \quad \hat{q}_{4(\text{del2\_2})} = 0.47791; \quad \hat{q}_{6(\text{del3\_2})} = 0.47320; \quad \hat{q}_{8(\text{del4\_2})} = 0.48272;$$

The weighted, full sample estimate of the mean is also computed:

$$\hat{q} = \frac{\sum_{i=1}^{n} y_i \cdot w_i}{\sum_{i=1}^{n} w_i} = 0.47375$$

JRR Step 4: Using the $r = 1, \ldots, R$ replicate estimates and the full sample estimate, the sampling variance is estimated using the following simple formula (which assumes the simplified case of two PSUs sampled per stratum, as in our example):

$$\begin{aligned} var_{JRR\_DelOne}(\hat{q}) &= \sum_{h=1}^{H} \frac{a_h - 1}{a_h} \sum_{\alpha=1}^{a_h} \left(\hat{q}_{(h\alpha)} - \hat{q}\right)^2 \\ &= 0.5 \cdot \sum_{r=1}^{8} \left(\hat{q}_r - \hat{q}\right)^2 \\ &= 0.5 \cdot \sum_{r=1}^{8} \left(\bar{y}_r - \bar{y}\right)^2 \\ &= .0000801449 \\ se_{JRR}(\bar{y}) &= \sqrt{.0000801449} = .008952 \end{aligned}$$

If only R replicate weights are provided with the data set, the JRR estimate of variance is the slightly simpler expression:

$$\begin{aligned} var_{JRR}(\hat{q}) &= \frac{R-1}{R} \sum_{r=1}^{R} \left(\hat{q}_r - \hat{q}\right)^2 \\ df &= R - 1 \end{aligned}$$

We note that when using the JKn method, it is entirely possible that different strata will have different counts of PSUs sampled within them. This means that the "coefficients" for each stratum in the first variance estimation formula above, defined by $(a_h - 1) / a_h$, will vary across the strata. Some data producers may provide data users with both replicate weights *and* the values of these coefficients for each replicate (where several replicate weights from the same deletion stratum will have the same coefficient), which are then input to the software for variance estimation purposes. Software procedures enabling JRR for variance estimation will allow users to input these coefficients corresponding to specific replicate weights, and data users in this situation should consult the data user guides carefully for the correct syntax to use in these cases. The slightly simpler expression for the JRR variance used above does not provide an unbiased estimate of the sampling variance for such designs, but still represents a reasonable approximation.

JRR Step 5: A 100(1 – α)% confidence interval for the population parameter is then constructed:

$$\begin{aligned} CI(q) &= \hat{q} \pm t_{1-\alpha/2,\, df}\sqrt{var_{JRR,JKn}(\hat{q})} \\ e.g.,\ CI(\bar{y}) &= 0.47375 \pm t_{1-\alpha/2,4} \cdot 0.008952 \\ &= 0.47375 \pm 2.7764 \cdot (0.008952) = (0.44890,\ 0.49860) \end{aligned}$$

Note that this variance estimation technique therefore only requires the final survey weights for the full sample and the R sets of **replicate weights** in order to perform the appropriate variance estimation. This is what enables survey organizations concerned about protecting respondent confidentiality (see Chapter 1) to produce public-use complex sample survey data sets that do not include stratum or cluster codes (which might be used, in theory, to identify a respondent), and only include the required weight variables for estimation of population parameters and replicated variance estimation. In contrast, stratum and cluster codes would be required to estimate variances using Taylor Series Linearization.

3.6.3.2 Balanced Repeated Replication (BRR)

The balanced repeated replication (BRR) method of variance estimation is a "half-sample" method that was developed specifically for estimating sampling variances under two PSU-per-stratum sample designs. The NCS-R and 2012 HRS data sets used in the example analyses presented in later chapters employ such a sampling error calculation model (see Section 4.3), with two PSUs (clusters) per stratum.

The evolution of the BRR method began with the concept of forming replicates by choosing one half of the sample. For a complex sample design with $h = 1, \ldots, H$ strata and exactly $a_h = 2$ PSUs per stratum, a **half sample replicate** could be formed by choosing 1 of the 2 PSUs from each stratum (e.g., in Figure 3.4, choose PSU 1 in strata 1, 2, and 3, and PSU 2 in stratum 4). By default, this choice of one PSU per stratum would define a **half-sample complement** (e.g., PSU 2 in strata 1, 2, and 3, and PSU 1 in stratum 4). For H strata and 2 PSUs per stratum, there are 2^H possible different half-samples that could be formed—a total of 16 for the simple sample in Figure 3.4, and 4,398,000,000,000 for the $H = 42$ strata in the NCS-R design. Since a complex sample design with H strata and 2 PSUs per stratum provides only H degrees of freedom for variance estimation, the formation of $R > H$ half sample replicates will not yield additional gains in efficiency for a replicated half-sample variance estimate.

BRR variance estimation proceeds using the following five steps.

BRR Step 1: What is the optimal procedure for selecting which half-sample replicates to employ in the variance estimation? McCarthy (1969) introduced the concept of **balanced repeated replication**, in which individual replicates are formed according to a pattern of "+" and "-" symbols that are found in the columns of a **Hadamard matrix.** The optimal efficiency of the BRR method for variance estimation based on half samples is due to its "balancing"—that is, complete algebraic cancellation of unwanted between-stratum cross-product terms, for example, $(y_{h1}-y_{h2})\cdot(y_{g1}-y_{g2})$, that enter the half-sample variance computation formula. Readers interested in a more complete mathematical development of this idea are referred to Wolter (2007).

Fortunately, contemporary software for survey data analysis makes it easy to apply the BRR variance estimation method. Using only sampling error stratum and cluster codes (see Section 4.3.1) provided with the survey data set, the software will invoke the correct form of the Hadamard matrix to construct the sample replicates. Figure 3.6 below illustrates the 4 × 4 Hadamard matrix than can be used to define BRR replicates for the 4-strata sample design in Figure 3.4. Each row of the matrix defines one BRR replicate. For BRR replicate 1, the "+" sign in the columns for Strata 1, 2, and 3 indicates that the 1st PSU in the stratum is assigned to the replicate. The "–" sign in the Stratum 4 column indicates the 2nd PSU is to be included in replicate 1. Likewise, BRR replicate 2 will include PSU 1 from stratum 1 and PSU 2 from strata 2, 3, and 4. Figure 3.7 illustrates the form of the 1st BRR replicate for the Figure 3.4 data set.

When a BRR half-sample replicate is created by choosing one PSU from each stratum, a "complementary" half-sample is also created from the unchosen paired PSUs in each stratum. There are several BRR variance estimation formulae that utilize both the BRR half-samples and their complementary half-samples to estimate variance (Rust, 1985). A special case of the BRR variance estimates that utilizes replicate sample estimates from both the original half samples and their complements is **Fay's method** (see below).

Hadamard matrices are defined only for dimensions that are multiples of four. Whenever the number of primary strata defined for a complex sample design is a multiple of four, exactly H BRR replicates are defined according to the patterns of "+/–" indicators in the rows and columns of the $H \times H$ Hadamard matrix. The corresponding BRR variance estimates are said to be **fully balanced.** If the number of primary strata in the complex sample

BRR Rep	**Stratum (h)**			
	1	**2**	**3**	**4**
1	+	+	+	–
2	+	–	–	–
3	–	–	+	–
4	–	+	–	–

FIGURE 3.6
Example Hadamard Matrix used to define BRR replicates for an $H = 4$ stratum sample design.

Stratum	PSU (Cluster)	Case	y_i	$w_{i,rep}$
1	1	1	.58	1×2
	1	2	.48	2×2
1	2	1	.	.
	2	2	.	.
2	1	1	.39	1×2
	1	2	.46	2×2
2	2	1	.	.
	2	2	.	.
3	1	1	.39	1×2
	1	2	.47	2×2
3	2	1	.	.
	2	2	.	.
4	1	1	.	.
	1	2	.	.
4	2	1	.47	2×2
	2	2	.50	2×2

FIGURE 3.7
Illustration of BRR Replicate 1 for the example data set.

design is not a multiple of four, the Hadamard matrix of dimension equal to the next multiple of 4 greater than H is used, and any additional columns beyond H are dropped. For example, a Hadamard matrix of dimension 44 × 44 is used to define the 44 half-sample replicates for the NCS-R, which has $H = 42$ strata for variance estimation, and the final two columns of this matrix are dropped. In such cases, the corresponding BRR variance estimates are said to be **partially balanced**.

BRR Step 2: A new replicate weight is then created for each of the $h = 1, \ldots, H$ BRR half sample replicates created in Step 1. Replicate weight values for cases in the complement half sample PSUs are assigned a value of "0" or "missing." The replicate weight values for the cases in the PSUs retained in the half sample are formed by multiplying their survey weights by a factor of 2.

BRR Step 3: Following the same procedure outlined for JRR and using the replicate weights developed for each BRR replicate sample, the weighted replicate estimates of the finite population parameter are computed. The full sample estimate of the population parameter is also computed:

$$\hat{q}_r = \bar{y}_r = \frac{\sum_{i \in rep}^{n_{rep}} y_i \cdot w_{i,rep}}{\sum_{i \in rep}^{n_{rep}} w_{i,rep}};$$

$$\hat{q}_1 = 0.4708; \quad \hat{q}_2 = 0.4633; \quad \hat{q}_3 = 0.4614; \quad \hat{q}_4 = 0.4692;$$

$$\bar{y}_w = \hat{q} = \frac{\sum_{i=1}^{n} y_i \cdot w_i}{\sum_{i=1}^{n} w_i} = 0.4737$$

BRR Step 4: The BRR estimate of the sampling variance of the weighted full sample estimate is computed using one of several simple formulas. Here we illustrate the computation using one of the more common half-sample variance estimation formulae:

$$var_{BRR}(\bar{y}_w) = var_{BRR}(\hat{q}) = \frac{1}{R}\sum_{r=1}^{R}(\hat{q}_r - \hat{q})^2$$
$$= \frac{1}{4}\sum_{r=1}^{4}(\bar{y}_r - \bar{y})^2$$
$$= 0.00007248$$
$$se_{BRR}(\bar{y}_w) = \sqrt{0.0000748} = 0.008515$$

Several software packages, such as WesVar PC, permit users to choose alternative half-sample variance estimation formulae, including **Fay's method** (Judkins, 1990). Interested users are referred to Rust (1985) or Wolter (2007) for more information on alternative half-sample variance estimation formulae.

BRR Step 5: A 100(1 – α)% confidence interval for the population parameter is then constructed (recall that in the case of BRR, $df = H$):

$$CI(q) = \hat{q} \pm t_{1-\alpha/2,\,df}\sqrt{var_{BRR}(\hat{q})}$$
$$\text{e.g., } CI(\bar{y}) = 0.4737 \pm t_{1-\alpha/2,4} \cdot 0.008515$$
$$= 0.4737 \pm 2.7764 \cdot (0.008515) = (0.45011, 0.49739)$$

3.6.3.3 Fay's BRR Method

As noted above, the literature on BRR variance estimation for complex sample survey data includes several alternatives to the estimators presented in Step 4 above. One alternative that deserves mention is **Fay's method**. Each Fay BRR replicate "blends" PSU contributions from the $a_h = 2$ PSUs in each stratum using a Fay factor or coefficient (ϵ) with $0 <= \epsilon < 1$. If an element (r = replicate, h = stratum) of the Hadamard matrix equals "+", then the survey weight for cases in the first PSU is multiplied by (2 – ϵ), and the survey weight for cases in the second PSU is multiplied by ϵ to form the replicate weight for replicate $r = 1, \ldots, R$. If element (r, h) of the Hadamard matrix equals "–", the scaling is reversed. The survey weights are multiplied by ϵ for cases in the first PSU in stratum h and by (2 – ϵ) in the second PSU to form the r-th replicate weight.

The replicated variance estimator for BRR replicates that are formed using the Fay factor in this fashion is:

$$var_{BRR,FAY}(\hat{q}) = \frac{1}{R \cdot (1-\varepsilon)^2}\sum_{r=1}^{R}(\hat{q}_r - \hat{q})^2 \tag{3.13}$$

Note that if the Fay factor, ϵ, is set to zero, this expression reduces to the standard BRR variance estimator in Step 4 above. In practice, Fay factors in the range of ϵ = (0.3–0.5) have been shown to work well (Judkins, 1990; Valliant et al., 2018).

All major software procedures for the analysis of complex sample survey data support BRR estimation of variances, including the optional Fay method.

3.6.3.4 The Bootstrap (Rao-Wu Rescaling Bootstrap)

Next, we consider the five steps involved in estimating variance using a popular variation of the bootstrap technique for complex sample survey data known as the **Rao-Wu Rescaling Bootstrap**.

Bootstrap Step 1: The Rao-Wu Rescaling Bootstrap (Rao et al., 1992) forms $b = 1, \ldots, B$ bootstrap replicates by sampling with replacement (SWR) m_h PSUs from each of the $h = 1, \ldots, H$ primary stage strata. Rust and Rao (1996) recommend setting $m_h = a_h - 1$ (one less than the number of sample PSUs in stratum h). In the example that we consider here, a total of $B = 100$ bootstrap samples are selected from the data displayed in Figure 3.4.

Bootstrap Step 2: For $m_h = a_h - 1$, the bootstrap weight for the each of the B selected replicates is:

$$w_{h\alpha i}^{(b)} = w_{h\alpha i} \cdot \frac{a_h}{(a_h - 1)} \cdot r_{h\alpha}^{(b)};$$

where: $r_{h\alpha}^{(b)}$ = the number of times PSU α is selected in replicate b.

Bootstrap Step 3: Using the replicate weights developed for each Bootstrap replicate sample, the weighted replicate estimates of the population parameter are computed. The weighted full sample estimate of the population parameter is also computed:

$$\hat{q}_{(b)} = \bar{y}_{(b)} = \frac{\sum_{i \in rep}^{n_{rep}} y_i \cdot w_{i,rep}}{\sum_{i \in rep}^{n_{rep}} w_{i,rep}}; \quad \hat{q}_{(b)} \in (0.4620, \ldots, 0.4900)$$

$$\bar{y}_w = \hat{q} = \frac{\sum_{i=1}^{n} y_i \cdot w_i}{\sum_{i=1}^{n} w_i} = 0.4737$$

Bootstrap Step 4: The variance estimate based on the $b = 1, \ldots, B$ bootstrap estimates is the Monte Carlo approximation:

$$var_{Boot}(\hat{q}) = \frac{1}{B} \sum_{b=1}^{B} (\hat{q}_{(b)} - \hat{q})^2$$

$$var_{Boot}(\bar{y}_w) = \frac{1}{100} \sum_{b=1}^{100} (\bar{y}_{(b)} - \bar{y})^2 = 0.0000844$$

$$se_{Boot}(\bar{y}_w) = \sqrt{0.0000844} = 0.0091865$$

Bootstrap Step 5: Symmetric confidence intervals can be constructed based on Bootstrap estimates of the standard error:

$$
\begin{aligned}
CI(q) &= \hat{q} \pm t_{1-\alpha/2,\,df}\sqrt{var_{Boot}\left(\hat{q}\right)} \\
\text{e.g., } CI(\bar{y}) &= 0.4737 \pm z_{1-\alpha/2}\cdot 0.0092 \\
&= 0.4737 \pm 1.9600\cdot(0.0092) = (0.4557, 0.4918)
\end{aligned}
$$

For large numbers of bootstrap replicates (e.g., $b > 1{,}000$), a histogram of the bootstrap replicate estimates simulates the sampling distribution of the estimator. This bootstrap "simulation" of the sampling distribution may be used to examine the asymptotic normality assumption or to derive asymmetric CIs for population values. See Kolenikov (2010) for additional discussion of these ideas.

Rao (2016) proposed a general framework for hypothesis testing with complex samples that relies on bootstrap replicates to approximate the distribution of a given *test statistic* (e.g., the Rao-Scott chi-square test statistic; see Chapter 6) under a null hypothesis. This approach has since been elaborated on by Kim et al. (2023). Given this approximate distribution of the test statistic under a null hypothesis (based on the bootstrap replicates), one would reject the null hypothesis if the weighted test statistic based on the full sample exceeds some critical value based on the approximated distribution.

3.6.3.5 Construction of Replicate Weights for Replicated Variance Estimation

At the foundation of replicate methods of variance estimation for complex samples is the concept that each JRR, BRR, or Bootstrap replicate is itself a weighted replication of the full sample design. Replicated variance estimation software by default requires the analyst to supply the stratum codes, cluster codes, and full sample weights that are needed to define the sample replicates and compute the replicate weights. Following this approach, replicate sample adjustments are applied to the full sample weight; post-selection weighting adjustments for nonresponse or post-stratification are *not* recomputed based only on the replicate samples. In theory, to capture the true variability introduced by these added weighting steps and minimize bias for estimated variances, these post-survey weighting adjustment steps should be repeated for each replicate sample (Rust & Rao, 1996). Since replicate re-weighting for nonresponse or post-stratification requires detailed and possibly confidential data (e.g., detailed characteristics of respondents *and* nonrespondents) that is not generally available to public data users, the re-weighting process is generally performed by the data producer and the resulting replicate weights are released with the data set.

What should an analyst do if the data producer does not supply replicate weights and only the stratum codes, cluster codes, and full sample weights are available? Analytically, it is impossible to accurately determine when replicate re-weighting will make a significant difference in reducing the bias in replicated variance estimates. A substantial amount of empirical evaluation based on large household survey data sets has found that replicate re-weighting for nonresponse and post-stratification produces few significant changes in estimated variances (Thompson, 2005; Saavedra & Harding, 2009; Chowdhury, 2013), and that any bias in variance estimates based on replicate weights that only apply the simple replicate weighting adjustments to the full sample weight will not be substantial.

The absence of major differences in these and other empirical examples should not be taken as evidence that re-computation of weighting adjustments for replicate samples does not improve the accuracy of complex sample variance estimates for survey statistics. In complex sample designs for populations such as school students and businesses, where extensive population data can be used effectively to achieve large post-stratification gains, re-weighted replicate creation on the part of the data producer will ensure that those gains are reflected in the variance estimates.

3.6.4 An Example Comparing Results from TSL, JRR, BRR, and the Bootstrap

The previous sections have presented simple illustrations of the TSL, JRR, BRR, and Bootstrap variance calculations based on the example data set in Figure 3.4. While these simple example calculations point out that estimated standard errors for the four calculation methods may in fact differ, these differences are magnified by the small sample size of the example data set. In this section, we present a more realistic comparison based on data from the HRS.

Several empirical and simulation-based studies have compared the properties of the TSL, JRR, and BRR methods of variance estimation for complex sample designs, including Kish and Frankel (1974), Rao and Wu (1985), Kovar et al. (1988), and Valliant (1990). Across a range of estimators and sample designs, these studies have shown few important differences in the results obtained using these three methods. The methods are unbiased and produce identical results in the special case where the estimator of interest is a linear statistic such as a weighted sample total.

For non-linear estimators commonly employed in survey analysis (e.g., regression coefficients), the TSL and JRR methods tend to have slightly lower bias (and smaller mean squared error) for estimates of sampling variance. However, confidence intervals constructed using BRR or Bootstrap estimates of standard errors provide better nominal coverage (e.g., 95 in 100 for a 95% CI) of true population parameters. For non-smooth functions of the sample data (e.g., sample quantiles), linearization cannot be applied directly (due to the need for smooth, continuous functions to compute derivatives), and the JRR method is known to result in biased estimates of the variance. As a result, BRR is often used in these situations to estimate variances (see Chapter 5 for more details).

Tables 3.2 and 3.3 present the results of simple analyses of the HRS Wave 1 (1992) data that illustrate the general equivalency of the TSL, JRR, BRR, and Bootstrap methods for sampling variance estimation. The comparison includes weighted estimates of the parameter of interest (identical for all methods) and estimated standard errors of the weighted estimates, computed using 7 methods: 1) SRS; 2–3) BRR and JRR estimates, computed using replicate weights that incorporate HRS nonresponse and post-stratification adjustments specific to each replicate sample; 4–6) TSL, BRR, and JRR estimates, using only replication adjustments to the full sample weight in the case of BRR and JRR; and 7) Bootstrap estimates from the contributed `survey` package in R that also use only the full sample weight to derive bootstrap weights.

The analysis summarized in Table 3.2 focuses on three descriptive estimates of characteristics of the HRS Wave 1 population: 1) mean years of schooling completed; 2) body weight in pounds; and 3) number (out of 10) of words recalled in a test of memory. Table 3.3 presents the estimated coefficients and comparative standard errors in a linear regression model for the number of words recalled, using respondent gender, race/ethnicity, and years of schooling as predictors.

Since each method employs the same overall weighted estimate, the point estimates of the population means (Table 3.2) or regression parameters (Table 3.3) are identical under the TSL, JRR, BRR, and Bootstrap methods (as they should be; the variance estimation

TABLE 3.2

Comparison of the TSL, JRR, BRR, and Bootstrap variance estimation methods for the estimation of descriptive population parameters

Statistic	n	Estimated Mean	Standard Error by Method						
			SRS	W-BRR	W-JRR	TSL	S-BRR	S-JRR	R-Boot
Years of School	9759	12.31	0.031	0.078	0.077	0.077	0.077	0.077	0.076
Body Weight (lbs)	9759	172.35	0.374	0.447	0.432	0.435	0.434	0.435	0.404
Words Recalled	9085	7.57	0.028	0.064	0.064	0.066	0.066	0.066	0.067

(*Source:* 1992 HRS)
Notes: W = With nonresponse and post-stratification adjustments to each replicate weight; S = Standard replication adjustments in the case of BRR and JRR; R = bootstrap variance estimation in the R `survey` package.

TABLE 3.3

Comparison of the TSL, JRR, BRR, and Bootstrap variance estimation methods for the estimation of a linear regression model for number of words recalled

Coefficient	$\hat{\beta}$	Standard Error by Method						
		SRS	W-BRR	W-JRR	TSL	S-BRR	S-JRR	R-Boot
Intercept	4.88	0.120	0.170	0.170	0.166	0.170	0.166	0.178
Male	−0.93	0.052	0.063	0.065	0.064	0.064	0.064	0.061
Black	−1.24	0.084	0.089	0.091	0.093	0.094	0.093	0.096
Hispanic	−0.51	0.107	0.140	0.143	0.141	0.141	0.141	0.139
School Years	0.27	0.009	0.011	0.012	0.011	0.011	0.011	0.012

(*Source:* 1992 HRS)
Notes: W = With nonresponse and post-stratification adjustments to each replicate weight; S = Standard replication adjustments in the case of BRR and JRR; R = bootstrap variance estimation in the R `survey` package.

method selected should not impact the point estimate). To two decimal places, estimated standard errors are virtually identical under the TSL, JRR, BRR, and Bootstrap methods. Any differences in the third decimal place are sufficiently small enough to be negligible in any practical survey analysis setting. Furthermore, comparison of the BRR and JRR estimates with replicate-specific adjustments for nonresponse to those that use only the full sample weight show few important trends or differences in the two series of standard error estimates. As discussed in Section 3.6.3.5 above, the absence of major differences in this empirical example should not be taken as evidence that re-computation of weighting adjustments for replicate samples does not improve the accuracy of complex sample variance estimates for survey statistics.

3.7 Hypothesis Testing in Survey Data Analysis

Throughout this book, we will emphasize the use of confidence intervals (or credible intervals in the case of Bayesian analysis) as the basis for making inferences concerning the population value of a descriptive parameter or to display the uncertainty associated with an estimated regression parameter. However, inference in the analysis of complex sample survey data is not limited to the construction of confidence (or credible) intervals for population parameters.

While we advise caution in using *p*-values to establish a "bright line" demarcation between significant and nonsignificant findings (Nuzzo, 2014), complex sample survey data may also be used to derive familiar test statistics (t, χ^2, F) for formal hypothesis tests of the form H_0: $\theta = \theta_0$ vs. H_A: $\theta \neq \theta_0$. In fact, as the discussion in Section 3.5.1 pointed out, the Student t test statistic is the **pivotal statistic** from which the expression for the $100(1 - \alpha)\%$ CI is derived. Table 3.4 compares the test statistics and hypothesis testing approaches for simple random sample (or independently and identically distributed) data to the modified approaches that have been developed for complex sample survey data.

Student t tests based on single weighted estimates of means, proportions, regression coefficients, and differences or linear combinations of these statistics incorporate the complex sample design by using design-adjusted standard errors in the computation of the test statistic. To test the null hypothesis, the test statistic is referred to a Student t distribution with degrees of freedom adjusted for the complex sample design. Chapter 5 will describe design-adjusted Student t tests for means, proportions, differences of means

TABLE 3.4

Comparisons of hypothesis testing procedures for simple random sample survey data vs. complex sample survey data

Simple Random Sample (i.i.d.) Survey Data	Complex Sample Survey Data
Student t-tests of hypotheses $(H_0 \mid H_A)$ for means, proportions, or single model parameters; for example, $\bar{Y} = \bar{Y}_0;\ \beta_j = 0$	Design-adjusted Student t-test • Correct standard error in denominator • Design-adjusted degrees of freedom
Student t-tests of simple hypotheses concerning differences or linear combinations of means; for example, $(\bar{Y}_1 - \bar{Y}_2) = 0;\ \sum_j a_j P_j = 0$	Design-adjusted Student t-test • Correct standard error in denominator, reflecting separate estimates of variances and covariances of component estimates • Design-adjusted degrees of freedom
X^2 test of independence (association) in bivariate and multi-way tables: • Pearson X^2, Likelihood Ratio G^2	Design-adjusted X^2 and F-tests • Rao-Scott first- and second-order corrections adjust for design effects in $\hat{\Sigma}(p)$ • X^2 transformed to F-test statistic
Full and partial F-tests of hypotheses for linear regression model goodness of fit and full vs. reduced model; for example, $H_0 : \beta = \{\beta_1, \beta_2, \ldots, \beta_p\} = 0;$ $H_0 : \beta_{(q)} = \{\beta_{p-q}, \ldots, \beta_p\} = 0.$	Design-adjusted Wald X^2 or F-test • Correct $\hat{\Sigma}(\hat{\beta})$ under complex sample design • Adjusted degrees of freedom
F-tests based on expected mean squares for ANOVA-type linear models	Linear regression parameterization of the ANOVA model. Design-adjusted Wald X^2 or F-tests as in linear regression above.
Likelihood Ratio X^2 tests for maximum likelihood estimates of parameters in generalized linear models; for example, $H_{0,MLE} : \beta = \{\beta_1, \beta_2, \ldots, \beta_p\} = 0;$ $H_{0,MLE} : \beta_{(q)} = \{\beta_{p-q}, \ldots, \beta_p\} = 0.$	Design-adjusted Wald X^2 or F-test • Correct $\hat{\Sigma}(\hat{\beta})$ under complex sample design • Adjusted degrees of freedom Design-adjusted Likelihood Ratio Test (Lumley & Scott, 2014)

and proportions, subpopulation estimates, and general linear combinations of estimates. Design-adjusted t tests for single parameters in linear regression or generalized linear models are addressed in Chapters 7, 8 and 9.

For tests of association between two cross-classified categorical variables, the standard Pearson X^2, Likelihood Ratio G^2, and Wald X^2 are replaced by design-adjusted forms of these traditional test statistics. Like their SRS counterparts, under the null hypothesis, the Rao-Scott "adjusted" X^2 and G^2 statistics and the adjusted Wald X^2 (computed with a design-based variance-covariance matrix) are expected to follow a central Chi-square distribution. Software packages offer the user the option of using the standard χ^2 version of each test statistic or a modified version that under H_0 is referred to an F distribution. More details on these design-adjusted tests of association for cross-classified data with accompanying examples are provided in Chapter 6.

F-tests are used in standard normal linear regression to test the overall fit of the model as well as to perform joint tests concerning the significance of subsets of the model parameters: for example, $H_0 : \boldsymbol{\beta} = \{\beta_1, \beta_2, ..., \beta_p\} = \mathbf{0}; H_0 : \boldsymbol{\beta}_{(q)} = \{\beta_{p-q}, ..., \beta_p\} = \mathbf{0}$. In the analysis of complex sample survey data, these conventional F-tests are replaced with a Wald X^2 test statistic that is provided both as a chi-square test or as a transformed F-test statistic. Chapter 7 addresses the use of the Wald test statistic for joint hypothesis tests involving multiple parameters in estimated linear regression models.

Standard approaches to analyses involving generalized linear models (e.g., logistic, probit, and Poisson regression models) typically employ maximum likelihood methods to estimate parameters and standard errors. Tests of hypotheses concerning the significance of a nested subset of the full vector of model parameters, that is, $H_0 : \boldsymbol{\beta}_{(q)} = \{\beta_{p-q}, ..., \beta_p\} = \mathbf{0}$, are therefore performed using the standard **likelihood ratio test (LRT)**. When generalized linear models (GLMs) are fitted to complex sample survey data, multi-parameter tests of this type are generally conducted using a design-adjusted Wald X^2 or F statistic. Lumley and Scott (2014) have outlined design-adjusted versions of likelihood ratio tests as well. These design-adjusted tests for GLM analyses are described in more detail in Chapters 8 and 9.

We note that all the approaches described thus far have been specific to the type of analysis being performed. Kim et al. (2023) have recently developed a unified and general approach to testing hypotheses based on analyses of complex samples using the type of bootstrapping approach discussed in Section 3.6.3.4. This approach, in which a variety of common test statistics can be computed using bootstrap replicate estimates of quantities of interest, does not require the use of design effects or estimated variance-covariance matrices accounting for complex sampling features, and can be easily applied with software that accepts bootstrap replicate weights. We view this new development as an attractive solution for data producers generating replicate weights for data users.

3.8 Bayesian Approaches to Finite Population Inference

Sections 3.3 to 3.7 of this chapter primarily focus on **design-based approaches** to making inference about finite populations. Models certainly play an important role in defining these approaches, but in each case, the estimation and inference for finite population parameters is based on the final survey weights (incorporating unequal probabilities of selection into the sample, nonresponse adjustments, and calibration adjustments) and the large sample properties associated with the expected sampling distribution of estimates under repeated sampling using the chosen probability sample design. As described in Section 2.2, the "tried and true" design-based techniques for the analysis of complex sample survey data are appealing because they require minimal assumptions, produce consistent estimates of the target population parameters, and provide robust measures of the sampling variability of the estimates.

Most survey practitioners accept these design-based methods as the method of choice for simple descriptive estimation of population characteristics. However, *model-driven* approaches to the descriptive analysis of survey data, including **Bayesian analysis methods**, can result in more efficient estimates (Si et al., 2015; Liu & Chen, 2020). Recent years have seen significant developments in terms of the methods and software for implementing Bayesian approaches to descriptive inference (Dong et al., 2014; Zhou et al., 2016b; Liu & Chen, 2020; Williams & Savitsky, 2021; Hornby et al., 2023), and we provide an overview of such approaches in this section.

There has been some controversy over time about the best analytic approaches for multivariable models, such as linear regression and logistic regression (see Section 3.4 and Theory Box 7.3). As a practical choice, most survey analysts have employed design-based approaches for multivariable modeling, due to their ease of implementation and the widespread availability of software procedures implementing such approaches. However, software implementing Bayesian approaches for making more efficient finite population inferences regarding multivariable model parameters is now more widely available, and we discuss these approaches in this section as well.

In general, recent years have seen a significant amount of research into model-based approaches for making finite population inferences based on survey data, the majority of which rely on Bayesian methods. Analysts of survey data should now consider this Bayesian inferential framework as an important alternative to classical design-based approaches. In the later example-focused chapters of this third edition of *Applied Survey Data Analysis*, we replicate as many analysis examples as possible following this general Bayesian framework and demonstrate available software procedures implementing these approaches. For now, we begin our discussion of Bayesian approaches to making finite population inferences with an overview of the basic ideas of Bayesian analysis.

3.8.1 Overview of Bayesian Analysis

We assume that readers of this third edition have some prior working knowledge of Bayesian analysis methods. We introduce terminology that might seem foreign to some readers without rigorous definitions. We recommend Gelman and Hill (2006) and Gelman et al. (2013) for readers who are interested in more general coverage of Bayesian data analysis techniques outside of the complex sampling setting.

Many leading survey statisticians see a model-based approach as providing a necessary principled framework for addressing problems in survey inference:

> "Bayesian methods provide a unified framework for addressing all of the problems of survey inference, including inferences for descriptive or analytical estimands, small or large sample sizes, inference under planned ignorable sample selection methods such as probability sampling, and problems where modeling assumptions play a more central role such as missing data or measurement error."
>
> **(R.J.A. Little in Chambers & Skinner, 2003)**

Briefly, **Bayesian inference** is focused on the **posterior distribution** of some parameter of interest θ (e.g., a finite population mean), denoted by $p(\theta \mid y)$, under an assumed model for the observed data y, denoted by the **likelihood function** $f(y \mid \theta)$, and a posited **prior distribution** for the parameter θ, denoted by $p(\theta)$. A Bayesian analysis of complex sample survey data involves defining 1) a prior distribution for the parameter(s) of interest, possibly using external sources (see Wiśniowski et al., 2020, for an example of the benefits of deriving informative priors using non-probability samples), and 2) a likelihood function for the data, given the model and the parameters of interest, which combines with the specified prior distribution to define a posterior distribution for the parameter(s) of interest. One then analyzes the data to simulate draws of the parameters of interest from this posterior distribution. These draws are then used to either 1) make inference about the parameter(s) of interest, or 2) determine a **posterior predictive distribution** for the unobserved elements of the population. Once a posterior predictive distribution is available, predictions on survey variables for elements not included in the sample can be computed, and inferences can be made regarding the finite population.

Inferences are generally based on medians or means of the draws from a given posterior distribution (which are the Bayesian "point estimates"), along with **credible intervals** for the parameter(s) of interest. These intervals are formed based on selected percentiles from the distribution of a large sample of draws from the (possibly simulated) posterior distribution. For example, given 1,000 draws from a posterior distribution, the 2.5 and 97.5 percentiles from this distribution of draws can be used to define a 95% credible interval for the parameter of interest. Importantly, while these credible intervals can be used to make inference about the value of a finite population parameter, they have a very different interpretation from *confidence* intervals. Credible intervals are literally probability statements; for example, if a 95% credible interval for a mean is (3.6, 7.2), we would say that the probability that the finite population mean lies between 3.6 and 7.2 is 0.95. This is quite different from confidence intervals, which need to be interpreted in the context of repeated sampling from a finite population (see Section 3.3).

Bayesian analysis methods are mathematically and computationally intensive, and often require simulations to determine posterior distributions. A popular choice for estimating a posterior distribution is the **Gibbs sampling** technique (see Gelman et al., 2013, for an overview). Complex sample design effects, which primarily arise from stratification, cluster sampling, and weighting due to differential inclusion probabilities/nonresponse adjustment/calibration adjustment, can be directly modeled or not, depending on the degree to which they are **informative** about the variables of interest in the analysis; we discuss this in more detail in Section 3.8.2.

Compared to a strictly design-based approach to inference, this is a distinguishing feature of more model-driven approaches. Design-based methods always incorporate weights and design variables such as strata and clusters in the analysis, which ensures robust variance estimation but sometimes leads to losses in efficiency (i.e., larger standard errors). Bayesian methods for survey data analysis do not rely on large-sample asymptotic theory; but in exchange, they are heavily model-driven and require many assumptions, especially concerning choices of prior distributions and distributions for the observed data.

In addition to the unifying theoretical framework cited in the opening quote in this section, Little (2003) identifies several features that should make the Bayesian approach appealing to survey analysts:

- Standard design-based inference (confidence intervals, test statistics) can be derived in the Bayesian framework using reasonable assumptions concerning the distribution of the data and the prior distribution for the parameter of interest.

- Survey applications typically assume noninformative priors; however, information gleaned from repeated and past surveys could be used to develop informative priors—e.g., information on the tails of the distribution that generates outlier values in a single survey.
- When sample sizes are large and diffuse priors apply, Bayesian estimates and credible intervals based on (possibly simulated) posterior distributions are quite like design-based estimates and confidence intervals, and ideally have frequentist properties that are in-line with design-based estimates (Little, 2015).

3.8.2 Accounting for Complex Sampling Features in Bayesian Analysis

As noted above, a Bayesian approach to inference from survey data can, if necessary, directly incorporate the features of the complex sample design in estimation and inference. **Ignorable sampling mechanisms** are important when making inferences based on Bayesian methods. If the sampling mechanism (e.g., stratification variables, measures of size in probability proportionate to size (PPS) sampling, cluster definitions) and the parameter of interest (e.g., mean household net worth) are *a priori* independent, and the sampling indicator (or equivalently the probability of inclusion for a population element in the sample) is independent of the variables of interest, then the sample design is **noninformative.** If the sampling indicator depends only on the available values of auxiliary variables used to design the sample (e.g., stratification variables on the sampling frame), then the sampling mechanism is **ignorable.** In other words, if the unobserved survey variables of interest are not associated with the probability of inclusion in the sample *after* conditioning on these auxiliary variables, then the design can be labeled as ignorable. Ignorable sampling mechanisms are closely related to the missing at random (MAR) concept (see Chapters 11 and 12).

Ignorable mechanisms are important because inferences about the parameter of interest can be made without explicitly modeling the sampling mechanism. So how does one make a complex design "ignorable" when using a Bayesian approach to making finite population inferences? Incorporating stratification and cluster sampling is relatively straightforward in Bayesian models and is similar to how design effects are handled in multilevel modeling approaches (see Chapter 13). Effects of sampling strata are generally treated as fixed (or parameters to be estimated), and effects of sampling clusters are treated as random. Introducing these effects in Bayesian models makes the complex design ignorable.

However, there are many modeling issues that can arise. There could be a loss of degrees of freedom if there are many strata; one could decide to only include "significant" stratum effects; one could choose to use covariates on which the strata were based rather than indicators for the strata themselves; and one must posit a distribution for the random cluster effects, including an assumption of whether the random effects are independent of the residual errors. None of these issues are truly unique to Bayesian approaches, and they are all shared with more model-based multilevel modeling approaches (see Section 11.2.3.1). Finally, how should the final survey weights be handled in a Bayesian approach (especially if they are informative about the parameters of interest)?

The hardest aspect of performing Bayesian analyses of complex sample survey data is properly accounting for unequal probabilities of selection via survey weights (Gelman, 2007). This is currently an active area of research, and we review important new developments in this area below. Early suggestions in the literature for maintaining the ignorability assumption when selection probabilities are unequal included allowing model parameters to vary as a function of selection probabilities (i.e., including interactions of predictors with selection probabilities in models), or stratifying by selection probabilities (Holt & Smith, 1979). Zheng and Little (2005) proposed an estimation method based on penalized splines for estimating totals from probability proportionate to size (PPS) sampling designs, where selection probabilities are known for all population members, and showed promising properties of their method compared to more standard model-based approaches.

Si et al. (2015) presented a novel approach that employs Bayesian methodology to make robust finite population inferences about means or proportions of interest in the scenario where only final survey weights (and no stratum or cluster codes) are provided for the respondents in a public-use survey data set (e.g., the Monitoring the Future Study; see Miech et al., 2016). This model-based approach simultaneously predicts the distribution of the weights among *nonsampled* cases in the population of interest and the values of the survey variable of interest for these cases (as a function of the weights), enabling simulations of the full population means or proportions based on posterior distributions for these descriptive parameters (given the sampled cases and their data). These authors demonstrated the advantages of this approach relative to classical design-based methods (e.g., Chapter 5) in terms of the efficiency of descriptive finite population estimates, for both full populations and subpopulations. The authors implemented this approach in the popular Stan software for Bayesian analysis, and we have provided annotated example code implementing this approach on the ASDA website.

Little (2015) outlined a more general paradigm for finite population inference that he refers to as "Calibrated Bayes." This approach involves using Bayesian methods for finite population inference that explicitly incorporate complex sampling features into the models, while making sure that the resulting estimates have good repeated-sampling (or "frequentist") properties. Savitsky and Toth (2016), León-Novelo and Savitsky (2019), Williams and Savitsky (2020), and Williams and Savitsky (2021) have since built on the ideas of Little (2015) and described pseudo-Bayesian and fully Bayesian approaches to making finite population inference in the presence of informative sampling features. The basic idea of these approaches is to produce survey-weighted "pseudo" posterior distributions, where draws from this distribution correctly reflect variability in the quantity of interest due to the complex sample design (including any dependencies introduced by cluster sampling) with large enough sample sizes.

In forming the survey-weighted posterior distribution for a given quantity of interest (e.g., a mean, or a regression coefficient), the final survey weights in a data set are first used to define a *weighted* likelihood function for the sample. This is like the idea of pseudo maximum likelihood estimation introduced in Chapter 8, which produces design-consistent estimates; see Savitsky and Toth (2016) and Williams and Savitsky (2020). Unfortunately, the survey-weighted posterior distribution that results from combining this weighted likelihood with a (preferably informative) prior distribution (see Wiśniowski et al., 2020) behaves like it arises from a simple random sample and needs to be adjusted to correctly account for the design effects of other complex sampling features on sampling variability. Per Williams and Savitsky (2021), this adjustment is made *after* the posterior has been formed (given the weighted likelihood and any prior distribution specified) and some form of Markov Chain Monte Carlo (MCMC) sampling from it has occurred, resulting in posterior draws of the parameter(s) of interest. These draws can still yield design-consistent *estimates* of the parameter(s) of interest (Williams & Savitsky, 2020), but additional adjustment of the draws is needed for appropriate variance estimation.

Each individual draw from this survey-weighted posterior is adjusted such that the distribution of the posterior draws matches the asymptotic normal distribution of the survey-weighted maximum likelihood estimates of the parameter(s) of interest under the pseudo maximum likelihood approach (Binder, 1981). The exact form of the adjustment relies on the original draw, the mean of the original draws, and a multivariate estimate of the design effect for the parameter(s) of interest (see the general discussion of design effects in Chapter 2). The adjusted draw is computed by multiplying the difference between the original draw and the mean by the estimated design effect, and then adding back the original mean. The key remaining component for computing these adjustments is therefore computation of the estimated design effect.

One possible approach to computing this design effect, outlined by Williams and Savitsky (2021), is to first estimate the components of the variance of the survey-weighted maximum likelihood estimate under pseudo maximum likelihood estimation (Binder, 1981), and compute this variance based on those components (this would be the "design-adjusted" variance needed for the design effect). Then, this variance is compared to the variance of the survey-weighted posterior distribution, which as we noted earlier behaves like a simple random sample (meaning that this would be the "simple random sample" variance needed for the design effect). Williams and Savitsky (2021) propose a "hybrid" approach to estimating the first variance, which combines 1) Taylor Series Linearization, to first approximate more complicated expressions defining this variance in simpler linear forms and enable the use of plug-in estimates (as described earlier in Section 3.6.2), and 2) replication techniques to estimate the variance of the linearized expressions, where each replicate would yield the replicate plug-in estimate(s) of the parameter(s) of interest needed to compute the first variance. Their approach also allows for the formation of replicate weights from existing stratum and cluster codes; see Section 3.6.3.

The second "SRS" variance can be estimated using MCMC draws from the survey-weighted posterior distribution, with formulas shown in Williams and Savitsky (2021). Given these two variances, the estimated multivariate design effect can be computed, and the adjustments described above can then be applied to each individual draw from the survey-weighted posterior distribution prior to computing design-sensitive credible intervals for the parameter of interest and making finite population inferences.

Hornby et al. (2023) have now made the Bayesian approaches described here possible in the `csSampling` contributed package for R. This package combines functions from the `survey` package, the `rstan` package, and the `brms` package to implement the complex approaches described above. We demonstrate the use of functions in this package in many of the later example chapters when presenting alternative Bayesian approaches to making finite population inferences and contrasting the results with those arising from design-based approaches.

We end this subsection by reiterating that Bayesian analysis of complex sample survey data is an active area of research at the time that this third edition of ASDA is being written. We anticipate additional extensions or enhancements of these Bayesian approaches in the near future, along with additional software development. We will strive to keep readers aware of any new developments in this area on the ASDA website.

3.8.3 Bayesian Approaches to Weight Trimming

Because fully weighted estimators reduce bias at the cost of increasing sampling variance, a compromise is often required, and various **weight trimming** approaches (Potter, 1990), or methods of reducing extreme weights without introducing severe bias (i.e., keeping the mean squared error relatively small), have been developed for the Bayesian analysis framework.

Holt and Smith (1979) proposed a Bayesian approach to estimating means where the means of strata defined by sampling weights (or inverses of probabilities of selection) were treated as random variables (or random effects) and showed that this approach had the effect of "smoothing" the sampling weights (as opposed to weight trimming). Elliott and Little (2000) built on this approach for estimation of means, showing improved performance of a weight smoothing model based on a nonparametric spline function for the underlying means of the strata based on the sampling weights. Elliott (2007) expanded this "weight smoothing" method to linear and generalized linear models.

More recently, Chen et al. (2017) provided a comprehensive overview of possible approaches for increasing the efficiency of weighted estimators, including various model-based weight trimming and weight smoothing strategies. Yang et al. (2023) evaluated nine different Bayesian approaches to making predictive inference, some of which involved the use of trimmed weights, via simulation, and found that the use of trimmed weights when applying Bayesian methods had advantages in terms of the root mean squared error (RMSE) of estimated proportions. Software implementing the "weight smoothing" approaches discussed above is freely available,[1] and we encourage data producers responsible for producing survey weights to consider these approaches prior to releasing weights for public use, as several studies have now demonstrated their ability to produce more efficient estimates.

3.9 Total Survey Error and Its Impact on Survey Estimation and Inference

This chapter has clarified how probability sampling provides a theoretical basis for unbiased estimation of finite population parameters and the associated sampling variance of the sample estimates. However, a successful scientific survey depends not only on the careful control of sampling variances but also other sources of survey error. The collection of all sources of error that can influence the precision and accuracy of survey estimates of population parameters is termed **total survey error** (TSE; Groves, 2004; Lessler & Kalsbeek, 1992). The TSE for a survey estimate is measured as the mean squared error (MSE), $Mean\ Squared\ Error = Variance + Bias^2$, or the variance of the estimate plus the square of the bias in the estimate. Table 3.5 provides a typical taxonomy of survey errors.

We provide a variety of standard definitions related to TSE in this section. The American Association for Public Opinion Research (AAPOR) also provides extensive documentation for researchers regarding standard reporting of the components of TSE,[2] such as response rates.[3] Standard definitions for many of these terms can also be found as part of ISO 3534–4 on Survey Sampling definitions and commentary.[4]

3.9.1 Variable Errors

The sources of error that cause sample estimates to disperse randomly about the true and unknown population value of interest across replications of the survey process are termed **variable errors**.

Sampling variances, the square roots of which yield the standard errors that have been described at length in the preceding sections of this chapter, derive from the statistical fact that only a subset of the full target population is observed in any given sample. As sample sizes increase, the sampling variance decreases. The sampling variance disappears entirely if a complete census of the target population is conducted. Probability sampling theory provides well-defined guidance for estimating sampling variances for survey estimates.

Interviewer variance and **response variance** (Fuller, 1987; Groves, 2004) enter the data during recruitment and interviewer- or self-administration of the actual survey, and may be attributed to random variability in the way that interviewers recruit respondents, ask survey questions, or record the survey answers (see West & Blom, 2017, for a review of the literature explaining interviewer effects), or the way that the respondents report their responses. In scientific surveys, interviewer variance and response variance are minimized by carefully training interviewers and designing and pretesting questions so that they are clearly worded and have comprehensive, easy-to-interpret response categories. Readers interested in an in-depth treatment of interviewer variance and other measurement errors are referred to Biemer et al. (1991).

Coding variance is primarily a technical source of random error in the data set. Research staff responsible for coding the survey data and transcribing the information to a data file for analysis may make errors in a random manner. As the technology for computer-assisted interviewing and data collection has advanced, this source of error has generally been reduced.

Each of these sources of variance—sampling, interviewer, response, and coding—contributes to the combined total variability in observed survey data. Short of conducting elaborate experiments, it will not be possible for the data producer, let alone the survey analyst, to decompose the total variance of a survey estimate into components attributable to each of these sources. Nevertheless, it is important to note that the combined influences of these multiple sources of variability are captured in the estimated standard errors that are computed from the survey data.

3.9.2 Biases in Survey Data

The English word *bias* has many meanings, but in a statistical context it refers to the tendency of a statistical estimate to deviate systematically from the true value it is intended to estimate across replications of the survey process. Students of statistics, including survey researchers, have long been taught that "unbiasedness"—the absence of bias—is one of the most desirable properties of a statistical estimator or procedure. In survey practice, bias should certainly be avoided; however, elimination of all sources of bias in survey data is probably neither practical nor even efficient when both the costs and benefits of reducing bias are considered. In fact, survey analysts tend to be less interested in purely unbiased estimators and more interested in estimators that have the property of being a *consistent* estimator of the population parameter of interest. **Consistent estimators** converge to the true population value as the sample size n increases to the population size N. Survey biases listed in the survey error taxonomy given in Table 3.5 above may be collapsed into two major types: **sampling bias** and **nonsampling bias** (Kish, 1965).

In probability samples, the greatest potential for **sampling bias** can be attributed to noncoverage of survey population elements. **Sample frame noncoverage** occurs when population elements are systematically excluded from the population registry or the data sources (maps and census counts) used to develop area probability sampling frames, and therefore have no probability of being included in the sample. Sample noncoverage bias may also occur in the process of screening selected dwelling units to identify and select eligible survey respondents. Survey producers minimize sampling bias through careful design and attention to detail in the sample selection process, field testing of screening and respondent selection procedures prior to the data collection period, and rigorous training and

TABLE 3.5

A Taxonomy of Survey Errors

Variable Errors	Biases
Sampling variance	Sample selection bias
Interviewer variance	Frame coverage bias
Response (Measurement) variance	Measurement bias
Coding variance	Nonresponse bias

on-site supervision of the field staff for the actual survey data collection. After the survey is complete, post-stratification weighting may be employed to attenuate any remaining sampling bias in the survey data (see Section 2.7).

Survey data are also vulnerable to **nonsampling bias** from two primary sources: **measurement bias**, or systematic bias in the way that respondents interpret and respond to questions, and **survey nonresponse**. Measurement bias may be deliberate on the part of the respondent, or it may be the unconscious result of poor questionnaire construction or interviewer training. Survey respondents who are asked to report their household income may underreport or fail to mention sources of income such as the sale of a parcel of land. Survey questions that ask about participation in elections, educational activities, or religious observances may be subject to over-reporting of participation, a phenomenon termed "social desirability" bias. Poorly worded or "leading" questions, or questionnaires that place questions out of context, may yield biased measures for the constructs in which the research investigator is truly interested. Across disciplinary areas, survey methodologists work very hard to understand the survey measurement process, and to design questions and questionnaire formats that accurately measure the underlying constructs of interest.

Survey nonresponse is another potential source of bias in sample-based estimates of population characteristics. The failure to obtain any data on a sample household or individual is termed **unit nonresponse**. A missing response to one or more individual variable items in an otherwise complete interview questionnaire is termed **item nonresponse** (Little & Rubin, 2019). Nonresponse to voluntary surveys, including those conducted by universities and other scientific research organizations, has become a major problem in countries in Western Europe and North America. Nonresponse bias in survey estimates of simple parameters such as population means or proportions is a function of the response rate and the difference in the values of the parameter for responding and nonresponding members of the sample. For estimates of a population proportion, for example, the nonresponse bias can be expressed using the following "deterministic" formula (which assumes that population elements can be divided into those that will always respond to a survey request and those that will never respond):

$$Bias_{NR}(p) = (1 - RRate) \times (p_R - p_{NR}) \tag{3.14}$$

In (3.14), *RRate* is the expected value of the population response rate, p_R is the value of the proportion for respondents and p_{NR} is the value of the proportion for nonrespondents. The absolute value of the expected nonresponse bias increases with the product of the nonresponse rate and the difference in p for respondents and nonrespondents. If the response rate is high or proportions for respondents and nonrespondents do not differ, the nonresponse bias will be very small. Unlike sampling variance, which decreases as sample size increases, nonresponse bias is persistent. Its expected value is not a function of sample size but remains unchanged regardless of how large or small the size of the survey sample.

The potential impact of nonresponse bias on the analysis of survey data may be best illustrated through a simple example. Assume a researcher is interested in studying parents' views on the need for increased government spending (hence potential increases in taxes) for elementary science education. Among parents who agree to participate in the survey, the expected proportion that support increased spending on elementary science education is $p_R = 0.6$, while for non-cooperating parents, $p_{NR} = 0.4$—a major difference between respondents and nonrespondents. If only 50% of the original random sample of n = 1,000 parents agreed to participate, the expected nonresponse bias for the proportion of interest would be:

$$Bias_{NR}(\hat{p}) = (1 - RR) \times (p_R - p_{NR}) = (1 - 0.5) \times (0.6 - 0.4) = 0.10.$$

Assuming for simplicity that the original sample of parents was selected using simple random sampling (SRS), the researcher would develop the following 95% confidence interval for this proportion:

$$\hat{p}_R \pm 1.96 \cdot [\hat{p} \cdot (1 - \hat{p}) / (n)]^{1/2} \cong \hat{p}_R \pm 1.96 \cdot [0.6(1 - 0.6) / (0.5 \cdot 1000)]^{1/2} = \hat{p}_R \pm 0.043$$

In this case, the size of the expected nonresponse bias is relatively large in comparison to the size of the 95% confidence interval half width—a result of the low response rate and the major difference in expected proportions for respondent and nonrespondent cases.

The purpose of this example is not to magnify the potential seriousness of nonresponse bias in survey estimation—high response rates and/or smaller differences in the expected statistics for respondents and nonrespondents would decrease the size of the expected bias. Past research has suggested that we may be overly concerned about the seriousness of nonresponse bias for certain types of survey measures, and especially measures of respondent attitudes and expectations (Keeter et al., 2000; Curtin et al., 2000). However, the example makes the point that we cannot be complacent and ignore the potential for nonresponse bias in the survey estimation process.

As described in Section 2.7, detailed data on respondents and nonrespondents and a model for the nonresponse mechanism are used to develop nonresponse adjustments to the survey analysis weights in actual survey practice. To the extent that the estimated model accurately describes the underlying nonresponse process, these adjustments to the survey weights serve to attenuate potential nonresponse bias and its impact on estimates and the corresponding inferences that the analyst draws from the survey data.

3.9.3 Analytic Error

The overall objective of this book is primarily to prevent researchers performing secondary analyses of complex sample survey data from making another type of error that is crucial to the TSE framework (Smith, 2011): **analytic error**. With this type of error, the researchers responsible for designing the survey, selecting the sample, cleaning the data, preparing the weights and sample design information, and generating final data sets for public use do their very best to minimize all the sources of TSE. However, independent researchers who then perform secondary analyses of the public-use survey data may either ignore the survey weights in estimation, ignore the other sample design information when estimating variances, or (in the worst case) completely ignore both key features of the sample design when performing analyses.

These types of analytic errors can lead to biased estimates, inefficient estimates, or both, painting an incorrect picture of the finite population of interest and effectively negating all the time, effort, and resources that the survey data producer dedicated to minimizing the other sources of error. Unfortunately, numerous studies of the published literature that have been conducted to date suggest that analytic error is quite common when researchers are performing secondary analyses of complex sample survey data (Briesacher et al., 2012; West et al., 2016, 2017; Khera et al., 2017; Ward, 2018; West & Sakshaug, 2018). These studies also clearly demonstrate that the implications of these errors for finite population inference can be alarming. While improved analytic guidelines and documentation for public-use survey data sets may go a long way in helping to prevent these types of analytic errors (Kolenikov et al., 2020), not all data users may read this documentation carefully, and other strategies are needed.

As a result, a new set of standards has been proposed for journal editors and other research outlets that disseminate the results of secondary analyses of survey data: Preferred Reporting Items for Complex Sample Survey Analysis, or PRICSSA (Seidenberg et al., 2023). In short, PRICSSA requires researchers presenting secondary analyses of complex sample survey data to provide clear responses to a small set of items asking about how they approached the data analysis (e.g., what survey weight variable was used for estimation, what design variables were used for variance estimation, how were missing data handled, is the code used for the data analysis available, etc.). The PRICSSA reporting requirement has already been adopted by the *Journal of Survey Statistics and Methodology*, and our hope is that more journals and research outlets will adopt these standards in the future to ensure transparency, reproducibility, and most importantly, correct data analysis.

Notes

1 https://github.com/mrelliot/weighttimming/blob/main/README.md
2 https://aapor.org/standards-and-ethics/standard-definitions/
3 https://aapor.org/response-rates/
4 https://www.iso.org/standard/56154.html

4

Preparation for Complex Sample Survey Data Analysis

4.1 Introduction

This chapter guides the survey analyst through a sequence of steps that are important to properly prepare for the analysis of a complex sample survey data set. The steps in this chapter should only be considered *after* steps 1 and 2 in Chapter 1 (defining the research problem, stating the research objectives, and understanding the sample design). These steps include reviewing the weights and the sampling error calculation model for the survey data set, examining the amount of missing data for key variables, and preparing for the analysis of sample subclasses. The chapter concludes with a short checklist to remind the reader of these critical preliminary steps.

We remind readers about two important terms. The term **data producer** applies to the statistical agency or the research team that is responsible (with the assistance of survey statisticians) for designing a probability sample from a larger population, collecting the survey data, developing final survey weights, and producing information about the complex sample design that can be used by research analysts for statistical analysis of the data. For example, the National Center for Health Statistics (NCHS) along with its contractor, Westat, Inc., was responsible for producing the 2011–2012 NHANES data that are analyzed throughout this book. The term **data user** is applied to the survey analyst or researcher who will use the survey data to develop and test scientific hypotheses concerning the survey population.

In most cases, the data producer has access to far more detailed information about the complex sample design and personal information concerning study respondents than the data user. To maintain confidentiality for survey respondents and minimize **disclosure risk**, the data producer is responsible for making optimal use of detailed and often confidential information on the individual sample cases to develop sampling error codes and final survey weights for the survey data set. In addition, the data producer is responsible for providing the data user with a cleaned data set, a codebook for the survey variables, and technical documentation describing the sampling and survey procedures. Many survey programs are now providing data users with a guide for statistical analysis of their survey data set as well (Kolenikov et al., 2020). For example, NCHS provides analytic guidelines online[1] that include detailed instructions for how to correctly analyze the NHANES data sets.

DOI: 10.1201/9781003472094-5

4.2 Final Survey Weights: Review by the Data User

Experience as survey statisticians and consultants has taught us that many survey data analysts struggle with the correct use of final survey weights in survey estimation and inference. For whatever reason, many otherwise sophisticated data users wish to place a "black box" around the process of weight development and the application of weights in their analysis. As described in Section 2.7, the final survey weights provided in survey data sets are generally the product of a sample selection weight (w_{sel}), a nonresponse adjustment factor (w_{nr}), and a post-stratification factor (w_{ps}): $w_{final,i} = w_{sel,i} \cdot w_{nr,i} \cdot w_{ps,i}$. For the reasons outlined above, the data producer is responsible for developing final survey weights for each survey respondent and linking the final survey weight variable to each respondent in the survey data file.

The final survey weight assigned to each survey respondent is a measure of the number of population members represented by that respondent, or alternatively, the relative share of the population that the respondent represents. When weights are applied in the statistical analysis of survey data, weighted calculations simply expand each respondent's contribution to reflect its representative share of the target population. Because the process of weight development has been discussed extensively in Section 2.7, the aim of this section is to remove any remaining mystique surrounding the weights that are provided with a survey data set.

Although data analysts will not typically be responsible for the actual weight calculations, they should familiarize themselves with the survey weight variable(s) and how weighted analysis may influence estimation and inference from their survey data. Key steps in this process of verification and familiarization include:

- Verifying the variable name for the appropriate weight for the intended analysis.
- Checking and reviewing the scaling and general distribution of the weights.
- Evaluating the impacts of the weights on key survey statistics.

The subsequent sections describe these three activities in more detail.

4.2.1 Identification of the Correct Weight Variable(s) for the Analysis

The data user will need to refer to the survey documentation (technical report, codebook, analytic guidelines, etc.) to identify the correct variable name for the final survey weight. Unfortunately, there are no standard naming conventions for weight variables, and we recommend great caution in this step as a result. Many years ago, a student mistakenly chose a variable labeled WEIGHT and produced a wonderful paper based on the NHANES data in which each respondent was weighted by their body weight in kilograms. The correct survey weight variable in the student's data file was stored under a different, less obvious, variable label.

Depending on the variables to be analyzed, there may be more than one weight variable provided with the survey data set. The 2012 HRS data set includes one weight variable (NWGTHH) for the analysis of financial unit (single adult or couple) variables, such as home value or total net worth, and a separate weight variable (NWGTR) for individual-level analysis of variables such as health status or earnings from a job. The 2011–2012 NHANES documentation instructs analysts to use the weight variable WTINT2YR for analyses of the medical history interview variables, and another weight variable (WTMEC2YR) for analyses of data collected from the medical examination phase of the study. The larger sample of NCS-R Part I mental health screening data (n = 9,282) is to be analyzed using one weight variable (NCSRWTSH), while another weight variable (NCSRWTLG) is the correct weight for analyses involving variables measured only for the in-depth Part II subsample (n = 5,692). Finally, the ESS Round 6 Russian Federation data set includes the "design weight" variable DWEIGHT and the "post-stratification" weight variable PSPWGHT. We use the post-stratified weight (PSPWGHT), incorporating post-stratification adjustments, for all analysis examples in this book.

Some public-use data sets may contain a large set of weight variables known as **replicate weights**. For example, the 1999–2000 NHANES public use data file included the replicate weight variables WTMREP01, WTMREP02, . . ., WTMREP52. As was discussed in detail in Chapter 3, replicate weights are used in combination with software that employs a replicated method of variance estimation, such as Balanced Repeated Replication (BRR), Jackknife Repeated Replication (JRR), or bootstrapping. When a public-use data set includes replicate weights, design variables for variance estimation (stratum and cluster codes) will generally *not* be included (see Section 4.3), and the survey analyst will need to specify the replicated variance estimation approach (BRR, JRR, BRR-Fay, bootstrapping, etc.) and identify the sequence of variables that contain the replicate weight values in their software procedure of choice (see Appendix A for more details on these software options).

4.2.2 Determining the Distribution and Scaling of the Weight Variable(s)

In everyday practice, it is always surprising to learn that an analyst who is struggling with the weighted analysis of survey data has never actually looked at the distribution of the weight variable. This is a critical step in preparing for analysis. Assessing a simple univariate distribution of the final survey weight variable provides information on: 1) the scaling of the weights; 2) the variability and skew in the distribution of the weights across sample cases; 3) extreme weight values; and 4) (possibly) missing data on the survey weight. Scaling of the weights is important for interpreting estimates of totals, and in older versions of software procedures may affect variance estimates. The variance and distribution of the weights may influence the precision loss for sample estimates (see Section 2.7). Extreme weights, especially when combined with outlier values for variables of interest, may produce instability in estimates and standard errors for complete sample or subclass estimates. Missing data or zero (0) values on weight variables may indicate an error in building the data set or a special feature of the data set. For example, 2011–2012 NHANES respondents that completed the medical history interview but did not participate in the medical examination phase of the study will have a positive, non-zero weight value for WTINT2YR, but will have a zero value for WTMEC2YR (see Table 4.1).

Table 4.1 provides simple distributional summaries of the final survey weight variables for the NCS-R, 2012 HRS, 2011–2012 NHANES, and ESS Russian Federation (Round 6) data sets. Inspection of these weight distributions quickly reveals that the scale of the weight values is quite different from one study to the next. For example, the sum of the NCS-R Part I weights is 9,282 (equal to the sample size), while the sum of the 2012 HRS individual weights is 90,698,760 (which would seem to approximate the size of the target HRS population in 2012).

Except for weighted estimates of population totals, weighted estimation of population parameters and standard errors should be invariant to a linear scaling of the weight values, that is, $w_{scale,i} = k \cdot w_{final,i}$, where k is an arbitrary constant. That is, the data producer may choose to multiply or divide the weight values by any constant and except for estimates of

TABLE 4.1

Descriptive Statistics for the Final Survey Weights in the Data Sets Analyzed in this Book

	NCS-R: NCSRWTLG	NCS-R: NCSRWTSH	NHANES: WTMEC2YR***	NHANES: WTINT2YR***	HRS: NWGTR	HRS: NWGTHH	ESS Russian Federation: DWEIGHT	ESS Russian Federation: PSPWGHT
n	5,692	9,282	5,864	5,864	20,554	20,554	2,484	2,484
Sum	5,692	9,282	232,002,539	231,898,519	90,698,760	89,174,512	2,484	2,484
Mean	1.00*	1.00*	39,653.87	39,546.13	4,412.71	4,338.55	1.00*	1.00*
SD	0.96	0.52	40,104.81	38,741.87	3,923.57	3,654.94	0.58	0.59
Min	0.11	0.17	0**	4,300	0**	0**	0.16	0.10
Max	10.10	7.14	220,579.78	220,233.32	19,866	17,875	4.00	4.00
Pctls.								
1%	0.24	0.36	0	6,800.60	0	0	0.27	0.22
5%	0.32	0.49	6,857.04	9,048.96	0	0	0.33	0.31
25%	0.46	0.69	15,096.34	15,497.84	1,738	1,746	0.57	0.57
50%	0.64	0.87	23,508.42	23,599.80	3,479	3,358.50	0.88	0.88
75%	1.08	1.16	43,508.42	42,892.68	5,692	5,536	1.28	1.28
95%	2.95	1.85	133,985.24	130,577.03	12,370	11,819	2.08	2.12
99%	4.71	3.17	169,896.24	167,711.39	19,618	17,855	3.06	3.08

Notes: * Suggests that the final survey weights have been normalized to sum to the sample size.
** Cases with weights of zero will be dropped from analyses and usually correspond to individuals who were not eligible to be in a particular sample.
*** Adults age 18+.

population totals, weighted estimates of population parameters and their standard errors should not change.

For many surveys, such as the 2011–2012 NHANES and the 2012 HRS, the final survey weights for individual cases will be **population-scale weights**, and the expected value of the sum of the weights will be the population size: $E\left(\sum_{i=1}^{n} w_i\right) = N$. For other survey data sets, a *normalized* version of the final survey weight is provided with the survey data. To "normalize" the final overall survey weights, data producers divide the final population-scale weight for each respondent by the mean final weight for the entire sample: $w_{norm,i} = w_i / \left(\sum_i w_i / n\right) = w_i / \bar{w}$. Many public-use data sets such as the NCS-R will have **normalized weights** available as the final overall survey weights. The resulting normalized weights will have a mean value of $\bar{w}_{norm} = 1.0$, and the normalized weights for all sample cases should add up to the sample size: $\sum_i w_{norm,i} = n$.

Normalizing final survey weight values is a practice that has its roots in the past, when software procedures for the analysis of survey data often misinterpreted the "sample size" for the weighted estimates of variances and covariances required in computations of standard errors, confidence intervals, or test statistics. As illustrated in Section 3.5.2, the degrees of freedom for variance estimation in analyses of complex sample survey data are determined by the sampling features (stratification, cluster sampling) and not the nominal sample size. Also, some data analysts feel more comfortable with weighted frequency counts that closely approximate the nominal sample sizes for the survey. However, there is a false security in assuming that a weighted frequency count of $\sum w_i = 1{,}000$ corresponds to an effective sample size of $n_{eff} = 1{,}000$. As discussed in Section 2.7, the effective sample size for 1,000 nominal cases will be determined in part by the weighting loss, L_w, that arises due to the variability in the weights and the correlations of the weights with the values of the survey variables of interest.

Fortunately, normalizing weights is *not necessary* when analysts use modern software procedures capable of incorporating any available complex sample design information into analyses of the survey data. This is especially important when estimation of population totals is a primary analytic objective; the use of normalized weights in these analyses will result in severe under-estimation of population totals, and population-scale weights need to be used instead. We will return to this issue in Chapter 5.

4.2.3 Weighting Applications: Sensitivity of Survey Estimates to the Weights

A third step that we recommend survey analysts consider the first time that they work with a new survey data set is to conduct a simple investigation of how the application of the survey weights affects the estimates and standard errors for several key parameters of interest.

To illustrate this step, we consider data from the NCS-R data set, where the documentation indicates that the final survey weight to be used for the *subsample* of respondents responding to both Part I and Part II of the NCS-R survey ($n = 5{,}692$) is NCSRWTLG. A simple univariate descriptive analysis of these survey weights in Stata reveals a mean of 1.00, a standard deviation of 0.96, a minimum of 0.11, and a maximum of 10.10 (see Table 4.1). These values indicate that the weights have been normalized and have moderate variance. In addition, we note that some survey weight values are below 0.50. Many standard statistical software procedures will round non-integer weights and set the weight to 0 if the normalized weight is less than 0.5—excluding such cases from certain analyses. This is an important consideration that underscores the need to use specialized software that incorporates the final survey weights correctly (and does not round them).

We first consider unweighted estimation of the proportions of NCS-R Part II respondents with lifetime diagnoses of either a major depressive episode (MDE), measured by a binary indicator equal to 1 (yes) or 0 (no), or alcohol dependence (ALD, also a binary indicator), in Stata:

```
mean mde ald if ncsrwtlg != .
```

Variable	Mean	Standard Error	95% Confidence Interval	
MDE	0.3155	0.0062	0.3035	0.3276
ALD	0.0778	0.0036	0.0709	0.0845

Note that we explicitly limit the unweighted analysis to Part II respondents (who have a non-missing value on the Part II survey weight variable NCSRWTLG). The unweighted estimate of the MDE proportion is 0.316, suggesting that almost 32% of the NCS-R population has had a lifetime diagnosis of MDE. The unweighted estimate of the ALD proportion is 0.078, suggesting that almost 8% of the NCS-R population has had a diagnosis of alcohol dependence at some point in their lifetime.

We then request *weighted* estimates of these proportions in Stata, first identifying the final survey weight to Stata with the svyset command, and then requesting weighted estimates by using the svy: mean command:

```
svyset [pweight = ncsrwtlg]
svy: mean mde ald
```

Variable	Mean	Linearized Standard Error	95% Confidence Interval	
MDE	0.1918	0.0054	0.1813	0.2023
ALD	0.0541	0.0031	0.0481	0.0601

The weighted estimates of population prevalence of MDE and ADL are 0.192 and 0.054, respectively. The unweighted estimates for MDE and ALD therefore have a positive bias for the true population proportions (there would be a big difference in reporting a population estimate of 32% for lifetime MDE versus an estimate of 19%).

In this simple example, the weighted estimates differ significantly from the unweighted means of the respondent measures. We would say that the survey weights in this case are **informative** about the two variables of interest (MDE and ALD). This is not always the case. Depending on the sample design and the nonresponse adjustment factors that contributed to the computation of individual weight values, weighted and unweighted estimates may or may not show significant differences. When this simple comparison of weighted and unweighted estimates of key population parameters shows a significant difference, the survey analyst should aim to understand why this difference occurs. Specifically, what are the factors contributing to the case-specific weights that would cause the weighted population estimates to differ from an unweighted analysis of the nominal set of sample observations?

Consider the NCS-R example. We know from the Chapter 1 description that according to the survey protocol, all Part I respondents reporting symptoms of a mental health disorder and a random subsample of symptom-free Part I respondents continued to complete the Part II in-depth interview. Therefore, the unweighted Part II sample contains an "enriched" sample of persons who qualify for one or more mental health diagnoses. Consequently, when the adjusted survey weight is applied to the Part II data, the unbiased weighted estimate of the true population value is substantially lower than the simple unweighted estimate.

Likewise, a similar comparison of estimates of the prevalence of physical functioning limitations using the 2011–2012 NHANES data would yield weighted population estimates that are lower than the simple unweighted prevalence estimates for the observed cases. The explanation for that difference lies in the fact that persons who self-report a disability are oversampled for inclusion in the NHANES, and the application of the weights adjusts for this initial over-sampling.

Repeating this exercise for several key variables of research interest should provide the user with confidence that they understand both how and why the application of the survey weights will influence estimation and inference for the population parameters that they will estimate from the sample survey data. We note that these examples were designed only to illustrate the calculation of weighted estimates; standard errors for the weighted estimates were not appropriately estimated to incorporate the complex design features of the NCS-R sample. Chapter 5 considers estimation of descriptive statistics and their standard errors in more detail.

4.3 Understanding and Checking the Sampling Error Calculation Model

The next step in preparing to work with a complex sample survey data set is to identify, understand, and verify the **sampling error calculation model** that the data producer has developed and encoded for the survey data set. Information about the sampling error calculation model can often be found in sections of the technical documentation for survey data sets entitled "sampling error calculations" or "variance estimation" (to name a couple of possibilities). In this section, we discuss the notion of a sampling error calculation model, and how to identify the appropriate variables in a survey data set that represent the model for variance estimation purposes.

A sampling error calculation model is an approximation or "model" for the actual complex sample design that permits practical estimation of sampling variances for survey statistics. Such models are necessary because in many cases, the most practical, cost-efficient sample designs for survey data collection pose analytical problems for complex sample variance estimation. Examples of sample design features that complicate direct analytic approaches to variance estimation include:

- Multi-stage sampling of survey respondents.
- Sampling units without replacement (WOR) at each stage of sample selection.
- Sampling of a single PSU from non-self-representing (NSR) primary stage strata.
- Small primary stage clusters that are not optimal for subclass analyses or pose an unacceptable disclosure risk.

The data producer has the responsibility of creating a sampling error calculation model that retains as much essential information about the original complex sample design as possible while eliminating the analytical problems and disclosure risks that the original design might pose for variance estimation.

4.3.1 Stratum and Cluster Codes in Complex Sample Survey Data Sets

The specification of the sampling error calculation model for a complex sample design entails the creation of a **sampling error stratum** and a **sampling error cluster** variable.

These **sampling error codes** identify the strata and "clusters" to which the survey respondents belong, approximating the original sample design as closely as possible while at the same time conforming to the analytical requirements of several methods for estimating variances from complex sample survey data (see Chapter 3). Because these codes *approximate* the stratum and cluster codes that would be assigned to survey respondents based on the original complex sample design, they will not be equal to the original design codes, and the approximate codes are often scrambled (or masked) to prevent the possibility of identifying the original design codes in any way. Sampling error stratum and cluster codes are *essential* for data users who elect to use software procedures that employ the Taylor Series Linearization (TSL) method for variance estimation. Software for replicated variance estimation (BRR, JRR, or bootstrapping) does not require these codes, provided that the data producer has generated replicate weights. If replicate weights are not available, software enabling replicated variance estimation can use the sampling error stratum and cluster codes to create sample replicates and develop the corresponding replicate weights.

In some complex sample survey data sets where confidentiality is of the utmost importance, variables containing sampling error stratum and cluster codes may not be provided directly to the research analyst (even if they represent approximations of the original design codes). In their place, the data producer's survey statisticians calculate **replicate weights** which reflect the sampling error calculation model and can be used for repeated replication methods of variance estimation (see Chapter 3). We provide examples of analyses using replicate weights in later chapters and on the ASDA website, and interested readers can refer to Appendix A to see how replicate weights would be identified in various software procedures.

Unfortunately, as is true for final survey weights, data producers have not adopted standard conventions for naming the sampling error stratum and cluster variables in public-use survey data sets. The variable containing the stratum code often has a label that includes some reference to stratification. Two examples from this book include the variable SDMVSTRA from the 2011–2012 NHANES data set, and the variable SESTRAT from the NCS-R data set.

The variable containing the sampling error cluster codes is sometimes referred to as a **sampling error computation unit (SECU)** variable, or it may be generically called the "cluster" code, the primary sampling unit (PSU) code, or pseudo primary sampling unit (PPSU) variable. Two examples that will be used throughout the example exercises in this book include the SDMVPSU variable from the 2011–2012 NHANES data set and the SECLUST variable from the NCS-R data set.

As was discussed in Section 2.5.2, some large survey data sets are still released without sampling error codes or replicate weight values that are needed for complex sample variance estimation. In such cases, analysts may be able to apply to the data producer for restricted access to the sampling error codes, or they may elect to use the generalized design effect that is advocated by these data producers. For a limited number of survey data sets, online, interactive analysis systems such as the SDA system[2] provide survey analysts with the capability of performing analysis without gaining direct access to the underlying sampling error codes for the survey data set. Analysts who are working with older public-use data sets may find that these data sets were released with the sampling error stratum and cluster codes concatenated into a single code (e.g., 3001 for stratum = 30 and cluster = 01). In these cases, the variable containing the unified code will need to be divided into two variables (stratum and cluster) by the survey data analyst for variance estimation purposes.

4.3.2 Building the NCS-R Sampling Error Calculation Model

We now consider the sample design for the National Comorbidity Survey Replication (NCS-R) as an illustration of the primary concepts and procedures for constructing a sampling error calculation model for a complex sample survey data set. Table 4.2 presents a side-by-side comparison of the features of the original NCS-R complex sample design and the procedures employed to create the corresponding sampling error calculation model. Interested readers can refer to Kessler et al. (2004) for more details on the original sample design for the NCS-R.

Note from Table 4.2 that in the NCS-R sampling error calculation model, the assumption of with-replacement sampling of ultimate clusters described in Chapter 3 is employed to address the analytic complexities associated with the multi-stage sampling and the without-replacement selection of NCS-R PSUs. The assumption that ultimate clusters are selected with replacement within primary stage strata ignores the finite population correction factor (see Section 2.4.2) and therefore results in a slight overestimation of true sampling variances for survey estimates.

To introduce several other features of the NCS-R sampling error calculation model, Table 4.3 illustrates the assignment of the sampling error stratum and cluster codes for six of the NCS-R sample design strata. In self-representing (SR) design strata 15 and 16, the

TABLE 4.2

Original Sample Design and Associated Sampling Error Calculation Model for the NCS-R

Original Sample Design	Sampling Error Calculation Model
The sample is selected in multiple stages. Primary stage units (PSUs), second stage units (SSUs), and third stage units are selected without replacement (WOR).	The concept of **ultimate clusters** is employed (see Chapter 3). Under the assumption that PSUs (ultimate clusters) are sampled with replacement, only PSU-level statistics (totals, means) are needed to compute estimates of sampling variance. The ultimate clusters are assumed to be **sampled with replacement (SWR)** at the primary stage. Finite population corrections are ignored, and simpler SWR variance formulas may be used for variance estimation.
Sixteen of the primary stage strata are self-representing (SR) and contain a single PSU. True sampling begins with the selection of SSUs within the SR PSU.	**Random groups** are formed for sampling error calculation. Each SR PSU becomes a sampling error stratum. Within the SR stratum, SSUs are randomly assigned to a pair of sampling error clusters.
Forty-six of the primary stage strata are non-self-representing (NSR). A single PSU is selected from each NSR stratum.	**Collapsed strata** are formed for sampling error calculation. Two similar NSR design strata (e.g., Strata A and B) are collapsed to form one sampling error computation stratum. The Stratum A PSU is the first sampling error cluster in the stratum, and the Stratum B PSU forms the second sampling error cluster.

TABLE 4.3

Illustration of NCS-R Sampling Error Code Assignments

	Original Sample Design		Sampling Error Calculation Model	
	Stratum	**PSU***	**Stratum**	**Cluster**
SR	15	1 2 3 4 5 6 7 8 9 10 11 12	15	1 = {1, 3, 5, 7, 9, 11} 2 = {2, 4, 6, 8, 10, 12}
	16	1 2 3 4 5 6 7 8 9 10 11 12	16	1 = {1, 3, 5, 7, 9, 11} 2 = {2, 4, 6, 8, 10, 12}
	.....			
NSR	17	1701	17	1 = 1701
	18	1801		2 = 1801
	19	1901	18	1 = 1901
	20	2001		2 = 2001

Note: * Recall from Section 2.8 and Table 4.2 that in self-representing (SR) strata, sampling begins with the selection of the smaller area segment units. Hence, in the NCS-R, the sampled units (coded 1–12) in each SR stratum (serving as its own PSU) are actually secondary sampling units. We include them in the PSU column because this was the first stage of non-certainty sampling in the SR strata. Unlike the SR PSUs, which serve as both strata and PSUs (each SR stratum is a PSU, that is, they are one in the same), the NSR strata can include multiple PSUs. In the NCS-R, one PSU was randomly selected from each NSR stratum (e.g., PSU 1701). NSR strata were then collapsed to form sampling error strata with two PSUs each, for the purpose of facilitating variance estimation.

"area segments" constitute the first actual stage of non-certainty sample selection—hence they are the ultimate cluster units with these two strata. To build the sampling error calculation model within each of these two SR design strata, the **random groups method** is used to randomly assign the area segment units to two sampling error clusters. This is done to simplify the calculations required for variance estimation. As illustrated, NCS-R nonself-representing (NSR) strata 17–20 contain a single PSU selection. The single PSU selected from each of these NSR strata constitutes an ultimate cluster selection. Because a minimum of two sampling error clusters per stratum is required for variance estimation, pairs of NSR design strata (e.g., 17 and 18, 19 and 20) are collapsed to create single sampling error strata with two sampling error computation units (PSUs) each.

Randomly grouping PSUs to form a sampling error cluster does not bias the estimates of standard errors that will be computed under the sampling error calculation model. However, forming random clusters by combining units does forfeit degrees of freedom, so the "variance of the variance estimate" may increase slightly.

If the collapsed stratum technique is used to develop the sampling error calculation model, slight overestimation of standard errors occurs because the collapsed strata ignore the true differences in the design strata that are collapsed to form the single sampling error calculation stratum. The following section provides interested readers with more detail on the combined strata, random groups, and collapsed strata techniques used in building sampling error calculation models for complex sample survey data.

4.3.3 Combining Strata, Randomly Grouping PSUs, and Collapsing Strata

Combining strata in building the sampling error calculation model involves the combination of PSUs from two or more different strata to form a single stratum with larger pooled sampling error clusters. Consider Figure 4.1. The original complex sample design involved

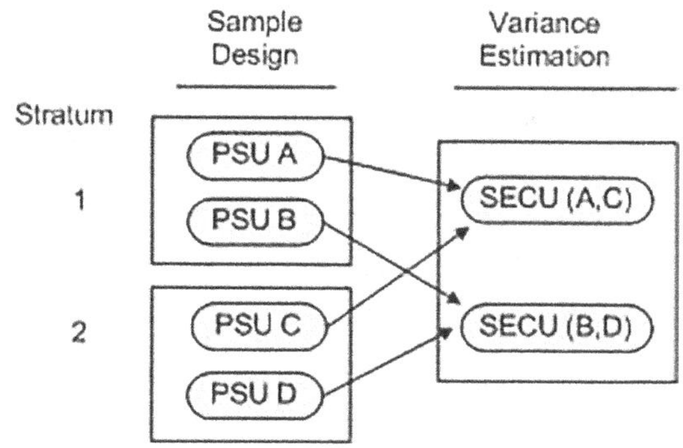

FIGURE 4.1
Illustration of the Combined Stratum Method.

paired selection of two PSUs within each primary stage stratum. For variance estimation, PSUs from the two design strata are combined, with the PSUs being randomly assigned into two larger sampling error clusters.

The technique of combining strata for variance estimation is typically used for one of two reasons: 1) the sample design has large numbers of primary stage strata and small numbers of observations per PSU, which could lead to variance estimation problems (especially when subclasses are being analyzed); or 2) the data producer wishes to group design PSUs to mask individual PSUs as part of a disclosure protection plan. The sampling error calculation model for the NHANES has traditionally employed combined strata to mask the identity of individual PSUs.

The **random groups method** randomly combines multiple clusters from a single design stratum to create two or more larger clusters for sampling error estimation. This is the technique illustrated for NCS-R self-representing design strata 15 and 16 in Table 4.3. Figure 4.2 presents an illustration of forming random groups of clusters for variance estimation purposes. In this illustration, there are eight area segment selections within an original self-representing design stratum. (The New York Metropolitan Statistical Area (MSA) is a common example of a geographic area that forms a self-representing stratum in national area probability samples of the United States population.) Figure 4.2 shows how the eight area segment clusters are randomly assigned into two larger sampling error clusters containing four of the original segments each.

The random groups method is typically used when many small design PSUs have been selected within a single stratum and the data producer wishes to simplify the calculation of variance for that stratum or to minimize disclosure risk that could result from releasing the original stratification and cluster coding. This technique is commonly used to create sampling error clusters in self-representing primary stage strata of multi-stage sample designs.

Collapsing strata is another technique that data producers may employ in building a sampling error calculation model for a complex sample design. This technique is generally used when a single primary sampling unit is selected from each primary stage design stratum (which precludes direct estimation of sampling variance for the stratum). The technique involves "collapsing" or "erasing" the boundary between *adjacent* strata, so that similar strata become one larger pseudo- or "collapsed" stratum with multiple primary sampling units. In contrast to the combined stratum method, the original design PSUs remain distinct as sampling error clusters in the collapsed sampling error stratum. We illustrate the idea of collapsing strata in Figure 4.3.

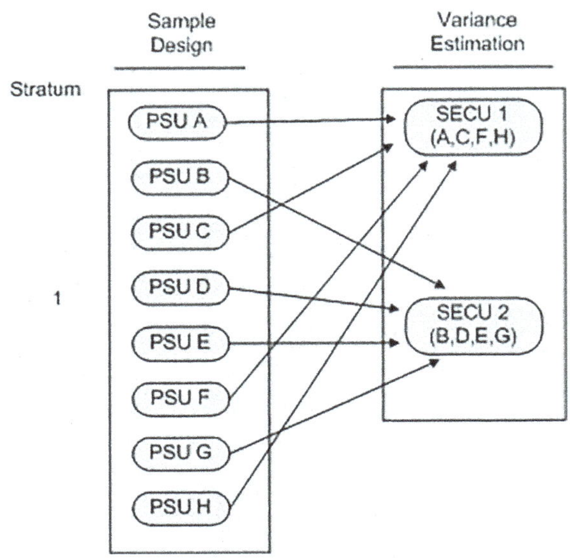

FIGURE 4.2
Illustration of the Random Groups Method.

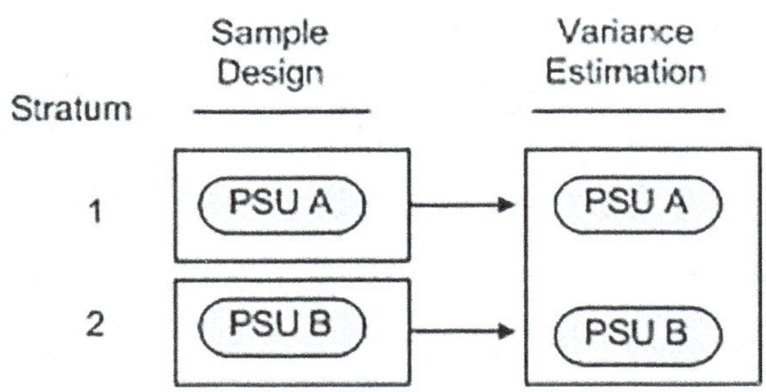

FIGURE 4.3
Illustration of the Collapsed Stratum Method.

4.3.4 Checking the Sampling Error Calculation Model for the Survey Data Set

As discussed previously in this chapter, identification of the variables identifying the sampling error strata and clusters is an essential step in preparing for the analysis of a complex sample survey data set. When these variables have been identified, a useful first step is to cross-tabulate these variables to get a sense of the sample sizes in each stratum and cluster.

The Stata software provides the `svyset` and `svydes` commands that enable users to define these key sample design variables (including final survey weights) for a data set and then describe the distributions of the cases to sampling error strata and clusters. These commands need to be submitted only once in each Stata session, unlike other software procedures that require these variables to be identified for each individual analysis (e.g., PROC SURVEYMEANS in SAS, PROC DESCRIPT in SUDAAN); readers can refer to Appendix A for details.

We illustrate the use of the `svyset` and `svydes` commands to show how these key design variables are identified and described in Stata for the NCS-R. First, the correct design variables were determined based on the documentation for the survey data set. Next, the following two Stata commands (`svyset` and `svydes`) were submitted, which produced the output that follows the commands:

```
svyset seclustr [pweight=ncsrwtlg], strata(sestrat) ///
vce(linearized) singleunit(missing)
svydes
```

Stratum	#Units included	#Units omitted	#Obs with complete data	#Obs with missing data	#Obs per included Unit		
					min	mean	max
1	2	0	44	0	22	22.0	22
2	2	0	53	0	26	26.5	27
3	2	0	77	0	31	38.5	46
4	2	0	87	0	42	43.5	45
5	2	0	72	0	32	36.0	40
6	2	0	68	0	31	34.0	37
7	2	0	127	0	61	63.5	66
8	2	0	84	0	38	42.0	46
9	2	0	89	0	44	44.5	45
10	2	0	78	0	23	39.0	55
11	2	0	57	0	28	28.5	29
12	2	0	61	0	24	30.5	37
13	2	0	63	0	26	31.5	37
14	2	0	45	0	20	22.5	25
15	2	0	73	0	36	36.5	37
16	2	0	66	0	30	33.0	36
17	2	0	41	0	18	20.5	23
18	2	0	55	0	26	27.5	29
19	2	0	62	0	29	31.0	33
20	2	0	159	0	70	79.5	89
21	2	0	187	0	75	93.5	112
22	2	0	178	0	82	89.0	96
23	2	0	186	0	84	93.0	102
24	2	0	197	0	94	98.5	103
25	2	0	264	0	124	132.0	140
26	2	0	155	0	58	77.5	97
27	2	0	172	0	79	86.0	93
28	2	0	117	0	45	58.5	72

```
                                                                     #Obs per included
                                            #Obs with    #Obs with         Unit
             #Units        #Units          complete     missing    ---------------------
Stratum     included       omitted          data         data       min    mean     max
--------   -----------   ----------   -----------   -----------   ----   -------   ----
     29         2             0              199           0          99     99.5    100
     30         2             0              121           0          57     60.5     64
     31         2             0              267           0         131    133.5    136
     32         2             0              191           0          86     95.5    105
     33         2             0               82           0          37     41.0     45
     34         2             0              236           0         108    118.0    128
     35         2             0              236           0          94    118.0    142
     36         2             0              203           0         100    101.5    103
     37         2             0              218           0         106    109.0    112
     38         2             0              197           0          84     98.5    113
     39         2             0              215           0          76    107.5    139
     40         2             0              164           0          66     82.0     98
     41         2             0              211           0          94    105.5    117
     42         2             0              235           0         116    117.5    119
--------   -----------   ----------   -----------   -----------   ----   -------   ----
     42        84             0             5692           0          18     67.8    142
                          --------------------------

                                            5692
                                            3590 = #Obs with missing values in the
                                       ---------   survey characteristics
                                            9282
```

The Stata output above illustrates the frequency distributions of the design variables that are produced by running the `svyset` and `svydes` commands for the NCS-R data set. The sampling error calculation model for the NCS-R sample design recognizes stratification of the population, and the variable containing the stratum codes for sampling error calculations is SESTRAT. The sampling error cluster (SECLUSTR) and overall survey weight (NCSRWTLG) variables are also identified, in addition to the method of variance estimation (Taylor Series Linearization). We also indicate what Stata should do if strata containing a single sampling error cluster are encountered in an analysis (variance estimates should be set to missing). Additional options for the `svyset` command are described in Appendix A. We note that if Stata users wish to reset the design variables declared in the `svyset` command, the `svyset, clear` command can be submitted.

The output from the `svydes` command displayed above indicates that the NCS-R sampling error calculation model has been specified as a paired selection design, with two sampling error clusters (# Units included) in each of 42 sampling error strata (for 84 sampling error clusters total). Sample sizes in the clusters (denoted in the Stata output by "# Obs per included Unit") range from 18 to 142, with an average of 67.8 sample respondents per ultimate cluster. This "two per stratum" sampling error coding is common in many public-use survey data sets and enables survey analysts to employ any of the variance estimation techniques (TSL, BRR, JRR, bootstrapping) discussed in Chapter 3. Stata also reports that 3,590 observations have missing data on at least one key design variable. This does not come as a surprise, because the Part II survey weight variable for NCS-R cases (NCSRWTLG) takes on missing values for cases that were not selected or subsampled for Part II of the NCS-R survey.

These simple initial tabulations help to familiarize data users with the sampling error calculation models that will be used for variance estimation and can be useful for describing complex sample design information and methods of variance estimation in scientific and technical publications. Analysts using other statistical software packages (e.g., SAS, SPSS, and SUDAAN) can easily perform these simple descriptive analyses using standard procedures for cross-tabulation.

4.4 Addressing Item Missing Data in Analysis Variables

Many survey data analysts, initially excited to begin the analysis of a new survey data set, are disappointed to find that many of the key survey variables they hope to analyze suffer from problems of missing data. As they move from simple univariate analysis to multivariable analysis, the missing data problem is compounded due to the **listwise deletion** from analysis by statistical software of any cases with missing values on one or more variables.

The first fact that all data users should recognize is that no survey data set is immune to some form of missing data problem. Therefore, addressing missing data should be as commonplace an analytic activity as choosing the form of a regression model to apply to a dependent variable of interest. Section 2.7 has already introduced weighting adjustments for unit nonresponse—one major source of missing data in sample surveys.

Here we will focus briefly on the implications of ignoring item missing data in analysis and then describe a convenient method for investigating the rates and patterns of missing data in a chosen set of analysis variables. Chapter 12 will examine methods for statistical imputation of item missing data using today's powerful new software tools.

4.4.1 Potential Bias Due to Ignoring Missing Data

Many analysts simply choose to ignore the missing data problem when performing analyses. If the rates of missing data on key analysis variables are very low (say < 1–2% of cases), the penalty for not taking active steps (i.e., weighting or imputation) to address the missing data is probably small. Consider a univariate analysis to estimate the population mean of a variable y. Under a simple deterministic assumption that the population is comprised of "responders" and "nonresponders," the expected bias in the respondent mean of the variable y is defined as follows:

$$Bias(\bar{Y}_R) = \bar{Y}_R - \bar{Y} = P_{NR} \times (\bar{Y}_R - \bar{Y}_{NR}) \tag{4.1}$$

In (4.1), $\bar{Y}$ is the true population mean, $\bar{Y}_R$ is the population mean for responders in the population, $\bar{Y}_{NR}$ is the population mean for nonresponders in the population, and P_{NR} is the expected proportion of nonrespondents in a given sample. For the bias to be large, the rate of missing data must be sizeable, and respondents must differ in their characteristics from nonrespondents. In statistical terms, the potential bias due to missing data depends on both the missing data pattern and the missing data mechanism. Chapter 12 will provide a more detailed review of missing data patterns and mechanisms.

4.4.2 Exploring Rates and Patterns of Missing Data Prior to Analysis

The Stata software provides data analysts with the `misstable patterns` and `misstable summarize` commands to display the patterns and rates of missing data across a set of variables that will be included in an analysis. The following example uses these commands to explore the missing data for six 2011–2012 NHANES variables that will be included in the multiple imputation regression example that is presented later in Section 12.6. The variables are diastolic blood pressure (BPXDI1_1), marital status (MARCAT), gender (RIAGENDR), race/ethnicity (RIDRETH1), age (AGEC and AGECSQ), body mass index (BMXBMI), and family poverty index (INDFMPIR). The command syntax to request the missing data summary for this set of six variables (where AGECSQ is simply AGEC squared) is:

```
misstable summarize bpxdi1 _ 1 bmxbmi indfmpir marcat ///
riagendr ridreth1 agec agecsq if age18p==1 & ///
wtmec2yr > 0, all
```

The command syntax to request the missing data patterns for the same six variables is:

```
misstable patterns bpxdi1 _ 1 bmxbmi indfmpir ///
marcat riagendr ridreth1 agec agecsq, freq
```

The actual Stata output produced by the `misstable patterns` and `misstable summarize` commands is provided below. In the `summarize` output, the first portion of the output lists the number of observed and missing values for each variable that has at least one missing value, and also variables with full data (since the `all` option was specified). The missing-value patterns output summarizes the frequencies of various patterns of missing data across the four variables with missing data, using a coding of '1' for observed and '0' for missing. For example, 4,416 observations have no missing data for these four variables, and a total of 9 observations have missing data only for the BMXBMI (body mass index) variable.

```
misstable summarize bpxdi1_1 bmxbmi indfmpir marcat riagendr ridreth1 agec agecsq, all

                                                            Obs<.
              |                              +------------------------------------
              |                              | Unique
     Variable |   Obs=.    Obs>.    Obs<.    | values           Min          Max
--------------+------------------------------+------------------------------------
     bpxdi1_1 |     503             5,112    |     49            10          120
     bpxdi1_1 |      90             5,525    |    385          13.4         82.1
     indfmpir |     487             5,128    |    406             0            5
       marcat |     300             5,315    |      3             1            3
     riagendr |                     5,615    |      2             1            2
     ridreth1 |                     5,615    |      5             1            5
         agec |                     5,615    |     63     -28.35516     33.64484
       agecsq |                     5,615    |     63       1261386     1131.975
```

```
Missing-value patterns
     (1 means complete)
                  |
                  |    Pattern
                  |
    Frequency     |  1  2  3  4
  ----------------+------------
        4,416     |  1  1  1  1
                  |
          386     |  1  1  1  0
                  |
          369     |  1  1  0  1
                  |
          230     |  1  0  1  1
                  |
           62     |  1  1  0  0
                  |
           48     |  0  1  1  1
                  |
           31     |  1  0  0  1
                  |
           22     |  1  0  1  0
                  |
           18     |  0  1  1  0
                  |
           12     |  0  1  0  1
                  |
            9     |  1  0  0  0
                  |
            6     |  0  0  1  1
                  |
            4     |  0  1  0  0
                  |
            2     |  0  0  1  0
  ----------------+------------
        5,615     |
                  |
Variables are (1) bmxbmi (2) marcat (3) indfmpir (4) bpxdi1_1
```

With recent advances in the theory of statistical analysis with missing data (Little & Rubin, 2019; Raghunathan, 2016) and today's improved software (e.g., Raghunathan et al., 2001; Berglund & Heeringa, 2014; StataCorp, 2023), data producers or data users often consider different methods for the **imputation** (or prediction) of missing data values. Depending on the patterns of missing data and the underlying process that generated the missingness, statistically sound imputation strategies can produce data sets with complete cases for analysis.

In many large survey programs, the data producer may perform imputation for key variables before the survey data set is released for public use. Typically, methods for **single imputation** are employed using techniques such as **hot deck imputation, regression imputation**, and **predictive mean matching.** Some survey programs may choose to provide data users with **multiply imputed** data sets (Kennickell, 1998; Schafer, 1996). When data producers perform imputation of item missing data, a best practice in data dissemination is to provide data users with both the imputed version of the variable (e.g., I_INCOME_AMT) and an indicator variable (e.g., I_INCOME_FLG) that identifies the values of the variables that have been imputed. Data users should expect to find general documentation for the imputations in the technical report that accompanies the survey data. The survey codebook should clearly identify the imputed variables and the imputation "flag" variables.

When imputations are not provided by the data producer (or the analyst chooses to employ their own imputation methods and models), the task of imputing item missing data will fall to the data user. Details on practical methods for user imputation of item-missing data are provided later in Chapter 12.

4.5 Preparing to Analyze Data for Sample Subpopulations

The analysis of complex sample survey data sets often involves separate estimation and inference for **subpopulations** or **subclasses** of the full population. For example, an analyst may wish to use the NHANES data to estimate the prevalence of diabetes separately for male and female adults, or compute HRS estimates of retirement expectations only for the population of the U.S. Census South Region. An NCS-R analyst may choose to estimate a separate logistic regression model for predicting the probability of past-year depression status for African American women only.

When analyses are focused in this way on these specific subclasses of the survey population, special care must be taken to correctly prepare the data and specify the subclass analysis in the code that is input to software procedures. Proper analysis methods for subclasses of survey data have been well-established in the survey methodology literature (Lohr, 2021; West et al., 2008; Rao, 2003; Korn & Graubard, 1999; Fuller et al., 1989; Cochran, 1977; Kish, 1965), and interested readers can consult these references for more general information on estimation of survey statistics and related variance estimation techniques for subclasses. A short summary of the theory underlying subclass analysis of survey data is also provided in Theory Box 4.2.

4.5.1 Subpopulation Distributions across Sample Design Units

Although most current survey data analysis software is programmed to correctly account for subclasses in analysis, a useful step in preparing for data analysis is to examine the distribution of the targeted subpopulation sample with respect to the sampling error strata and clusters that have been defined under the sampling error calculation model. Figure 4.4 illustrates the different distributional patterns that might be observed in practice.

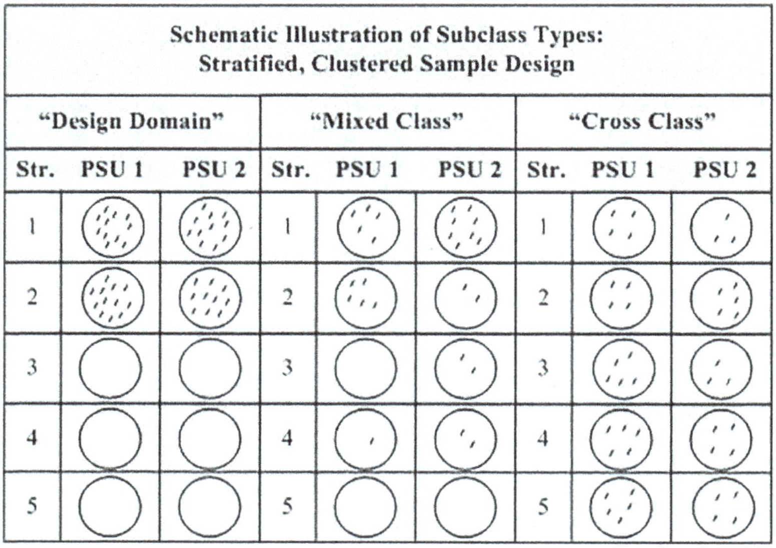

FIGURE 4.4
Subclass Types in Stratified Cluster Sample Designs.

Design domains, or subclasses that are restricted to only a subset of the primary stage strata (e.g., adults in the Census South Region, or residents of urban counties), constitute a broad category of analysis subclasses. In general, the analysis of design domain subclasses should not be problematic in most contemporary survey analysis software. Analysts should recognize that sampling errors estimated for domain subclasses will be based on fewer degrees of freedom than estimates for the full sample or cross-classes, given that design domains are generally restricted to specific sampling strata.

A second pattern that may be observed is a **mixed class**, or a population subclass that is not isolated in a subset of the sample design strata but is unevenly and possibly sparsely distributed across the strata and clusters of the sampling error calculation model. Experience has shown that many survey analysts often "push the limits" of sample designs, focusing on rare or highly concentrated subclasses of the population—e.g., Hispanic adults with asthma in the HRS. While such analyses are by no means precluded by the sample design, survey analysts are advised to exercise care in approaching subclass analyses of this type for several reasons: 1) nominal sample sizes for mixed classes may be small, but design effects due to weighting and cluster sampling may still be substantial; 2) a highly uneven distribution of cases to strata and clusters will introduce instability or bias in the variance estimates (especially those based on the Taylor Series Linearization method); and 3) software approximations to the complex sample design degrees of freedom (df = # clusters – # strata) may significantly overstate the true precision of the estimated sampling errors and confidence intervals. Survey analysts who intend to analyze data for mixed classes are encouraged to consult with a survey statistician, but we would recommend that these subclasses be handled using unconditional subclass analysis approaches (see Section 4.5.2).

We label the third pattern as a **cross-class** (Kish, 1987). Cross-classes are subclasses of the survey population that are broadly distributed across all strata and clusters of a complex sample design. Examples of cross-classes in a national area probability sample survey of adults might include males, or individuals aged 40 and above. Properly identified to the software, subclasses that are true cross-classes present very few problems in survey data analysis—sampling error estimation and degrees of freedom determination for confidence intervals and hypothesis tests should be straightforward.

We now consider two alternative approaches to subclass analysis that analysts of complex sample survey data sets could take in practice and discuss applications where one approach might be preferred over another.

4.5.2 The Unconditional Approach for Subclass Analysis

Recall from Chapter 3 that the estimated standard error of an estimated survey statistic expresses the degree of variability in the statistic that we would expect from one hypothetical sample to another around the true population parameter of interest. If the subclass analysis were restricted only to sample cases that belonged to the subclass (sometimes referred to as the **conditional subclass analysis** approach, because it conditions subclass inference on the observed sample), sampling error estimates would in effect assume that each hypothetical sample would have the *same* fixed number of subclass members (say m) and that the distribution of the fixed subclass sample size, m, across design strata and clusters would remain unchanged from sample to sample. When analyzing complex sample survey data sets in practice, this would only be the case when analyzing design domains (as defined above) with sample sizes that are fixed by design.

In all other applications, the subclass sample size is a *random variable*—varying both in its size and its distribution across the design strata and clusters. By performing an

unconditional subclass analysis (i.e., one that does not condition on the observed distribution of subclass cases to particular strata and clusters of the full sample design, as described in West et al., 2008), we take into account these added sources of true sampling variability. Point estimates of population parameters will be identical under both subclass analysis approaches. However, because a conditional analysis does *not* incorporate this latter source of variance, it tends to result in *underestimates* of standard errors, which cause analysts to overstate the precision of estimated survey statistics for subclasses. Therefore, simply using the data management capabilities of statistical software to delete or filter out those cases that do not fall into the subclass of interest can produce biased inferences. This requires some additional preparation on the part of the data analyst (see Section 4.5.3 below).

We note that this distinction between the conditional and unconditional subclass analysis approaches is only relevant when Taylor Series Linearization is the technique being employed for variance estimation (see Chapter 3 for details). This is because linearization relies on the sampling error strata and clusters defined by the sampling error calculation model to compute variance estimates. When replication methods (e.g., JRR) are employed for variance estimation, only the weighted estimates from each replicate sample are used for variance estimation purposes, and the distribution of the subclass sample across the sampling error strata and clusters becomes irrelevant.

To make this distinction between the variance estimation methods in the case of subclass analysis clear, we consider an example. In a complex sample with a sampling error calculation model that defines two clusters per stratum, suppose that a subclass of analytic interest has no sample members in one of the two clusters in a particular stratum. If one were to employ linearization and attempt the *conditional* subclass analysis approach, deleting cases that did not belong to the subclass, this cluster would be deleted from the data set, and the software would encounter problems attempting to determine the contribution of between-cluster variance within this stratum to the overall variance estimate (given that it only has one cluster present for variance estimation, when at least two are required). The *unconditional* approach would allow linearization to proceed, as the cluster with no sampled elements from the subclass would now contribute totals of zero to the linearized variance estimates (reflecting sampling zeroes). If one were to employ replication approaches, the cluster in question would simply not contribute to any of the replicate sample estimates used for variance estimation.

We urge analysts to carefully consider the distinction between these two approaches, given that most software for the analysis of complex sample survey data uses linearization as the default variance estimation technique. Different software procedures will react differently if linearization is requested for variance estimation and only one sampling error computation unit (SECU) is detected within a sampling error stratum (possibly due to the use of the conditional approach; this is sometimes referred to as a "**lonely PSU**"). For example, the Stata `svy:` commands (by default) will not proceed with computing linearized standard errors and only report missing values, while the SAS SURVEY procedures will simply omit those strata with only a single SECU present from variance estimation and provide a note in the SAS log. Users will need to specify the appropriate options to implement the correct unconditional subclass analysis approach, and we will emphasize this in all of our chapters presenting examples.

Finally, we note that secondary analysts may be interested in quantifying which subclass differences are largest when conducting subclass analyses. This approach is known as **dominance analysis**, and Luchman (2015) describes how to perform this type of analysis when analyzing complex sample survey data in Stata. The inputs to this analysis need to come from the correct type of unconditional analysis approach. We revisit this idea in Chapter 5.

4.5.3 Preparation for Subclass Analyses

To properly prepare for a correct unconditional subclass analysis of a complex sample survey data set, data users should generate indicator variables for each subclass S that they are interested in analyzing:

$$I_{S,i} = \begin{cases} 1 \text{ if case } i \text{ is a member of the subclass of interest} \\ 0 \text{ if case } i \text{ is not a member of the subclass of interest} \end{cases}$$

Each case in the complete sample should be assigned a value of 1 or 0 for this indicator variable (if information used to define the subclass of interest is available for each case). This indicator variable will be used to identify the subclass members to the analysis software.

Once these indicator variables have been created for subclasses of interest, data users need to utilize appropriate software options for unconditional subclass analyses. For example, Stata procedures offer the `subpop()` option, where the subclass indicator variable can be specified in the parentheses; SUDAAN offers the SUBPOPN keyword; SPSS (in the Complex Samples Module) offers the SUBPOP and DOMAIN subcommands; and SAS offers the DOMAIN keyword for PROC SURVEYMEANS, PROC SURVEYREG, and PROC SURVEYLOGISTIC specifically, and allows users to indicate domains using the first variable listed in the TABLES statement of PROC SURVEYFREQ. Similar options and functions are available in other software, such as the `survey` package in R; see Appendix A for details.

THEORY BOX 4.2: THEORETICAL MOTIVATION FOR UNCONDITIONAL SUBCLASS ANALYSES

To illustrate the importance of following the unconditional subclass analysis approach mathematically, we consider the variance of a sample total (the essential building block for variance estimation based on Taylor Series Linearization). We denote design strata by h (h = 1, 2, . . ., H), first-stage PSUs within strata by α (α = 1, 2, . . ., a_h), and sample elements within PSUs by i (i = 1, 2, . . ., $n_{h\alpha}$). The weight for element i, taking into account factors such as unequal probability of selection, nonresponse, and possibly post-stratification, is denoted by $w_{h\alpha i}$. We refer to specific subclasses using the notation S. An estimate of the total for a variable Y in a subclass denoted by S is computed as follows (Cochran, 1977):

$$\hat{Y}_S = \sum_{h=1}^{H}\sum_{\alpha=1}^{a_h}\sum_{i=1}^{n_{h\alpha}} w_{h\alpha i} I_{S,h\alpha i} y_{h\alpha i} \tag{4.2}$$

In this notation, I represents an indicator variable, equal to 1 if sample element i belongs to subclass S, and 0 otherwise. The closed-form analytic formula for the variance of this subclass total can be written as follows:

$$Var(\hat{Y}_S) = \sum_{h=1}^{H} \frac{a_h}{(a_h - 1)} \left[\sum_{\alpha=1}^{a_h} \left(\sum_{i=1}^{n_{h\alpha}} w_{h\alpha i} I_{S,h\alpha i} y_{h\alpha i} \right)^2 - \frac{\left(\sum_{\alpha=1}^{a_h} \sum_{i=1}^{n_{h\alpha}} w_{h\alpha i} I_{S,h\alpha i} y_{h\alpha i} \right)^2}{a_h} \right] \tag{4.3}$$

This formula shows that the variance of the subclass total is calculated by summing the between-cluster variance in the subclass totals within strata, across the H sample strata. The formula also shows how the indicator variable is used to ensure that all sample elements (and their design strata and PSUs) are recognized in the variance calculation; this emphasizes the need for the software to recognize all of the original design strata and PSUs. Analysts should note that if all $n_{h\alpha}$ elements within a given stratum denoted by h and PSU denoted by α do not belong to the subclass S (although elements from that subclass theoretically *could* belong to that PSU in any given sample), that PSU will still contribute to the variance estimation: The PSU helps to define the total number of PSUs within stratum h (a_h), and contributes a value of 0 to the sums in the variance estimation formula. In this way, sample-to-sample variability in the estimation of the total because the subclass sample size is a random variable is captured in the variance calculation.

4.6 A Final Checklist for Data Users

We conclude this chapter with a summary of the important preparation steps that data users should follow prior to beginning the analysis of a complex sample survey data set.

1. Review the documentation for the data set provided by the data producer, specifically focusing on sections discussing the development of the final survey weights and sampling error (standard error) estimation. Contact the data producer if any questions arise.
2. Identify the correct survey weight variable for the analysis, keeping in mind that many survey data sets include separate weights for different types of analyses. Perform simple descriptive analyses of the weight variable, noting the general distribution of the weights, whether the weights have been normalized, and whether there are missing or 0 weight values for some cases. Select a few key variables from the survey data set and compare weighted and unweighted estimates of descriptive parameters (e.g., means, proportions) for these variables.
3. Identify the variables in the data set containing the sampling error calculation codes (for strata and clusters) that define the sampling error calculation model. Examine how many clusters were selected from each sampling stratum (according to the sampling error calculation model), and whether clusters have small sample sizes. If only a single sampling error cluster is identified in a sampling stratum, contact the data producer, or consult the documentation for the data set for guidance on recommended variance estimation methods. Determine whether replicate weights are present if sampling error calculation codes are not available, and make sure that the statistical software is capable of accommodating replicate weights (see Section 4.2.1).
4. Create a final analysis data set containing only the analysis variables of interest (including the final survey weights, sampling error calculation variables, and case identifiers). Examine univariate and bivariate summaries for the key analysis variables to determine possible problems with missing data or unusual values on the individual variables.

5. Review the documentation provided by the data producer to understand the procedure (typically nonresponse adjustment) used to address unit nonresponse or nonresponse to a wave or phase of the survey data collection. Analyze the rates and patterns of item-missing data for all variables that will be included in the analysis. Investigate the potential missing data mechanism by defining indicator variables flagging missing data for the analysis variables of interest. Use statistical tests (e.g., chi-square tests, two-sample t-tests) to see if there are any systematic differences between respondents providing complete responses and respondents failing to provide complete responses on important analysis variables (e.g., demographics). Choose an appropriate strategy for addressing missing data using the guidance provided in Section 4.4 and Chapter 12.
6. Define indicator variables for important analysis subclasses. *Do not delete cases that are not a part of the primary analysis subclass.* Assess a cross-tabulation of the stratum and cluster sampling error calculation codes for the subclass cases to identify the distribution of the subclass across the strata and clusters defined by the sampling error calculation model. Consult a survey statistician prior to analysis of subclasses that exhibit the "mixed class" characteristic illustrated in Figure 4.4. Make sure to employ appropriate software options for unconditional subclass analyses if using Taylor Series Linearization for variance estimation.

Based on many years of consulting on the analysis of survey data, the authors firmly believe that following these steps before performing any analyses of complex sample survey data sets will eliminate the most common mistakes that result in wasted time and effort or (even worse) unknowingly result in biased, incorrect analyses of the survey data.

Notes

1 http://www.cdc.gov/nchs/data/series/sr_02/sr02_161.pdf

2 The SDA system is available at the website of the University of Michigan Inter-University Consortium for Political and Social Research (ICPSR) and is produced by the Computer-Assisted Survey Methods Program at the University of California—Berkeley. Visit www.icpsr.umich.edu for more details.

5

Descriptive Analysis for Continuous Variables

"In 2023, 22.5 percent of people with a disability were employed, compared with 65.8 percent of those without a disability. The unemployment rate for people with a disability (7.2 percent) was little changed in 2023. The rate for people without a disability (3.5 percent) was unchanged."

U.S. Bureau of Labor Statistics, News Release, February 22, 2024, https://www.bls.gov/news.release/pdf/disabl.pdf.

5.1 Introduction

Estimation of totals, means, variances, and other distributional statistics for survey variables may be the primary objective of an analysis plan or possibly an exploratory step on a path to a more multivariable treatment of the survey data. Major reports of survey results can be filled with detailed tabulations of descriptive estimates. Scientific papers and publications that may emphasize a more analytical multivariable treatment of the data typically include initial descriptive estimation of population characteristics as part of the overall analysis plan.

This chapter describes techniques for generating population estimates of descriptive parameters for continuous survey variables. Section 5.2 examines features that distinguish descriptive analysis of complex sample survey data from more standard descriptive analyses of simple random samples or convenience samples. Basic methods and examples of estimating important descriptive parameters for univariate, continuous distributions are described in Section 5.3, followed in Section 5.4 by methods for studying simple bivariate relationships between two continuous survey variables. The chapter concludes with coverage of methods for descriptive estimation for subpopulations (Section 5.5) and estimation of simple linear functions of distributional statistics such as differences of means (Section 5.6).

5.2 Special Considerations in Descriptive Analysis of Complex Sample Survey Data

Descriptive analysis of complex sample survey data shares much in common with the standard statistical methods that are taught in any introductory statistics course, but there are several subtle differences that are important to keep in mind.

DOI: 10.1201/9781003472094-6

5.2.1 Weighted Estimation

Descriptive analysis of both continuous and discrete survey measures is intended to characterize the distributions of variables over the full survey population. If population elements have differing probabilities of inclusion in the sample (deliberately introduced by the sampling statistician or induced by a nonresponse mechanism), unbiased estimation for the population distribution requires weighting of the sample data. Depending on the distribution of the weights and the correlations between the weights and the survey variables of interest, unweighted sample estimates may yield a very biased picture of the true population distribution (Little & Vartivarian, 2005).

It is informative to plot the value of the survey weight against the value of the survey variable of interest. Figure 5.1 illustrates a plot of this type, with the 2012 HRS household-level survey weight value on the vertical axis and the value of 2012 HRS total household net worth on the horizontal axis. Diagnostic plots of this type inform the user of the following important properties of the weights and their relationship to the variable of interest: 1) the variability of the weights, which can affect the sampling variability of descriptive estimates (see Section 2.7.5); 2) any strong functional relationship between individual weight values and survey observations, signaling that the weights are highly "informative" for the estimation of population parameters; and 3) the potential for "weighted outliers"—points representing large values on both the survey variable and the survey weight that in combination will have a major influence on the estimated distributional statistics and their standard errors.

For example, Figure 5.1 suggests that there is considerable variability in the 2012 HRS household weights, but there is little evidence of a functional (e.g., linear, quadratic) relationship of the weights with the measures of total household net worth. Therefore, for the total household net worth variable, weighted estimation will result

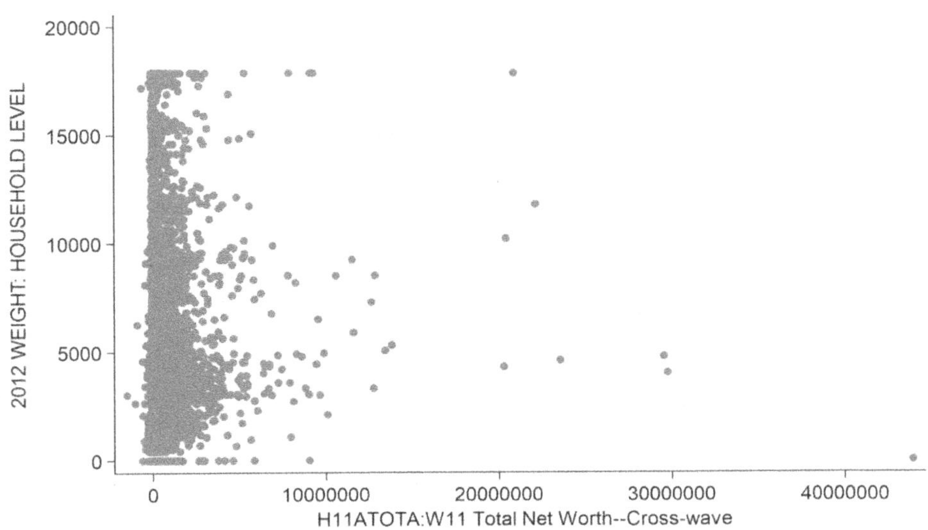

FIGURE 5.1
Examining the relationship of a survey variable (total household net worth) and the final household-level survey weight in the 2012 HRS data set.

in increased sampling variances for estimates, but the weighting may not have major impacts on the estimates of population parameters (see Section 2.7). The plot also identifies eight HRS cases with exceptionally large values for total household net worth (e.g., >$20,000,000). The survey weights for three of these "large net worth" cases are also large, each above the mid-point of the weight range and one close to the maximum for all 2012 HRS household weight values. These three points warrant investigation (see Chapter 4) because they may have significant influence on the estimates of descriptive parameters in addition to substantial leverage in the estimation of multivariable models.

5.2.2 Design Effects for Descriptive Statistics

Estimates of descriptive parameters (or descriptive statistics) derived from complex sample survey data can be subject to substantial design effects, due to the stratification, cluster sampling, and weighting associated with the design. Empirical research has consistently demonstrated that complex sample design effects are largest for weighted estimates of population means and totals for single survey variables, smaller for estimates of subpopulation means and proportions, and substantially reduced for regression coefficients and other statistics that are functions of two or more variables (Skinner et al., 1989; Kish et al., 1976). Throughout this chapter, the example analyses will include estimation of design effect ratios to provide the reader with a sense of the magnitudes of the design effects and their variability across survey variables and estimated statistics.

5.2.3 Matching the Method to the Variable Type

Survey responses are recorded on a variety of scales, including simple binary choices (yes, no), nominal categories (Democrat, Republican, Independent), ordinal scales (1 = excellent, 2 = very good, 3 = good, 4 = fair, 5 = poor), ordinal counts (years of education, weeks not employed), interval scale variables (age in years), fully continuous variables (systolic blood pressure in mmHg), grouped or "interval censored" continuous variables (income categories, for example, $10,000–$24,999), and censored or "semi-continuous" variables (the value of household wealth in stocks and bonds). With the obvious exception of nominal categorical variables and interval censored measures, it is common for survey analysts to apply descriptive analysis techniques appropriate for continuous data to survey variables that are measured on scales that are not strictly continuous.

The descriptive analytic methods described in this chapter may certainly be applied to ordinal, interval-scale, and semi-continuous survey measures; however, analysis results should be presented in a way that acknowledges the "non-continuous" character of the underlying data. The classical interpretation of ordinal scales as a measure of an underlying continuous latent variable recognizes that the mapping of that construct to the ordinal metric may not be linear—moving from poor to fair on a health rating scale may not reflect the same degree of health improvement as moving from fair to good. The estimated 75th percentile of the distribution of household mortgage debt for HRS households may be a misleading statistic if the percent of households with no mortgage is not also reported. In short, as is true in all statistical summarization, it is the survey analyst's responsibility to ensure that the reporting and interpretation of even the simplest descriptive analyses are accurate and transparent for the intended audience.

5.3 Simple Statistics for Univariate Continuous Distributions

This section outlines methods for estimating simple statistics that describe the population distributions of survey variables, including graphical tools and statistical estimators for population totals, means, variances, and percentiles. Weighted estimators of the most important distributional parameters for continuous survey variables are described in this section, and with a few exceptions, most of the major statistical software packages support weighted estimation for these statistics. When available, design-based estimation of these univariate descriptive parameters and the corresponding standard errors is illustrated in Stata using the NCS-R, the 2011–2012 NHANES, and the 2012 HRS data sets. We also present alternative Bayesian approaches to these descriptive analyses for selected examples using contributed packages in the R software (see Section 3.8). Examples of code for estimating these same descriptive quantities using the newest versions of other software packages can be found on the ASDA website.

5.3.1 Graphical Tools for Descriptive Analysis of Survey Data

The authors' experience suggests that statistical graphics are currently underutilized as tools for the exploratory analysis of survey data (Tukey, 1977), which includes effective visual display of estimated population distributions of single variables or the population relationships among two or more independent variables (Cleveland, 1993; Tufte, 1983). A thorough treatment of the many graphical tools available to today's analysts would require a separate volume and, as a general subject, is well-covered in any number of existing publications (Mitchell, 2022; SAS Institute Inc., 2023a; Maindonald & Braun, 2007). In this chapter, we will use selected tools in Stata and R to illustrate the concepts of graphical presentation and exploration of survey data.

The key requirement in any graphical treatment of complex sample survey data is that the method and the software must enable the analyst to incorporate the influences of the survey weights in the construction of the final display. Just as an unweighted estimate of a mean may be a biased estimate of the true population value, a statistical graphic developed from unweighted survey data may present a misleading image of the true survey population that the analysis is intended to represent.

The graphical procedures in today's statistical software packages differ in their capability to incorporate weights in graphical analyses. Stata is one package that can incorporate the survey weights in the development of statistical graphics; however, as the examples in this chapter will illustrate, the different graphical procedures may require specification of the survey weight as a standard probability weight (`pweight =`), a population frequency weight (`fweight =`), a relative weight (`aweight =`) that Stata proportionately scales to the sample size, or a flexible, user-defined weight (`iweight =`) designed for specific forms of estimation in Stata commands. Stata graphics include a number of options for displaying the estimated form of the population distribution for a single variable, including histograms, boxplots, spikeplots, and kernel density (kdensity) plots. The following two examples demonstrate Stata's capability to build histogram and boxplot graphics that incorporate survey weights.

EXAMPLE 5.1: A WEIGHTED HISTOGRAM OF TOTAL CHOLESTEROL (SOURCE: 2011–2012 NHANES)

After first opening the 2011–2012 NHANES data set in Stata, we generate an indicator variable for adults (respondents with age greater than or equal to 18), for use in the analyses:

```
gen age18p = 1 if age >= 18 & age != .
replace age18p = 0 if age < 18
```

The Stata software considers system missing values as being larger than the largest non-missing value for a variable in a data set. This is the reason for the condition that the value on the AGE variable must be greater than or equal to 18 *and* not missing for the indicator variable to be equal to 1. System missing values are important to consider in *all* software packages when manually creating indicator variables, so that missing values are not mistakenly coded into either a 1 or 0.

The following Stata commands then generate the weighted histogram of NHANES total cholesterol measures presented in Figure 5.2.

```
generate int_wtmec2yr = int(WTMEC2YR)
histogram lbxtc if age18p [fweight = int_wtmec2yr]
```

Note that the Stata `histogram` command requires that the weight be specified as an integer `fweight` (frequency weight) value; Stata does not allow for the use of `pweight` variables in the creation of histograms. Since these frequency weights must be integer values, a new weight is created by dropping the decimal places from the original values of the 2011–2012 NHANES MEC analysis weight, WTMEC2YR. Since WTMEC2YR is a population-scale weight, trimming the decimal places will have no significant effect on the graphical representation of the estimated distribution of total cholesterol.

Figure 5.2 displays the histogram generated by Stata. Since the 2011–2012 NHANES data have been weighted, the distributional form of the histogram represents an estimate of the distribution of total cholesterol for the NHANES survey population (U.S. adults age 18 and older) during this time period.

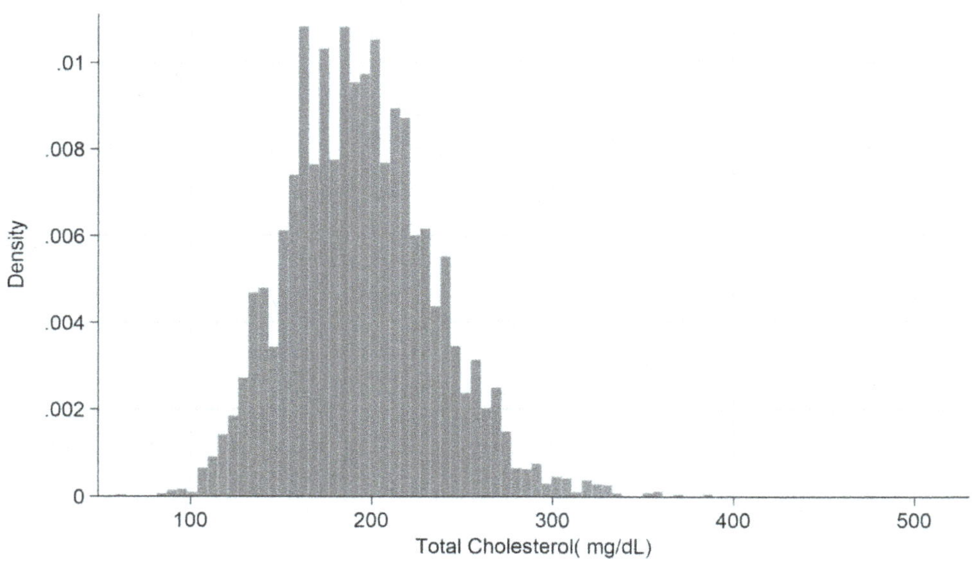

FIGURE 5.2
A Weighted Histogram of Total Cholesterol

Source: 2011–2012 NHANES.

EXAMPLE 5.2: WEIGHTED BOXPLOTS OF TOTAL CHOLESTEROL FOR U.S. ADULT MEN AND WOMEN (SOURCE: 2011–2012 NHANES)

For scientific reports and papers, boxplots are a very useful tool for the graphical presentation of estimated (weighted) population distributions of single survey variables. The following Stata command requests a pair of boxplots that represent the estimated gender-specific population distributions of total serum cholesterol in U.S. adults during the 2011–2012 time period.

```
gen female = (riagendr == 2)
graph box lbxtc [pweight = WTMEC2YR] if age18p==1, by(female)
```

Note that the Stata `graph box` command permits the specification of the analysis weight as a `pweight` (or probability weight) variable—no conversion to integer values is needed.

The pair of boxplots generated by the Stata command is displayed in Figure 5.3. These boxplots suggest that male and female adults in the 2011–2012 NHANES survey population have very similar distributions on this particular continuous variable.

5.3.2 Estimation of Population Totals

The estimation of a population total and its sampling variance has played a central role in the development of probability sampling theory. In survey practice, optimal methods for estimation of population totals are extremely important in government and academic surveys that focus on agriculture (e.g., acres in production), business (e.g., total employment), and organizational research (e.g., hospital costs). Acres of corn planted per year, total natural gas production for a 30-day period, and total expenditures attributable to a prospective change in benefit eligibility are all quantities of research interest that require efficient estimation of

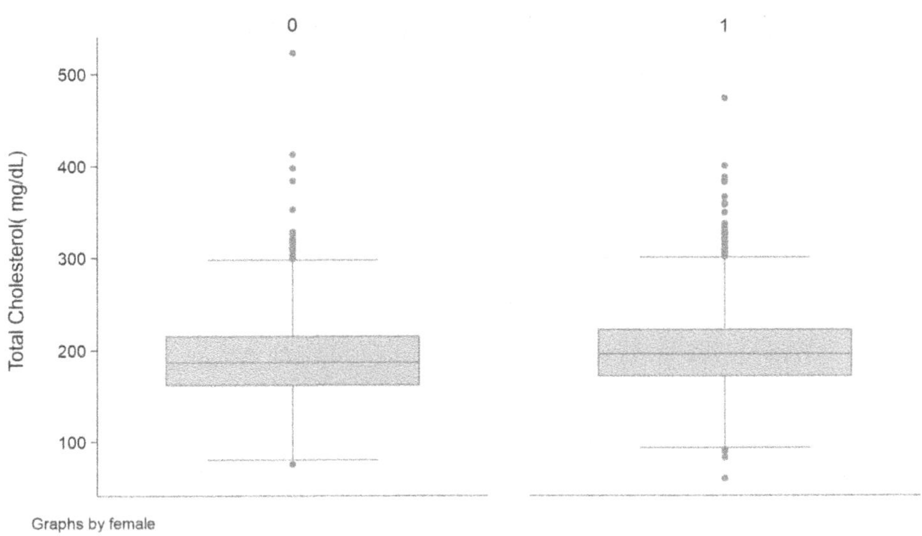

FIGURE 5.3
Boxplots of the gender-specific weighted distributions of total serum cholesterol for U.S. adults

Source: 2011–2012 NHANES.

population totals. In those agencies and disciplines where estimation of population totals plays a central role, advanced "model-assisted" estimation procedures and specialized software are the norm. Techniques such as the **generalized regression (GREG) estimator** or **calibration estimators** (e.g., Andersson, 2019) integrate the survey data with known population controls on the distribution of the population weighting factors to produce efficient weighted estimates of population totals (Deville & Sarndal, 1992; Valliant et al., 2000).

Most statistical software packages do not currently support advanced techniques for estimating population totals such as the GREG or calibration methods. Valliant et al. (2018) provide a variety of R code enabling such tasks. Stata and other software packages that support complex sample survey data analysis do provide the capability to compute simple weighted or **expansion estimates** of finite population totals, and also include a limited set of options for including population controls in the form of post-stratified estimation (see Theory Box 5.1).

In the case of a complex sample design including stratification (with strata indexed by $h = 1, \ldots, H$) and cluster sampling (with clusters within stratum h indexed by $\alpha = 1, 2, \ldots, a_h$), the simple weighted estimator for the population total can be written as follows:

$$\hat{Y}_w = \sum_{h=1}^{H}\sum_{\alpha=1}^{a_h}\sum_{i=1}^{n_{h\alpha}} w_{h\alpha i}\, y_{h\alpha i} \tag{5.1}$$

A closed-form, unbiased estimate of the variance of this estimator is:

$$var(\hat{Y}_w) = \sum_{h=1}^{H}\frac{a_h}{(a_h-1)}\left[\sum_{\alpha=1}^{a_h}\Big(\sum_{i=1}^{n_{h\alpha}} w_{h\alpha i}\, y_{h\alpha i}\Big)^2 - \frac{\Big(\sum_{\alpha=1}^{a_h}\sum_{i=1}^{n_{h\alpha}} w_{h\alpha i}\, y_{h\alpha i}\Big)^2}{a_h}\right] \tag{5.2}$$

Recall from Section 3.6.2 that this simple estimator for the variance of a weighted total plays an important role in the computation of Taylor Series Linearization estimates of sampling variances for more complex estimators that can be approximated as linear functions of estimated totals.

This simple weighted estimator of the finite population total is often labeled the **Horvitz-Thompson estimator** or "H-T" estimator (Horvitz & Thompson, 1952). Practically speaking, this labeling is convenient, but the estimator in (5.1) and the variance estimator in (5.2) make additional assumptions beyond those that are explicit in Horvitz and Thompson's original derivation. Theory Box 5.1 provides interested readers with a short summary of the theory underlying the H-T estimator.

THEORY BOX 5.1: THE HORVITZ-THOMPSON ESTIMATOR OF A POPULATION TOTAL

The **Horvitz-Thompson estimator** (Horvitz & Thompson, 1952) of the population total for a variable Y is written as follows:

$$\hat{Y} = \sum_{i=1}^{N}\frac{\delta_i Y_i}{p_i} = \sum_{i=1}^{n}\frac{y_i}{p_i} \tag{5.3}$$

In (5.3), $\delta_i = 1$ if element i is included in the sample and 0 otherwise, and p_i is the probability of inclusion in the sample for element i. The H-T estimator is an unbiased estimator for the population total Y, because the only random variable defined in the estimator is the indicator of inclusion in the sample (the y_i and p_i values are fixed in the population):

$$E(\hat{Y}) = E\sum_{i=1}^{N}\frac{\delta_i Y_i}{p_i} = \sum_{i=1}^{N}\frac{E(\delta_i)Y_i}{p_i} = \sum_{i=1}^{N}\frac{p_i Y_i}{p_i} = \sum_{i=1}^{N} Y_i = Y \tag{5.4}$$

An unbiased estimator of the sampling variance of the H-T estimator is:

$$var(\hat{Y}) = \sum_{i=1}^{n} y_i^2 \frac{(1-p_i)}{p_i^2} + \sum_{i=1}^{n}\sum_{j\neq i}^{n}\frac{y_i y_j}{p_{ij}}\left(\frac{p_{ij}-p_i p_j}{p_i p_j}\right) \tag{5.5}$$

In this expression, p_{ij} represents the probability that both elements i and j are included in the sample; these joint inclusion probabilities must be supplied to statistical software to compute these variance estimates.

The H-T estimator weights each sample observation inversely proportionate to its sample selection probability, $w_{HT,i} = w_{sel,i} = 1/p_i$, and does not explicitly consider nonresponse adjustment or post-stratification (Section 2.7). In fact, when the final survey weight incorporates all three of these conventional weight factors, the variance estimator (5.2) does not fully reflect the stochastic sample-to-sample variability associated with the nonresponse mechanism, nor does it capture true gains in precision that may have been achieved through the post-stratification of the weights to external population controls.

Because survey nonresponse is a stochastic process that operates on the selected sample, the variance estimator could (in theory) explicitly capture this added component of sample-to-sample variability (Valliant, 2004). This method assumes that the data user can access the individual components of the survey weight, in addition to unique codes for the post-strata. Stata does provide the capability to directly account for the reduction in sampling variance due to the post-stratification using the `poststrata()` and `postweight()` options for the `svyset` command, as illustrated in Chapter 2.

The effects of nonresponse adjustment and post-stratification weighting on the sampling variance of estimated population totals and other descriptive statistics may also be captured through the use of replicate weights, in which the nonresponse adjustment and the post-stratification controls are separately developed for each BRR or JRR replicate sample of cases (Valliant et al., 2018). See Chapter 3 for more details on this process.

Two major classes of totals can be estimated using (5.1). If $y_{h\alpha i}$ is a binary indicator for an attribute (e.g., 1 = has the disease, 0 = disease free), the result is an estimate of the size of the subpopulation that shares that attribute: $\hat{Y}_w = \sum_{h=1}^{H}\sum_{\alpha=1}^{a_h}\sum_{i=1}^{n_{h\alpha}} w_{h\alpha i} y_{h\alpha i} = \hat{M} \leq \sum_{h=1}^{H}\sum_{\alpha=1}^{a_h}\sum_{i=1}^{n_{h\alpha}} w_{h\alpha i}\cdot 1 = \hat{N}.$

Alternatively, if y is a continuous measure of an attribute of the sample case (e.g., acres of corn, monthly income, annual medical expenses), the result is an estimate of the population total of y, $\hat{Y}_w = \sum_{h=1}^{H}\sum_{\alpha=1}^{a_h}\sum_{i=1}^{n_{h\alpha}} w_{h\alpha i} y_{h\alpha i} = \hat{Y}$. Example 5.3 will illustrate the estimation of a subpopulation total, and Example 5.4 will illustrate the estimation of a population total of a variable Y.

EXAMPLE 5.3: ESTIMATING THE TOTAL COUNT OF U.S. ADULTS HAVING EXPERIENCED A MAJOR DEPRESSIVE EPISODE (MDE) IN THEIR LIFETIME

The MDE variable in the NCS-R data set is a binary indicator (1 = yes, 0 = no) of whether an NCS-R respondent reported a major depressive episode at any point in their lifetime. The aim of this example analysis is to estimate the total number of individuals who have experienced a lifetime major depressive episode along with the standard error of the estimate (and the 95% confidence interval).

For this analysis, the NCS-R survey weight variable NCSRWTSH is selected to analyze all respondents completing the Part I survey (n = 9,282), where the lifetime diagnosis of MDE was assessed. Because the NCS-R data producers normalized the values of the NCSRWTSH variable so that the weights would sum to the sample size, the weight values must be expanded back to the population scale in order to obtain an unbiased estimate of the population total. This is accomplished by multiplying the Part I weight for each case by the ratio of the NCS-R survey population total (N = 209,128,094 U.S. adults age 18+, Source: July 2001 Current Population Survey) divided by the count of sample observations (n = 9,282). The SECLUSTR variable contains the codes representing NCS-R sampling error clusters, while SESTRAT is the sampling error stratum variable.

```
gen ncsrwtsh_pop = ncsrwtsh * (209128094 / 9282)
svyset seclustr [pweight = ncsrwtsh_pop], strata(sestrat)
```

Once the complex design features of the NCS-R sample have been identified using the `svyset` command, the `svy: total` command is issued to obtain an unbiased weighted estimate of the population total along with a standard error for the estimate. The `mat list r(table)` post-estimation command then displays the actual estimates and confidence limits (rather than the default scientific notation that is used for large values), and the `estat effects` command computes an estimate of the design effect for this estimated total.

```
svy: total mde
mat list r(table)
estat effects
```

n	df	$\hat{Y}_w$	$se(\hat{Y}_w)$	$CI_{.95}(\hat{Y}_w)$	$d^2(\hat{Y}_w)$
9,282	42	40,092,207	2,567,488	(34,910,806, 45,273,607)	9.03

The resulting Stata output indicates that 9,282 observations have been analyzed and that there are 42 design-based degrees of freedom. The weighted estimate of the total population of U.S. adults who have experienced an episode of major depression in their lifetime is $\hat{Y} = 40,092,207$. The estimated value of the design effect for the weighted estimate of the population total is $d^2(\hat{Y}_w) = 9.03$, suggesting that the NCS-R variance of the estimated total is approximately nine times greater than that expected for a simple random sample of the same size.

Weighted estimates of population totals can also be computed for subpopulations. Consider subpopulations of NCS-R adults classified by marital status (married, separated/widowed/divorced, and never married). Under the complex NCS-R sample design, correct unconditional subpopulation analyses (see Section 4.5.2) can be specified in Stata by adding the `over()` option to the `svy: total` command:

```
svy: total mde, over(mar3cat)
mat list r(table)
estat effects
estat size
```

Subpopulation	n	Estimated Total Lifetime MDE	Standard Error	95% CI	$d^2(\hat{Y})$
Married	5,322	20,304,191	1,584,109	(17,107,330, 23,501,051)	6.07
Sep./Wid./Div.	2,017	10,360,671	702,601	(8,942,723, 11,778,618)	2.22
Never Married	1,943	9,427,345	773,137	(7,867,090, 10,987,600)	2.95

Note that the MAR3CAT variable is included in parentheses, to request estimates for each subpopulation defined by the levels of that variable. We once again include the `mat list r(table)` post-estimation command, which prints the full estimates of the totals and their confidence limits (rather than scientific notation). We also include the command `estat size` to display the unweighted sample sizes for each subclass in the output.

EXAMPLE 5.4: USING THE 2012 HRS DATA TO ESTIMATE TOTAL HOUSEHOLD WEALTH

Next, we consider the example problem of estimating the total value of household wealth for the 2012 HRS target population (U.S. households with adults born prior to 1960). The 2012 HRS constructed variable that measures total wealth for HRS household financial units is H11ATOTA. It is equal to the total value of reported household assets minus the household liabilities (debts), exclusive of assets/liabilities related to ownership of second homes.

To begin the analysis, we first identify the HRS variables containing the sampling error computation units, or ultimate clusters (SECU), and the sampling error stratum codes (STRATUM). We also specify the NWGTHH variable (for 2012) as the survey weight variable for the analysis, because we are performing an analysis at the level of the HRS household financial unit. The HRS data set includes an indicator variable (NFINR for 2012) that identifies the individual respondent who is the financial reporter for each HRS sample household. This variable is used to create a subpopulation indicator (FINR) that restricts the estimation to only sample members who are financial reporters for their HRS household unit. We then apply the `svy: total` command to the H11ATOTA variable, measuring the total value of household wealth.

```
gen finr = 1
replace finr = 0 if nfinr != 1
svyset secu [pweight = nwgthh], strata(stratum)
svy, subpop(finr): total H11ATOTA
```

n	df	$\hat{Y}_w$	$se(\hat{Y}_w)$	$CI_{.95}(\hat{Y}_w)$
13,657	56	$\$2.53 \times 10^{13}$	$\$1.35 \times 10^{12}$	$(2.26 \times 10^{13}, 2.80 \times 10^{13})$

The Stata output indicates that the 2012 HRS target population includes approximately 58,970,000 households (not shown). In 2012, these 59 million estimated households owned household wealth valued at an estimated $\hat{Y}_w = \$2.53 \times 10^{13}$, with a 95% CI of ($\$2.26 \times 10^{13}$, $\$2.80 \times 10^{13}$).

5.3.3 Means of Continuous, Binary, or Interval Scale Data

The estimation of the population mean, $\bar{Y}$, for a continuous, binary, or interval scale variable is a very common task for survey researchers seeking to describe the central tendencies of these variables in target finite populations of interest. An estimator of the population mean, $\bar{Y}$, can be written as a non-linear ratio of two estimated finite population totals:

$$\bar{y}_w = \frac{\sum_{h=1}^{H}\sum_{\alpha=1}^{a_h}\sum_{i=1}^{n_{h\alpha}} w_{h\alpha i} y_{h\alpha i}}{\sum_{h=1}^{H}\sum_{\alpha=1}^{a_h}\sum_{i=1}^{n_{h\alpha}} w_{h\alpha i}} = \frac{\hat{Y}}{\hat{N}} \tag{5.6}$$

Note that if y is a binary variable coded 1 or 0, the weighted mean estimates the population proportion or prevalence, P, of "1s" in the population (see Chapter 6):

$$\bar{y}_w = \frac{\sum_{h=1}^{H}\sum_{\alpha=1}^{a_h}\sum_{i=1}^{n_{h\alpha}} w_{h\alpha i} y_{h\alpha i}}{\sum_{h=1}^{H}\sum_{\alpha=1}^{a_h}\sum_{i=1}^{n_{h\alpha}} w_{h\alpha i}} = \frac{\hat{M}}{\hat{N}} = p \tag{5.7}$$

Since $\bar{y}_w$ is not a linear statistic, a closed form formula for the variance of this estimator does not exist. As a result, variance estimation methods such as Taylor Series Linearization (TSL), BRR, JRR, or the Bootstrap (see Chapter 3) must be used to compute an estimate of the sampling variance of this estimator of the population mean. An application of Taylor Series Linearization to the weighted mean in (5.6) results in the following TSL variance estimator for the ratio mean:

$$var(\bar{y}_w) \doteq \frac{var(\hat{Y}) + \bar{y}_w^2 \times var(\hat{N}) - 2 \times \bar{y}_w \times Cov(\hat{Y}, \hat{N})}{\hat{N}^2} \tag{5.8}$$

Closed form formulas are available to estimate the variances and covariances of the estimators of the population totals included in this TSL expression (see Section 3.6.2), and BRR, JRR, and Bootstrap replicated variance estimation methods can also be used to estimate the standard errors of the estimated means (see Section 3.6.3). We begin the following example using a design-based approach to make inference about a finite population mean, and then turn to an alternative Bayesian approach to making the same inference for comparison purposes.

EXAMPLE 5.5: ESTIMATING THE MEAN VALUE OF HOUSEHOLD INCOME USING THE 2012 HRS DATA

We start with a design-based approach to this analysis. In this example, we analyze the mean of the household income variable (H11ITOT) for the 2012 HRS survey population.

Because this is a household-level measure, we once again apply the 2012 HRS household weight variable, NWGTHH.

```
svyset secu [pweight = nwgthh], strata(stratum)
```

The `svy: mean` command is then submitted in Stata to compute the weighted estimate of the mean household income in the 2012 HRS population (based on the subclass of financial reporters; see Example 5.4 above), along with a linearized standard error (the default variance estimation technique in Stata) and a design-based 95% confidence interval for the population mean:

```
svy, subpop(finr): mean H11ITOT
```

n	df	$\bar{y}_w$	$se(\bar{y}_w)$	$CI_{.95}(\bar{y}_w)$
13,657	56	\$71,382	\$1937	(\$67,502, \$75,263)

We see that the estimated mean of total household income for households with individuals in the HRS survey population is \$71,382, with an associated 95% confidence interval of (\$67,502, \$75,263).

Alternative Bayesian Approach

We now turn to an alternative Bayesian approach, making use of the contributed **`csSampling` package** in the R software (see Section 3.8 for more details about the general approach that functions in this package enable). We present the details on how to obtain, install, and load this package as part of this example, and then assume that the package has already been installed for later examples. At the time of this writing, the `csSampling` package needs to be obtained from GitHub. We anticipate that the package will eventually become part of the CRAN archive, meaning that all of these steps may not be necessary to install and load the package, but for now we illustrate this process. First, we load the `devtools` package in R, and then we install the contributed package from its GitHub repository:

```
library(devtools)
install_github("RyanHornby/csSampling")
```

After this installation process has been completed, we then load all of the packages that we will need to implement the Bayesian approaches discussed in Section 3.8:

```
library(csSampling)
library(rstan)
library(brms)
library(survey)
rstan_options(auto_write = TRUE)
```

The last function helps to enable automatic generation of code for the Stan software (utilized via the `rstan` package when using `csSampling`; see Section 3.8) for the Bayesian modeling and simulation.

Once all of these packages (and their dependencies) have been loaded, we proceed with some initial data management. We start by loading the full 2012 HRS data set from the directory in which it has been saved:

```
# read in the full HRS data set for the third edition
load("X:/. . ./hrs12.rdata")
```

Next, we need to create a reduced data frame that has complete data for our variables of interest and our subpopulation of interest. Note that since this is a model-based approach, we do not need to worry about the unconditional vs. conditional subclass analysis distinction from Chapter 4.

```
hrs.red <- hrs12[hrs12$nfinr == 1, c("h11itot", "secu",
"stratum", "nwgthh")]
hrs.red <- hrs.red[complete.cases(hrs.red),]
```

Next, we normalize the 2012 HRS household weights, which is required to match the approach the Stan uses for handling the weights in the modeling:

```
# need to normalize HRS weights to match Stan
hrs.red$wtsc <- hrs.red$nwgthh / mean(hrs.red$nwgthh)
```

We now create an HRS design object for the analysis functions in the `survey` package, declaring all of the complex sampling features (and specifically the normalized weights):

```
# survey design object
hrs.des <- svydesign(id = ~secu, strata = ~stratum, weights = ~wtsc,
nest = T, data = hrs.red)
```

Now, we carry out the design-adjusted Bayesian analysis, executing the approach described in Section 3.8 to adjust the draws from the survey-weighted posterior distribution for the mean for the HRS complex sample design effect. We first set a seed so that we can replicate the results if needed (given that the procedure involves random draws from the posterior). Then, we define a model formula that declares the (normalized) survey weights and our parameter of interest; in this case, we frame the analysis as focusing on the intercept in an intercept-only model for total household income (which is the mean of interest). Next, we use the `cs _ sampling _ brms()` function to carry out the approach outlined in Section 3.8, with help from the `brms` package and Stan for the actual modeling and simulation. We assume that total household income follows a normal distribution (`family = gaussian()`).

```
# Bayesian approach, flat prior by default
set.seed(41279)
model _ formula <- formula("h11itot|weights(wtsc) ~ 1")
mod.brms <- cs _ sampling _ brms(svydes = hrs.des, brmsmod =
brmsformula(model _ formula), data = hrs.red, family =
gaussian())
```

After the function has finished running, the results of the Bayesian analysis can be printed as follows:

```
# Print results
mod.brms$stan _ fit
```

```
              mean   se_mean       sd       2.5%   . . .      97.5%
Intercept  71338.32     50.82  1071.99   69280.69   . . .   73460.27
```

In the output, a portion of which is shown above, the mean of the adjusted draws from the survey-weighted posterior distribution is $71,338.32, and the 95% credible interval for the mean (based on the 2.5 percentile and 97.5 percentile of 1,000 adjusted draws from the survey-weighted posterior) is ($69,280.69, $73,460.27). This design-adjusted Bayesian approach is therefore *more efficient* than the standard design-based approach, while still fully accounting for all of the complex sampling features, and the point estimate is nearly identical.

We note in this example that we used the default `brms` prior for this intercept parameter, which is non-informative. In many repeated cross-sectional or longitudinal surveys, prior waves or repeated cross-sectional samples have the potential to yield useful prior information about parameters of interest when applying this Bayesian approach. **Prior distributions** for parameters of interest can easily be specified as part of the `csSampling` code above. For example, suppose that we knew from prior waves of the HRS that the mean total household income (the intercept in this model) followed a normal distribution with a mean of $70,000 and a standard deviation of $1,000 (possibly derived from previous analyses of standard errors). This prior could be specified as follows:

```
# With Normal Prior, mean = 70000, SD (SE) = 1000
mod.brms2 <- cs_sampling_brms(svydes = hrs.des, brmsmod =
brmsformula(model_formula), prior =
c(set_prior("normal(70000,1000)", class = "Intercept")),
data = hrs.red, family = gaussian())
```

This analysis results in even more efficient inference, with a posterior mean of $70,639.61 (pulled down somewhat by the informative prior) and a 95% credible interval of ($69,248.30, $72,034.69). Informative prior distributions can certainly be helpful in this regard.

EXAMPLE 5.6: ESTIMATING MEAN SYSTOLIC BLOOD PRESSURE USING THE 2011-2012 NHANES DATA

In this example, the 2011–2012 NHANES data are used to estimate the mean systolic blood pressure (mmHg) for the U.S. adult population age 18 and older. The survey weight is the NHANES medical examination weight (WTMEC2YR). The following Stata command sequence declares the variables describing the complex sampling features of the 2011–2012 NHANES to Stata (where SDMVPSU contains masked PSU codes and SDMVSTRA contains masked stratum codes), generates the subpopulation indicator (AGE18P), and then computes the weighted estimate of the mean systolic blood pressure, the linearized estimate of the standard error of the weighted mean (based on an appropriate unconditional subclass analysis), a design-based 95% confidence interval for the mean, and a design effect for the estimate:

```
svyset sdmvpsu [pweight = WTMEC2YR], strata(sdmvstra)
gen age18p = 1 if age >= 18 & age != .
replace age18p = 0 if age < 18
svy, subpop(age18p): mean BPXSY1
estat effects
```

n	df	$\bar{y}_w$	$se(\bar{y}_w)$	$CI_{.95}(\bar{y}_w)$	$d^2(\bar{y}_w)$
5,132	17	122.03	0.62	(120.73, 123.33)	8.42

Based on the sample of n = 5,132 adults, the 2011–2012 NHANES estimate of mean systolic blood pressure for U.S. adults age 18 and older is $\bar{y}_w$ = 122.03 mmHg. As observed in the examples of estimation of population totals, the design effect is fairly large when estimating a descriptive population parameter for the entire target population. For this estimate of mean systolic blood pressure, the estimated design effect ratio indicates that the variance of the estimated mean for the full NHANES sample is over eight times greater than would be expected from an SRS of the same size. For this estimate of mean systolic blood pressure in U.S. adults, the NHANES effective sample size is roughly $n_{eff} \sim 5132 / 8.42 = 610$.

Alternative Bayesian Approach

Assuming that the `csSampling` package and all of its dependencies have been loaded in R (see Example 5.5), we start with the necessary data management (including a condition that defines the subclass of interest) and creation of the appropriate 2011–2012 NHANES `survey` design object:

```
# read in the full NHANES data set for the third edition
load("X:/. . ./nhanes1112.rdata")

# create complete data set with variables of interest
nhanes.red <- nhanes1112[nhanes1112$age >= 18 &
!is.na(nhanes1112$age), c("bpxsy1", "sdmvpsu", "sdmvstra", "wtmec2yr")]
nhanes.red <- nhanes.red[complete.cases(nhanes.red),]

# need to normalize NHANES weights to match Stan
nhanes.red$wtsc <- nhanes.red$wtmec2yr /
mean(nhanes.red$wtmec2yr)

# survey design object
nhanes.des <- svydesign(id = ~sdmvpsu, strata = ~sdmvstra,
weights = ~wtsc, nest = T, data = nhanes.red)
```

Next, we carry out the design-adjusted Bayesian analysis using the approach described in Section 3.8, assuming a non-informative prior for the mean of interest:

```
# Bayesian approach, flat prior
set.seed(41279)
model_formula <- formula("bpxsy1|weights(wtsc) ~ 1")
mod.brms <- cs_sampling_brms(svydes = nhanes.des, brmsmod =
brmsformula(model_formula), data = nhanes.red, family =
gaussian())

# Print results
mod.brms$stan_fit
```

The mean of the adjusted draws from the survey-weighted posterior distribution for the mean systolic blood pressure is 122.02, which is nearly identical to the design-based point estimate. The 95% credible interval for the mean is (121.54, 122.46), which like Example 5.5 reflects an increase in efficiency from the model-based Bayesian approach, even when the simulated posterior draws of the mean are fully adjusted for the complex sampling features. One could also incorporate a prior distribution for the mean based on analyses of NHANES data from previous cycles, as was illustrated in Example 5.5.

EXAMPLE 5.7: ESTIMATING THE MEAN VALUE OF TOTAL HOUSEHOLD WEALTH USING THE 2012 HRS DATA

Next, the 2012 HRS data are used to compute an estimate of the mean of total household wealth for U.S. households with at least one adult born prior to 1960. As in Examples 5.4 and 5.5, the estimate is restricted to the subpopulation of HRS respondents who are the financial reporters for their household unit. The Stata commands required to perform this analysis follow:

```
svyset secu [pweight=nwgthh], strata(stratum)
svy, subpop(finr): mean H11ATOTA
estat effects
```

n	df	$\bar{y}_w$	$se(\bar{y}_w)$	$CI_{.95}(\bar{y}_w)$	$d^2(\bar{y}_w)$
13,657	56	$428,471	$17,354	($393,707, $463,234)	3.21

Based on this analysis of reports from $n = 13{,}657$ HRS financial units, the mean household wealth in the 2012 HRS survey population is $\bar{y}_w = \$428{,}471$. The estimated design effect for the estimated mean, $d^2(\bar{y}_w) = 3.21$, indicates that while there is still a loss of precision in the HRS estimate relative to simple random sampling, the loss in effective sample size is not as severe as in the previous example based on the 2011–2012 NHANES data.

Alternative Bayesian Approach

Following the same basic approach from Example 5.5, we carry out the Bayesian analysis, only this time specifying an informative normal prior distribution for the mean household wealth based on analyses of HRS data from prior waves, with mean $400,000 and standard deviation $20,000:

```
# read in the full HRS data set for the third edition
load("X:/. . ./hrs12.rdata")

# create complete data set with variables of interest
hrs.red <- hrs12[hrs12$nfinr == 1, c("h11atota", "secu",
"stratum", "nwgthh")]
hrs.red <- hrs.red[complete.cases(hrs.red),]

# normalize HRS household weights to match Stan
hrs.red$wtsc <- hrs.red$nwgthh / mean(hrs.red$nwgthh)

# survey design object
hrs.des <- svydesign(id = ~secu, strata = ~stratum, weights = ~wtsc,
nest = T, data = hrs.red)

# With Normal Prior, mean = 400000, SD (SE) = 20000
set.seed(41279)
model_formula <- formula("h11atota|weights(wtsc) ~ 1")
mod.brms <- cs_sampling_brms(svydes = hrs.des, brmsmod =
brmsformula(model_formula), prior =
c(set_prior("normal(400000,20000)", class = "Intercept")),
data = hrs.red, family = gaussian())
```

```
# Print results
mod.brms$stan _ fit
```

The resulting mean of the design-adjusted draws from the survey-weighted posterior distribution is $423,606.38, which is not far off from the point estimate from the design-based approach, but is shifted downward somewhat by the prior. We also see a 95% credible interval of ($407,200, $439,998), once again reflecting much more efficient inference compared to the design-based approach.

The `csSampling` package also offers a variety of plotting options for visualizing the posterior distributions of parameters of interest, before and after the design adjustments to the draws from the survey-weighted posterior. Figure 5.4 [generated in R using the simple function `plot(mod.brms, varnames = "Intercept")`] shows how the design adjustments result in draws from the survey-weighted posterior with more variability, reflecting the effects of the 2012 HRS complex sampling features. This plot gives us a sense of what the (adjusted) posterior distribution of the mean correctly incorporating the sampling variability looks like, and how inferences based on the unadjusted posterior would be overly liberal. We also note how both posterior distributions are centered at roughly the same value, as they both incorporate the 2012 HRS household weights.

5.3.4 Standard Deviations of Continuous Variables

Although experience suggests that it is not a common task, survey analysts may wish to compute an unbiased estimate of the population standard deviation of a continuous variable. Just as weights are required to obtain unbiased (or nearly unbiased) estimates of the population mean, weights must also be employed to obtain consistent estimates of the standard deviation of a random variable in a designated target population. A weighted estimator of the population standard deviation of a variable y can be written as follows:

$$s(y)_w = \sqrt{\frac{\sum_{h=1}^{H}\sum_{\alpha=1}^{a_h}\sum_{i=1}^{n_{h\alpha}} w_{h\alpha i}(y_{h\alpha i} - \bar{y}_w)^2}{\sum_{h=1}^{H}\sum_{\alpha=1}^{a_h}\sum_{i=1}^{n_{h\alpha}} w_{h\alpha i} - 1}} \quad (5.9)$$

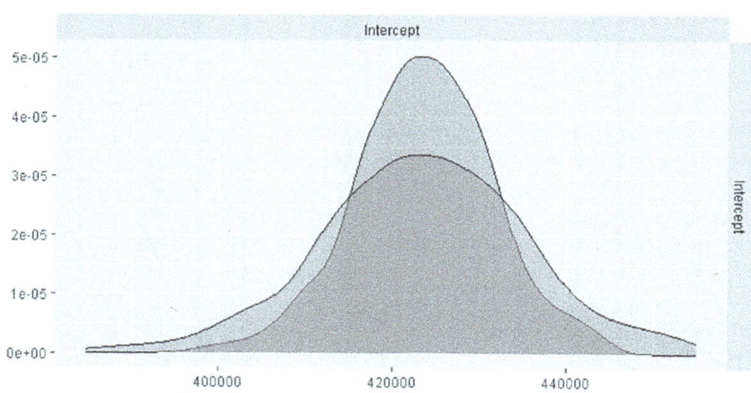

FIGURE 5.4
Overlaid density plots showing the unadjusted (but survey-weighted) posterior distribution for the mean total household assets (taller peak, less spread), versus the fully design-adjusted posterior distribution for the mean (shorter peak, more spread), incorporating complex sample design effects from the 2012 HRS.

In (5.9), the estimate of the population mean for the variable y is calculated as in (5.6). In Stata, analysts may use the `summarize` command in combination with an "`aweight`" specification for the survey weight variable to generate this unbiased estimate of the population standard deviation.

EXAMPLE 5.8: ESTIMATION OF THE POPULATION STANDARD DEVIATION OF TOTAL HOUSEHOLD NET WORTH BASED ON THE 2012 HRS

For this example, if we wish to estimate the population standard deviation of total household net worth based on the 2012 HRS, we could use the following command in Stata:

```
summarize H11ATOTA [aweight = nwgthh]
```

This will yield a point estimate of the population standard deviation, which in this case is $1,143,716. Unfortunately, there are no straightforward commands in Stata for making inference about this quantity, via estimated standard errors and design-adjusted confidence intervals. Users of the SAS software can use PROC UNIVARIATE or PROC MEANS with a WEIGHT statement to generate this weighted estimate of the population standard deviation S_y.

Alternative Bayesian Approach

Inference for these types of quantities, where design-based analysis commands may not be readily available in the software, may be more straightforward using Bayesian approaches. In this example, we can specify the standard deviation of interest as the standard deviation of the residuals in an intercept-only linear regression model for the total household assets variable, and make inference about this standard deviation based on design-adjusted draws from the survey-weighted posterior distribution for the parameters defining this model. These draws were automatically generated as part of the analysis performed in R for Example 5.7, given that this residual standard deviation was one of the parameters in that specified model. This parameter was labeled as "sigma" in the output from that analysis.

The posterior mean of the adjusted draws for this "sigma" parameter was $1,113,687, which is quite close to the design-based parameter estimate, but as part of the Bayesian approach, we can also compute a 95% credible interval for this standard deviation: ($1,101,343, $1,125,997). This approach would therefore enable inferences about the likely values of this parameter in the 2012 HRS target population.

5.3.5 Estimation of Percentiles, Medians, and Measures of Inequality in Population Distributions for Continuous Variables

5.3.5.1 Estimation of Distribution Quantiles

Estimation of quantiles, such as the median (Q_{50}) or the 95th percentile (Q_{95}) of the population distribution of a continuous variable, can play an important role in analyses of survey data. A sociologist may wish to compare sample estimates of percentiles of household income for a regional survey population to nationally defined poverty criteria. An epidemiologist may wish to estimate the 95th percentile of prostate specific antigen (PSA) levels in a metropolitan sample of men over the age of 40.

The **ungrouped method** of quantile estimation (Loomis et al., 2005) builds on results related to the weighted estimator for totals presented earlier in this chapter (see

Section 5.3.2), employing a weighted sample estimate of the population **cumulative distribution function (CDF)** of a survey variable. Specifically, the CDF for a variable y in a given finite population of size N is defined as follows:

$$F(x) = \frac{\sum_{i=1}^{N} I(y_i \leq x)}{N} \tag{5.10}$$

In (5.10), $I(y_i \leq x)$ is an indicator variable equal to 1 if y_i is less than or equal to a specified value of y (denoted here by x), and 0 otherwise. The weighted estimator of the CDF from a complex sample of size n from this population is then written as follows:

$$\hat{F}(x) = \frac{\sum_{h=1}^{H}\sum_{\alpha=1}^{a_h}\sum_{i=1}^{n_{h\alpha}} w_{h\alpha i} I(y_{h\alpha i} \leq x)}{\sum_{h=1}^{H}\sum_{\alpha=1}^{a_h}\sum_{i=1}^{n_{h\alpha}} w_{h\alpha i}} \tag{5.11}$$

The q-th quantile (e.g., $q = 0, 0.25, 0.50, 0.75, 1$) for a variable y is the smallest value of y such that the population CDF is greater than or equal to q. For example, the median would be the smallest value of y at which the CDF is greater than or equal to 0.5. The ungrouped method of estimating a quantile first considers the order statistics (the sample values of y ordered from smallest to largest), denoted by $y_1, \ldots, y_n$, and finds the value of j ($j = 1, \ldots, n$) such that

$$\hat{F}(y_j) \leq q < \hat{F}(y_{j+1}). \tag{5.12}$$

Then, the estimate of the q-th population quantile Y_q is calculated as follows:

$$\hat{Y}_q = y_j + \frac{q - \hat{F}(y_j)}{\hat{F}(y_{j+1}) - \hat{F}(y_j)}(y_{j+1} - y_j) \tag{5.13}$$

Kovar et al. (1988) report results of a simulation study that suggests that Balanced Repeated Replication (BRR) performs well for variance estimation and construction of confidence intervals when working with estimators of non-smooth functions like quantiles. SAS, the `survey` package in R, and the WesVar software currently implement the BRR variance estimation approach for quantiles, and we recommend the use of this variance estimation approach for estimated quantiles in practice (when feasible, given the sampling error calculation model provided). A variance estimate for the estimated quantile in (5.13) can also be computed using Taylor Series Linearization (Binder, 1991). The SAS software currently uses the linearized variance estimator by default. The JRR approach to variance estimation is known to be badly biased for these types of estimators (Miller, 1974), but modifications to the Jackknife approach addressing this problem have been developed (Shao & Wu, 1989).

EXAMPLE 5.9: ESTIMATING POPULATION QUANTILES FOR TOTAL HOUSEHOLD WEALTH USING THE 2012 HRS DATA

This example considers the total household wealth variable collected from the 2012 HRS sample and aims to estimate the 0.25 quantile, the median, and the 0.75 quantile of household wealth in the HRS target population. The SAS software is used first in this example because Stata does not currently support commands specifically dedicated to the estimation of standard errors of weighted estimates of quantiles in complex sample survey data sets. Stata users can download and install the user-written `epctile` command (using `findit epctile` in web-aware Stata) to compute weighted estimates of quantiles along with design-adjusted standard errors and approximate 95% confidence intervals.

The following SAS code generates the quantile estimates and standard errors, using an unconditional subclass analysis approach:

```
proc surveymeans data=hrs12 q1 median q3;
weight nwgthh;
stratum stratum;
cluster secu;
var h11atota;
domain finr;
run;
```

Table 5.1 summarizes the results provided in the SAS output, and compares the estimates with those generated using the BRR approach to variance estimation in the SAS software. In SAS, BRR variance estimation is requested by simply adding the VARMETHOD=BRR specification to the PROC SURVEYMEANS command statement.

The estimated median of the total household wealth for the HRS target population is \$141,907. The estimate of the mean total household wealth from Example 5.7 was \$428,471, suggesting that the distribution is highly skewed toward higher dollar values.

From the side-by-side comparison in Table 5.1, SAS BRR estimates of the corresponding standard errors differ slightly from the TSL standard errors, as expected. The resulting inferences about the population quantiles would not differ substantially in this example as a result, but we do note that the BRR estimates would yield more conservative inferences.

The following Stata code can be used to perform the same analysis using the aforementioned user-written command:

```
svyset secu [pweight = nwgthh], strata(stratum)

epctile H11ATOTA, percentiles(25 50 75) subpop(if nfinr==1) svy
```

TABLE 5.1

Estimation of percentiles of the distribution of total household wealth

	SAS (TSL)		SAS (BRR)		Stata (TSL)	
Percentile	$\hat{Q}_p$	$se(\hat{Q}_p)$	$\hat{Q}_p$	$se(\hat{Q}_p)$	$\hat{Q}_p$	$se(\hat{Q}_p)$
Q_{25}	\$21,953	\$2,215	\$21,953	\$2,287	\$22,000	\$2,200
Q_{50} (Median)	\$141,907	\$7,754	\$141,907	\$7,761	\$142,000	\$8,000
Q_{75}	\$439,965	\$18,652	\$439,965	\$19,940	\$440,000	\$18,869

(Source: 2012 HRS)

We note how unconditional subpopulation analyses can also be performed using this command in Stata, where the subpopulation is specified as an option. We also note that the svy option is needed to refer to the complex sample design variables identified in the svyset command. The Stata results in Table 5.1 indicate that while some rounding of the estimates is performed, the results agree with those generated using TSL in SAS.

Finally, we note that quantile estimation for finite populations is also possible using functions in the contributed survey package for R. Here are the functions that we would need to execute to replicate this example, using the default variance estimation approach of TSL and being careful to specify the unconditional subpopulation analysis via subset():

```
hrs.des <- svydesign(id = ~secu, strata = ~stratum, weights = ~nwgthh,
nest = T, data = hrs12)
svyquantile(~h11atota, subset(hrs.des, nfinr == 1), c(0.25, 0.5, 0.75))
```

```
$h11atota
          quantile    ci.2.5    ci.97.5         se
0.25         22000     18000      26900   2221.401
0.5         142000    127000     158000   7737.463
0.75        440000    404500     480000  18844.465
```

We see the same output generated by the other procedures. If one wanted to use BRR instead, given the fact that each HRS stratum has two sampling error computation units available, we could create BRR replicate weights "on the fly" as follows:

```
hrs.des2 <- as.svrepdesign(hrs.des, type = "BRR")
```

The new design object could then be input to the svyquantile() function to use BRR for variance estimation:

```
svyquantile(~h11atota, subset(hrs.des2, nfinr == 1), c(0.25, 0.5, 0.75))
```

Just like the SAS approach to this example, we remind readers that the BRR replicate weights created "on the fly", starting with the HRS sampling error codes, have not been individually calibrated to external population control totals. This may result in some bias in the individual replicate estimates.

5.3.5.2 Estimation of Measures of Inequality in Population Distributions

Estimation of individual quantiles for the distribution of a continuous variable enables the survey analyst to address questions such as "What is the median value of total household wealth in U.S. households with members age 65 and older?" Furthermore, by estimating quantiles at multiple points over the (0, 1) range of $F(y)$, for example, $q = \{.01,.05,.10,.25,.50,.75,.90,.95,.99\}$, the analyst can gain insight into how the variable of interest is distributed across the elements in the population. This is especially important when the characteristic of interest is highly skewed across the survey population and simple measures of central tendency (means) and dispersion (standard deviation) may not be sufficient to describe the underlying distribution. This is true in household surveys where financial measures of income, wealth, debts and liabilities, and medical expenditures are highly skewed in the direction of small subpopulations with large unit values. It is also true in surveys of businesses and farms where small fractions of the full survey population account for a major share of the production of a good, service, or commodity (e.g., refined gasoline, mobile telephone service, livestock).

To answer questions regarding non-uniformity in the distribution of such variables across the elements of the target population, economists and other social scientists have developed a growing class of graphical displays and indices as tools to describe distributional inequality (Lorenz, 1905; Nygard & Sandstrom, 1981). Complex sample survey data are often used as the basis for deriving the statistical measures of inequality for target populations, and over the past thirty years there has been steady development in the theory for correctly producing weighted estimates and robust standard errors of these statistics (Binder & Kovacevic, 1995; Bhattacharya, 2007; Langel & Tille, 2013). Software procedures for estimating inequality statistics from complex sample survey data are also now available as user-written commands or functions (e.g., Stata .ado files or R packages) available for the major statistical software packages (Jenkins, 2006; Alfons & Templ, 2013).

EXAMPLE 5.10: ESTIMATING THE LORENZ CURVE AND GINI COEFFICIENT FOR THE 2012 HRS POPULATION DISTRIBUTION OF TOTAL HOUSEHOLD WEALTH

This example will illustrate an application of the user-written Stata command `svylorenz` (Jenkins, 2006) to estimate the **Gini coefficient** for the distribution of total household wealth in the HRS target population. The Gini coefficient takes values in the (0, 1) range, and in this example measures the degree of inequality in the distribution of total wealth across the population of households that is studied in the 2012 HRS. A Gini coefficient value of $G = 0$ indicates perfect equality in the wealth distribution, with larger values signifying increasingly greater inequality in the wealth distribution.

Following the estimating equation formulation of Binder and Kovacevic (1995), an estimator of the Gini coefficient is:

$$\hat{G}(y) = \frac{2 \cdot \sum_{h=1}^{H}\sum_{\alpha=1}^{a_h}\sum_{i=1}^{n_{h\alpha}} w^{*}_{h\alpha i} \cdot \hat{F}_{h\alpha i} \cdot y_{h\alpha i} - 1}{\bar{y}_w} \tag{5.14}$$

where:

$w^{*}_{h\alpha i} = w_{h\alpha i} / \sum_h \sum_\alpha \sum_i w_{h\alpha i}$,

$\hat{F}_{h\alpha i}$ = the estimate of the CDF, F(y), at the ordered value of y for case i in cluster α and stratum h;

$\bar{y}_w$ = the weighted estimate of the mean of y.

After installing the `svylorenz` command (via the command `ssc install svylorenz`), the following sequence of Stata commands will generate the estimates and standard errors for the Gini coefficient, the share (proportion) of the total held by the quantile group, the cumulative share by quantile group, and the Lorenz curve ordinates for total household wealth based on the 2012 HRS.

```
svyset secu [pweight = nwgthh], strata(stratum)
gen poshllatota = hllatota
replace poshllatota = 0 if hllatota < 0
gen finr = 1
replace finr = 0 if nfinr != 1
svylorenz poshllatota, ngp(10) subpop(finr)
```

TABLE 5.2

Estimation of wealth inequality measures for the 2012 HRS population of households

Statistic	Estimate	SE	95% Confidence Interval
Gini Coefficient	0.727	0.009	(0.709, 0.745)
	Group Share (Percent) of Total Household Wealth		
1st Decile	0.00%	—	(0.00%, 0.00%)
2nd Decile	0.06%	0.01%	(0.04%, 0.08%)
3rd Decile	0.50%	0.05%	(0.43%, 0.62%)
4th Decile	1.42%	0.08%	(1.25%, 1.60%)
5th Decile	2.57%	0.13%	(2.30%, 2.83%)
6th Decile	4.21%	0.18%	(3.86%, 4.57%)
7th Decile	6.50%	0.24%	(5.99%, 6.92%)
8th Decile	10.30%	0.34%	(9.60%, 10.90%)
9th Decile	17.20%	0.48%	(16.29%, 18.18%)
10th Decile	57.24%	1.39%	(54.52%, 59.97%)

(*Source:* 2012 HRS)

For this example, all negative reported values of H11ATOTA are set to a minimum value of 0 since the `svylorenz` command requires that all values be greater than or equal to zero. Table 5.2 summarizes the results of this analysis of asset inequality in the 2012 HRS household population. The estimated Gini coefficient is $\hat{G}(y) = 0.727$ (SE = 0.009), which suggests a high degree of concentration of total wealth among households in the upper deciles of the total net worth distribution. The degree of concentration is evident in the estimates of the group shares of total wealth for the ten deciles of the household distribution. The upper decile is estimated to hold 57.2% (SE = 1.4%) of all wealth owned by the 2012 HRS household population.

5.4 Bivariate Relationships between Two Continuous Variables

There are four basic analytic approaches that can be used to examine bivariate relationships between two continuous survey variables: 1) generation of a scatter plot; 2) computation of a pairwise correlation coefficient, *r*; 3) estimation of the ratio of two continuous variables, $\hat{R} = \hat{Y} / \hat{X}$; and 4) estimation of the coefficients of the simple linear regression of one variable on another, $\hat{Y} = \hat{\beta}_0 + \hat{\beta}_1 \cdot X$. The first three of these techniques are reviewed in this section. Chapter 7 will address the estimation of linear regression models for continuous dependent variables in detail.

5.4.1 X-Y Scatter Plots

A comparison of the 2011–2012 NHANES MEC measures of HDL cholesterol (the *y* variable) and total serum cholesterol (the *x* variable) illustrates how a simple scatter plot can be used to gain insight into the bivariate relationship of two survey variables. Figure 5.5 presents an unweighted scatter plot examining the relationship of these two variables. The figure suggests a positive relationship between HDL cholesterol and total serum cholesterol,

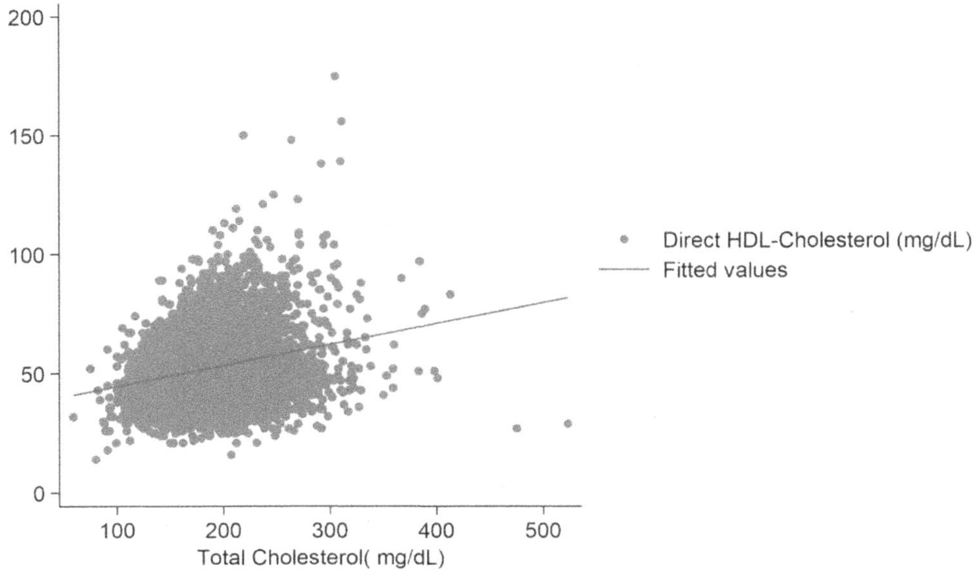

FIGURE 5.5
HDL Cholesterol vs. Total Serum Cholesterol (mg/dl) in U.S. Adults (unweighted points, with the fit of a survey-weighted regression line included).

Source: 2011–2012 NHANES.

but also illustrates considerable variability in the relationship, along with the presence of a few points that stand out as potential outliers.

A drawback to the simple X-Y scatter plot summary is that there is no practical way to incorporate the survey weights and the corresponding population frequency associated with each point in the two-dimensional X-Y space. Stata does provide a unique display in which the area of the dot representing an X-Y point is proportionate to its survey weight, and other packages (e.g., the `svyplot()` function in the `survey` package in R) offer these capabilities as well. Unfortunately, in survey data sets with hundreds and even thousands of data points, it is not possible to obtain the resolution needed to evaluate single points or to detect patterns in these types of weighted scatter plot displays.

One technique for introducing information on the effect of the weights in the plotted X-Y relationship is to overlay on the scatter plot the line representing the weighted regression of y on x, $\hat{y}_{wls} = \hat{\beta}_0 + \hat{\beta}_1 \cdot x$. To do this in Stata, we use the `twoway` graphing command, where the first command in parentheses defines the scatter plot and the second command overlays the weighted estimate of the simple linear regression model relating total cholesterol to HDL cholesterol (note that the probability weight variable is defined in square brackets for the second command). We have added the fit of this survey-weighted linear regression model to Figure 5.5.

```
twoway (scatter lbdhdd lbxtc if age18p==1) ///
(lfit lbdhdd lbxtc if age18p==1 [pweight = WTMEC2YR])
```

Another useful visualization technique for understanding the nature of the association between two continuous variables is **locally weighted scatter plot smoothing (LOWESS).**

This is a nonparametric technique used to fit a smooth curve to a set of data on two continuous variables, where the smooth curve is defined by the best local fit within a specific window of data points. The survey package within R provides a convenient function for adding a smooth LOWESS curve to a given scatter plot (the svysmooth() function) in a manner that accounts for the survey weights associated with the individual data points. The attractive feature of adding this type of curve to a scatter plot is that this technique does not assume that the relationship between the two variables is linear, and can give the analyst a sense of whether a relationship between two variables is actually non-linear in the target population.

Here is how we would re-create the same plot from Figure 5.5 in R, and add the smooth LOWESS curve to the plot:

```
# read in the full NHANES data set for the third edition
load("X:/. . ./nhanes1112.rdata")

nhanes.des <- svydesign(id = ~sdmvpsu, strata = ~sdmvstra,
weights = ~wtmec2yr, nest = T, data = nhanes1112)

# unweighted scatter plot
plot(lbdhdd ~ lbxtc, subset(nhanes1112, age >= 18),
xlab="Total Cholesterol", ylab="HDL Cholesterol", col="lightblue")

# survey-weighted LOWESS smoother (bandwidth = window)
smoother <- svysmooth(lbdhdd ~ lbxtc, subset(nhanes.des, age >= 18),
bandwidth=10)

# add smooth curve to plot
lines(smoother, col="black", lwd=4, lty=2)
```

Figure 5.6 presents the resulting plot. The smoothing function is clearly sensitive to the unusual data points identified previously, but for the bulk of the data points in the sample, the survey-weighted relationship appears to be roughly linear.

5.4.2 Product Moment Correlation Statistic (r)

The product moment correlation (r) is a standard measure of the strength of the linear association between two continuous variables. Kish and Frankel (1974) included this pairwise correlation statistic in their simulation studies and found that TSL, BRR, and JRR performed similarly in the estimation of standard errors for estimates of pairwise correlation statistics. Unfortunately, few current software systems provide the capability to estimate single correlations (or weighted correlation matrices) and confidence intervals for r that account for complex sample design features.

A weighted estimator of the product moment correlation statistic is written as follows:

$$r_w = \frac{s_{xy,w}}{s_{x,w} \cdot s_{y,w}} = \frac{\sum_{h=1}^{H}\sum_{\alpha=1}^{a_h}\sum_{i=1}^{n_{h\alpha}} w_{h\alpha i}(y_{h\alpha i} - \bar{y}_w)(x_{h\alpha i} - \bar{x}_w)}{\sqrt{\sum_{h=1}^{H}\sum_{\alpha=1}^{a_h}\sum_{i=1}^{n_{h\alpha}} w_{h\alpha i}(y_{h\alpha i} - \bar{y}_w)^2} \cdot \sqrt{\sum_{h=1}^{H}\sum_{\alpha=1}^{a_h}\sum_{i=1}^{n_{h\alpha}} w_{h\alpha i}(x_{h\alpha i} - \bar{x}_w)^2}} \quad (5.15)$$

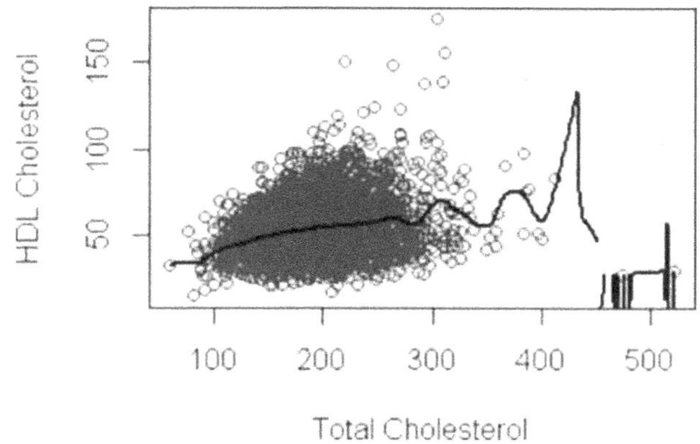

FIGURE 5.6
HDL Cholesterol vs. Total Serum Cholesterol (mg/dl) in U.S. Adults (unweighted points, with the fit of a survey-weighted LOWESS curve included).

Source: 2011–2012 NHANES.

If a survey analyst does require weighted estimates and standard errors for the pairwise correlation statistic, an alternative to direct estimation is to first standardize the two variables for which the correlation is desired, and then compute a weighted estimate of the slope parameter in a simple linear regression model relating the two standardized variables (see Chapter 7).

EXAMPLE 5.11: ESTIMATION OF THE CORRELATION OF ADULTS' TOTAL AND HIGH-DENSITY CHOLESTEROL MEASURES IN THE 2011–2012 NHANES

In this example, we will use the 2011–2012 NHANES data and illustrate the indirect method of using linear regression and standardized survey measures to estimate the correlation of total and high-density cholesterol levels in the U.S. adult population. To begin, we use the Stata `summarize` command with the `[aweight =]` specification to obtain weighted estimates of the population mean and standard deviation for the two cholesterol measures (see Example 5.8 for estimation of the means and standard deviations):

```
sum lbdhdd lbxtc [aweight = WTMEC2YR] if age18p == 1
```

The Stata `correlate` command with the `[aweight =]` specification produces a weighted estimate of the population correlation, $r_w = 0.241$:

```
correlate lbdhdd lbxtc [aweight = WTMEC2YR] if age18p == 1
```

Next, the original values for the 2011–2012 NHANES high-density (LBDHDD) and total (LBXTC) cholesterol measures are "standardized" by subtracting from each value the weighted estimate of the mean and dividing that difference by the weighted estimate of the standard deviation:

```
gen stdlbxtc = (lbxtc - 194.4355) / 41.05184
gen stdlbdhdd = (lbdhdd - 52.83826) / 14.93157
sum stdlbdhdd stdlbxtc [aweight = WTMEC2YR] if age18p == 1
```

The `summarize` command produces output that confirms (not shown) that the standardized cholesterol measures do, in fact, have a mean of zero and a standard deviation of 1.0. Using Stata's `svy: regress` command, the standardized measure for high-density cholesterol is then regressed on the standardized value for total cholesterol. Using the standard `svyset` command, the regression is weighted by the NHANES MEC weight, and the estimated (linearized) standard error of the weighted estimate of the total cholesterol coefficient accounts for the primary stage stratification and cluster sampling in the complex sample design.

```
svyset sdmvpsu [pweight = WTMEC2YR], strata(sdmvstra)
svy, subpop(age18p): regress stdlbdhdd stdlbxtc
```

Variable	n	$\hat{\beta}$	$se(\hat{\beta})$	df	$CI_{.95,df}(\beta)$
STDLBXTC	5,187	0.241	0.014	17	(0.211, 0.271)

The coefficient, $\hat{\beta}$, in the survey-weighted regression of `stdlbdhdd` on `stdlbxtc` is a population-weighted estimate of the correlation of the two cholesterol measures. The estimated value, $\hat{\beta} = r_w = 0.241$, matches the weighted estimate for the original (not standardized) variable generated above using the `correlate` command. The design-adjusted standard error of the correlation estimate is $se(\hat{\beta}) = se(r_w) = 0.014$, and the corresponding design-based 95% confidence interval is $CI_{.95,17}(\beta)$ = (0.211, 0.271).

5.4.3 Ratios of Two Continuous Variables

Occasionally, survey analysts need to estimate the ratio of two continuous survey variables (e.g., the ratio of HDL cholesterol level to total cholesterol). The ratio estimator of the population mean (5.6) of a single variable can be generalized to an estimator of the population ratio of two survey variables:

$$\hat{R} = \frac{\hat{Y}}{\hat{X}} = \frac{\sum_{h=1}^{H}\sum_{\alpha=1}^{a_h}\sum_{i=1}^{n_{h\alpha}} w_{h\alpha i}\, y_{h\alpha i}}{\sum_{h=1}^{H}\sum_{\alpha=1}^{a_h}\sum_{i=1}^{n_{h\alpha}} w_{h\alpha i}\, x_{h\alpha i}} \tag{5.16}$$

The following TSL approximation provides estimates of the sampling variance of ratio estimates:

$$var(\hat{R}) \doteq \frac{var(\hat{X}) + \hat{R}^2 \cdot var(\hat{Y}) - 2 \cdot \hat{R} \cdot cov(\hat{Y}, \hat{X})}{\hat{X}^2} \tag{5.17}$$

BRR, JRR, and bootstrap variance estimation options now available in the major statistical software packages provide appropriate alternatives to TSL for estimating standard errors of $\hat{R}$.

EXAMPLE 5.12: ESTIMATING THE POPULATION RATIO OF HIGH DENSITY TO TOTAL CHOLESTEROL FOR U.S. ADULTS

A weighted estimate of this ratio based on 2011–2012 NHANES respondents age 18+ is obtained in Stata using the `svy: ratio` command. Note that we once again define the subpopulation indicator (AGE18P) first before running the analysis, assuming that the variable has not already been generated:

```
svyset sdmvpsu [pweight = WTMEC2YR], strata(sdmvstra)
gen age18p = 1 if age >= 18 & age != .
replace age18p = 0 if age < 18
svy, subpop(age18p): ratio (lbdhdd/lbxtc)
estat effects
```

n	df	$\hat{R}$	$se(\hat{R})$	$CI_{.95}(\hat{R})$	$d^2(\hat{R})$
5,187	17	0.272	0.003	(0.266, 0.278)	8.030

Note that the two variables defining the numerator and denominator of the ratio are indicated in parentheses in the `svy: ratio` command, with the numerator variable listed before a forward slash (/), and the denominator variable listed after the slash.

THEORY BOX 5.2: RATIO ESTIMATORS OF POPULATION TOTALS

If two survey variables x and y are highly correlated and a control total for the variable x, X_{pop}, is known from an auxiliary data source, then the ratio estimator of the population total, $\hat{Y}_R = \hat{R} \cdot X_{pop}$, may provide a more precise estimate than the simple weighted estimator of the population total, $\hat{Y}_w$, described in Section 5.3.2. Because X_{pop} is assumed to be free of sampling variability, the standard error of the ratio estimator of the population total is $se(\hat{Y}_R) = se(\hat{R}) \cdot X_{pop}$.

5.5 Descriptive Statistics for Subpopulations

Section 4.5.3 discussed the important analytical differences between *conditional* and *unconditional* subpopulation analyses. Section 5.3.2 presented examples of unconditional subpopulation analyses in the estimation of totals. Stata's `svy` commands provide two options for correctly specifying unconditional subpopulation analyses. The `over()` option requests unconditional subclass analyses for subpopulations defined based on *all* levels of a categorical variable. This option is only available in Stata's descriptive analysis procedures. The more general `subpop()` option can be used with all `svy` commands in Stata. This option requests analysis for a specific subpopulation identified by an indicator variable, coded 1 if the respondent belongs to the subpopulation of interest and 0 otherwise. Procedures for survey data analysis in other software packages will have similar command options available for these unconditional subpopulation analyses (see the ASDA website for examples).

EXAMPLE 5.13: ESTIMATING THE PROPORTIONS OF MALES AND FEMALES AGE ≥ 70 WITH DIABETES BASED ON THE 2012 HRS DATA

This first subpopulation analysis example uses the 2012 HRS data set to estimate the prevalence of diabetes for U.S. males and females age 70 years and older. Because the HRS variable DIABETES is equal to 1 for respondents with diabetes and 0 otherwise, estimating the mean of this binary variable will result in the prevalence estimates of interest. The following example command uses a logical condition in the `subpop()` option to specify the age range and then the `over()` option to perform separate subpopulation analyses for males and females.

```
svyset secu [pweight = nwgtr], strata(stratum)
svy, subpop(if nage >= 70): mean diabetes, over(gender)
```

Gender	df	n	$\bar{y}_w$	$se(\bar{y}_w)$	$CI_{.95}(\bar{y}_w)$
Male	54	3,495	0.274	0.007	(0.259, 0.289)
Female	54	4,369	0.227	0.009	(0.210, 0.244)

We see that more elderly men (27.4%) are estimated to have diabetes compared to elderly women (22.7%).

It might be tempting for an analyst to make the mistake of taking a *conditional* approach to these subpopulation analyses, using a Stata command like the following:

```
svy: mean diabetes if age > 70 & gender == 1
```

Use of the sub-setting `if` modifier to define a subpopulation for design-based analyses of stratified samples, along with the use of Taylor Series Linearization for variance estimation, is inappropriate. All cases not satisfying the condition are temporarily deleted from the analysis, along with their sample design information. This effectively fixes the stratum sample sizes for variance estimation calculations (when in fact the subpopulation sample sizes for each of the strata should be treated as random variables). This also introduces the risk of entire sampling error clusters with no subpopulation members present in the sample being deleted and ignored in the variance estimation, rather than treated as sampling zeroes. See Chapter 4 for more details. Taking a Bayesian approach to these analyses, per the earlier examples in this chapter, does not introduce the same complexities.

EXAMPLE 5.14: ESTIMATING MEAN SYSTOLIC BLOOD PRESSURE FOR MALES AND FEMALES AGE > 45 USING THE 2011–2012 NHANES DATA

Consider an example based on the 2011-2012 NHANES data of estimating the mean systolic blood pressure for male and female adults over the age of 45. We once again illustrate a combination of options in Stata to set up this analysis, using a logical condition in the `subpop()` option to specify the age range and then the `over()` option to perform subpopulation analyses for males and females in this age range separately. Assuming that the complex sample design variables have already been specified using `svyset`, we also request design effects for each subpopulation estimate using the `estat effects` post-estimation command.

```
svy, subpop(if age > 45): mean BPXSY1, over(riagendr)
estat effects
estat size
```

Gender	n	df	$\bar{y}_w$	$se(\bar{y}_w)$	$CI_{.95}(\bar{y}_w)$	$d^2(\bar{y}_w)$
Male	1,329	17	128.30	0.87	(126.47, 130.13)	3.53
Female	1,343	17	128.12	0.95	(126.19, 130.18)	4.27

Survey analysts should be aware that restricting analysis to a subpopulation of the full sample may result in a reduction in the effective degrees of freedom for variance estimation. Since Stata employs the variable degrees of freedom method discussed in Section 3.5.2, its commands ignore any original design strata that do not contain one or more observations from the subpopulation of interest. Stata signals that complete strata have been dropped by including a note in the output indicating that one or more design strata were "`omitted because they contain no subpopulation members`." Approaches to this issue are not currently uniform across the different major statistical software packages.

The greatest reductions in effective degrees of freedom for variance estimation can occur when survey analysts are interested in estimation for rare subpopulations that comprise only a small percent of the survey population, or the subpopulation of interest is defined by a domain of sample strata, such as a single Census region. (Refer to Figure 4.5 in Chapter 4 for an illustration of how subpopulations may be distributed across the strata and clusters of a complex sample design.)

Software procedures for correct unconditional subpopulation analyses are currently also available in R (the `survey` package), SUDAAN, WesVar PC, Mplus, SPSS Complex Samples, IVEware, and SAS (the SURVEY procedures). Examples of subpopulation analyses using these other software packages are available on the ASDA website. Analysts could also take a more model-based Bayesian approach to these types of descriptive analyses for subpopulations; see the earlier examples in this chapter.

5.6 Linear Functions of Descriptive Estimates and Differences of Means

The capability to estimate *functions* of descriptive statistics, especially differences of means or proportions, is an important feature of many survey analyses. In general, many important functions of descriptive statistics can be written as **linear combinations** of the descriptive statistics of interest. Examples of such linear combinations of estimates include differences of means, weighted sums of means used to build economic indices, or computation of a moving average of three monthly survey means, for example:

$$\begin{aligned}\hat{\Delta} &= \bar{y}_1 - \bar{y}_2\\ \hat{I} &= .25\cdot\bar{y}_1 + .40\cdot\bar{y}_2 + 1.5\cdot\bar{y}_3 + 1.0\cdot\bar{y}_4\\ \bar{y}_{moving} &= (1/3)\cdot\bar{y}_{t1} + (1/3)\cdot\bar{y}_{t2} + (1/3)\cdot\bar{y}_{t3}\end{aligned}$$

Consider the general form of a linear combination of $j = 1, \ldots, J$ descriptive statistics (e.g., estimates of means for J subpopulations):

$$f(\theta_1,\ldots,\theta_J) = \sum_{j=1}^{J} a_j\theta_j \tag{5.18}$$

In (5.18), θ_j represents the statistic of interest for the j-th subpopulation, while the a_j terms represent constants defined by the analyst. This function is estimated by simply substituting estimates of the descriptive statistics into the expression:

$$f(\hat{\theta}_1,\ldots,\hat{\theta}_J) = \sum_{j=1}^{J} a_j\hat{\theta}_j \tag{5.19}$$

The variance of this estimator is then calculated as follows:

$$var\left(\sum_{j=1}^{J} a_j\hat{\theta}_j\right) = \sum_{j=1}^{J} a_j^2 var(\hat{\theta}_j) + 2\times\sum_{j=1}^{J-1}\sum_{k>j}^{J} a_j a_k cov(\hat{\theta}_j,\hat{\theta}_k) \tag{5.20}$$

Note that the variance of the linear combination incorporates the variances of the individual component estimates as well as the covariances of the estimates.

Covariance between descriptive estimates can arise in several ways. First, the statistics of interest may represent overlapping subpopulations (see Kish, 1987). An example would be a contrast comparing the mean systolic blood pressure for men to that for the total population, for example, $\hat{\Delta} = \bar{y}_{male} - \bar{y}_{total}$, or a longitudinal analysis in which the mean blood pressures of a sample panel of individuals are compared at two points in time, $\hat{\Delta} = \bar{y}_{time2} - \bar{y}_{time1}$. Due to the intraclass correlation among the sample elements within the clusters that may be defined by complex sample designs, a degree of covariance is possible even when estimates are based on non-overlapping samples, for example, $\hat{\Delta} = \bar{y}_{female} - \bar{y}_{male}$.

Under conditions where samples are overlapping or the complex sample design itself induces a covariance in the finite population estimates, statistical software must compute and store the covariances of estimates in order to correctly compute the variance of a linear combination of means or other descriptive statistics. Stata's `lincom` command is one example of a post-estimation command that enables analysts to correctly compute estimates and standard errors for linear combinations of estimates.

5.6.1 Differences of Means for Two Subpopulations

Analysts of survey data are frequently interested in making inferences about differences of descriptive statistics for two subpopulations. The inference can be based on a 95% confidence interval for the difference of the two means, for example,

$$CI = (\bar{y}_{female} - \bar{y}_{male}) \pm t_{df,.975} \cdot se(\bar{y}_{female} - \bar{y}_{male}), \tag{5.21}$$

or it might employ a **two-sample *t*-test** based on the Student *t* statistic,

$$t = \left(\bar{y}_{female} - \bar{y}_{male}\right) / \, se\left(\bar{y}_{female} - \bar{y}_{male}\right). \tag{5.22}$$

Applying the general formula for the variance of a linear combination (5.20 above), the standard error of the difference in the mean of the two subpopulations samples can be expressed as:

$$se(\bar{y}_1 - \bar{y}_2) = \sqrt{var(\bar{y}_1) + var(\bar{y}_2) - 2cov(\bar{y}_1, \bar{y}_2)} \tag{5.23}$$

Under simple random sampling, the covariance of estimates from distinct subpopulation samples is zero; however, for clustered samples or samples that share elements, the covariance of the two sample means may be non-zero and will generally be positive in value. The following example estimates the contrast in the mean value of total household wealth for two subpopulations of HRS households: one subpopulation in which the household head has less than a high school education and a second subpopulation of households headed by a college-educated adult.

EXAMPLE 5.15: ESTIMATING DIFFERENCES IN MEAN TOTAL HOUSEHOLD WEALTH BETWEEN 2012 HRS SUBPOPULATIONS DEFINED BY EDUCATIONAL ATTAINMENT LEVEL

Testing differences of subpopulation means using the `svy: mean` procedure in Stata is a two-step process. First, the mean of total household wealth is estimated for subpopulations defined by the levels of the EDCAT variable in the HRS data set. The estimates are further restricted to HRS household financial respondents (FINR = 1), and the 2012 HRS household weight (NWGTHH) is used to compute the estimated means:

```
gen finr = 1
replace finr = 0 if nfinr != 1
svyset secu [pweight = nwgthh], strata(stratum)
svy, subpop(finr): mean h11atota, over(edcat)
```

Education of Head	Stata Label	$\bar{y}_w$	$se(\bar{y}_w)$	$CI_{.95}(\bar{y}_w)$
0–11 yrs	1	\$122,089	\$10,596	(\$100,863, \$143,314)
12 yrs	2	\$259,027	\$9,802	(\$239,391, \$278,664)
13–15 yrs	3	\$336,309	\$17,202	(\$301,849, \$370,768)
16+ yrs	4	\$834,141	\$46,478	(\$741,035, \$927,247)

Stata automatically saves these estimated means along with their estimated variances and covariances in working memory until the next command is issued. Stata assigns the four subpopulations of household heads with 0–11, 12, 13–15 and 16+ years of education to internal reference labels 1, 2, 3 and 4.

Following the computation of the subpopulation means, the `lincom` post-estimation command is used to estimate the difference in means for household heads with 0–11 years of education (subpopulation 1) vs. 16+ years of education (subpopulation 4):

```
lincom c.H11ATOTA@1.edcat - c.H11ATOTA@4.edcat
```

Education of Head	$\bar{y}_{0-11}-\bar{y}_{16+}$	$se(\bar{y}_{0-11}-\bar{y}_{16+})$	$CI_{.95}(\bar{y}_{0-11}-\bar{y}_{16+})$
0–11 vs. 16+	–\$712,052	\$48,886	(–\$809,983, –\$614,122)

After this post-estimation command is submitted, Stata outputs the estimated difference of the two subpopulation means ($\hat{\Delta}=\bar{y}_{0-11}-\bar{y}_{16+}=-\$712{,}052$). The 95% confidence interval for the population difference does not include 0, suggesting that households headed by a college graduate have a significantly higher mean total household wealth compared to households in which the head does not have a high school diploma.

To display the estimated variance-covariance matrix for the subpopulation estimates of mean household wealth, the Stata post-estimation command `vce` is used:

```
vce
```

The output produced by this command is shown in the symmetric 4 × 4 matrix in Table 5.3. Note that in this example, the covariance of the estimated means for the 0–11 and 16+ household heads is negative (-0.587×10^8).

Applying the formula for the standard error of the contrast:

$$
\begin{aligned}
se(\bar{y}_{0-11}-\bar{y}_{16+}) &= \sqrt{var(\bar{y}_{0-11})+var(\bar{y}_{16+})-2\cdot cov(\bar{y}_{0-11},\bar{y}_{16+})} \\
&= \sqrt{[1.123+21.600-2\cdot(-0.587)]\times10^8} \\
&\doteq \$48{,}886
\end{aligned}
$$

TABLE 5.3

Estimated variance-covariance matrix for estimated subpopulation means of total household wealth (estimates by education level of household head)

Subpopulation	1	2	3	4
1	1.123×10^8	0.257×10^8	0.564×10^8	-0.587×10^8
2		0.961×10^8	0.640×10^8	0.194×10^8
3			2.959×10^8	1.572×10^8
4				2.160×10^9

Source: 2012 HRS

The result of this direct calculation exactly matches the output from the `lincom` command in Stata.

Alternative Bayesian Approach

One can also frame a comparison of two subpopulation means as the regression coefficient associated with an indicator for one of the two groups in a simple linear regression model for the continuous variable of interest. In the case of this example, we prepare a data frame object in R containing complete cases for the two subpopulations of interest and include the variable of interest, an indicator for being in the lowest education category (1 = yes, 0 = no), and the complex sampling features:

```
# read in the full HRS data set for the third edition
load("X:/. . ./hrs12.rdata")

# create complete data set with variables of interest
hrs.red <- hrs12[(hrs12$edcat == 1 | hrs12$edcat == 4) &
hrs12$nfinr == 1, c("h11atota", "edcat", "secu", "stratum", "nwgthh")]
hrs.red <- hrs.red[complete.cases(hrs.red),]
hrs.red$edcat1[hrs.red$edcat == 1] <- 1
hrs.red$edcat1[hrs.red$edcat == 4] <- 0
```

Next, we normalize the household weights and create the survey design object:

```
# need to normalize HRS weights to match Stan
hrs.red$wtsc <- hrs.red$nwgthh / mean(hrs.red$nwgthh)

# survey design object
hrs.des <- svydesign(id = ~secu, strata = ~stratum, weights
= ~wtsc, nest = T, data = hrs.red)
```

Then, we apply the Bayesian approach discussed in Section 3.8 to simulate design-adjusted draws from the survey-weighted posterior distribution for the intercept and the regression coefficient in a normal linear regression model, where the coefficient for the indicator of lowest education represents the difference in means between these two subpopulations. In this case, the covariance of the two means will be captured in the variance of the estimated coefficient:

```
# Bayesian approach, flat prior
set.seed(41279)
model_formula <- formula("h11atota|weights(wtsc) ~ edcat1")
```

```
mod.brms <- cs _ sampling _ brms(svydes = hrs.des,
brmsmod = brmsformula(model _ formula), data = hrs.red,
family = gaussian())

# Print results
mod.brms$stan _ fit
```

In this analysis, the mean of the design-adjusted draws of the regression coefficient from the survey-weighted posterior distribution is –$713,272.66, which is quite similar to the point estimate from the design-based approach. The 95% credible interval for the regression coefficient is (–$783,901.77, –$635,888.63), which is narrower than the 95% confidence interval from the design-based approach. We therefore see an increase in the efficiency of the estimated mean difference when using the model-based Bayesian approach that is similar to that demonstrated for earlier examples.

In this example, we assumed a flat (non-informative) prior for the regression coefficients in this simple model. If one had prior information (say, from earlier HRS waves) regarding the size of this coefficient and its standard error, it could easily be specified as part of this analysis. For example, suppose that based on a prior analysis, the mean of this coefficient was –$700,000, the standard error was $40,000, and the mean is assumed to follow a normal distribution. We could specify this prior as follows:

```
# With informative prior for coefficient
mod.brms <- cs _ sampling _ brms(svydes = hrs.des, brmsmod =
brmsformula(model _ formula),
prior = c(set _ prior("normal(-700000,40000)", class = "b",
coef="edcat1")), data = hrs.red, family = gaussian())
```

The resulting posterior mean is –$706,887.93, and the 95% credible interval is (–$762,729.71, –$650,491.54). While we would still conclude that there is a significant difference between these groups, the informative prior would clearly shift our inference somewhat.

5.6.2 Comparing Means over Time

Analysts working with longitudinal survey data are often interested in comparing the means of a longitudinal series of survey measures. Chapter 11 will address the various types of longitudinal data and introduce more sophisticated tools for longitudinal analysis of survey data.

When comparing means on single variables measured at two or more "waves" or points in time, an approach similar to that used in Section 5.6.1 can be applied. However, since longitudinal data are often released as a separate file for each time period with distinct weight values for each time point, special data preparation steps may be needed. We now consider an example of this approach using two waves of data from the HRS study.

EXAMPLE 5.16: ESTIMATING DIFFERENCES IN MEAN TOTAL HOUSEHOLD WEALTH FROM 2010 TO 2012 USING DATA FROM THE HRS STUDY

To estimate the difference in 2010 and 2012 mean household wealth for households in the HRS target population, the first step is to "stack" the 2010 and 2012 HRS data sets, combining them into a single data set. Provided that they responded to the survey in both HRS waves, each panel household has two records in the stacked data set—one for their 2010 observation and another for the 2012 interview. We provide more details on how to deal with attrition when analyzing longitudinal data in Chapter 11.

Each pair of household records includes the wave-specific measure of total household wealth and the wave-specific survey weight. For this example, when stacking the data sets, we assigned these values to two new common variables, TOTWEALTH and HHWEIGHT. Each record also includes the permanently assigned stratum and cluster codes. As described in Example 5.4, the HRS public-use data sets include an indicator variable for each wave of data that identifies the respondents who are the household financial reporters (MFINR for 2010, NFINR for 2012). Using the `over(year)` option, estimates of mean total household wealth are computed separately for 2010 and 2012. The `subpop(finr2010_2012)` option restricts the estimates to the financial reporters for each of these two data collection years. The post-estimation `lincom` command then estimates the difference of means for the two time periods, its linearized standard error, and a 95% confidence interval for the difference.

```
gen hhweight = mwgthh
replace hhweight = nwgthh if year == 2012
gen totwealth = h10atota
replace totwealth = h11atota if year == 2012
gen finr2010 = 0
replace finr2010 = 1 if (year == 2010 & mfinr == 1)
gen finr2012 = 0
replace finr2012 = 1 if (year == 2012 & nfinr == 1)
gen finr2010_2012 = 0
replace finr2010_2012 = 1 if finr2010 == 1 | finr2012 == 1
svyset secu [pweight = hhweight], strata(stratum)
svy, subpop(finr2010_2012): mean totwealth, over(year)
```

Year	$\bar{y}_{year}$	$se(\bar{y}_{year})$
2010	\$432,829	\$16,011
2012	\$437,807	\$17,016

```
lincom c.totwealth@2010.year - c.totwealth@2012.year
```

Contrast	$\bar{y}_{2010}-\bar{y}_{2012}$	$se(\bar{y}_{2010}-\bar{y}_{2012})$	$CI_{.95}(\bar{y}_{2010}-\bar{y}_{2012})$
2010 vs. 2012	–\$4,978	\$7,936	(–\$20,877, \$10,921)

Note that the `svyset` command has been used again to specify the recoded sampling weight variable (HHWEIGHT) in the stacked data set. The `svy: mean` command is then used to request weighted estimates and linearized standard errors (and the covariance of the estimates, which is saved internally) for each subpopulation defined by the YEAR variable. The resulting survey-weighted estimate of the difference of means is $\hat{\Delta}=\bar{y}_{2010}-\bar{y}_{2012}=-\$4{,}978$, with a linearized standard error of \$7,936. The analysis does not provide evidence that the mean total household wealth increased significantly from 2010 to 2012, a period in which U.S. households were still recovering from the impacts of the global recession that began in 2008.

5.7 Exercises

5.1) Download the European Social Survey—Russian Federation data set from the ASDA website. Using any statistical software, explore the data set by obtaining a list of variables and identify the complex sample design and final survey weight (post-stratification weight) variables. Answer these questions:

a. What are the names of each variable?

b. How many strata and clusters does this data set have?

c. Examine the distributional characteristics of the weight variable. Fill in the table below. Does the weight sum to the sample size or a population total?

Summary of Weight Variable (PSPWGHT)

N	Sum	N Missing	Mean	Minimum	Maximum

d. How many design degrees of freedom will be used during analysis and how did you calculate this number?

5.2) Identify the two survey variables measuring 1) Satisfaction with Life and 2) Trust in the Police. Next, answer the following questions:

a. Prepare a survey-weighted (if possible in your software of choice) histogram for each variable STFLIFE and TRSTPLC. What conclusions can be made based on the histograms? Are the population distributions skewed, or symmetrical? Do there seem to be any outliers?

b. What measurement scale is used for each variable? How can you justify use of a continuous variable analysis approach for these variables?

c. Use a statistical software procedure of your choice and fill in the table below. Ignore the complex sampling features for this initial descriptive analysis.

Variable Name	n	Mean	SE	Minimum	Maximum

d. Perform a survey-weighted analysis (with no design adjustment of the standard errors) for the same two variables. Complete the table below.

Variable Name	n	Mean	SE	95% CI

e. Repeat the analysis from Part d., but perform a fully design-based analysis, using the weights and the stratum and cluster codes for variance estimation. Use both the Taylor Series Linearization method and the Jackknife Repeated Replication methods for variance estimation. Fill in the table below.

Variable Name	n	Mean	TSL SE	JRR SE	95% CI (TSL)	95% CI (JRR)

f. Repeat the analysis from Part e., now using the design-adjusted Bayesian approach implemented in the R software (use a non-informative prior distribution in each case). Add these results to the table below.

Variable Name	n	Posterior Mean	Posterior SD	95% Credible Interval

g. How do the standard errors differ between the 4 analyses of parts c, d, e, and f? What impacts do the weights and the design variables have on the standard errors? Is there a difference between the standard errors using the TSL versus the JRR variance estimation method? What changes in the Bayesian analysis?

h. What are the estimated design effects for each estimated mean? Complete the table below.

Variable Name	n	Mean	Design Effect

5.3) Estimate the proportion of people in the Russian Federation target population who voted in the last election. Use the variable called VOTED_LASTELECTION, the TSL method for variance estimation, and the PSPWGHT variable for weighted estimation.

a. What is the estimated proportion of people in the ESS Russian Federation who voted in the last election? What is the estimated standard error and 95% confidence interval for this proportion? Is this proportion of voters significantly different from a hypothesized proportion of 0.55?

b. Re-estimate the proportion of voters among the subclasses of the variable called MALE (0=Female, 1=Male). Make sure to use an unconditional approach for this analysis. Fill in the table below.

Subclass	Variable	Mean (Linearized SE)	95% CI
Females	Voted Last Election		
Males	Voted Last Election		

c. Next, estimate the difference in proportions between men and women and discuss if the difference is significant at the alpha = 0.05 level.

	Variable	Mean (SE), *p* value
Difference in Proportions	Voted Last Election	

What conclusions would you make about the voting behavior among men and women?

d. Now, carry out the alternative Bayesian analysis of this difference in proportions (using the Bernoulli family for this binary voting variable, and specifying gender as a predictor in the model). Does anything change about our inference when using a Bayesian approach to the analysis?

5.4) Using a software procedure of your choice (remember that not all software enables percentile estimation in addition to estimation of standard errors for weighted estimates of percentiles) and the 2011–2012 NHANES data set, estimate the 25th percentile, the median, and the 75th percentile of Body Mass Index (BMXBMI) for U.S. adults aged 50 and older. You will need to create a subpopulation indicator of respondents aged 50+ for this analysis. Remember to perform an appropriate unconditional subpopulation analysis for this population subclass. Compute 95% confidence intervals for each percentile.

6

Categorical Data Analysis

> "If the same group of individuals is classified in two or more different ways, as persons may be classified as inoculated and not inoculated, and also may be attacked and not attacked by disease, then we may require to know if the two classifications are independent."
>
> **—R.A. Fisher (1925)**

6.1 Introduction

A casual perusal of the codebook and variable descriptions for most public-use survey data sets quickly leads to the observation that the responses to the majority of survey questions in the social sciences, public health, and related fields of research are measured as a binary choice (e.g., yes, no), a selection from multinomial response categories (e.g., ethnicity), a choice from an ordinal scale (e.g., strongly agree to strongly disagree), or possibly a discrete count of events. This chapter covers procedures for simple univariate, bivariate, and selected multivariate analyses for such categorical survey responses, focusing on the adaptation of established analytic techniques to complex sample survey data. For readers interested in a more thorough discussion of these basic techniques of categorical data analysis, we recommend Agresti (2012).

The outline of this chapter parallels that of Chapter 5. Section 6.2 highlights several important considerations in categorical data analysis for complex sample surveys. Basic methods for analyzing a single categorical variable are described in Section 6.3, including estimation of category proportions, goodness of fit tests to compare the sample estimates to hypothesized population values, graphical display of the estimated population distribution across the K categories of the single variable, and alternative Bayesian approaches. Section 6.4 extends the discussion to bivariate analyses, including statistics that measure association and tests of hypotheses concerning independence of two categorical variables. The chapter concludes in Section 6.5 with coverage of two techniques for multivariate categorical data: the Cochran-Mantel-Haenszel test, and simple log-linear models for cross-tabulated data. Multivariable regression modeling and related methods for categorical data will be introduced later in Chapters 8 and 9.

6.2 A Framework for Analysis of Categorical Survey Data

We begin this chapter by introducing a framework for the analysis of categorical data collected in complex sample surveys. We introduce methods for accommodating complex

DOI: 10.1201/9781003472094-7

sample design features and important considerations for both univariate and bivariate analyses of categorical data. Sections 6.3 and 6.4 go into more detail about these analysis approaches.

6.2.1 Incorporating the Complex Sample Design and Pseudo-Maximum Likelihood

The simplicity of categorical survey responses can belie the range of sophistication of categorical data analysis techniques—techniques that range from the simplest estimation of category proportions to complex multivariable and even multilevel regression models. The majority of the standard estimators and test statistics for categorical data analysis are derived under the method of **maximum likelihood**, and assume that the data are independent and identically distributed (i.i.d.) according to a discrete probability distribution. Under simple random sampling assumptions, categorical variables are assumed to follow one of several known probability distributions or **sampling models** (Agresti, 2012), that is, the binomial, the multinomial, the Poisson, the product-multinomial or, more rarely, a hypergeometric model. Unfortunately, due to sample weighting, cluster sampling, and stratification, the true likelihood of the sample survey data is generally difficult to specify analytically. Therefore, the simple elegance of maximum likelihood methods for estimation and inference does not easily transfer to complex sample survey data.

This chapter will introduce the methods and software that survey statisticians have developed to adjust standard analyses for complex sample design effects, including weighted estimation of proportions, design-based estimation of sampling variance, and generalized design effect corrections for key test statistics. Later, in Chapters 8 and 9, different forms of the generalized linear model and **pseudo maximum likelihood** techniques will be discussed for regression modeling of categorical data that follow approximate binomial, multinomial, Poisson, or negative binomial distributions.

6.2.2 Proportions and Percentages

In Section 6.3, we discuss estimation of proportions and their standard errors. In this chapter, we denote estimates of population proportions as p and population proportions as π. Many software procedures choose to output estimates of proportions and standard errors on the percentage scale. Translation between estimates and standard errors for proportions and percentages simply involves the following multiplicative scaling:

$$percent = 100 \cdot p, \quad se(percent) = 100 \cdot se(p), \quad var(percent) = 100^2 \cdot var(p).$$

While these relationships may be obvious to experienced analysts, our experience suggests that it is easy for mistakes to be made, especially in translating standard errors from the percentage to the proportion scale.

6.2.3 Cross-Tabulations, Contingency Tables, and Weighted Frequencies

Categorical data analysis becomes more complex (and also more interesting) when the analysis is expanded from a single variable to include two or more categorical variables. With two categorical variables, the preferred summary is typically a two-way data display with $r = 1, \ldots, R$ rows and $c = 1, \ldots, C$ columns, often referred to in statistical texts and literature as a **cross-tabulation** or a **contingency table**. Crosstabs and contingency tables are not limited to two dimensions but may include a third (or higher) dimension, that is, $l = 1, \ldots, L$ layers or sub-tables based on categories of a third variable. For simplicity of presentation, the majority of the examples in this chapter will be based on the cross-tabulation of two categorical variables (see Section 6.4).

Consider the simple $R = 2$ by $C = 2$ (2×2) tabular display of observed sample frequencies shown in Figure 6.1. Under simple random sampling (SRS), one-, two-, three-, or higher-dimension arrays of unweighted sample frequencies like the 2×2 array illustrated in Figure 6.1 can be used directly to estimate statistics of interest, such as the proportion of individuals in category 1 of variable 2 that respond in category 1 of variable 1, $p_{1|1} = n_{11} / n_{1+}$, or derive tests of hypotheses for the relationships between categorical variables, for example, the Pearson chi-square (χ^2) test.

However, because individuals can be selected into a sample with varying probabilities, estimates and test statistics computed from the unweighted sample frequencies may be biased for the true properties of the survey population. Therefore, it is necessary to translate from unweighted sample counts to weighted frequencies, as shown in Figure 6.2. For example, the weighted frequency (or estimated population count) in cell (0,1) is $\hat{N}_{01} = \sum_{h=1}^{H} \sum_{\alpha=1}^{a_h} \sum_{i \in (0,1)} w_{h\alpha i}$. The weighted proportions estimated from these weighted sample frequencies, for example, $p_{rc} = \hat{N}_{rc} / \hat{N}_{++}$, will reflect the relative size of the total population in the corresponding cell, row margin, or column margin of the cross-tabulation.

We discuss bivariate analyses in more detail in Section 6.4.

Variable 2	Variable 1: 0	Variable 1: 1	Row Margin
0	n_{00}	n_{01}	n_{0+}
1	n_{10}	n_{11}	n_{1+}
Column Margin	n_{+0}	n_{+1}	n_{++}

FIGURE 6.1
Bivariate Distribution of Observed Sample Frequencies.

Variable 2	Variable 1: 0	Variable 1: 1	Row Margin
0	$\hat{N}_{00}$	$\hat{N}_{01}$	$\hat{N}_{0+}$
1	$\hat{N}_{10}$	$\hat{N}_{11}$	$\hat{N}_{1+}$
Column Margin	$\hat{N}_{+0}$	$\hat{N}_{+1}$	$\hat{N}_{++}$

FIGURE 6.2
Bivariate Distribution of Weighted Sample Frequencies.

6.3 Univariate Analysis of Categorical Data

Simple descriptive analyses of a single categorical variable in a survey data set can take a number of forms, including: estimation of a simple population proportion π for binary responses; estimation of the vector of population proportions for each of the $k = 1, \ldots, K$ categories of a multi-category nominal or ordinal response variable; and, with care, estimation of the means for ordinally-scaled responses (see Section 5.3.3). To draw inferences about the population parameters being estimated, analysts can construct $100(1 - \alpha)\%$ confidence intervals for the parameters or perform Student t or simple χ^2 hypothesis tests. Recent methodological and software developments have also enabled analysts to take Bayesian approaches to making finite population inferences about proportions (see Section 3.8). Properly weighted graphical displays of the frequency distribution of the categorical variable are also very effective tools for presentation of results.

6.3.1 Estimation of Proportions for Binary Variables

Estimation of a single population proportion, π, for a binary response variable requires only a straightforward extension of the ratio estimator (Section 5.3.3) for the population mean of a continuous random variable. By recoding the original response categories to a single indicator variable y_i with possible values 1 and 0 (e.g., yes = 1, no = 0), the ratio mean estimator estimates the proportion or **prevalence**, π, of "1s" in the population:

$$p = \frac{\sum_{h=1}^{H}\sum_{\alpha=1}^{a_h}\sum_{i=1}^{n_{h\alpha}} w_{h\alpha i} I(y_i = 1)}{\sum_{h=1}^{H}\sum_{\alpha=1}^{a_h}\sum_{i=1}^{n_{h\alpha}} w_{h\alpha i}} = \frac{\hat{N}_1}{\hat{N}} \tag{6.1}$$

Most software procedures for survey data analysis provide the user with at least two alternatives for calculating ratio estimates of proportions. The first alternative is to use the single variable option in commands such as `svy: prop` and `svy: tab` in Stata, or PROC SURVEYFREQ in SAS. These procedures are designed to estimate the univariate proportions for the discrete categories of a single variable as well as total, row, and column proportions for cross-tabulations of two or more categorical variables. The second alternative for estimating single proportions is to first generate an indicator variable for the category of interest (with value 1 if a case falls into the category of interest, or 0 if a case does not fall into the category of interest) and then apply a procedure designed for estimation of means (e.g., `svy: mean` in Stata or PROC SURVEYMEANS in SAS). Both approaches will yield identical estimates of the population proportion and the standard error, but as explained below, estimated confidence intervals developed by the different procedures may not agree exactly due to differences in the methods used to derive the lower and upper confidence limits.

An application of Taylor Series Linearization (TSL) to the ratio estimator of π in (6.1) results in the following TSL variance estimator for the ratio estimate of a simple proportion:

$$v(p) \doteq \frac{V(\hat{N}_1) + p^2 \cdot V(\hat{N}) - 2 \cdot p \cdot Cov(\hat{N}_1, \hat{N})}{\hat{N}^2} \tag{6.2}$$

Replication techniques (JRR, BRR, Bootstrap) can also be used to compute estimates of the variances of these estimated proportions and the corresponding design-based confidence intervals and test statistics.

If the analyst estimates a proportion as the mean of an indicator variable (e.g., using Stata's `svy: mean` procedure), a standard design-based confidence interval is constructed for the proportion: $CI(p) = p \pm t_{1-\alpha/2,df} \cdot se(p)$. One complication that arises when proportions are viewed simply as the mean of a binary variable is that the true proportion, π, is constrained to lie in the interval (0,1). When the estimated proportion of interest is extreme (i.e., close to 0 or close to 1), the standard design-based confidence limits may be less than 0 or exceed 1. In addition, these symmetric "Wald" confidence intervals for proportions have repeatedly been shown in simulation studies to have less than nominal coverage (meaning, they do not cover the true population proportion at the stated rate; Dean & Pagano, 2015; Franco et al., 2019). To address this problem, alternative computations of design-based confidence intervals for proportions have been proposed, including (among others) the logit transformation procedure and the modified Wilson procedure (Rust & Hsu, 2007; Dean & Pagano, 2015; Franco et al., 2019).

Stata's `svy: tab` command uses the logit transformation procedure by default. Implementation of this procedure for constructing a confidence interval is a two-step process. First, using the weighted estimate of the proportion, one constructs a 95% CI for the logit transform of p:

$$\text{CI}\,[\text{logit}(p)] = \{\text{A,B}\} = \left\{ \ln\left(\frac{p}{1-p}\right) - \frac{t_{1-\alpha/2,df} \cdot se(p)}{p \cdot (1-p)}, \ln\left(\frac{p}{1-p}\right) + \frac{t_{1-\alpha/2,df} \cdot se(p)}{p \cdot (1-p)} \right\} \quad (6.3)$$

where:

p= the weighted estimate of the proportion of interest;

$se(p)$= the design-based Taylor Series approximation to the standard error of the estimated proportion.

Next, the two confidence limits on the logit scale, A and B, are transformed back to the original (0, 1) scale:

$$CI(p) = \left\{ \frac{e^A}{1+e^A}, \frac{e^B}{1+e^B} \right\} \quad (6.4)$$

Although procedures such as `svy: tab` in Stata and PROC CROSSTAB in SUDAAN default to the logit transformation formula for estimating CI(p) for all values of p, the adjusted CIs generally do not differ from the standard symmetric CIs unless $p < 0.10$ or $p > 0.90$. Via empirical simulation studies, Dean and Pagano (2015) found support for the use of adjusted CIs based on the Agresti-Coull and Clopper-Pearson methods of forming the intervals when $p < 0.10$ or $p > 0.90$, and these authors also find that the logit transformation and modified Wilson procedure tend to work well in these settings (assuming that the standard error of the estimated proportion is not zero). Interested readers can find a description of the modified Wilson procedure in Theory Box 6.1.

Wu (2016) advocated for the computation of **empirical likelihood confidence intervals** for finite population proportions (Wu & Rao, 2006), demonstrating empirically that they have improved coverage relative to other competing methods when proportions are extremely large or small. Unfortunately, outside of R code provided in Appendix Section A4 of Wu (2005), this approach is not yet widely implemented in standard software for the analysis of complex sample survey data. We will provide any updates in this regard on the ASDA website.

The most recent empirical work in this area has been performed by Franco et al. (2019), who find that it is crucial to compute confidence intervals for proportions using improved estimators of *effective* sample sizes that account for the complex sampling features (see Chapter 2). Via simulation studies, these authors empirically evaluated the coverage of additional alternatives to the confidence intervals evaluated by Dean and Pagano (2015) and others, including those based on Bayesian approaches. They ultimately recommend computing confidence intervals based on the Wilson method (see Theory Box 6.1) or a Bayesian approach with a uniform prior. In addition, these authors provided R code enabling the calculations of each alternative interval that they evaluated in secondary analyses (in addition to the necessary effective sample size calculations). We illustrate the use of this contributed R code in selected examples below.

EXAMPLE 6.1: ESTIMATING THE PROPORTION OF THE U.S. ADULT POPULATION WITH AN IRREGULAR HEART BEAT

In this example, data from the 2011–2012 NHANES are used to estimate the proportion of U.S. adults with an irregular heartbeat (defined by BPXPULS = 2 in the NHANES data set). To enable a direct comparison of the results from analyses performed using `svy: tab`, `svy: prop`, and `svy: mean` in Stata, we generate a binary indicator, equal to 1 for respondents reporting an irregular heartbeat, and 0 otherwise. This recoding of the original response categories for BPXPULS would not be necessary for estimation performed using either `svy: tab` or `svy: prop`. Note that as in the Chapter 5 analysis examples, we create an indicator for the adult subpopulation age 18 and older, enabling appropriate unconditional subpopulation analyses (see Section 4.5.2). As in previous examples, during the recoding steps, missing data on the original variable are explicitly assigned to a missing data code on the recoded variable.

```
gen irregular = .
replace irregular = 1 if bpxpuls == 2
replace irregular = 0 if bpxpuls == 1

gen age18p = 1 if age >= 18 & age != .
replace age18p = 0 if age < 18

svyset sdmvpsu [pweight = WTMEC2YR], strata(sdmvstra)

svy, subpop(age18p): tab irregular, se ci col deff
svy, subpop(age18p): proportion irregular
svy, subpop(age18p): mean irregular
estat effects
```

Table 6.1 below provides the weighted estimate of the proportion, the estimated standard error, the 95% confidence interval, and the estimated design effect produced by each of these three alternative commands. (Note that `svy: tab` and `svy: prop` produce estimated proportions and standard errors for each category; only the results for IRREGULAR = 1 are shown.)

TABLE 6.1

Estimated Proportion of Adults in the U.S. Population with an Irregular Heartbeat, with Alternative 95% Confidence Interval Calculations (n = 5,374, design degrees of freedom = 17)

Method	Estimate	Linearized SE	95% CI	Design Effect
`svy: tab, logit CI`	0.0164	0.0017	(0.0132, 0.0204)	1.2107
`svy: prop, logit CI`	0.0164	0.0017	(0.0132, 0.0204)	1.2107
`svy: mean, Wald CI`	0.0164	0.0017	(0.0129, 0.0200)	1.2107
`surveyci() in R, Wald`	—	—	(0.0133, 0.0196)	—
`surveyci() in R, Jeffreys Prior`	—	—	(0.0135, 0.0198)	—
`surveyci() in R, Uniform Prior`	—	—	(0.0136, 0.0199)	—
`surveyci() in R, Clopper-Pearson`	—	—	(0.0134, 0.0199)	—
`surveyci() in R, Wilson`	—	—	(0.0136, 0.0199)	—
`surveyci() in R, Agresti-Coull`	—	—	(0.0135, 0.0199)	—
`surveyci() in R, arcsine square root`	—	—	(0.0135, 0.0198)	—
`surveyci() in R, logit`	—	—	(0.0136, 0.0199)	—
`confint(svymean()) in R, Wald`	0.0164	0.0017	(0.0129, 0.0200)	—
`confint(svyciprop()) in R, logit`	0.0164	0.0017	(0.0132, 0.0204)	—

Source: 2011–2012 NHANES

For comparison, we also compute the alternative 95% confidence intervals evaluated in Franco et al. (2019). The following R code defines a function that computes all of these alternative intervals at once:

```
library(survey)
library(haven)

# Load key function from Franco et al. (2019)
source("http://umich.edu/~bwest/CIarrFcn.R")

# Define function for computing the Franco et al. intervals
# Credit: Sofi Sinozich (sofi.sinozich@gmail.com)
surveyci <- function(formula, design, deff) {
  if(length(all.vars(formula))>1) {
    message("Can only be used for single variables.")
    return()
  }
  # compute design-adjusted estimate of p and DEFF
  p <- svymean(formula, design, na.rm=TRUE, deff=deff)
  # compute the effective sample size
  n_eff <- dim(design$variables[all.vars(formula)])[1]/deff(p)
  # Compute and return the Franco et al. intervals
  CIarrFcn(p*n_eff, n_eff, 0.05)
}
```

Next, we apply this function to the subpopulation of interest in the 2011–2012 NHANES data (where the `subset()` function implements correct unconditional subclass analyses):

```
# Read NHANES data file
load("X:/. . ./nhanes1112.rdata")
```

```
# Create survey design object
nhanes.des <- svydesign(id = ~sdmvpsu, strata = ~sdmvstra,
weights = ~wtmec2yr, nest = T, data = nhanes1112)

# Compute adjusted Franco et al. (2019) intervals
surveyci(~irregular, subset(nhanes.des, age >= 18),
deff="replace")

# Comparison with more standard intervals
irr_p <- svymean(~irregular, subset(nhanes.des, age >= 18),
na.rm=TRUE, deff="replace")

# Wald
confint(irr_p, df=17)

# Logit
confint(svyciprop(~irregular, subset(nhanes.des, age >= 18),
method="logit", df=17))
```

The results in Table 6.1 suggest that for the period 2011–2012, 1.64% of U.S. adults were estimated to have an irregular heartbeat, with a 95% CI of (1.32%, 2.04%) based on the logit transform method in the `svy: tab` and `svy: prop` commands, and a 95% CI of (1.29%, 2.00%) based on the `svy: mean` command. Although the two confidence intervals are not substantially different, due to the small value of p, the logit transform 95% confidence interval for the proportion of adults with an irregular heartbeat is not symmetric about the point estimate of 1.64% prevalence. We also note that the various alternative intervals evaluated by Franco et al. (2019) would largely agree in terms of inference, but would generally produce more *efficient* inference than the standard design-based intervals would provide (as the confidence intervals are narrower).

THEORY BOX 6.1: THE MODIFIED WILSON PROCEDURE FOR CI(*P*)

The modified Wilson procedure (Kott & Carr, 1997; Korn & Graubard, 1999; Dean & Pagano, 2015; Franco et al., 2019) for complex samples constructs the confidence limits for the proportion using the expression:

$$CI(p)_{Wilson} = \frac{(2n^{*}p + t^{2}) \pm (t\sqrt{(t^{2} + 4p(1-p)n^{*})})}{2(n^{*} + t^{2})}$$

where:

p= the weighted estimate of the population proportion, π;

$t = t_{1-\alpha/2,df}$ = the $(1-\alpha/2)$ critical value of the Student t distribution with df design degrees of freedom; and

$n^{*} = p\cdot(1-p)/var_{des}(p) = n/d^{2}(p)$= the effective sample size for the estimate of π.

From the 2011–2012 NHANES data, the modified 95% Wilson confidence interval for the proportion of U.S. Adults with an irregular heartbeat (computed in R) is equal to (0.0136, 0.0199), which is quite similar to the design-based confidence interval in Table 6.1 computed by Stata using the logit transformation procedure. Interested readers can review Dean and Pagano (2015) and Franco et al. (2019) for additional empirical evaluations of the performance of these alternative approaches to forming intervals for proportions using survey data.

6.3.2 Estimation of Category Proportions for Multinomial Variables

Estimation of multinomial proportions and standard errors for multi-category survey response variables (e.g., race/ethnicity, measured by 1 = Mexican, 2 = Other Hispanic, 3 = White, 4 = Black, 5 = Other) is a direct extension of the method for estimating a single proportion. The ratio estimator in (6.1) is simply applied to indicator variables for each distinct category k:

$$p_k = \frac{\sum_{h=1}^{H}\sum_{\alpha=1}^{a_h}\sum_{i=1}^{n_{h\alpha}} w_{h\alpha i} I(y_i = k)}{\sum_{h=1}^{H}\sum_{\alpha=1}^{a_h}\sum_{i=1}^{n_{h\alpha}} w_{h\alpha i}} = \frac{\hat{N}_k}{\hat{N}} \tag{6.5}$$

The result is a vector of estimated population proportions for each category: $p = \{p_1,...,p_K\} = \{\hat{N}_1 / \hat{N},...,\hat{N}_K / \hat{N}\}$. Standard errors for the weighted estimates of the individual category proportions can be computed using Taylor Series Linearization (as presented in (6.2)) or by BRR, JRR, or Bootstrap (replication) methods. The most convenient method of simultaneously estimating the K category proportions for a multinomial categorical variable is through the use of the single variable option in procedures such as `svy: tab` or `svy: prop` in Stata, or PROC SURVEYFREQ in SAS.

EXAMPLE 6.2: ESTIMATING THE PROPORTIONS OF U.S. ADULTS BY RACE AND ETHNICITY

The following Stata commands estimate the proportions of the 2011–2012 NHANES adult survey population in each category defined by self-reported race and ethnicity:

```
svyset sdmvpsu [pweight = WTMEC2YR], strata(sdmvstra)
svy, subpop(age18p): prop RIDRETH1
estat effects
```

Table 6.2 presents the estimated proportions of U.S. adults for the five NHANES race/ethnicity categories along with the linearized standard errors of the estimates, the 95% CIs, and design effects for the estimated proportions.

As a brief aside, the large design effects for the sampling variance of the estimated proportions are due to the large sample sizes in the primary stage clusters of the 2011–2012 NHANES probability sample. However, the NHANES samples are not primarily designed to generate precise *overall* estimates of these types of proportions; they are designed to generate precise estimates for *subpopulations* of the U.S. population defined by demographic characteristics such as race/ethnicity and gender (Mohadjer & Curtin, 2008). NHANES design effects for estimates of subpopulation proportions are considerably smaller.

TABLE 6.2

Estimated Proportions of Adults in the U.S. Population by Race/Ethnicity.

Approach	Race/Ethnicity	Estimated Proportion	Linearized SE	95% CI	Design Effect
Design-Based	Mexican	0.079	0.017	(0.050, 0.124)	29.36
	Other Hispanic	0.066	0.015	(0.041, 0.106)	26.85
	White	0.659	0.039	(0.573, 0.736)	48.44
	Black	0.117	0.023	(0.076, 0.176)	37.98
	Other	0.078	0.011	(0.058, 0.104)	11.92
	Race/Ethnicity	**Posterior Mean**	**Posterior SD**	**95% Credible Interval**	
Bayesian (Survey-Weighted Dirichlet Multinomial Model)	Mexican	0.08	<0.01	(0.07, 0.09)	
	Other Hispanic	0.07	<0.01	(0.06, 0.07)	
	White	0.66	0.01	(0.65, 0.67)	
	Black	0.12	<0.01	(0.11, 0.13)	
	Other	0.08	<0.01	(0.07, 0.09)	

Source: 2011–2012 NHANES (n = 5,615)

Alternative Bayesian Approach

We now consider an alternative Bayesian approach to making inferences about these five population proportions in the target NHANES population. Here, we apply the methodology outlined in Section 3.8 to generate a survey-weighted posterior distribution for the five proportions defining the distribution of this multinomial race/ethnicity variable, and apply appropriate design adjustments to the simulated draws from this survey-weighted posterior that reflect that complex sample design features of the 2011–2012 NHANES sample. More specifically, we apply the survey-weighted Dirichlet multinomial model described in Hornby et al. (2023), using the following sequence of R code:

```
# read in the full NHANES data set for the third edition
load("X:/. . ./nhanes1112.rdata")

# create complete data set with variables of interest
nhanes.red <- nhanes1112[nhanes1112$age >= 18 &
!is.na(nhanes1112$age), c("ridreth1", "sdmvpsu",
"sdmvstra", "wtmec2yr")]
nhanes.red <- nhanes.red[complete.cases(nhanes.red),]

# need to normalize NHANES weights to match Stan
nhanes.red$wtsc <- nhanes.red$wtmec2yr /
mean(nhanes.red$wtmec2yr)

# survey design object
nhanes.des <- svydesign(id = ~sdmvpsu, strata = ~sdmvstra,
weights = ~wtsc, nest = T, data = nhanes.red)
# Create jackknife replicate weights
nhanes.rep <- as.svrepdesign(nhanes.des)

# Define the Dirichlet multinomial model for Stan
mod_dm <- stan_model(model_code = load_wt_multi_model())
```

```
# Prepare the data
y <- as.factor(nhanes.rep$variables$ridreth1)
yM <- model.matrix(~y -1)
n <- dim(yM)[1]
K <- dim(yM)[2]
alpha <- rep(1,K)
weights <- nhanes.rep$pweights
data_stan <-
list("y"=yM,"alpha"=alpha,"K"=K,"n"=n,"weights"=weights)
ctrl_stan <-
list("chains"=1,"iter"=2000,"warmup"=1000,"thin"=1)

# Run the Stan model with cs_sampling
mod1 <- cs_sampling(svydes = nhanes.rep, mod_stan = mod_dm,
data_stan = data_stan, ctrl_stan = ctrl_stan, rep_design = TRUE)

# View the results
mod1$stan_fit
```

This code, which uses a Uniform prior on the proportions, generates output that refers to the estimated proportions as "theta" values. Table 6.2 presents the posterior means for each of these proportions, in addition to 95% credible intervals. We see agreement with the design-based approach in terms of the point estimates, along with more efficient inferences for the proportions based on the narrower credible intervals.

EXAMPLE 6.3: ESTIMATING THE PROPORTIONS OF U.S. ADULTS BY BLOOD PRESSURE CATEGORY USING THE 2011–2012 NHANES DATA

A common technique in survey data analysis is to classify the population into discrete categories based on the values of one or more continuous survey variables. This example illustrates the variable recoding steps and use of the `svy: tab` command in Stata that are required to estimate the proportions of U.S. adults falling into each of the following four categories of blood pressure status: 1) normal (systolic blood pressure < 120 and diastolic blood pressure < 80); 2) pre-hypertension (SBP 120–139, or DBP 80–89); 3) Stage 1 Hypertension (SBP 140–159, or DBP 90–99); and 4) Stage 2 Hypertension (SBP 160+, or DBP 100+).

The analysis begins with the recoding steps. Based on the NHANES mean values of diastolic (DBP) and systolic (SBP) blood pressure (averaging several replicate measures), respondents are assigned to one of the four blood pressure categories.

```
generate bp_cat = .
replace bp_cat = 1 if (sbp < 120 & dbp < 80)
replace bp_cat = 2 if (missing(bp_cat) & ///
((sbp >= 120 & sbp < 140) | (dbp >= 80 & dbp < 90)))
replace bp_cat = 3 if (missing(bp_cat) & ///
((sbp >= 140 & sbp < 160) | (dbp >= 90 & dbp < 100)))
replace bp_cat = 4 if (missing(bp_cat) & ///
((160 <= sbp & sbp !=. ) | (dbp >= 100 & dbp != .)))
```

The NHANES sample design is identified through the `svyset` command and the `svy: tab` command is then applied to generate weighted estimates of the proportions of the adult population (AGE18P = 1) falling into each of these four categories:

```
svyset sdmvpsu [pweight = WTMEC2YR], strata(sdmvstra)
```

```
svy, subpop(age18p): tab bp _ cat, obs se ci
svy, subpop(age18p): tab bp _ cat, deff
```

The output generated by these commands is summarized in Table 6.3 below.

TABLE 6.3

Estimated Proportions of U.S. Adults by Blood Pressure Status

Blood Pressure Category	Estimated Proportion	Linearized SE	95% CI	Design Effect
Normal	0.472	0.016	(0.440, 0.505)	6.70
Pre-Hypertension	0.428	0.012	(0.403, 0.454)	4.10
Stage 1 Hypertension	0.080	0.006	(0.068, 0.093)	3.19
Stage 2 Hypertension	0.020	0.004	(0.013, 0.032)	6.79

Source: 2011–2012 NHANES; n = 5,356

These results suggest that an estimated 47.2% of the adult population has "normal" blood pressure (95% CI = 44.0%, 50.5%), while roughly 43% of the adult population is at the pre-hypertension stage (95% CI = 40.3%, 45.4%). Approximately 10% of the adult population is estimated to have either Stage 1 or Stage 2 hypertension. As in Example 6.1, the asymmetry of the 95% CIs for the smaller estimated proportions is due to Stata's use of the logit transform method to derive the CI limits. One could also apply the model-based Bayesian approach outlined in Example 6.2 to possibly increase the efficiency of the inferences related to these four proportions.

6.3.3 Testing Hypotheses Concerning a Vector of Population Proportions

When testing a null hypothesis concerning multinomial proportions of the form H_0: $\{\pi_1 = 0.5, \pi_2 = 0.25, \pi_3 = 0.25\}$, the standard **chi-square goodness of fit (GOF) test statistic,** $X^2 = n_{++} \cdot \sum_k (p_k - \pi_k)^2 / \pi_k$ (with degrees of freedom equal to the number of categories minus 1), can be considerably biased for complex sample survey data. Jann (2008) has extended the GOF test to the complex sample survey data setting, and this procedure has been implemented in the user-written mgof command for Stata (we omit details of the approach here; interested readers can review the article).

EXAMPLE 6.4: A GOODNESS OF FIT TEST FOR PROPORTIONS OF RUSSIANS AGE 15+ BY MARITAL STATUS

Table 6.4 summarizes a descriptive analysis of the ESS-Russian Federation data set in which proportions for the multinomial distribution of the Russian population age 15 years and older across marital status categories are estimated (see the Stata commands below). Based on these same data, assume the analyst wishes to test the null hypothesis that the true population proportions of Russian adults in the three categories are: married (50%), previously married (25%), and never married (25%).

```
use "X:\. . .\ess6 _ russia.dta", clear
svyset psu [pweight = pspwght], strata(stratify)
svy: tab marcat, obs se ci
svy: tab marcat, deff
```

After installing the user-written mgof command (web-aware Stata users can type `findit mgof` for help finding the command), a new variable, pi, is created using the

TABLE 6.4

Estimated Proportions of Russian Adults by Marital Status

Marital Status	Estimated Proportion	Linearized SE	95% CI	Design Effect
Married	0.504	0.013	(0.479, 0.529)	1.62
Previously Married	0.230	0.012	(0.208, 0.254)	1.84
Never Married	0.266	0.013	(0.241, 0.293)	2.25

Source: ESS Round 6 (2012)—Russian Federation (n = 2,444)

recode command that contains the null hypothesis values for each category of the MARCAT variable:

```
recode marcat (1=.5) (2=.25) (3=.25), generate (pi)
```

Next, the mgof command is submitted to test the null hypothesis. Note that the svy option is added to the command to ensure that the complex sampling features previously specified using svyset are considered:

```
mgof marcat = pi, svy
```

Submitting this command produces the following design-adjusted test statistics and p-values for the goodness of fit hypothesis test:

Goodness-of-fit Statistic	F	$P(F > \mathcal{F}_{1.92,337.97})$
Rao-Scott $X^2_{Pearson}$	1.68	0.19
$X^2_{LR, Jann}$	1.69	0.19

Interpreting the results of these tests, the null hypothesis is not rejected by either the design-adjusted Rao-Scott version of Pearson's chi-square test statistic (see Section 6.4.4 below) or the likelihood ratio chi-square statistic developed by Jann (2008). The weighted estimates computed using the collected sample data therefore appear to "fit" the hypothesized population distribution of category proportions.

6.3.4 Graphical Display for a Single Categorical Variable

Categorical survey variables naturally lend themselves to graphical display and analytic interpretation. As discussed in Chapter 5, it is important that the software used to generate graphics for analysis or publication has the capability to incorporate the survey weights, so that the display accurately reflects the estimated distribution of the variable in the survey population.

EXAMPLE 6.5: PIE CHARTS AND VERTICAL BAR CHARTS OF THE ESTIMATED PROPORTIONS OF RUSSIANS AGE 15+ BY MARITAL STATUS

Using Stata graphics, simple survey-weighted pie charts (Figure 6.3) or survey-weighted vertical bar charts (Figure 6.4) produce an effective display of the estimated population distribution across categories of a single categorical variable. The Stata commands used to generate these two figures follow; note the specification of a pweight in each command. Note also that both the graph pie and graph bar commands require

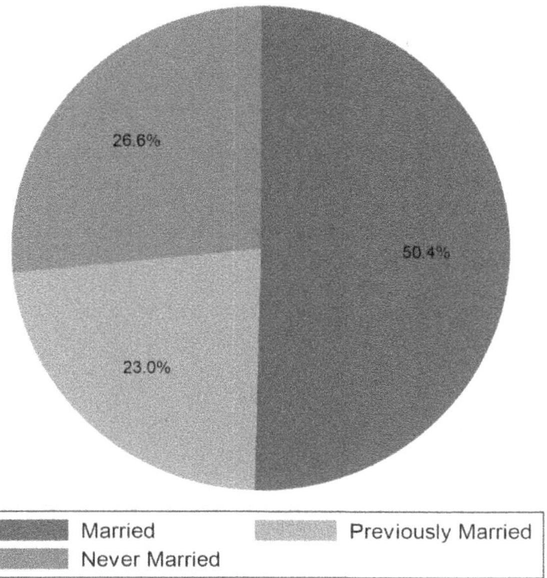

FIGURE 6.3
Pie Chart of the Estimated Distribution of the Marital Status of Russian Adults.

Source: ESS Round 6 (2012)—Russian Federation.

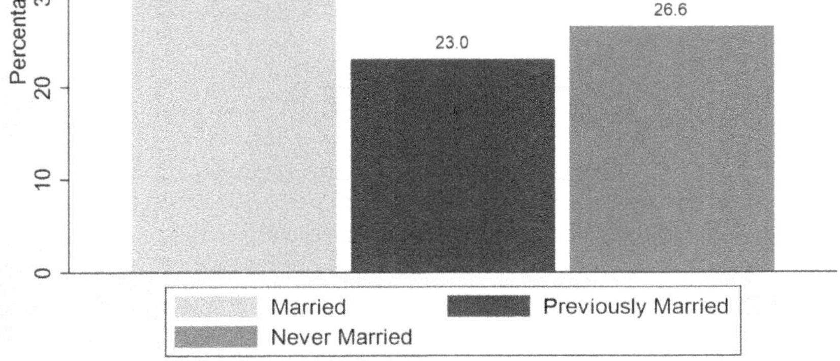

FIGURE 6.4
Bar Chart of the Estimated Distribution of the Marital Status of Russian Adults.

Source: ESS Round 6 (2012)—Russian Federation.

the initial creation of indicator variables for the individual marital status categories (marcat1, marcat2, and marcat3). This is accomplished using the generate(marcat) option in the tabulate command:

```
tabulate marcat, generate(marcat)
label var marcat1 "Married"
label var marcat2 "Previously Married"
label var marcat3 "Never Married"

* Pie Chart (one long command).
graph pie marcat1 marcat2 marcat3 ///
[pweight=pspwght], plabel( _all percent, ///
format(%9.1f)) scheme(s2mono) legend ///
(label(1 "Married") label(2 "Previously Married") ///
label(3 "Never Married"))

* Vertical Bar Chart (one long command).
graph bar (mean) marcat1 marcat2 marcat3 ///
[pweight=pspwght], percentages ///
bar(1,color(gs12)) bar(2,color(gs4)) ///
bar(3,color(gs8)) blabel(bar, format(%5.1f)) ///
bargap(7) scheme(s2mono) ///
legend (label(1 "Married") ///
label(2 "Previously Married") label(3 "Never Married")) ///
ytitle ("Percentage")
```

6.4 Bivariate Analysis of Categorical Data

Bivariate analysis of categorical data may take a number of forms, ranging from estimation of proportions for the joint distribution of the two variables (total proportions) or estimation of "conditional" proportions based on levels of the second categorical variable (i.e., row proportions or column proportions) to techniques for measuring and testing the bivariate association between two categorical variables. Bivariate categorical data analyses and reporting are important in their own right, and they are also important as exploratory tools in the development of more complex multivariable models (see the regression model building steps in Section 8.3).

6.4.1 Response and Factor Variables

Unlike simple linear regression for continuous survey variables, there is no requirement to differentiate variables by type (dependent or independent) or postulate a cause/effect relationship between two or more categorical variables that are being analyzed simultaneously. An example where assigning the variables to dependent or independent status is not necessary would be a bivariate analysis of the population distribution by education levels and Census region. Nevertheless, in many categorical data analysis problems, it is natural to think of one categorical variable as the **response variable** and the other(s) as a **factor variable** that may be associated with the discrete outcome for the response variable.

This section will introduce examples centered around the analysis of the joint distribution of U.S. adults' experience with a lifetime episode of major depression (yes, no) and their gender (male, female), based on the NCS-R data set. In these examples, it is convenient to label the NCS-R indicator variable for a lifetime major depression episode (MDE) as the response variable and the respondent's gender (SEX) as the factor. Assigning response and factor labels to the categorical variables also sets up the transition to our later discussion of regression modeling of categorical data where dependent and independent variables are clearly specified.

6.4.2 Estimation of Total, Row, and Column Proportions for Two-Way Tables

Based on the weighted frequencies illustrated in Figure 6.2, estimates of population proportions can be computed as the ratio of the weighted sample frequency for the cell to the appropriate weighted total or marginal frequency value. For example, Figure 6.5 illustrates estimation of the **total proportions** of the population in each cell and margin of the table. Note that the numerator of each estimated proportion is the weighted total frequency for the cell, for example, $\hat{N}_{A1}$, and the denominator is the weighted total population frequency, $\hat{N}_{++}$.

Statistical software also enables the user to condition estimates of population proportions on the sample in particular rows or columns of the cross-tabulation. Figure 6.6 illustrates the calculations of weighted estimates of **row proportions** for the estimated population distribution in Figure 6.2.

The following example uses the NCS-R data on lifetime major depression episode (MDE) and gender (SEX) to illustrate the Stata commands to estimate total proportions, row proportions, standard errors, and confidence intervals.

EXAMPLE 6.6: ESTIMATION OF TOTAL AND ROW PROPORTIONS FOR THE CROSS-TABULATION OF GENDER AND LIFETIME MAJOR DEPRESSION STATUS (SOURCE: NCS-R)

The first of two `svy: tab` commands in Stata (shown below) requests the default estimates of the total proportions for the SEX × MDE cross-tabulation along with the corresponding standard errors, 95% CIs, and design effects. The `row` option in the second `svy: tab` command specifies that the estimates, standard errors, CIs, and design effects will be for the row proportions.

```
svyset seclustr [pweight = ncsrwtsh], strata(sestrat)
svy: tab sex mde, se ci deff
svy: tab sex mde, row se ci deff
```

Factor	Response		
	0	1	
A	$p_{A0}=\frac{\hat{N}_{A0}}{\hat{N}_{++}}$	$p_{A1}=\frac{\hat{N}_{A1}}{\hat{N}_{++}}$	$p_{A+}=p_{A0}+p_{A1}$
B	$p_{B0}=\frac{\hat{N}_{B0}}{\hat{N}_{++}}$	$p_{B1}=\frac{\hat{N}_{B1}}{\hat{N}_{++}}$	$p_{B+}=p_{B0}+p_{B1}$
	$p_{+0}=p_{A0}+p_{B0}$	$p_{+1}=p_{A1}+p_{B1}$	$p_{++}=1.0$

FIGURE 6.5
Estimation of Overall (Total) Population Proportions (Multinomial Sampling Model).

<table>
<tr><th rowspan="2">Factor</th><th colspan="2">Response</th><th rowspan="2"></th></tr>
<tr><th>0</th><th>1</th></tr>
<tr><td>A</td><td>$p_{0|A} = \frac{\hat{N}_{A0}}{\hat{N}_{A+}}$</td><td>$p_{1|A} = \frac{\hat{N}_{A1}}{\hat{N}_{A+}}$</td><td>$p_{A+} = 1.0$</td></tr>
<tr><td>B</td><td>$p_{0|B} = \frac{\hat{N}_{B0}}{\hat{N}_{B+}}$</td><td>$p_{1|B} = \frac{\hat{N}_{B1}}{\hat{N}_{B+}}$</td><td>$p_{B+} = 1.0$</td></tr>
</table>

FIGURE 6.6
Estimation of Population Row Proportions (Product Multinomial Sampling Model).

TABLE 6.5
Estimated Proportions of U.S. Adults by Gender and Lifetime Major Depression Status

Description	Parameter	Estimated Proportion	Linearized SE	95% CI	Design Effect
		Total Proportions			
Male, no MDE	π_{A0}	0.407	0.007	(0.393, 0.421)	1.87
Male, MDE	π_{A1}	0.072	0.003	(0.066, 0.080)	1.64
Female, no MDE	π_{B0}	0.402	0.005	(0.391, 0.413)	1.11
Female, MDE	π_{B1}	0.120	0.003	(0.114, 0.126)	0.81
		Row Proportions			
No MDE | Male	$\pi_{0\|A}$	0.849	0.008	(0.833, 0.864)	2.08
MDE | Male	$\pi_{1\|A}$	0.151	0.008	(0.136, 0.167)	2.08
No MDE | Female	$\pi_{0\|B}$	0.771	0.006	(0.759, 0.782)	0.87
MDE | Female	$\pi_{1\|B}$	0.229	0.006	(0.218, 0.241)	0.87

Source: NCS-R

The estimated proportions, standard errors, 95% CIs, and design effect output from these two commands are summarized in Table 6.5, where we see evidence of a higher prevalence of lifetime MDEs among females.

6.4.3 Estimating and Testing Differences in Subpopulation Proportions

Estimates of row proportions in two-way tables (e.g., $\hat{p}_{1|B} = 0.229$ in Table 6.5) are in fact subpopulation estimates in which the subpopulation is defined by the levels of the factor variable. Analysts interested in testing differences in response category proportions between two levels of a factor variable can use methodology similar to that discussed in Section 5.6 for comparison of subpopulation means.

EXAMPLE 6.7: COMPARING THE PROPORTIONS OF U.S. ADULT MEN AND WOMEN WITH LIFETIME MAJOR DEPRESSION

This example uses data from the NCS-R to test a null hypothesis that there is no difference in the proportions of U.S. adult men and women with a lifetime diagnosis of a major depressive episode. To compare male and female row proportions for MDE = 1, the `svy: prop` command with the `over()` option is used to estimate the vector of row proportions (see Table 6.5 above) and their design-based variance-covariance matrix. The `lincom` command is then executed to estimate the contrast of the male and female proportions and the standard error of this difference. The relevant Stata commands follow:

```
svyset seclustr [pweight=ncsrwtsh], strata(sestrat)
svy: proportion mde, over(sex)
lincom 1.mde@1.sex - 1.mde@2.sex
```

The output of the `lincom` command provides the following estimate of the male-female difference, its standard error, and a 95% CI for the contrast of proportions.

$\hat{\Delta} = p_{male} - p_{female}$	$se(\hat{\Delta})$	$CI_{.95}(\Delta)$
−0.079	0.010	(−0.098, −0.059)

Because the design-based 95% CI for the difference in proportions does not include 0, the data suggest that the prevalence of lifetime major depression episodes for women is in fact significantly higher than that for men.

Alternative Bayesian Approach

To implement the alternative Bayesian approach to testing this difference in proportions for significance, we frame the analysis as a logistic regression model with a single predictor (an indicator for being male) and a binary dependent variable (MDE). We then simulate draws from the survey-weighted posterior distribution for the intercept and the regression coefficient for being male in this model, and adjust them for the NCS-R complex sample design effects (see Section 3.8 for details). Our focus is on the 95% credible interval for the regression coefficient (assuming a flat prior for the coefficient). We carry out this analysis with the following R code (note the use of `family = bernoulli()` to fit the logistic model):

```
# read in the full NCS-R data set for the third edition
load("X:/. . ./ncsr.rdata")

# create complete data set with variables of interest
ncsr.red <- ncsr[, c("mde", "sexm", "seclustr", "sestrat", "ncsrwtsh")]
ncsr.red <- ncsr.red[complete.cases(ncsr.red),]

# need to normalize the NCS-R weights to match Stan
ncsr.red$wtsc <- ncsr.red$ncsrwtsh /
mean(ncsr.red$ncsrwtsh)

# survey design object
ncsr.des <- svydesign(id = ~seclustr, strata = ~sestrat,
weights = ~wtsc, nest = T, data = ncsr.red)

# Bayesian approach, flat prior
set.seed(41279)
model_formula <- formula("mde|weights(wtsc) ~ sexm")
mod.brms <- cs_sampling_brms(svydes = ncsr.des,
brmsmod = brmsformula(model_formula), data = ncsr.red, family = bernoulli())
mod.brms$stan_fit
```

The resulting mean of the adjusted draws of the coefficient for being male from the survey-weighted posterior distribution is −0.52, which is on the log-odds scale. The 95% credible interval for the coefficient is (−0.62, −0.42), which leads to the same inference: males have significantly reduced odds of an MDE. If we exponentiate this mean, we get an estimate of the ratio of odds of having an MDE for males relative to females: 0.595. This is consistent with the ratio of odds based on the weighted point estimates from the design-based approach: (0.151 / 0.849) / (0.229 / 0.771) = 0.599. See Example 5.15 for the necessary code to specify a prior distribution for this coefficient if one is available.

6.4.4 Chi-Square Tests of Independence of Rows and Columns

For a 2 × 2 table, the contrast of estimated subpopulation proportions examined in Example 6.7 above is equivalent to a test of whether the response variable (MDE) is independent of the factor variable (SEX). More generally, under SRS, two categorical variables are independent of each other if the following result for the expected proportion in row r, column c of the cross-tabulation holds:

$$\hat{\pi}_{rc} = \text{Expected proportion in row } r\text{, column } c = \frac{n_{r+}}{n_{++}} \cdot \frac{n_{+c}}{n_{++}} = p_{r+} \cdot p_{+c} \tag{6.6}$$

Under SRS, formal testing of the null hypothesis that two categorical variables are independent can be conducted by comparing the expected cell proportion under the independence assumption, $\hat{\pi}_{rc}$, to the observed proportion from the survey data, p_{rc}. Intuitively, if the differences ($\hat{\pi}_{rc}$-p_{rc}) are large, then there is evidence that the independence assumption does not hold and that there is an association between the row and column variables. In standard practice, two statistics are commonly used to test the null hypothesis of independence in two-way tables:

$$\text{Pearson's Chi-square Test Statistic: } X^2_{Pearson} = n_{++} \cdot \sum_r \sum_c (p_{rc} - \hat{\pi}_{rc})^2 / \hat{\pi}_{rc} \tag{6.7}$$

$$\text{Likelihood Ratio Test Statistic: } G^2 = 2 \cdot n_{++} \cdot \sum_r \sum_c p_{rc} \times \ln\left(\frac{p_{rc}}{\hat{\pi}_{rc}}\right) \tag{6.8}$$

Under the null hypothesis of independence for rows and columns of a two-way table, these test statistics both follow a central χ^2 distribution with $(R - 1) \times (C - 1)$ degrees of freedom.

Ignoring the complex sample designs that underlie most survey data sets can introduce bias in the estimated values of these test statistics. To correct the bias in the estimates of the population proportions used to construct these test statistics, weighted estimates of the cell, row, and column proportions are substituted, for example, $p_{rc} = \hat{N}_{rc} / \hat{N}_{++}$. To correct for the design effects on the sampling variances of these proportions, two general approaches have been introduced in the survey statistics literature. Both approaches involve scaling the standard $X^2_{Pearson}$ and G^2 test statistics by dividing them by an estimate of a **generalized design effect factor**, GDEFF. Theory Box 6.2 provides a mathematical explanation of how the generalized design effect adjustments are computed.

Fay (1979, 1985) and Fellegi (1980) were among the first to propose such a correction based on a generalized design effect. Rao and Scott (1984) and later Thomas and Rao (1987) extended the theory of generalized design effect corrections for these test statistics. The

Rao-Scott method requires the computation of **generalized design effects** that are analytically more complicated than the Fellegi approach. The Rao-Scott corrections are now the standard in procedures for the analysis of categorical survey data in software packages such as Stata and SAS. The design-adjusted Rao-Scott Pearson and Likelihood Ratio chi-square test statistics are computed as follows:

$$X^2_{R-S} = X^2_{Pearson} / GDEFF, \qquad G^2_{R-S} = G^2 / GDEFF \tag{6.9}$$

Under the null hypothesis of independence of rows and columns, both of these adjusted test statistics can be referred to a χ^2 distribution with $(R - 1) \times (C - 1)$ degrees of freedom. Thomas and Rao (1987) showed that a transformation of the design-adjusted X^2_{R-S} and G^2_{R-S} values produced a more stable test statistic that under the null hypothesis closely approximated an F distribution. Table 6.6 below defines the F-transformed versions of these two chi-square test statistics and the corresponding F reference distributions to be used in testing the independence of rows and columns:

TABLE 6.6

F-Transformations of the Rao-Scott Chi-square Test Statistics

F-Transformed Test Statistics	F Reference Distribution under H_0
$F_{R\text{-}S,Pearson} = X^2_{R-S} / df$ *where* : df is the design degrees of freedom	$\mathcal{F}_{(R-1)(C-1),(R-1)(C-1)\cdot df}$
$F_{R-S,LRT} = G^2_{R-S} / (C-1)$ *where* : C is the number of columns in the crosstab	$\mathcal{F}_{(C-1),df}$

THEORY BOX 6.2: FIRST- AND SECOND-ORDER DESIGN EFFECT CORRECTIONS

The Fellegi (1980) method for a generalized design effect correction to the chi-square test statistic is best summarized as a three-step process. First, the average of the design effects for the $R \times C$ (unweighted) proportions involved in the computation of the chi-square statistic is computed. The standard Pearson or Likelihood Ratio chi-square statistic computed under simple random sampling is then divided by the average design effect. The resulting adjusted chi-square test statistic is referred to a χ^2 distribution with degrees of freedom equal to $(R - 1) \times (C - 1)$ to test the null hypothesis of independence.

Rao and Scott (1984) built on this method by advocating the use of weighted estimates of the proportions in the construction of the standard chi-square statistics. Under the Rao-Scott method, the **generalized design effect** is defined as the mean of the eigenvalues of the following matrix, $\boldsymbol{D}$:

$$\boldsymbol{D} = \boldsymbol{V}_{Design} \boldsymbol{V}^{-1}_{SRS} \tag{6.10}$$

In (6.10), $\boldsymbol{V}_{Design}$ is the matrix of design-based (e.g., linearized) variances and covariances for the $R \times C$ vector of estimated proportions used to construct the chi-square test statistic, and $\boldsymbol{V}_{SRS}$ is the matrix of variances and covariances for the estimated proportions given a simple random sample of the same size. The Rao-Scott generalized design effect factor, *GDEFF*, for two-way tables can then be written as follows:

$$GDEFF = \frac{\sum_r \sum_c (1-p_{rc}) \cdot d^2(p_{rc}) - \sum_r (1-p_{r+}) \cdot d^2(p_{r+}) - \sum_c (1-p_{+c}) \cdot d^2(p_{+c})}{(R-1)(C-1)} \quad (6.11)$$

The design-adjusted test statistics introduced in (6.9) are computed based on **first-order design corrections** of this type.

Thomas and Rao (1987) derived **second-order design corrections** to the test statistics, which incorporate variability in the eigenvalues of the $\boldsymbol{D}$ matrix. These second-order design corrections can be implemented by dividing the adjusted test statistic based on the first-order correction (e.g.,$X^2_{R-S} = X^2_{Pearson} / GDEFF$) by the quantity $(1 + a^2)$, where a represents the coefficient of variation of the eigenvalues of the $\boldsymbol{D}$ matrix. The F-transformed version of this second-order design-corrected version of the Pearson chi-square statistic is currently the default test statistic reported by Stata's `svy: tab` command for analyses of two-way tables. Thomas and Rao (1987) used simulations to show that this second-order correction controls Type I error rates much better when there is substantial variance in the eigenvalues of $\boldsymbol{D}$.

In theoretical terms, the aim in applying the first- or second-order complex sample design corrections to the standard Chi-square test statistic is to transform the test statistic such that under the null hypothesis of independence, H_0, the moments (i.e., mean, variance) of its distribution will match those of the central χ^2 reference distribution with $(R - 1) \times (C - 1)$ degrees of freedom (Lohr, 2021).

A third form of the chi-square test statistic that may be used to test the null hypothesis of independence of rows and columns in a cross-tabulation of two categorical variables is the **Wald chi-square test statistic** (see Theory Box 6.3). We will see in later chapters that Wald statistics play an important role in hypothesis testing for linear and generalized linear models. However, simulation studies have shown that the standard Pearson chi-square test statistic and its design-adjusted forms proposed by Rao and Scott (1984) and Rao and Thomas (1988) perform best for both sparse and non-sparse tables (Sribney, 1998), and that these tests are more powerful than the Wald test statistic, especially for larger tables. As a result, Stata makes the $F_{R\text{-}S,Pearson}$ test statistic (with a second-order design correction incorporated) the default test statistic reported by its `svy: tab` command, and this is also the default test statistic reported by PROC SURVEYFREQ in SAS (with a first-order design correction incorporated).

THEORY BOX 6.3: THE WALD CHI-SQUARE TEST OF INDEPENDENCE FOR CATEGORICAL VARIABLES

The Wald chi-square test statistic for the null hypothesis of independence of rows and columns in a two-way table is defined as follows:

$$Q_{Wald} = \hat{Y}'(H\hat{V}(\hat{N})H')^{-1}\hat{Y}, \tag{6.12}$$

$$\text{where } \hat{Y} = (\hat{N} - E) \tag{6.13}$$

is a vector of $R \times C$ differences between the observed and expected cell counts, that is, $\hat{N}_{rc} - E_{rc}$, where under the independence hypothesis, $E_{rc} = \hat{N}_{r+} \cdot \hat{N}_{+c} / \hat{N}_{++}$. The matrix term $H\hat{V}(\hat{N})H'$ represents the estimated variance-covariance matrix for the vector of differences. In the case of a complex sample design, the variance-covariance matrix of the weighted frequency counts, $\hat{V}(\hat{N})$, is estimated using a TSL, BRR, JRR, or bootstrapping approach that captures the effects of stratification, cluster sampling, and weighting.

Under the null hypothesis of independence, Q_{Wald} follows a χ^2 distribution with $(R-1) \times (C-1)$ degrees of freedom. An F-transformation of the Wald chi-square test statistic reported in SUDAAN and other software programs is:

$$F_{Wald} = Q_{Wald} \times \frac{df - (R-1)(C-1) + 1}{(R-1)(C-1)df} \sim \mathcal{F}_{(R-1)(C-1),\, df-(R-1)(C-1)+1} \text{ under } H_0. \tag{6.14}$$

EXAMPLE 6.8: TESTING THE INDEPENDENCE OF MDE AND GENDER IN U.S. ADULTS USING THE NCS-R DATA

In Example 6.7, based on both the estimated 95% confidence interval for the difference in proportions of males and females with a lifetime major depression episode and the 95% credible interval for the same difference (according to the Bayesian approach), we concluded that U.S. adult females appear to have significantly higher rates of lifetime major depression episodes (MDEs) than their male counterparts. An equivalent way of interpreting that result is to say that adult prevalence of MDE is not independent of gender. A second approach to address the question of gender-specific differences in the prevalence of lifetime major depression is to perform the Rao-Scott Chi-square test of the null hypothesis H_0: MDE prevalence is independent of gender. The following Stata command can be used to perform a design-adjusted test of that null hypothesis:

```
svyset seclustr [pweight=ncsrwtsh], strata(sestrat)
svy: tab sex mde, se ci deff
```

A summary of the results of this analysis is presented in Table 6.7. By default, Stata reports the standard uncorrected Pearson chi-square test statistic ($X^2_{Pearson} = 92.15$, $p < 0.0001$) and then reports the (second-order) design-adjusted Rao-Scott F-test statistic ($F_{R\text{-}S,Pearson} = 57.98$, $p < .0001$). In this case, both the unadjusted X^2 and the design-corrected Rao-Scott F-test statistics provide strong evidence to reject the null hypothesis and conclude that there is a significant association between MDE prevalence and gender.

As in the example, applying the design adjustment to the test statistic may not cause the analyst to alter their conclusion regarding the significance of an association between two categorical variables. However, as Example 6.9 will show, a failure to properly account for complex sample design effects can lead to erroneous conclusions—the error almost always being to assume that an association exists when in fact correct treatment of the data do not support that finding.

TABLE 6.7

Design-based Analysis of the Association between Gender and Lifetime Experience of Major Depression

Tests of Independence			
Unadjusted X^2	$P(\chi^2_{(1)} > X^2_{Pearson})$	Rao-Scott F	$P(F_{1,42} > F_{R\text{-}S})$
$X^2_{Pearson} = 92.15$	$p < 0.0001$	$F_{R\text{-}S,Pearson} = 57.98$	$p < 0.0001$
Parameters of the Rao-Scott Design-Adjusted Test			
$n = 9{,}282$	Design $df = 42$	$GDEFF = 1.37$	$a = 0.00$

(*Source:* NCS-R)

EXAMPLE 6.9: TESTING THE INDEPENDENCE OF ALCOHOL DEPENDENCE AND EDUCATION LEVEL IN YOUNG ADULTS (AGES 18–28) USING THE NCS-R DATA

This example uses the `svy: tab` command in Stata to compute the Rao-Scott F-statistics and test the independence of two categorical variables that are available in the NCS-R data set (for Part II respondents, requiring use of the Part II weight NSCRWTLG): ALD, an indicator of receiving a diagnosis of alcohol dependence in the lifetime, and ED4CAT, a categorical variable measuring educational attainment (1 = less than high school, 2 = high school, 3 = some college, 4 = college and above). The analysis is restricted to the subpopulation of NCS-R Part II respondents 18–28 years of age. After identifying the complex design features to Stata, we request the cross-tabulation analysis and any related design-adjusted test statistics by using the `svy: tab` command:

```
svyset seclustr [pweight=ncsrwtlg], strata(sestrat)
svy, subpop(if 18 <= age & age < 29): tab ED4CAT ald, ///
row se ci deff
```

ED4CAT is specified as the row (factor) variable and ALD as the column (response) variable. Weighted estimates of the row proportions are requested using the `row` option. Table 6.8 summarizes the estimated row proportions and standard errors for the ALD × ED4CAT cross-tabulation, along with the Rao-Scott F-test of independence.

TABLE 6.8

Design-based Analysis of the Association between Alcohol Dependence and Education Level for Young Adults Ages 18–28

Education Level (Grades)	**Alcohol Dependence (ALD) Row Proportions (Linearized SE)** 0 = No	1 = Yes	Total
0–11	0.909 (0.029)	0.091 (0.029)	1.000
12	0.951 (0.014)	0.049 (0.014)	1.000
13–15	0.951 (0.010)	0.049 (0.010)	1.000
16+	0.931 (0.014)	0.069 (0.014)	1.000
Total	0.940 (0.009)	0.060 (0.009)	1.000
Tests of Independence			
Unadjusted X^2	$P(\chi^2_{(3)} > X^2_{Pearson})$	Rao-Scott F	$P(F_{2.75,\,115.53} > F_{R\text{-}S})$
$X^2_{Pearson} = 27.21$	$p < 0.0001$	$F_{R\text{-}S,Pearson} = 1.65$	$p = 0.186$
Parameters of the Rao-Scott Design-Adjusted Test			
$n_{18\text{–}28} = 1{,}275$	Design $df = 42$	$GDEFF = 6.63$	$a = 0.56$

(*Source:* NCS-R)

An estimated 9.1% of young adults in the lowest education group have been diagnosed with alcohol dependence at some point in their lifetime (95% CI = 4.7%, 17.0%), while an estimated 6.9% of young adults in the highest education group have been diagnosed with alcohol dependence (95% CI = 4.6%, 10.2%). By default, Stata reports the standard uncorrected Pearson chi-square test statistic ($X^2_{Pearson} = 27.21, p < 0.0001$) and then reports the (second-order) design-adjusted Rao-Scott *F*-test statistic ($F_{R-S,Pearson} = 1.65, p = 0.186$) (see Table 6.8).

The standard Pearson X^2 test rejects the null hypothesis of independence at $\alpha = 0.05$; however, when the corrections for the complex sample design are introduced, the Rao-Scott design-adjusted test statistic fails to reject a null hypothesis of independence between education and a lifetime diagnosis of alcohol dependence in this younger population. The appropriate inference in this case would thus be that there is no evidence of a bivariate association between these two categorical factors in this subpopulation. Multivariable analyses examining additional potential predictors of alcohol dependence could certainly be examined at this point (see Chapter 8 for examples).

We remind readers that Stata is using a second-order design correction for the test statistic, which is why the results of these analyses may differ from those found using other software packages (note the decimal degrees of freedom for the design-adjusted *F*-statistic, due to the second-order correction). If a user specifies the `deff` option in the `svy: tab` command, Stata also reports both the mean generalized design effect (*GDEFF* = 6.63) used in the first-order correction and the coefficient of variation of the generalized design effects (*a* = 0.56) used in the second-order correction.

Additional test statistics, including design-adjusted likelihood ratio and Wald test statistics, can be requested in Stata by using the `lr` and `wald` options for the `svy: tab` command. These options do not lead to substantially different conclusions in this illustration and will generally not lead to different inferences about associations between two categorical variables. As mentioned previously, Stata developers advocate the use of the second-order design-adjusted Pearson chi-square statistic (or the Rao-Scott chi-square statistic and its *F*-transformed version) in all situations involving cross-tabulations of two categorical variables measured in complex sample surveys (Sribney, 1998).

6.4.5 Odds Ratios and Relative Risks

The **odds ratio**, which we denote by ψ, can be used to quantify the association between the levels of a **response variable** and a categorical factor. Figure 6.7 displays NCS-R weighted estimates (row proportions) of the prevalence of one or more lifetime episodes of major depression by gender (from Table 6.5).

SEX	MDE		
	0	1	
A - Male	$p_{0\|A} = \frac{\hat{N}_{A0}}{\hat{N}_{A+}} = 0.849$	$p_{1\|A} = \frac{\hat{N}_{A1}}{\hat{N}_{A+}} = 0.151$	$p_{A+} = 1.0$
B - Female	$p_{0\|B} = \frac{\hat{N}_{B0}}{\hat{N}_{B+}} = 0.771$	$p_{1\|B} = \frac{\hat{N}_{B1}}{\hat{N}_{B+}} = 0.229$	$p_{B+} = 1.0$

FIGURE 6.7
Estimates of Row Proportions for MDE, by Gender.

We briefly introduced the odds ratio in Example 6.7. The odds ratio compares the odds that the response variable takes a specific value across two levels of the factor variable. If the response variable is truly independent of the chosen factor, then $\psi = 1.0$. For example, from Figure 6.7, the estimated male (A)/female (B) odds ratio for MDE is:

$$\hat{\psi} = \frac{Odds(MDE=1\,|\,Male)}{Odds(MDE=1\,|\,Female)} = \frac{p_{1|A}/(1-p_{1|A})}{p_{1|B}/(1-p_{1|B})} = \frac{p_{1|A}/p_{0|A}}{p_{1|B}/p_{0|B}} = \frac{0.151/0.849}{0.229/0.771} = 0.599$$

Note that although this estimate of ψ is computed using the estimated row proportions for the SEX × MDE table, the same estimate would be obtained if the estimated total proportions had been used (Table 6.5):

$$\hat{\psi} = \frac{p_{A1}/p_{A0}}{p_{B1}/p_{B0}} = \frac{0.072/0.407}{0.120/0.402} = 0.599$$

Since this odds ratio is estimated with no additional controls for other factors such as age or education, it would be labeled as an estimate of an **unadjusted odds ratio** in any scientific reports. Note that a correct description of this result is the following: "The *odds* that adult men experience major depression in their lifetime are estimated to be only 59.9% as large as the odds for women." A common mistake in reporting results for estimated odds ratios is to make a statement like the following: "The *probability* that a man experiences an episode of major depression in their lifetime is 59.9% of that for women."

The latter statement above is confusing the odds ratio statistic with a related, yet different, comparative measure: the **relative risk** (computed here using the estimates in Table 6.5):

$$\hat{RR} = \frac{\text{Prob}(MDE=1\,|\,Male)}{\text{Prob}(MDE=1\,|\,Female)} = \frac{p_{1|A}}{p_{1|B}} = \frac{0.151}{0.229} = 0.659$$

The relative risk is the ratio of two conditional probabilities: the probability of MDE for males and the probability of MDE for females. Although both the odds ratio and the relative risk measure the association of a categorical response and a factor variable, they should be distinguished. Only in instances where the prevalence of the response of interest is very small for all levels of the factor (i.e.,$p_{1|A}$ and $p_{1|B} < 0.01$) will the odds ratio and relative risk statistics converge to similar numerical values.

A related quantity that is often of interest to epidemiologists is the population **attributable fraction (AF)** associated with a particular risk factor when analyzing categorical outcomes (e.g., disease status). The AF, which is a function of estimated relative risks associated with different levels of a categorical risk factor (all relative to some baseline category, usually representing no exposure to the risk factor), represents the proportional reduction in population disease (or mortality) that would occur if exposure to a risk factor were reduced to some baseline level, ideally representing no exposure to the risk factor. Heeringa et al. (2015a) reviewed the literature on estimation of AFs, and presented straightforward approaches to estimating (and making inferences about) AFs based on complex sample survey data (see also Section 8.6).

If the response and factor variables are independent, then $\psi = 1.0$ (and RR = 1.0). Therefore, to test if categorical response and factors variables are independent, it would be reasonable to construct a confidence interval of the form $\hat{\psi} \pm t_{1-\alpha/2,df} \cdot se(\hat{\psi})$, and establish whether (or not) the null value of ψ=1 is contained within the interval. Although a TSL approximation to $se(\hat{\psi})$ can be derived directly, a CI for ψ is generally obtained from the technique of simple logistic regression (see below).

6.4.6 Simple Logistic Regression to Estimate the Odds Ratio

Logistic regression for binary dependent variables will be covered in depth in Chapter 8. Here, the logit function and simple logistic regression models are briefly introduced to demonstrate their application to estimation of the unadjusted odds ratio and its confidence interval (see also Example 6.7).

The natural logarithm of the odds is termed a **logit function** (arising from the term "**log**-odds un**its**"). Again, using the NCS-R MDE example in Table 6.5, the logits of the probabilities of MDE for the male and female factor levels are:

$$\text{logit}(p_{1|A}) = \ln(Odds(MDE=1 \mid Male)) = \ln\left(\frac{p_{1|A}}{1-p_{1|A}}\right) = \ln\left(\frac{0.151}{0.849}\right) = -1.727$$

$$\text{logit}(p_{1|B}) = \ln(Odds(MDE=1 \mid Female) = \ln\left(\frac{p_{1|B}}{1-p_{1|B}}\right) = \ln\left(\frac{0.229}{0.771}\right) = -1.214$$

Consider a single indicator variable, I_{male}, coded 1 = male and 0 = female, that distinguishes the two levels of SEX. The outcome MDE is coded 1 = yes, 0 = no. A simple logistic regression model for these data is written as follows:

$$\text{logit}(P(MDE=1 \mid SEX)) = \beta_0 + \beta_1 \cdot I_{male}$$

Then, we can derive the following result:

$$\hat{\psi} = \frac{p_{1|A}/(1-p_{1|A})}{p_{1|B}/(1-p_{1|B})} = \frac{\exp(\text{logit}(p_{1|A}))}{\exp(\text{logit}(p_{1|B}))} = \frac{\exp(\hat{\beta}_0 + \hat{\beta}_1 \cdot 1)}{\exp(\hat{\beta}_0 + \hat{\beta}_1 \cdot 0)} = \exp(\hat{\beta}_1)$$

Therefore, given an estimate of the parameter β_1 that would be obtained by fitting the simple logistic regression model to the survey data, we can easily compute an estimate of the odds ratio. A 100(1-α)% CI for ψ can then be developed by exponentiating the lower and upper bounds of the standard CI for β_1: $CI(\psi) = \left(\exp(\hat{\beta}_1 - t_{1-\alpha/2,df} \cdot se(\hat{\beta}_1)), \exp(\hat{\beta}_1 + t_{1-\alpha/2,df} \cdot se(\hat{\beta}_1))\right)$. The resulting confidence interval is not symmetric about the estimated odds ratio but provides more accurate coverage of the true population value for a specified level of Type I error (α).

EXAMPLE 6.10: SIMPLE LOGISTIC REGRESSION TO ESTIMATE THE NCS-R MALE/FEMALE ODDS RATIO FOR LIFETIME MAJOR DEPRESSIVE EPISODE

As mentioned previously, logistic regression will be covered in detail in later chapters. Here, a simple logistic regression of the NCS-R MDE variable on the indicator of male gender (SEXM) is used to illustrate the technique for estimating the unadjusted Male/Female odds ratio for MDE and a 95% CI for that odds ratio:

```
svyset seclustr [pweight = ncsrwtsh], strata(sestrat)
svy: logistic mde sexm
```

From the output provided by the `svy: logistic` command, the estimated odds ratio and a 95% CI for the population odds ratio are:

$\hat{\psi}_{MDE}$(SE)	$CI_{.95}(\psi)$
0.597 (0.041)	(0.520, 0.685)

Based on this analysis, the odds that an adult male has experienced a lifetime MDE are only 59.7% as large as the odds of MDE for adult females, which agrees (allowing for some rounding error) with the simple direct calculation above. Since the 95% CI does not include $\psi = 1$, we would reject the null hypothesis that MDE status is independent of gender.

6.4.7 Bivariate Graphical Analysis

Graphical displays are also useful tools to describe the bivariate distribution of two categorical variables. The Stata graphics command below produces gender-specific vertical bar charts for the ESS-Russia Federation MARCAT1-MARCAT3 variables generated in Example 6.3 above (note that the `pweight` option is used to specify the survey weights, and the `over()` option is used to generate a plot for each level of gender). The resulting graphic is shown in Figure 6.8.

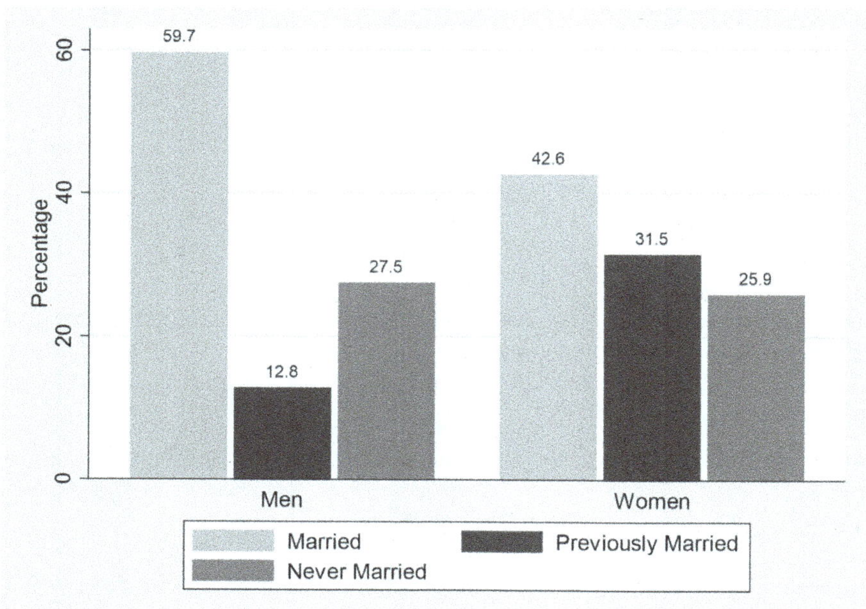

FIGURE 6.8
Bar Chart of the Estimated Distributions of Marital Status of Russian Adult Men and Women. ESS Round 6 (2012)—Russian Federation.

```
label define gndr 1 "Men" 2 "Women"
label values gndr gndr

graph bar (mean) marcat1 marcat2 marcat3 ///
[pweight=pspwght], percentages ///
 bar(1,color(gs12)) bar(2,color(gs4)) ///
 bar(3,color(gs8)) blabel(bar, format(%5.1f)) bargap(7) ///
 scheme(s2mono) over(gndr) ///
 legend (label(1 "Married") ///
 label(2 "Previously Married") ///
 label(3 "Never Married")) ytitle("Percentage")
```

6.5 Analysis of Multivariate Categorical Data

During the past 25 years, multivariate analysis involving three (or more) categorical variables has increasingly shifted to regression-based methods for generalized linear models (Chapters 8 and 9). The regression framework provides the flexibility to estimate the association of categorical responses and factors as well as the ability to adjust for continuous covariates. In this section, we briefly review the adaptation of two long-standing techniques for the analysis of multivariate categorical data to complex sample survey data: the Cochran-Mantel-Haenszel test, and simple log-linear modeling of the expected proportions of counts in multi-way tables defined by cross-classifications of categorical variables.

6.5.1 The Cochran-Mantel-Haenszel Test

Commonly used in epidemiology and related health sciences, the Cochran-Mantel-Haenszel (CMH) test permits tests of the association between two categorical variables while controlling for the categorical levels of a third variable (Cochran, 1954; Mantel & Haenszel, 1959). For example, an analyst may be interested in testing the association between a lifetime diagnosis of major depressive episode and gender while controlling for age categories. Although not widely available in software procedures that support complex sample survey data analysis, design-based versions of the CMH test are available in SUDAAN's CROSSTAB procedure.

PROC CROSSTAB in SUDAAN supports two alternative methods for estimating the adjusted or **common odds ratio** and **common relative risk** statistics: the Mantel-Haenszel (M-H) method and the logit method. Both methods adjust for the complex sample design and generally result in very similar estimates of common odds ratios and relative risks.

EXAMPLE 6.11: USING THE NCS-R DATA TO ESTIMATE AND TEST THE ASSOCIATION BETWEEN GENDER AND DEPRESSION IN THE U.S. ADULT POPULATION WHEN CONTROLLING FOR AGE

In Examples 6.7, 6.8 and 6.10, we employed multiple approaches to analyzing the relationship of sex and MDE. In each case, we found evidence of a significant overall association between gender and a diagnosis of lifetime depression when analyzing the NCS-R data, where females had greater odds of receiving a diagnosis of depression at some point in their lives. This example is designed to test whether or not this association holds in the U.S. adult population when controlling for age. Given that the CMH test can be applied when the control variable is a categorical variable, a four-category age variable named AGECAT is constructed: 1 = ages 18–29, 2 = ages 30–39, 3 = ages 40–49, and 4 = ages 50+. The SUDAAN CROSSTAB procedure is then run to derive the CMH test and to use both the Mantel-Haenszel and logit methods to estimate the common age-adjusted male/female odds ratio and relative risk for MDE.

```
title "ASDA, Example 6.11: Test of Association between
Gender and MDE When Controlling for Age (NCS-R)";
options ls=120 ps=64;
proc crosstab data=ncsr filetype=sas deft1;
nest sestrat seclustr;
weight ncsrwtsh;
class agecat mde sexm / nofreq;
tables agecat*sexm*mde;
risk mhor mhrrl lor lrrl;
test cmh chisq;
print nsum wsum rowper serow colper secol / stest=default
atest=default adjrisk=all style=nchs;
run;
```

Note that the sampling error codes (NEST statement) and the final survey weights (WEIGHT statement) are identified first, with the cluster codes nested within the strata. Next, the CLASS statement specifies that all three variables are categorical. In the TABLES statement, the AGECAT variable is identified first, defining it as the categorical control variable for this analysis. We define SEXM as the row variable, and MDE as the column variable. The RISK statement then requests estimates of the Mantel-Haenszel common odds ratio and relative risk ratio, and the logit-based common odds and relative risk ratios. Finally, the TEST statement requests the overall design-adjusted CMH test, in addition to the Wald chi-square tests which will be performed for each age stratum. Table 6.9 summarizes the key elements of the SUDAAN output.

The value of the design-adjusted CMH test statistic in this case is $X^2_{CMH} = 94.26$, which has $p < 0.0001$ on one degree of freedom, suggesting that there is still a strong association between gender and lifetime depression after adjusting for age. This is supported by the design-adjusted Wald chi-square statistics produced by SUDAAN for each age group (see Equation 6.12). In each age stratum, there is evidence of a significant association of gender with lifetime depression. After adjusting for age, both the M-H and logit estimates of the common male/female odds ratio are about $\hat{\psi}_{MH|Age} \approx 0.60$. This can be

TABLE 6.9

SUDAAN Output for the Cochran-Mantel-Haenszel (CMH) Test of MDE vs. SEX, Controlling for AGECAT

Cochran-Mantel-Haenszel Test Results				
$X^2_{CMH} = 94.26$			$df = 1$	$p < 0.0001$
Age Category-Specific Wald Tests of Independence for MDE and SEX				
Age Category	18–29	30–39	40–49	50+
Q_{Wald}	23.47	23.22	9.12	38.28
$P(Q_{Wald} > \chi^2_1)$	$p < 0.0001$	$p < 0.0001$	$p = 0.0043$	$p < 0.0001$
Age-Adjusted Estimates of Common Male/Female Odds Ratios and Relative Risks				
Statistic	ψ_{M-H}	ψ_{Logit}	RR_{M-H}	RR_{Logit}
Point estimate	0.59	0.60	0.91	0.91
95% CI	(0.51, 0.67)	(0.52, 0.68)	(0.88, 0.93)	(0.89, 0.93)

compared to the estimated odds ratio, $\hat{\psi}_{Unadj} \approx 0.60$ from Example 6.10 when no adjustment for age is included. Males appear to have roughly 40% lower odds of having a lifetime diagnosis of depression compared to females, when adjusting for age. The M-H and logit method estimates of the common risk ratios also suggest a significant difference in the *probability* of depression for the two groups, with the expected probability about 10% lower for males when adjusting for age.

6.5.2 Log-Linear Models for Contingency Tables

A text on the analysis of survey data would not be complete without a mention of log-linear models for multi-way contingency tables (Bishop et al., 1975; Agresti, 2012; Stokes et al., 2002). Log-linear models permit analysts to study the **association structure** among categorical variables. In a sense, a log-linear model for categorical data is analogous to an ANOVA model for the cell means of a continuous dependent variable. The dependent variable in the log-linear model is the natural logarithm of the expected counts (or multinomial probabilities) for cells of a multi-way contingency table. The model parameters are estimated effects associated with the categorical variables and their interactions.

For example, the following is the log-linear model under the null hypothesis of independence for three categorical variables X, Y, and Z:

$$\log(p_{ijk}) = \mu + \lambda_i^X + \lambda_j^Y + \lambda_k^Z \tag{6.15}$$

where: p_{ijk} = *the expected cell proportion under the model;*

$$\mu = \log(p_0) = 1/(\#\,cells)$$

A model that includes a first order interaction between the X and Y variables would be written as:

$$\log(p_{ijk}) = \mu + \lambda_i^X + \lambda_j^Y + \lambda_k^Z + \lambda_{ij}^{XY} \tag{6.16}$$

The cell proportions are assumed to follow a multinomial distribution and the model parameters are estimated using the method of maximum likelihood or iterative procedures such as **iterative proportional fitting (IPF)** or the **Newton-Raphson method**. Tests of nested models (note that the model in (6.15) is nested in the model in (6.16)) are performed using the **likelihood ratio test**.

Log-linear models for SRS data can be analyzed in virtually every major software package (e.g., PROC CATMOD in SAS). Presently, most software packages such as Stata, SAS, and SPSS do not include a procedure to perform the traditional log-linear modeling of complex sample survey data. The `svyloglin()` function in the R `survey` package (Lumley, 2010) does enable log-linear modeling for complex sample survey data. The IVEWare® (Raghunathan et al., 2001) %SASMOD command employs a generic jackknife repeated replication (JRR) algorithm that in combination with PROC CATMOD can be used for log-linear model estimation and inference with data from complex sample surveys (see Example 6.12 below).

There are two major explanations for why programs for log-linear modeling are not included in most software packages for survey data analysis. The first is that the common, general structure of the input data (grouped cell counts of individual observations) does not lend itself readily to design-based estimation and inference where individual cases may have varying weights. In Skinner et al. (1989), Rao and Thomas discuss the extension of their design-based adjustment for chi-square test statistics to the conventional likelihood ratio tests for log-linear models, but this technique would take on considerable programming complexity as the dimension and number of variations on the association structure in the model increases.

Grizzle et al. (1969) introduced the weighted least squares (GSK) method of estimating log-linear and other categorical data models. This generalized technique was programmed in the GENCAT software (Landis et al., 1976) and required the user to input a design-based variance-covariance matrix for the vector of cell proportions in the full cross-tabular array. Under the GSK method, tests of hypothesis are performed using Wald statistics.

A second explanation for the scarcity of log-linear modeling software for complex sample survey data is that log-linear models for expected cell counts or multinomial proportions can be re-parameterized as logistic regression models (Agresti, 2012), and all of the major software packages have more advanced procedures for fitting logistic and other generalized linear models to complex sample survey data. These models will be considered in detail in Chapters 8 and 9.

EXAMPLE 6.12: A SIMPLE LOG-LINEAR MODEL TO TEST THE ASSOCIATION BETWEEN LIFETIME MAJOR DEPRESSION EPISODE AND GENDER

To illustrate an application of log-linear modeling to complex sample survey data, we return again to the now familiar question of whether the NCS-R data provide evidence of a significant association between gender and a lifetime experience of an episode of major depression. The log-linear model that we will estimate is:

$$\log(p_{ijk}) = \mu + \lambda_i^{MDE} + \lambda_j^{SEX} + \lambda_{ij}^{MDE,SEX}$$

where: p_{ijk} = *the expected proportion in cell (i,j) under the model (multinomial sampling is assumed).*

The IVEWare® %SASMOD program syntax used to fit the model (which requires running IVEware within the SAS software) is:

```
%sasmod(dir=P:\ASDA, setup=new, name=SASMOD _ CATMOD2);
datain ncsr _ a;
cluster seclustr;
stratum sestrat;
weight ncsrwtsh;

proc catmod;
model mde*sexm= _ response _ ;
loglin mde sexm mde*sexm;
run;
```

Note that %SASMOD utilizes the NCS-R strata, cluster, and weight variables to form JRR replicates of the full sample. The program then calls PROC CATMOD in SAS to estimate the model for each replicate and the full sample. The standard JRR formula (Section 3.6.3) is then used to estimate the variances of the model parameter estimates.

Using the R `survey` package (Lumley, 2010), the following R code imports the NCS-R data from Stata, estimates the log-linear model, and generates a test of the interaction between gender and MDE:

TABLE 6.10

IVEWare® %SASMOD/SAS® CATMOD Log-linear Model Analysis of the Association between Gender and Lifetime Experience of Major Depression

Description	Parameter	Parameter Estimate	JRR SE	t Statistic (H_0: λ=0)	Prob(t_{42}>t)
MDE	λ^{MDE}	0.735	0.017	42.326	<0.0001
Gender	λ^{Sex}	0.123	0.012	10.551	<0.0001
MDE x Gender	$\lambda^{MDE,Sex}$	−0.129	0.017	−7.439	<0.0001

(*Source:* NCS-R)

```
# load the haven package
library(haven)
ncsr <- read_dta("P:/. . ./ncsr.dta")
# load the survey package
library(survey)
# create survey design object
ncsrsvyp1 <- svydesign(strata=~sestrat, id=~seclustr,
weights=~ncsrwtsh, data=ncsr, nest=T)
# run null model with mde and sexm
null <- svyloglin(~mde+sexm, ncsrsvyp1)
summary(null)
# update null model with interaction
saturated <- update(null, ~.+mde:sexm)
summary(saturated)
```

Both %SASMOD and the R `svyloglin()` analysis produce identical estimates of the model parameters, with only negligible differences in the standard errors and test statistics, due to the different default methods for complex sample variance estimation—JRR in %SASMOD and Taylor Series Linearization in `svyloglin()`.

Table 6.10 summarizes the results of the analysis. For the 2 × 2 cross-classification of SEX × MDE, the estimated log-linear model for the log of the multinomial proportions in each of the four cells includes: a row effects parameter, $\hat{\lambda}^{sex}$; a column effects parameter, $\hat{\lambda}^{MDE}$; and a parameter for the interaction of row and column, $\hat{\lambda}^{sex,mde}$. Since there are four cell proportions and three parameters to estimate, this loglinear model is "saturated". In this simple 2 × 2 example table, the null hypothesis of interest is whether the cell probabilities are solely a function of the margin variables $\hat{\lambda}^{sex}$ and $\hat{\lambda}^{MDE}$ (i.e., the model of independence), versus the alternative that the interaction parameter, $\hat{\lambda}^{sex,mde}$, also plays a significant role. Therefore, in this simple 2 × 2 example, testing the null hypothesis of independence of MDE and gender is equivalent to testing H0: $\hat{\lambda}^{sex,mde} = 0$.

From Table 6.10, $\hat{\lambda}^{sex,mde} = -0.129$ with standard error $se(\hat{\lambda}^{sex,mde}) = 0.017$, corresponding to a Student t test statistic of $t = -7.439$ ($p < 0.0001$). Consistent with previous analyses, we would again reject the null hypothesis that U.S. adults' lifetime experience of major depression is independent of their gender.

6.6 Summary

Variables for which responses are measured on a binary, multinomial, ordinal, or count scale are the predominant form of data collected in complex sample surveys. In this chapter, we have introduced methods for conducting basic univariate and bivariate analyses of categorical variables from complex sample survey data sets. By introducing the Cochran-Mantel-Haenszel Test, log-linear models, and the logistic regression model, we have opened the door to a larger discussion (see Chapters 8, 9, and 13) of multivariable methods in the analysis of categorical data from complex sample surveys. The developments of these later chapters will focus on regression modeling of categorical responses using methods such logistic or probit regression, multinomial and ordinal logistic regression, and Poisson and negative binomial regression.

6.7 Exercises

6.1) Using the software procedure of your choice and the ESS-Russian Federation data set (see the ASDA website to download the data), estimate the row proportions in a two-way table where marital status (MARCAT, 1='MARRIED', 2='PREVIOUSLY MARRIED', 3='NEVER MARRIED') is the factor variable (or row variable) and Voted in the Last Election (VOTE, 1=YES, 0=NO) is the response variable (or column variable). Make sure to correctly identify and incorporate the sampling error stratum, sampling error cluster, and final sampling weight variables in the ESS Russian Federation data set. Use the weight that incorporates post-stratification and design features. Then, answer the following questions:

a. What is the value of the Rao-Scott F-statistic for the overall test of the null hypothesis that marital status category and voting status are not associated? What are the degrees of freedom for this design-adjusted F-statistic?

b. Under this null hypothesis, what is the p-value for the F reference distribution?

c. Based on this test, what is your statistical decision regarding the association between marital status and voting behavior in the ESS Russian Federation survey population?

d. Do you arrive at the same conclusion when taking a model-based Bayesian approach to the analysis? Make sure to show your code in detail.

6.2) Repeat the analysis from 6.1) above, only performing unconditional subclass analyses for men and women separately (GNDR: 1 = Male 2 = Female). Do your inferences change when restricting the target population to only men or only women?

6.3) Create an indicator variable of high satisfaction with life (called HIGH_SAT) using the STFLIFE variable, coded so that 1 = (STFLIFE=9 OR 10), 0 = (0<=STFLIFE <=8) and missing values are set to system-missing in your software of choice. Then answer the following questions, while using the appropriate complex sample design variables and weights:

a. What proportions of the total population, males, and females in the ESS-RF survey population are estimated to have high satisfaction with life? Make sure to provide three (3) weighted estimates of proportions and design-adjusted standard errors to fully answer this question.

b. Compute design-adjusted 95% CIs or credible intervals for each of the three proportions estimated in part a. Use your software of choice and indicate which of the approaches discussed in the chapter were used to compute the intervals (and why).

c. Is there a significant association between gender and high satisfaction with life? Provide statistical support for your inference.

6.4) Extend the analysis in **6.3)** by conducting an analysis of the association between Gender and High Satisfaction with Life separately for each of the four age groups defined by the variable AGECAT. Then, answer these questions:

a. When the age groups are analyzed separately, does the association (or lack thereof) between Gender and High Satisfaction with Life continue to hold?

b. Provide the design-adjusted *F*-statistic and *p*-value for each of the age-specific analyses to support your answer.

c. If the answer to part a. is "No," how do you explain this pattern of results in plain English?

6.5) Extra Credit Question (use SUDAAN or another software procedure that is able to perform this test): Perform a Cochran-Mantel-Haenszel trend test of the association of Gender and High Satisfaction with Life, controlling for age groups.

a. Prior to running the analysis, create a numeric version of the STRATIFY variable with values of 1–8 corresponding to the eight unique string values of STRATIFY, and name the new variable NSTRAT. If using SUDAAN, sort the data by NSTRAT and PSU before use of the SUDAAN command.

b. What are the values of the Wald-*F* test statistic, degrees of freedom, and *p*-value from the CMH general association Wald-*F* Test?

c. What is your conclusion regarding the significance of this association, controlling for the four age groups?

7

Linear Regression Models

7.1 Introduction

"Study regression. All of statistics is regression."

This quote came as a recommendation from a favorite professor to one of the authors while he was in the process of choosing a concentration topic for his comprehensive exam. The broader interpretation of the quote requires placing the descriptor in quotes, "regression"; but ask individuals with backgrounds as varied as social science graduate students or quality control officers in a paper mill to decipher the statement and they will think first of the linear regression model. Given the importance of the linear regression model in the history of statistical analysis, the emphasis that it receives in applied statistical training, and its importance in real world statistical applications, the narrower interpretation is quite understandable.

This chapter introduces linear regression modeling for complex sample survey data—its similarities and how it differs (theoretically and procedurally) from standard ordinary least squares (OLS) regression analysis. We assume that the reader is familiar with the basic theory and methods for simple (single predictor) and multiple (multiple predictor) linear regression analysis for continuous dependent variables. Readers interested in a comprehensive reference on the topic of linear regression are referred to Draper and Smith (1981), Kleinbaum et al. (1988), Neter et al. (1996), DeMaris (2004), Fox (2008), Faraway (2014), Harrell (2015), or many other excellent texts on the subject.

Focusing on practical approaches for complex sample survey data, we emphasize "aggregated" **design-based approaches** to the linear regression analysis of survey data (sometimes referred to as **population-averaged modeling**), where design-based variance estimates for weighted estimates of regression parameters in *finite* populations are computed using nonparametric methods such as the Taylor Series Linearization (TSL) method, Balanced Repeated Replication (BRR), Jackknife Repeated Replication (JRR), or bootstrapping. We will also consider state-of-the-art Bayesian approaches to fitting population-averaged linear regression models, which incorporate appropriate adjustments to draws from a survey-weighted posterior distribution for the regression parameters that reflect complex sample design effects (see Section 3.8).

More general **model-based approaches** to the linear regression analysis of complex sample survey data, which may explicitly include fixed effects of sampling strata and/or random effects of sampling clusters in **multilevel models** and may or may not utilize the survey weights (e.g., Skinner et al., 1989; Pfeffermann et al., 1998; Little, 2003; Pfeffermann, 2011), are introduced in Chapter 13. Over the years, there have been many contributions to the survey methodology literature comparing design-based and model-based approaches to the regression analysis of survey data, including papers by DuMouchel and Duncan (1983), Hansen et al. (1983), and Kott (1991, 2018).

DOI: 10.1201/9781003472094-8

7.1.1 A Brief History of Design-Based Linear Regression Modeling

We present a brief history of important statistical developments in design-based linear regression analysis of complex sample survey data to begin this chapter. Kish and Frankel (1974) were two of the first to empirically study and discuss the impacts of complex sample designs on inferences related to regression coefficients. Fuller (1975) derived a linearization-based variance estimator for multiple regression models with unequal weighting of observations and introduced variance estimators for estimated regression parameters under stratified and two-stage sampling designs. Shah et al. (1977) further discussed the violations of standard linear model assumptions when fitting linear regression models to complex sample survey data, discussed appropriate methods for making inferences about linear regression parameters estimated using survey data, and presented an empirical evaluation of the performance of variance estimators based on TSL.

Binder (1983) focused on the sampling distributions of estimators for regression parameters in finite populations and defined related variance estimators. Skinner et al. (1989, Sections 3.3.4 and 3.4.2) summarized estimators of the variances for regression coefficients that allowed for complex sample designs (including linearization estimators) and recommended the use of linearization methods or other robust methods (such as JRR) for variance estimation. Kott (1991) further discussed the advantages of using variance estimators based on TSL for estimates of linear regression parameters: protection against within-PSU correlation of random errors, protection against possible non-constant variance of the random errors, and the fact that a within-PSU correlation structure does not need to be identified to have a nearly unbiased estimator.

Fuller (2002) provided a modern summary of regression estimation methods for complex sample survey data, and Pfeffermann (2011) discussed the relative merits of a variety of possible design-based and model-based approaches to fitting linear regression models to complex sample survey data, presenting empirical support for the use of a "q-weighted" method. In this method, the modified survey weights used to compute estimates of the regression coefficients are the original (and possibly adjusted) probability-based weights divided by the *expectations* of the weights as function of the covariates in the model of interest (i.e., predicted values of the weights based on a model regressing the weights on the predictor variables in the model of interest). We consider this approach, which has demonstrated efficiency benefits, in the illustration presented later in this chapter.

More recently, most of the literature in this area has focused on design-adjusted linear regression model diagnostics (Li & Valliant, 2009, 2011a, 2011b, 2015; Liao & Valliant, 2012a, 2012b), and more efficient design-sensitive (Kott, 2018) and Bayesian (e.g., León-Novelo & Savitsky, 2019) approaches to fitting population-averaged linear regression models. We consider these alternative approaches and examine their efficiency benefits throughout this chapter.

7.2 The Linear Regression Model

Regression analysis is a study of the relationships among variables: a **dependent variable** and one or more **independent variables**. Figure 7.1 illustrates a simple linear regression model of the relationship of a dependent variable, y, and a single independent variable x. The regression relationship among the observed values of y and x is expressed as a regression model, for example, $y = \beta_0 + \beta_1 x + \varepsilon$, where y is the dependent variable, x is the

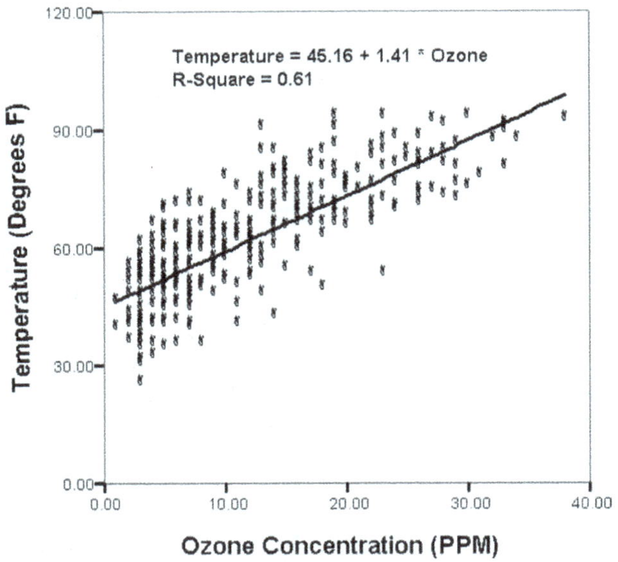

FIGURE 7.1
Linear regression of air temperature on ozone level
(Source: Breiman & Friedman, 1985).

independent variable, β_0 and β_1 are model parameters, and ε is an error term that reflects the difference between the observed value of y and its conditional expectation under the model, $\varepsilon = y - \hat{y} = y - \beta_0 - \beta_1 x$.

In statistical practice, a fitted regression model may be used to simply predict the expected outcome for the dependent variable based on a vector of independent variable measurements $\boldsymbol{x}$, $E(y \mid \boldsymbol{x}) = \hat{\beta}_0 + \hat{\beta}_1 x_1 + \ldots + \hat{\beta}_p x_p$, or to explore the functional relationship of y and x. Across the many scientific disciplines that use regression analysis methods, dependent variables may also be referred to as response variables, regressands, outcomes, or even "left-hand side variables." Independent variables may be labeled as predictors, regressors, covariates, factors, co-factors, explanatory variables, or "right-hand side variables." We primarily refer to response variables and predictor variables in this chapter, but other terms can be used interchangeably.

This chapter will focus on the broad class of regression models known as **linear models**, or models for which the conditional expectation of y given x, $E(y \mid x)$, is a linear function of the unknown parameters. Consider the following three specifications of linear models:

$$y = \beta_0 + \beta_1 x + \varepsilon \tag{7.1}$$

Note in this model that the dependent variable, y, is a linear function of the unknown parameters and the independent variable x. Next, consider this specification:

$$y = \beta_0 + \beta_1 x + \beta_2 x^2 + \varepsilon \tag{7.2}$$

In this model (7.2), the response variable y is still a linear function of the β parameters for x and x^2; however, the linear model defines a *non-linear relationship* between y and x. Finally, consider this third specification:

$$\begin{aligned} y &= \boldsymbol{x\beta} + \varepsilon \\ &= \sum_{j=0}^{p} \beta_j x_j + \varepsilon \\ &= \beta_0 + \beta_1 x_1 + \beta_2 x_2 + \cdots + \beta_p x_p + \varepsilon \end{aligned} \tag{7.3}$$

Here the linear model is first expressed in **vector notation**. Vector notation may be used as an abbreviation to represent a complex model with many parameters and to facilitate computations using the methods of matrix algebra.

When specifying linear regression models, it is useful to be able to reference specific observations on the subjects in a survey data set:

$$y_i = \boldsymbol{x}_i \boldsymbol{\beta} + \varepsilon_i \tag{7.4}$$

where:

$$\boldsymbol{x}_i = \begin{bmatrix} 1 & x_{1i} & \ldots & x_{pi} \end{bmatrix} \text{ and } \boldsymbol{\beta}^T = \begin{bmatrix} \beta_0 & \beta_1 & \ldots & \beta_p \end{bmatrix}$$

In this notation, i refers to sampled element (or respondent) i in a survey data set.

7.2.1 The Standard Linear Regression Model

Standard procedures for unbiased estimation and inference for the linear regression model involve the following assumptions:

1. The model for $E(y \mid x)$ is linear in the parameters [see (7.2) above];
2. Correct model specification—in short, the model includes the predictor variables and regression coefficients that accurately reflect the true model under which the data were generated;
3. $E(\varepsilon_i \mid \boldsymbol{x}_i) = 0$, or the expected value of the residuals given a set of values on the predictor variables is equal to 0;
4. Homogeneity of Variance—$Var(\varepsilon_i \mid \boldsymbol{x}_i) = \sigma^2_{y \cdot x}$, or the variance of the residuals given values on the predictor variables is a constant parameter equal to $\sigma^2_{y \cdot x}$;
5. Normality of residuals (and also y): for continuous outcomes, we assume that $\varepsilon_i \mid \boldsymbol{x}_i \sim N(0, \sigma^2_{y \cdot x})$, or that given values on the predictor variables, the residuals are independently and identically distributed (i.i.d.) as normal random variables with mean 0 and constant variance $\sigma^2_{y \cdot x}$; and
6. Independence of residuals: As a consequence of (5), $Cov(\varepsilon_i, \varepsilon_j \mid \boldsymbol{x}_i, \boldsymbol{x}_j) = 0, i \neq j$, or residuals on different subjects are *uncorrelated* given values on their predictor variables.

There are several implications of these standard model assumptions. First, we can write:

$$\hat{y} = E(y \mid \boldsymbol{x}) = E(\boldsymbol{x\beta}) + E(\varepsilon) = \boldsymbol{x\beta} + 0 = \boldsymbol{x\beta} = \beta_0 + \beta_1 x_1 + \ldots + \beta_p x_p \tag{7.5}$$

This equation for the predicted value of y is the **regression function**, or the expected value of the dependent variable y conditional on a set of values on the predictor variables (of which there are p). Further, we can write:

$$var(y_i \mid \mathbf{x}_i) = \sigma^2_{y \cdot x} \quad (7.6)$$

$$Cov(y_i, y_j \mid \mathbf{x}_i, \mathbf{x}_j) = 0 \quad (7.7)$$

These assumptions, therefore, imply that the dependent variable has constant variance given values on the predictors and that no two values on the dependent variable are correlated given values on the predictors. Putting all these implications together, we have

$$y_i \sim N(\mathbf{x}_i \boldsymbol{\beta}, \sigma^2_{y \cdot x}) \quad (7.8)$$

Values on the dependent variable, y, are therefore assumed to be i.i.d. normally distributed random variables, with a mean defined by the linear combination of the regression parameters and the predictor variables, and a constant variance.

7.2.2 Survey Treatment of the Regression Model

Since the late 1940s and early 1950s, when economists and sociologists (Kendall & Lazarsfeld, 1950; Klein & Morgan, 1951) first applied regression analysis to complex sample survey data, survey statisticians have sought to relate design-based estimation of regression relationships to the standard linear model. The result was the linked concepts of a finite population and the superpopulation model, which are described in more detail in Chapter 3 (see also Theory Box 7.1).

When survey analysts conduct a regression analysis using complex sample survey data, they choose between two targets—the finite population or a more universal superpopulation model—for their estimation and associated inference. Sometimes this is a conscious choice, and sometimes the choice is less conscious and only implicit in the analyst's presentation of the results and the inferences that are drawn.

In theory, weighted, design-based estimation of a regression model permits the survey analyst to make unbiased inferences concerning the coefficients in that model as they exist within the finite population that corresponds to the geographic, demographic, and temporal definitions of the survey population. To extend this inference beyond the survey population requires the analyst to make generalizing assumptions about the relationship of the finite population that has been studied to an overarching superpopulation model that may govern the relationships of the variables of interest. Theory Box 7.1 discusses the relationship of the superpopulation model and the finite population regression function in more detail. In practice, if the survey population is large and the regression model is correctly specified, then the survey data analyst may treat these two conceptual approaches as virtually equivalent (Korn & Graubard, 1999; Skinner et al., 1989).

THEORY BOX 7.1: FINITE POPULATIONS AND SUPERPOPULATION MODELS

In theory, weighted estimation of a linear regression model, $y = \beta_0 + \beta_1 x + \varepsilon$, from complex sample survey data results in unbiased estimates of the regression function,

$E(y \mid x) = B_0 + B_1 x$, where $B = [B_0, B_1]$ are **finite population regression parameters**. If instead of observing a sample of size n of the N population elements a complete census had been conducted, the finite population regression parameter B_1 for this simple "line" could be computed algebraically as follows:

$$B_1 = \frac{\sum_{i=1}^{N}(X_i - \bar{X})(Y_i - \bar{Y})}{\sum_{i=1}^{N}(X_i - \bar{X})^2}$$

The theory suggests that we distinguish between the true regression model parameters, $\boldsymbol{\beta}$, and the particular values, **B**, that characterize the current finite survey population that was sampled for the survey. For example, consider a simple linear regression model of the relationship of years of education with adults' earned income. The model can be interpreted as:

$$E(Income| \, Education) = \beta_0 + \beta_1 \cdot Education(years)$$

or

$$Income| \, Education \sim N(\beta_0 + \beta_1 \cdot Education(years), \sigma^2_{I \cdot E})$$

In contrast, a strict finite population regression interpretation is that for the particular population that has been sampled, the best prediction of adult income, given their education level, is found on the line $E(Income| \, Education) = B_0 + B_1 \times Education(years)$. The concept of a **superpopulation** links these two interpretations by introducing the assumption that although the surveyed population is finite (size N), the individual relationship between income and education in that finite population conforms to an underlying superpopulation model. Therefore, the fixed set of pairs of income and education values in the finite population of N elements is itself a sample from an infinite number of possible data pairs that could be generated by a stochastic superpopulation model, denoted by $\zeta : Income = \beta_0 + \beta_1 \cdot Education(years) + \varepsilon$.

Under the superpopulation model, the finite population parameters **B** will vary about the true model parameters $\boldsymbol{\beta}$. However, the bias of making inferences for $\boldsymbol{\beta}$ based on estimates, $\hat{\mathbf{B}}$ (which are unbiased for **B**), is small, and of the order $O(N^{-1/2})$ (meaning that the bias is a function of $N^{-1/2}$, and will thus become small as the population size becomes larger). Therefore, for large populations, an unbiased estimate of **B** (generally computed using weighted least squares) can serve as an unbiased estimate of $\boldsymbol{\beta}$. Model diagnostics can then be used to question the hypothesized superpopulation model or the suitability of **B** as a summary measure of the relationships.

7.3 Four Steps in Linear Regression Analysis

There are four basic steps that analysts should follow when fitting regression models to complex sample survey data: specification, estimation, evaluation (diagnostics), and inference. These steps apply to all types of regression models and not just those discussed in this chapter for continuous response variables. The following sections describe each step, considering the standard approach first, followed by the adaptation of the step for complex sample survey data.

7.3.1 Step 1: Specifying and Refining the Model

Survey data are typically observational data. The process of initial specification and subsequent refinement of a regression model for survey data involves multiple iterations of the four-step process. At the beginning of each cycle in this iterative process, it is important for the survey analyst to step back from the "number crunching" and critically evaluate the scientific interpretation and the plausibility of the emerging model.

One initially postulates a model based on subject matter knowledge and empirical investigation of the data. The specific aims of the analysis will often determine the choice of the dependent variable, y, and one or more independent variables of particular interest, x. Scientific subject matter knowledge and information gleaned from prior studies and publications can be used to identify additional independent variables (i.e., covariates that are known predictors of the dependent variable) or variables that may **mediate** or **moderate** (DeMaris, 2004) the relationships of the independent variables of primary interest with the dependent variable y. For example, an epidemiologist aiming to model the relationship of obesity with systolic blood pressure (in mmHg) decides to include age, gender, and race of the respondent as additional covariates. Based on prior research conducted by a colleague, she has evidence that advanced age moderates the relationship between systolic blood pressure level and obesity. She will therefore test for an interaction between age and the obesity measure (as well as other potential interactions—see Section 7.4).

Empirical results from simple descriptive and graphical statistical analyses of the survey data itself can also be used to identify independent variable candidates for the model. The epidemiologist in our example might conduct an exploratory analysis, plotting systolic BP against respondent age. At older ages, the resulting scatter plot shows a curvilinear relationship with systolic BP, suggesting the addition of a quadratic term to the initial model. See Cleveland (1993) for a good resource on data exploration.

Contemporary survey data sets may contain hundreds of variables, and there is a temptation to bypass the scientific review and empirical investigations and "see what works." Regression procedures in statistical software packages often include variable selection algorithms such as **stepwise regression**, **forward selection**, and **backward selection** that are capable of culling a set of significant predictors from a large input of independent variable choices. These algorithms may prove useful in the model exploration and fitting process, but they are numerical tools and should not substitute for the survey analyst's own scientific and empirical assessment of the model and its final form. Careless use of such techniques leaves the analyst exposed to problems of **confounding** or **spurious relationships** that can distort the model and its interpretation.

A variety of variable selection and model building approaches have been proposed for linear regression models. One such model building "recipe" commonly used in our teaching and consulting practice is described in Section 7.4.5.

7.3.2 Step 2: Estimation of Model Parameters

After the survey data analyst has specified a linear regression model (Step 1), the next step in the modeling process involves computation of estimates of the regression parameters in the specified model. This section describes mathematical methods that can be used for estimation of those parameters.

7.3.2.1 Estimation for the Standard Linear Regression Model

By far, the most popular method of estimating unknown parameters in linear regression models is **ordinary least squares (OLS) estimation.** This method focuses on estimating the unknown set of regression parameters $\boldsymbol{\beta}$ in a specified model by minimizing the residual sum of squares (or sum of squared errors, SSE) based on the model:

$$SSE = \sum_{i=1}^{n}(y_i - \hat{y}_i)^2 = \sum_{i=1}^{n}(y_i - \boldsymbol{x}_i\boldsymbol{\beta})^2 \tag{7.9}$$

Once the estimate of $\boldsymbol{\beta}$ has been obtained analytically (see 7.11), an estimate of the variance of the random errors in the model, $\sigma^2_{y \cdot x}$, is obtained as follows:

$$\hat{\sigma}^2_{y \cdot x} = \frac{\sum_{i=1}^{n}(y_i - \boldsymbol{x}_i\hat{\boldsymbol{\beta}})^2}{n - p} \tag{7.10}$$

Here, p is the number of regression parameters in the specified model.

The least squares estimate has several important properties. First, parameter estimates and their variances and covariances are analytically simple to compute, requiring only a single non-iterative algebraic or matrix algebra computation:

$$\begin{aligned} \hat{\boldsymbol{\beta}} &= (\boldsymbol{X}^T\boldsymbol{X})^{-1}\boldsymbol{X}^T\boldsymbol{y} \\ \boldsymbol{var}(\hat{\boldsymbol{\beta}}) &= \hat{\Sigma}\,(\hat{\boldsymbol{\beta}}) = (\boldsymbol{X}^T\boldsymbol{X})^{-1}\hat{\sigma}^2_{y \cdot x} \end{aligned} \tag{7.11}$$

Second, the estimator is unbiased:

$$E(\hat{\boldsymbol{\beta}} \mid \boldsymbol{X}) = \boldsymbol{\beta} \tag{7.12}$$

Third, the estimator has the lowest variance among all other unbiased estimators that are also linear functions of the response values, making it the **best linear unbiased estimator (BLUE).** Finally, assuming normally distributed errors, the least squares estimates are equal to estimates derived based on **maximum likelihood estimation** (see Chapter 8 for more details).

As described in Section 7.2, one key assumption of the standard linear regression model is homogeneity of the error variance:

$$var(\varepsilon) = var(y_i \mid \boldsymbol{x}_i) = \sigma^2_{y \cdot x} = \text{constant} \tag{7.13}$$

In practice, it is common to find that the variance of the residuals is heterogeneous, varying over the $i = 1, \ldots, n$ cases with differing values of y and x. For OLS estimation, the consequence of heterogeneity of variance is loss of efficiency (larger standard errors) in the estimation of the regression coefficients. **Weighted least squares (WLS) estimation** of the regression coefficients addresses this inefficiency by weighting each sample observation's contribution to the sum of squares by the reciprocal of its residual variance, $W_i = 1/\sigma^2_{y \cdot x, i}$. In matrix notation, the WLS estimator of the linear regression coefficients is:

$$\hat{\boldsymbol{\beta}} = (\boldsymbol{X}^T\boldsymbol{W}\boldsymbol{X})^{-1}\boldsymbol{X}^T\boldsymbol{W}\boldsymbol{y} \tag{7.14}$$

Here, W is an $n \times n$ diagonal matrix (with zeroes off the diagonal) and the n values of the inverse variance weights on the diagonal.

The standard linear regression model and the OLS estimator are statistically elegant, but the underlying assumptions are easily violated when analyzing real world data. To minimize the mean squared error of estimation, techniques such as transformation of the dependent variable, WLS estimation, and other approaches such as ridge regression (Hoerl & Kennard, 1970) and robust variance estimation (Judge et al., 1985) have been developed to address problems of non-normality, heterogeneity of variances, collinearity of predictors, and correlated errors.

7.3.2.2 Linear Regression Estimation for Complex Sample Survey Data

Estimation of regression relationships for complex sample survey data alters the standard approach to estimation of coefficients and their standard errors. We first discuss what changes with estimation of the parameters, and then address what changes in terms of variance estimation.

1. **Estimation of parameters.** The observed data from a complex sample survey are typically not "identically distributed." Due to variation in sample selection and sample inclusion probabilities, survey weights must generally be employed to develop unbiased estimates of the finite population regression parameters. Recall that in standard methods of regression analysis, WLS estimation incorporates a weight for each sample element inversely proportional to the residual variance. In the context of fitting regression models to complex sample survey data sets, where survey weights have been calculated to compensate for unequal probability of selection, unit nonresponse, and possibly post-stratification or calibration (see Section 2.7), the survey weights can be incorporated into the estimation of the regression parameters via the use of WLS estimation. The contribution of each case to the residual sum of squares is made proportional to its population weight. This results in the following analytic formula for the weighted least squares estimate of the finite population regression parameters:

$$\hat{B} = (X^T W X)^{-1} X^T W y \tag{7.15}$$

Here, W is an $n \times n$ diagonal matrix (with zeroes off the diagonal) and the n values of the final survey weights on the diagonal. Theory Box 7.2 provides mathematical motivation for the weighted survey estimator of B, and Theory Box 7.3 further considers the decision of whether to use survey weights when estimating regression parameters from complex sample survey data.

Although the motives for conventional WLS estimation (heterogeneity of residual variances) and weighted survey estimation (unbiased estimation of the finite population regression parameters) are very different, the WLS computational algorithms that have been built into regression software for several decades serve perfectly well for weighted survey estimation of the finite population regression model. Consequently, an analyst who naively specified the survey weight as the weight variable in a standard linear regression program (e.g., SAS PROC REG, `regress` in Stata, and SPSS Linear Regression) has obtained the correct design-based estimates of the population regression parameters. Unfortunately, the naïve use of weighted survey estimation in standard linear regression programs generally results in biased estimates of standard errors for the parameter estimates (see below). Stata explicitly recognizes the differences in weighting concepts and requires the user to declare probability weights, or "pweights." Stata also uses a robust estimator of variance when a "pweight" is specified in the standard linear regression command (`regress`).

Correct interpretation of estimated regression parameters is essential for effectively communicating the results of investigations involving regression analyses in scientific publications. In a simple linear regression model for a continuous dependent variable, after obtaining estimates of the regression parameters, we can calculate the following expected values for the response variable, associated with a one-unit change in a given predictor variable x:

$$E(y \mid x = x_0) = \hat{B}_0 + \hat{B}_1 x_0 \tag{7.16}$$

$$E(y \mid x = x_0 + 1) = \hat{B}_0 + \hat{B}_1 x_0 + \hat{B}_1 \tag{7.17}$$

We can therefore write the following about the estimate of the regression parameter β_1:

$$\hat{B}_1 = E(y \mid x = x_0 + 1) - E(y \mid x = x_0) \tag{7.18}$$

The parameter estimate therefore describes, on average, the expected change in the continuous response variable y for a one-unit change in the predictor variable x.

When an additional predictor variable z is added to the model, representing a theoretical control variable or possibly a confounding variable, a portion of the relationship between x and y is attributable to the variable z, and the interpretation of the regression parameter for the predictor variable x requires that the predictor variable z be fixed at a constant value, z_0:

$$E(y \mid x = x_0, z = z_0) = \hat{B}_0 + \hat{B}_1 x_0 + \hat{B}_2 z_0 \tag{7.19}$$

$$E(y \mid x = x_0 + 1, z = z_0) = \hat{B}_0 + \hat{B}_1 x_0 + \hat{B}_1 + \hat{B}_2 z_0 \tag{7.20}$$

$$\hat{B}_1 = E(y \mid x = x_0 + 1, z = z_0) - E(y \mid x = x_0, z = z_0) \tag{7.21}$$

We therefore interpret the estimate of the parameter B_1 in this case as the expected difference in y associated with a one-unit increase in x, holding the value of the predictor variable z constant. When multiple control variables are added to a linear regression model, we hold all of them at fixed values when interpreting the regression parameter for a primary predictor of interest. The interpretation of estimated regression parameters does not change at all when analyzing sample survey data sets collected from finite populations, aside from the fact that the regression parameters are describing relationships in a finite population of interest. We will consider interpretations of regression parameters in detail in all examples presented in this chapter.

THEORY BOX 7.2: A "WLS" ESTIMATOR FOR FINITE POPULATION REGRESSION MODELS

When fitting regression models to complex sample survey data collected from a finite population, we choose estimates of the finite population parameters $\boldsymbol{B}$ that minimize the following objective function:

$$f(\mathbf{B})=\sum_{i=1}^{N}(y_i-\mathbf{x}_i\mathbf{B})^2$$

We can think of this objective function $f(\mathbf{B})$ as a finite population "residual" sum of squares, SSE_{pop}. An unbiased sample estimate of this total incorporating the final survey weights can be written as follows:

$$W\hat{S}SE_{pop}=\sum_{h}^{H}\sum_{\alpha}^{a_h}\sum_{i=1}^{n_{h\alpha}}w_{h\alpha i}(y_{h\alpha i}-\mathbf{x}_{h\alpha i}\mathbf{B})^2$$

In this expression, h is a stratum index, α is a cluster (or primary sampling unit) index, and i is an index for elements within the α-th cluster. Weighted least squares estimation is used to derive estimates of $\mathbf{B}$ that minimize this sample estimate of the actual objective function for the finite population.

THEORY BOX 7.3: SHOULD SURVEY WEIGHTS BE USED TO ESTIMATE REGRESSION MODELS?

This is an important question that statisticians and data analysts working with complex sample survey data need to answer when fitting regression models. When employing the design-based approaches that we illustrate in this chapter and attempting to make inferences about finite populations, the use of survey weights ensures that the estimates of regression parameters will be unbiased with respect to the sample design. Weighted estimation does not, however, protect analysts from model misspecification; if an analyst is fitting a poorly specified model using survey weights, they will simply be computing unbiased estimates of the regression parameters in a model that does a poor job of describing relationships in the larger target population.

Statisticians and data analysts who advocate *model-based* approaches emphasize the importance of sound model specification. Those following this approach will generally argue that the use of survey weights in estimation should not be necessary if a model has been correctly specified; in these cases, the use of weights in estimation introduces the risk of computing inefficient estimates, with standard errors larger than they need to be (Korn & Graubard, 1999). If the computed survey weights are informative about a dependent variable of interest, then the variables used to compute the weights (or the weights themselves!) should be included in the model as covariates. Pfeffermann (2011) discusses alternative model-based and design-based approaches to fitting regression models to survey data, clarifying the differences between these approaches and finding general support (via simulation) for a variant of probability-weighted estimation.

Given that this is an area of healthy debate in the survey statistics literature (see Gelman, 2007, and Heeringa et al., 2015b), we encourage data analysts to not choose a side, *per se*, but rather to consider the sensitivity of their inferences to alternative choices of estimation approaches. Today's statistical software makes it very easy to fit a carefully-specified regression model with and without survey weights (holding the variance estimation approach constant), and examine the sensitivity of the results to the use of weights in estimation. In short, if the use of weights leads to substantially different parameter estimates and inferences, a model may be misspecified (e.g., Korn & Graubard, 1999), and the weighted estimates should be reported (as they will be unbiased). However, if the use of weights in estimation does not change parameter estimates substantially and only results in large increases in the standard errors of the estimates, a model may indeed be well-specified, and the use of weights in estimation may be unnecessary. One can also formally compare weighted and unweighted estimates of regression parameters to assess the significance of the differences, using a method described by Fuller (1984) that we will illustrate in an upcoming example. Bollen et al. (2016) provide a comprehensive review of related methods for comparing weighted and unweighted estimates of regression parameters to assess the need for using weights in regression analysis.

Overall, we feel that the design-based approaches illustrated in this and later chapters, where the use of weights in estimation results in unbiased finite population estimates of regression parameters, generally work well in practice and are easier to defend as having favorable properties. We are usually never certain that regression models have been correctly specified, despite our best efforts, and we are willing to accept slight losses in the efficiency of estimates as long as they are unbiased. See Heeringa et al. (2015b), Binder (2011), Pfeffermann (2011), or Korn and Graubard (1999) for more perspective on this issue. We consider the "q-weighted" approach for increasing the efficiency of weighted survey estimates of regression parameters that was proposed and evaluated by Pfeffermann (2011) in our illustration later in this chapter. We also consider the alternative Bayesian approach to fitting linear regression models discussed in Section 3.8 and evaluate any gains in efficiency that it achieves in the upcoming example.

2. **Estimation of variances of parameter estimates.** As described in Chapter 3, the complex designs of most large probability samples (stratification, cluster sampling, unequal selection probabilities) preclude the use of conventional variance estimators that can be derived for maximum likelihood estimation (MLE) for data that are presumed to be i.i.d. draws from a probability distribution (i.e., normal, binomial, Poisson). Instead, robust, nonparametric methods based on the Taylor Series Linearization (TSL) of the estimator or replication variance estimation methods (BRR, JRR, bootstrapping) are employed (Wolter, 2007).

The general approach to variance estimation using TSL for linear regression coefficients can be illustrated for the simple linear regression model (which involves a single predictor variable). Extension of the method to multiple linear regression is straightforward but algebraically more complex. In the case of a simple linear regression model with a single predictor x, an analytic formula for the calculation of the associated finite population regression parameter B (given all data for the finite population with size N) can be written as follows:

$$B = \frac{\sum_{i=1}^{N} y_i x_i}{\sum_{i=1}^{N} x_i^2} = \frac{T_{xy}}{T_{x^2}} \tag{7.22}$$

This formula can be written as a ratio of two totals: T_{xy} and T_{x^2}. We can calculate an estimate of this ratio by applying the survey weights to the observed sample data:

$$\hat{B} = \frac{\sum_h \sum_\alpha \sum_{i=1}^{n_{h\alpha}} w_{h\alpha i} y_{h\alpha i} x_{h\alpha i}}{\sum_h \sum_\alpha \sum_{i=1}^{n_{h\alpha}} w_{h\alpha i} x_{h\alpha i}^2} = \frac{t_{xy}}{t_{x^2}} \tag{7.23}$$

Note that the sample estimate $\hat{B}$ is also a ratio of sample totals. Under the TSL approximation method, an estimate of the sampling variance of this ratio of two sample totals can be written as follows:

$$var(\hat{B}) \cong \frac{var(t_{xy}) + \hat{B}^2 var(t_{x^2}) - 2\hat{B} cov(t_{xy}, t_{x^2})}{(t_{x^2})^2} \tag{7.24}$$

In multiple linear regression, TSL approximation methods require weighted sample totals for the squares and cross-products of all the y and $x = \{1\ x_1 \ldots x_p\}$ combinations. The computations are more complex, but the approach is a direct extension of the technique shown here for a simple linear regression model. Replication methods like JRR, BRR, or bootstrapping (see Section 3.6.3) can also be used to estimate sampling variances of estimated regression parameters (Kish & Frankel, 1974).

Statistical software designed for regression analysis of complex sample survey data applies the TSL, BRR, JRR, or bootstrapping methods to estimate the sampling variance of each parameter estimate, $\text{var}(\hat{B}_j)$, $j = 1, \ldots, p$, as well the $p(p+1)$ / 2 unique covariances, $\text{cov}(\hat{B}_j, \hat{B}_k)$, between the parameter estimates. These estimates of sampling variances and covariances are then assembled into the estimated **variance-covariance matrix** of the parameter estimates:

$$\boldsymbol{var(\hat{B})} = \hat{\Sigma}(\boldsymbol{\hat{B}}) = \begin{bmatrix} var(\hat{B}_0) & cov(\hat{B}_0, \hat{B}_1) & \cdots & cov(\hat{B}_0, \hat{B}_p) \\ cov(\hat{B}_0, \hat{B}_1) & var(\hat{B}_1) & \cdots & cov(\hat{B}_1, \hat{B}_p) \\ \vdots & \vdots & \ddots & \vdots \\ cov(\hat{B}_0, \hat{B}_p) & cov(\hat{B}_1, \hat{B}_p) & \cdots & var(\hat{B}_p) \end{bmatrix} \tag{7.25}$$

The estimated variances and covariances can then be used to develop Student t-statistics and Wald chi-square or Wald F-statistics required to test hypotheses concerning the population values of the regression parameters (see Section 7.3.4).

A variety of procedures exist in statistical software packages for fitting linear regression models to survey data using the methods discussed in this section; we consider procedures in some of the more common general-purpose statistical software packages here. In SAS, PROC SURVEYREG can be used; IBM SPSS Statistics offers the General Linear Model procedure in the Complex Samples Module (CSGLM); Stata offers the `svy: regress` command, which will be considered in this book; SUDAAN offers PROC REGRESS; IVEware offers the %REGRESS module; and the `survey` package in R offers the `svyglm()` function. We will also demonstrate the use of the contributed `csSampling` package in R (Hornby et al., 2023) to illustrate alternative Bayesian approaches to fitting linear regression models that fully account for complex sampling features. See Appendix A for more details on software options for linear regression analysis of complex sample survey data.

7.3.3 Step 3: Model Evaluation

The standard linear regression model is an "elegant" statistical tool, but its simplicity and "best" properties hinge on several model assumptions (see Section 7.2 above). Standard texts on linear regression modeling (e.g., Neter et al., 1996) provide detailed coverage of procedures designed to evaluate the model **goodness of fit (GOF)**, examine how closely the data match the basic model assumptions, and determine if certain observations are unduly influencing the fit of the model. The process does not change substantially when fitting finite population models to complex sample survey data sets. In this section, we briefly consider some of the model diagnostics that can be used to evaluate the model and discuss some current work on model diagnostics adapted for regression models fitted to complex sample survey data.

1. **Explained Variance and Goodness of Fit.** A standard measure of the "fit" of the regression model to the data is the **coefficient of multiple determination** or the **R^2 statistic**, which is interpreted as the proportion of variance in the dependent variable explained by regression on the independent variables:

$$R^2 = 1 - \frac{SSE}{SST} \tag{7.26}$$

In (7.26), SST refers to the total sum of squares, or the sum of squared differences between the response values and the mean of the response variable, and SSE is given in (7.9) above. The use of R^2 as a measure of explained variance carries forward to regression modeling of complex sample survey data, although the statistic that is output by the analysis software is generally a *weighted* version, where each squared difference contributing to the sums is weighted by the corresponding survey weight:

$$R^2_{weighted} = 1 - \frac{WSSE}{WSST} \tag{7.27}$$

Although in theory it could be argued that this weighted R^2 statistic estimates the proportion of population variance explained by the population regression of y on x, in practice it is safe to simply view it as the fraction of explained variance in y attributable to the regression on x. Analysts who are new to regression modeling of social science, education, or epidemiological data should not fret if the achieved

R^2 values are lower than those seen in their textbook training. Physicists may be disappointed with $R^2 < 0.98$–0.99 and chemists with $R^2 < 0.90$, but social scientists and others who work with human populations will find that their best regression model will often explain only 20%–40% of the variation in the dependent variable.

2. **Residual diagnostics.** In the standard regression context, analysis of the distributional properties of the residual terms, $\varepsilon_i = (y_i - \hat{y}_i)$, is used to evaluate how well the assumptions of the normal linear model are met (see Section 7.2 above). Despite the theoretical distinction between the concept of a finite population regression model and a broader superpopulation model, we recommend using standard residual analysis to evaluate regression models that are fitted to complex sample survey data. Li and Valliant (2015) describe advances in the study of residual diagnostics for regression models fitted to data from complex samples. These advances have now been implemented in a contributed package for the R software (`svydiags`), and we illustrate use of the functions in this package later in this chapter (see Section 7.5.4).
3. **Model specification and homogeneity of variance.** Two-way scatter plots of the residuals for the estimated model against the predicted values, $\hat{y}$, and the independent variables, x_j, can identify problems with lack of correct functional form (e.g., omitting a squared term for age when modeling blood pressure) or where a moderating variable or interaction has not been correctly included in the model.

   These same plots may be used to diagnose a problem with heterogeneity of residual variances. A pattern of residuals that spreads out in a fan-shaped pattern with increasing values for the predicted y or increasing values of an independent variable is a common observation. To address the problem of heterogeneity of variance and reduce standard errors for the estimated β coefficients, a standard approach in regression analysis is to employ weighted least squares (WLS), weighting each observation's contribution to the sum of squared errors inversely proportional to its residual variance, that is, $w_i^{(hv)} = 1/\sigma^2_{y\cdot x_i}$. With complex sample survey data, this becomes complicated, because weighted estimating equations for finite population regression parameters, $\boldsymbol{B}$, already include the final survey weights, say $w_i^{(survey)}$. While it is possible to create a composite weight, $w_i^* = w_i^{(survey)} \cdot w_i^{(hv)}$ (see Little, 2004 for a suggested approach), in cases of serious heterogeneity of variance it may be possible to identify a transformation of the dependent or independent variables (see below) that eliminates much of the residual variance heterogeneity but still permits the use of the survey weights in the estimation of the regression function.
4. **Normality of the residual errors.** This assumption can be assessed using standard diagnostic plots for the model-based residuals (e.g., normal quantile-quantile (Q-Q) plots or histograms). The Q-Q plot is a plot of quantiles for the observed residuals against those computed from a theoretical normal distribution having the same mean and variance as the distribution of observed residuals. A straight 45° line in this plot would therefore suggest that normality is a reasonable assumption for the random errors in the model. We present examples of these Q-Q plots in the application later in this chapter.

   In large survey data sets, formal tests of the normality of the residuals (e.g., the Kolmogorov-Smirnov and Shapiro-Wilks tests) tend to be extremely "powerful" and the null hypothesis of normality will be rejected due to the slightest deviation from normality. Our recommendation is to use the more informal visual methods illustrated in Section 7.5 below. Empirical research has provided evidence that as long as the residuals display symmetry about $E(e_i) = 0$, the regression estimates are quite robust against failure of strict normality. If the residual distribution is highly skewed or irregular (e.g., bimodal), the analyst should first determine that the model has been correctly specified (see Section 7.4.1), and no important predictor variables or interaction terms have been omitted in error.

Transformation of the dependent variable is a common method to address serious problems of non-normality of residuals and heterogeneity of residual variances. Analysts should be careful when making transformations, however, because they can destroy the straightforward interpretation of the parameters discussed above. A common transformation that is often used when violations of normality are apparent and residual distributions appear to be right-skewed is the natural (base *e*) log transformation of the response variable:

$$\ln(y) = B_0 + B_1 x + e \tag{7.28}$$

When this transformation is used, the regression parameters still have a somewhat-straightforward interpretation:

$$\frac{E(y \mid x = x_0 + 1)}{E(y \mid x = x_0)} = \frac{e^{(B_0 + B_1 x_0 + B_1)}}{e^{(B_0 + B_1 x_0)}} = e^{B_1} \tag{7.29}$$

That is, a one-unit change in the predictor variable will *multiply* the expected response by $\exp(B_1)$. The important issue to keep in mind when transforming the dependent variable is that predicted values on the response variable need to be back-transformed to the original scale of the response. For example, square-root transformations are often used to stabilize variance of the residuals, and predicted values based on this type of model would need to be squared to return to the original scale of the response.

5. **Collinearity diagnostics.** When specifying linear regression models, analysts need to be careful not to include predictor variables that are highly correlated with each other in the same model. This type of model misspecification can lead to problems of **multicollinearity**, which can have an adverse effect on the properties of the estimated regression parameters: standard errors of the estimated coefficients become inflated due to **variance inflation factors** (which are a function of the amount of variance in a given predictor explained by the other predictors), and the signs of estimated coefficients may even change artificially relative to what their values would be in the absence of multicollinearity. Analysts should always first examine the correlations of the different predictor variables that they are considering, and make sure that none of them are excessively high in absolute value (say, greater than 0.7). In addition, **condition indices** for fitted linear regression models should not exceed 30 in value, and variance inflation factors greater than 2 warrant consideration and further analysis. In these cases, analysts should consider either combining correlated predictors (for example, using principal components analysis or factor analysis) or omitting redundant predictors. See Faraway (2014) for more discussion of problems with multicollinearity.

Methodological developments in this area have considered the computation of these important diagnostic statistics for assessing multicollinearity when fitting linear regression models to complex sample survey data. Liao and Valliant (2012a, 2012b) have led the way in this area, developing the underlying theory for determining these diagnostic statistics in a manner that accounts for the complex sampling features and illustrating practical examples of their implementation. These design-adjusted diagnostic statistics have also been implemented in the `svydiags` package in R, and we will demonstrate their computation and interpretation later in this chapter.

6. **Outliers and influence statistics.** Finally, analysts should determine if the sample data include **outlier values** (observations poorly fitted by a given model) or **influential points** (observations that have a strong effect on the fit of a model). Tools such as standard residual plots, **studentized residuals, hat statistics** (measures of leverage), and **Cook's Distance (D) statistics** have been developed for this purpose. Regression models can be fitted with and without potential outliers and influential points to assess whether inferences based on the fitted model are changing substantially depending on the points in question. For more details on these diagnostic methods in the standard linear regression case, we refer readers to Neter et al. (1996) or Faraway (2014).

Influential points should still be investigated for their impact on the fit of a model when fitting linear regression models to complex sample survey data sets. What changes is the way that complex sampling features are incorporated when calculating the various statistics (e.g., Cook's D statistics) used for residual diagnostics. Li and Valliant (2009, 2011a) were the first to discuss the identification of influential points when sampling weights are involved in the estimation of linear regression models, and these authors also discussed how to adapt the forward search method for identifying influential cases to the complex sampling context (Li & Valliant, 2011b). More recent work by Li and Valliant (2015) discussed extensions of the adjustments to diagnostic statistics for samples involving stratification and cluster sampling. Collectively, the current literature in this area suggests that the final survey weights are most important for assessing the influence of individual observations on model fit. These design-adjusted diagnostic measures of outliers and influential points are also available in the contributed `svydiags` package for the R software.

THEORY BOX 7.4: MODEL EVALUATION FOR COMPLEX SAMPLE SURVEY DATA

When fitting regression models to complex sample survey data collected from a finite population, outlier statistics used for model evaluation are simply adapted by replacing unweighted values with weighted values. This theory box presents mathematical expressions for some of these adaptations, each of which have now been implemented in the contributed `svydiags` package for R. To keep the expressions as general as possible, we use notation from generalized linear model theory. Generalized linear models will be discussed in more detail in Chapters 8 and 9.

First, we consider computation of **Pearson residuals:**

$$r_{p_i} = (y_i - \mu_i(\hat{\boldsymbol{B}}_w))\sqrt{\frac{w_i}{V(\hat{\mu}_i)}}$$

In this notation, μ_i refers to the expected value of the outcome (y) for sampled case i, computed as a function of the weighted estimates of the regression parameters. The term w_i refers to the final survey weight for the i-th case based on the complex sample design, and the term $V(\mu_i)$ refers to the **variance function** for the outcome, which partly defines the variance of the outcome variable in a generalized linear model as a function of the expected value of the outcome.

The **hat matrix** is computed as

$$\boldsymbol{H} = \mathbf{W}^{1/2}\mathbf{X}(\mathbf{X}'\mathbf{W}\mathbf{X})^{-1}\mathbf{X}'\mathbf{W}^{1/2}$$

where

$$\mathbf{W} = diag\left\{\frac{w_1}{V(\mu_1)[g'(\mu_1)]^2}, \ldots, \frac{w_n}{V(\mu_n)[g'(\mu_n)]^2}\right\}.$$

This matrix is used for the computation of various diagnostic statistics, and specifically, measures of leverage. This matrix is an $n \times n$ diagonal matrix, with zeroes off the diagonal and diagonal elements defined by the final survey weights divided by a term that is a function of the variance function for the observation and the derivative of the **link function** g for the specific generalized linear model (in a linear regression model, the link function is the identity function). Note that the derivative of the link function is computed for the diagonal elements and is evaluated as a function of the expected value of the outcome according to the model. If a **canonical link** is used to define the generalized linear model (which is often the case in practice), the diagonal elements simplify to

$$\mathbf{W} = diag(w_1 V^{-1}(\mu_1), \ldots, w_n V^{-1}(\mu_n)).$$

Diagonal elements of the hat matrix (denoted by h_{ii}) updated with the survey weights are used to identify influential cases with high leverage. A common rule of thumb is to identify diagonal elements larger than $2(p + 1) / n$ (where n is the *effective* sample size), or to identify large gaps in leverage values. Removing cases with high leverage will generally have little impact on estimates of regression parameters, but will have a large impact on the uncertainty of the estimates (i.e., standard errors).

Next, we consider computation of **Cook's Distance (D) statistic** in the complex sample design setting. Cook's D statistic can be useful for identifying observations that have a large impact on estimates of the regression parameters when they are removed from the estimation, taking the precision of the estimates into account. Cook's D statistic is computed for an individual sample observation i as follows:

$$c_i = \frac{w_i^* w_i e_i^2}{p\phi V(\hat{\mu}_i)(1 - h_{ii})^2} \boldsymbol{x}_i'[V\hat{a}r(U_w(\hat{\boldsymbol{B}}_w))]^{-1}\boldsymbol{x}_i$$

where

w_i^* = survey weight

w_i = remainder of the diagonal element in the Hat matrix

(e.g., for a canonical link)

e_i = residual

p = number of parameters in the regression model

ϕ = dispersion paramter in the generalized linear model

$V\hat{a}r(U_w(\hat{B}_w))$ = linearized variance estimate of the **score equation**, which is used for pseudo mazimum likelihood estimation in generalized linear models fitted to complex sample survey data (see Chapter 9)

Once the value of Cook's D has been determined for an individual sample element, the following test statistic can be computed to assess the significance of the D statistic:

$$\frac{(df-p+1)\cdot c_i}{df} \doteq F(p, df-p)$$

where

df= number of PSUs-number of strata (design-based degrees of frredom)

This test statistic is approximately F-distributed and can be used to identify any unusual elements that are having a significant impact on the estimates of the regression parameters when taking the complex sample design features into account.

We illustrate the computation of these various diagnostic statistics using functions available in the `svydiags` package in R in the example presented later in this chapter.

This evaluation step of regression analysis may lead to a modification of the model structure. One then cycles back through the model fitting and model diagnosis process, with careful consideration of model structure and parsimony (i.e., finding the best-fitting model using the smallest number of relevant predictors possible).

7.3.4 Step 4: Inference

After following the three steps above, the final step in the model building process is making inferences about the regression parameters in the finite population of interest. This section describes that process.

7.3.4.1 Inference Concerning Model Parameters

After rigorously evaluating the fit of a model, an analyst can use the estimated parameters and their standard errors to characterize or infer about the conditional distribution of y given the predictor variables x. The analyst can perform a variety of hypothesis tests concerning the parameters being estimated, ranging from tests for a single regression parameter to tests for multiple regression parameters. We begin this section by considering tests for single regression parameters.

In the standard linear regression context, when the residuals follow a normal distribution, hypothesis tests for a single regression parameter associated with predictor variable k employ a t test statistic:

$$t = \frac{\hat{\beta}_k - \beta_k}{se(\hat{\beta}_k)} \sim t_{n-p} \tag{7.30}$$

Therefore, when performing a test of the null hypothesis H_0: $\beta_k = 0$ versus the alternative hypothesis H_A: $\beta_k \neq 0$, one can calculate a t-statistic by dividing the estimate of the regression parameter β_k by its standard error. For reasonably large samples, when H_0 is true, this test statistic is distributed as a variate from a Student t distribution with n - p degrees of freedom, where p is the number of parameters being estimated in the regression model (including the intercept). The analyst (or their software) refers the computed t-statistic to a Student t distribution with $n - p$ degrees of freedom. If the absolute value of the test statistic, t, exceeds a critical value of the Student t distribution (e.g., $t_{1-\alpha/2,\ n-p}$ for a two-sided test), H_0 is rejected with a Type I error probability of α. Pivoting on the value of t for the two-sided test, a 100(1 – α)% confidence interval for the true model parameter can be constructed as $\hat{\beta}_k \pm t_{1-\alpha/2,df} \cdot se(\hat{\beta}_k)$.

When constructing the confidence interval (or the pivotal hypothesis test statistic) for a single regression parameter estimated from complex sample survey data, two aspects of the inferential process change: 1) $se(\hat{B})$, or the correct standard error of the estimated regression parameter, is *estimated* using a nonparametric technique like Taylor Series Linearization, BRR, JRR, or bootstrapping; and 2) the degrees of freedom for the Student t reference distribution must be adjusted to reflect the reduced degrees of freedom for the complex sample estimate of $se(\hat{B})$. Recall from Chapter 3 that the design degrees of freedom are approximated as $df = \sum_h a_h - H$, or the number of primary stage ultimate clusters minus the number of primary stage strata. For example, in the NCS-R data set, the approximation to the design degrees of freedom is 84–42 = 42. The correct estimate of the standard error and degrees of freedom for the Student t reference distribution can be used to develop a design-based 100(1 – α)% confidence interval (corresponding to a Type I error rate of α) for the regression parameter of interest, as follows:

$$\hat{B} \pm t_{1-\alpha/2,\ df} \cdot se(\hat{B}) \tag{7.31}$$

The t-statistic for the comparable two-sided hypothesis test of H_0: $B = 0$ can be developed as: $t = \hat{B} / se(\hat{B})$.This is the form of the t test statistic that is routinely printed in tables of regression model output. The "p-values" generally printed alongside the test statistics are the probability that $t_{df} \geq t$.

In standard linear regression, F-tests are often used to test hypotheses about multiple parameters in the model. The **overall *F*-test** typically reported in the analysis of variance (ANOVA) table output generated by software procedures for fitting regression models using standard OLS methods tests the null hypothesis H_0: $\beta_1 = \beta_2 = \ldots = \beta_p = 0$, that is, that the fitted model predicts $E(y|x)$ no better than a model that only includes the intercept ($\beta_0 = \bar{Y}$). **Partial *F*-tests** can be used to test whether selected subsets of parameters in the model are not significantly different from 0. In this case, a "full" model is compared to a nested "reduced" model that contains a subset of the predictor variables in the "full" model. This type of hypothesis test is essentially a test of the null hypothesis that multiple parameters (the parameters omitted in the "reduced" model) are all equal to 0. This multi-parameter test can be extremely useful for testing hypotheses about categorical predictor variables represented by several indicator variables in a regression model (see Section 7.4.2.1). More formally, we can use the following notation to indicate the $p = p_1 + p_2$ predictor variables of interest:

$$\begin{aligned} x &= (x_1, x_2) \\ x_1 &= p_1 \text{ predictors} \\ x_2 &= p_2 \text{ predictors} \end{aligned}$$

Then, we can write the full model as follows:

$$full\text{: } y = x\beta + \varepsilon \tag{7.32}$$

The reduced model then omits the p_2 predictor variables:

$$reduced\text{: } y = x_1\beta_1 + \varepsilon \tag{7.33}$$

Then, to test the null hypothesis that the regression parameters associated with the p_2 predictor variables are all equal to 0, that is, H_0: $\boldsymbol{\beta_2} = \mathbf{0}$, we can calculate the following partial F-test statistic:

$$F = \frac{\dfrac{SSE_{reduced} - SSE_{full}}{(n-p_1)-(n-p_1-p_2)}}{\dfrac{SSE_{full}}{n-p_1-p_2}} = \frac{\dfrac{SSE_{reduced} - SSE_{full}}{p_2}}{\dfrac{SSE_{full}}{n-p}} \tag{7.34}$$

Under the null hypothesis and assuming normally distributed residuals, this F-statistic follows an F distribution with numerator degrees of freedom equal to p_2 and denominator degrees of freedom equal to $n - p$. This general result allows one to perform a variety of multi-parameter tests when comparing nested linear regression models, at least in the simple random sample setting. This F-statistic is also fairly robust to slight deviations from normality in the residuals.

In the complex sample survey data setting, these multi-parameter tests must be adapted to the complex design features of the sample. **Wald test statistics** (Judge et al., 1985) replace the overall F-test and the partial F-test. The equivalent multi-parameter Wald test statistics can be calculated as follows:

$$\begin{aligned} Overall\text{: } Modified\ X^2_{W,overall} &= \frac{\hat{\boldsymbol{B}}^T \hat{\Sigma}(\hat{\boldsymbol{B}})^{-1} \hat{\boldsymbol{B}}}{p} = F_{W,overall} \\ Partial\text{: } Modified\ X^2_{W,partial} &= \frac{\hat{\boldsymbol{B}}_2^T \hat{\Sigma}(\hat{\boldsymbol{B}}_2)^{-1} \hat{\boldsymbol{B}}_2}{p_2} = F_{W,partial} \end{aligned} \tag{7.35}$$

where: $\hat{\boldsymbol{B}}, \hat{\boldsymbol{B}}_2$ are vectors of estimated regression parameters; and
$\hat{\Sigma}(\hat{\boldsymbol{B}}), \hat{\Sigma}(\hat{\boldsymbol{B}}_2)$ are the estimated variance/covariance matrices.

Under the null hypothesis H_0: $\boldsymbol{B} = \boldsymbol{0}$, the overall modified Wald test statistic, $F_{W,overall}$, follows an *F* distribution with numerator degrees of freedom equal to *p* and denominator degrees of freedom equal to the design degrees of freedom (*df*). Likewise, to test H_0: $\boldsymbol{B_2} = \boldsymbol{0}$, or the null hypothesis that the p_2 parameters are all equal to 0 in the nested model, the modified Wald partial test statistic is referred to the critical value of the *F* distribution with p_2 and *df* degrees of freedom.

Wald tests can also be used to test more general hypotheses regarding linear combinations of regression model parameters. Consider the null hypothesis H_0: $\boldsymbol{CB} = \boldsymbol{0}$, where $\boldsymbol{C}$ is a matrix that defines specific linear combinations of the regression parameters in the vector $\boldsymbol{B}$. In this case, a version of the Wald test statistic that follows a chi-square distribution with degrees of freedom equal to the rank of the matrix $\boldsymbol{C}$ under the specified null hypothesis can be written as follows:

$$X_W^2 = [\boldsymbol{C}\hat{\boldsymbol{B}}]'[\boldsymbol{C}\hat{\Sigma}(\hat{\boldsymbol{B}})\boldsymbol{C}']^{-1}[\boldsymbol{C}\hat{\boldsymbol{B}}] \approx \frac{\text{contrast "squared"}}{\text{variance of contrast}} \tag{7.36}$$

We consider one example of this more general type of hypothesis test for linear combinations of multiple parameters. First, suppose that a specified linear regression model includes three parameters of interest, that is, $\boldsymbol{B}' = [B_1, B_2, B_3]$. Using this more general framework for the Wald test statistic, if one wished to test the null hypothesis H_0: $B_2 - B_3 = 0$ (or equivalently, H_0: $B_2 = B_3$), the $\boldsymbol{C}$ matrix would take the following form:

$$\boldsymbol{C} = \begin{bmatrix} 0 & 0 & 0 \\ 0 & 1 & -1 \\ 0 & 0 & 0 \end{bmatrix}$$

Then, the Wald test statistic specified in (7.36) above would follow a chi-square (χ^2) distribution with one degree of freedom (because the rank of the $\boldsymbol{C}$ matrix is 1). These more general hypothesis tests are available in a variety of software packages that can fit regression models to complex sample survey data sets, including Stata, SAS, and R. Dividing these more general chi-square test statistics by the number of parameters being tested will result in test statistics that follow *F* distributions, like (7.35). We consider examples of how to specify these tests for multiple parameters in Section 7.5.

7.3.4.2 Prediction Intervals

Predicting expected outcomes for populations and single individuals is an important scientific application of regression modeling (Neter et al., 1996). Even in the social and health sciences, where regression models are more often used to explore relationships among dependent and independent variables, prediction from fitted regression models still has a role. In the standard linear regression case, given a set of predictor values, $\boldsymbol{x}_{obs,i}$, the estimated regression model can be used to calculate a predicted value for *y*, in addition to **confidence intervals** and **prediction intervals** for the predicted value. First, we consider the expected value of the response variable *y* given an estimated model and a known vector of values on the predictor variables, $\boldsymbol{x}_{obs,i}$:

$$\hat{E}(y_i \mid \boldsymbol{x}_{obs,i}) = \boldsymbol{x}_{obs,i}\hat{\beta} \tag{7.37}$$

Given this expected value, we can calculate a confidence interval for the expected value once the variance of the expected value is calculated:

$$var(\hat{E}(y_i \mid \boldsymbol{x}_{obs,i})) = \boldsymbol{x}'_{obs,i} cov(\hat{\beta}) \boldsymbol{x}_{obs,i} \tag{7.38}$$

A 100(1 − α)% confidence interval for the expected value of y given $\boldsymbol{x}_{obs,i}$ (i.e., the average expected outcome for a population of cases with covariates $\boldsymbol{x}_{obs,i}$) can then be calculated as follows:

$$\boldsymbol{x}_{obs,i}\hat{\beta} \pm t_{1-\alpha/2,n-p}\sqrt{var(\hat{E}(y_i \mid \boldsymbol{x}_{obs,i}))} \tag{7.39}$$

Note that the confidence interval above does not consider the variance of the random errors that are also a part of the linear regression model. A **prediction interval** for a *single future value* of y does take this estimated variance ($\hat{\sigma}^2_{y \cdot x}$) into account:

$$\boldsymbol{x}_{obs,i}\hat{\beta} \pm t_{1-\alpha/2,n-p}\sqrt{var(\hat{E}(y_i \mid \boldsymbol{x}_{obs,i})) + \hat{\sigma}^2_{y \cdot x}} \tag{7.40}$$

Prediction intervals, therefore, are wider than more standard confidence intervals for the expected value because they also include variance in the prediction due to random error. Both intervals can give analysts a notion of the precision of the predicted values based on the fitted model if prediction of future values is an important objective of the modeling process.

Confidence intervals for predicted values can also be computed in the context of regression models for complex sample survey data. Standard errors for the predicted values can be computed using the **delta method** (essentially linearization), which is a technique that can accommodate a wide variety of general predictions, based on fixed values of the predictors and the estimated regression parameters. For computational details on this technique, which is an approximate method based on large samples, interested readers can refer to https://www.stata.com/manuals/rnlcom.pdf. The intervals change in that the degrees of freedom used to calculate the critical t-statistic are now based on the design degrees of freedom, and the variance-covariance matrix of the parameter estimates is estimated using approximate methods like TSL or replicated methods like JRR and BRR. Correct computation of these confidence intervals for predicted values when fitting regression models to complex sample survey data is currently implemented in Stata's `predictnl` post-estimation command.

One can also compute different forms of **marginal predicted values** based on regression models fitted to complex samples (along with confidence intervals for the marginal predicted values). Heuristically, the marginal predicted value of a dependent variable for a *given value* of a covariate of interest is computed by first obtaining predicted values of the dependent variable using design-based estimates of the regression parameters of interest, and assuming that every case in the data set has the *same given value* of the covariate of interest. The values of the other covariates (aside from the covariate of interest) are either left unchanged or fixed to constant values (e.g., their means) when computing these predicted values for each case. These predicted values are then averaged across all cases to compute the marginal predicted value, and a standard error for this marginal predicted value can be computed using the delta method. One can think of these marginal predicted values as overall average predictions for a given choice of values on the covariates.

Average marginal effects of specific changes in the covariate of interest on the dependent variable can also be computed for any given values of the other covariates by comparing marginal predicted values for different values of the covariate of interest (e.g., males vs. females) with selected other covariates fixed to particular values (e.g., everyone who is age 60; Bauer, 2015). These approaches are currently implemented in the Stata software via the `margins` command, and marginal predicted values computed using this command (and corresponding confidence intervals for the predicted values) can then be plotted using the subsequent `marginsplot` command. We illustrate the use of these commands in our illustration later in this chapter.

7.4 Some Practical Considerations and Tools

In this section, we discuss important considerations for data analysts fitting regression models to complex sample survey data sets in practice and provide practical guidance on steps to avoid potential pitfalls when fitting regression models and making inferences based on the fitted models.

7.4.1 Distribution of the Dependent Variable

We specifically focus our discussion in this chapter on linear regression models for **continuous dependent variables** (e.g., weight in kilograms, blood pressure in millimeters of mercury to the second decimal place, weekly household expenditures on food items). Surprisingly, many survey data sets include very few variables that are measured on a truly continuous scale. Many response variables may be **semi-continuous**, **censored**, or **grouped** (or "coarsened") in nature. We do not consider models for semi-continuous, censored, or grouped dependent variables in this chapter; Tobit regression models, Heckman selection models, and other forms of latent variable models might be considered by analysts for these types of response variables (Skrondal & Rabe-Hesketh, 2004). The Stata software currently provides versions of commands designed to handle dependent variables of these types (e.g., `svy: ivtobit`) in the setting of complex sample survey data.

Many survey variables of interest are measured as ordinal scale variables. Examples include age in years and education in years. Survey questions such as *"On a scale of 1 to 5, where 1 is excellent and 5 is poor, please rate your overall health"* produce a response on an ordinal scale. Over the years it has been common practice to fit linear regression models to ordinal scale dependent variables. DeMaris (2004) describes an ordinal scale variable as **approximately continuous** if it meets the following conditions: the number of sample observations, n, is large; measurement is at least on an ordinal scale; the response has at least five ordered levels; and the distribution of responses to the ordered categories is not skewed and ideally is approximately normal in appearance.

We agree that there are obvious cases where it may be acceptable to apply linear regression to an ordinal dependent variable; however, analysts will generally have a difficult time satisfying the underlying statistical assumptions for the models discussed in this chapter when working with ordinal outcomes. This could lead to highly inefficient or faulty inferences. Especially for ordinal variables with a small number of levels, a better approach is to choose a regression model that is more appropriate for the measurement scale of the dependent variable. Several regression models appropriate for ordinal and categorical response variables are discussed in detail in Chapter 9.

7.4.2 Parameterization and Scaling for Independent Variables

An extremely important aspect of fitting linear regression models is the treatment and coding of categorical *predictor* variables, which can definitely be considered in the linear regression models discussed in the chapter. When analysts consider nominal categorical predictor variables (e.g., race/ethnicity, region of the country) in linear regression models, they often generate **indicator variables** (a.k.a. "dummy" variables) to represent levels of the categorical predictor variables in the models. Regression parameters associated with these indicator variables (which are equal to 1 for cases falling into a specific category and 0 otherwise) represent changes in the expected value of the continuous outcome for a specific category relative to a **reference category**, which does *not* have an indicator variable included in the model. Alternative dichotomous specifications of these indicator variables are possible (e.g., 1/–1 "effect" coding, where estimated regression parameters represent changes in the expected value of the outcome relative to the *overall mean*), but we focus on the (1, 0) coding in this book for ease of interpretation.

Consider a categorical predictor variable measuring race/ethnicity with three possible values in a survey of a human population: 1 = Caucasian, 2 = African American, and 3 = other race/ethnicity. Analysts need to choose one of these three categories to be a reference category, and this choice is generally guided by contrasts of interest and research objectives (e.g., comparing African Americans and other groups to Caucasians). In cases where the choice of the reference category is not clear, choosing the most prevalent group in the sample data will suffice. In the case of the race/ethnicity variable, an analyst choosing "Caucasian" to be the reference category would need to create two indicator variables to include as predictors in a regression model: a variable indicating African Americans (1 = African American, 0 = Caucasian or other race/ethnicity), and a variable indicating the other race/ethnicity group (1 = other race/ethnicity, 0 = Caucasian/African American). Most modern statistical software capable of fitting regression models to survey data will perform this "dummy" coding automatically for categorical predictors, requiring the analyst to simply choose the reference category for the analysis.

Continuing with the race/ethnicity example, suppose that a regression model was fitted to a continuous response variable y, where race/ethnicity was the only predictor variable. The dummy coding described above would lead to the following regression function:

$$E(y \mid x) = B_0 + B_1 x_1 + B_2 x_2 \tag{7.41}$$

In this model, $x_1 = 1$ for African Americans and 0 otherwise, while $x_2 = 1$ for other racial/ethnic groups and 0 otherwise. The expected value on the continuous outcome variable y for African Americans would therefore be calculated as

$$E(y \mid x) = B_0 + B_1 \times 1 + B_2 \times 0 = B_0 + B_1, \tag{7.42}$$

and the expected value for individuals in other racial/ethnic groups would be calculated as

$$E(y \mid x) = B_0 + B_1 \times 0 + B_2 \times 1 = B_0 + B_2. \tag{7.43}$$

Because both indicator variables would be equal to 0 for Caucasians, the expected value on the outcome variable for Caucasians would simply be B_0. The regression parameters B_1 and B_2 therefore represent differences in the expected outcomes (i.e., means) between

African Americans and Caucasians (B_1) or other racial/ethnic groups and Caucasians (B_2), and hypothesis tests about differences between the groups could therefore be conducted by testing whether these parameters are equal to 0. Similarly, the difference in expected outcome values between the non-reference groups (African Americans and others) would be equal to $B_1 - B_2$, or the difference in the regression coefficients for these two groups.

When using statistical software to fit regression models to complex sample survey data, analysts have two choices: they can create indicator variables manually, or they can use special options in the different procedures to have the software automatically create indicator variables for the levels of categorical variables, given knowledge about a reference category. Users of the Stata software can employ **factor variable coding** to define the reference category of a given categorical predictor variable when specifying a regression model. For example, one could fit a linear regression model using this command:

```
regress depvar ib1.catvar
```

where `depvar` refers to the dependent variable in a linear regression model (fitted using the `regress` command), `catvar` refers to a categorical predictor variable, and `ib1.` refers to the specific value of `catvar` that is to be set as the reference category (in this case, the indicator variable (`i`) that will be omitted from the model as the baseline (`b`) category is 1). In general, when fitting any type of regression model, Stata users can include `i.` before the names of any categorical predictor variables, and Stata will by default set the *lowest alphanumeric category* as the reference category (if the user does not specify some form of `ib#.`). The following code presents an example of this, once again considering the race/ethnicity example (where OUTCOME is the continuous outcome variable and ETHNIC is the race/ethnicity variable):

```
regress outcome i.ethnic
```

We will consider several additional examples of factor variable coding in Stata later in this chapter.

THEORY BOX 7.5: ANOVA AND ANCOVA AS LINEAR REGRESSION ANALYSIS

Statistical texts on linear models often include separate chapters for linear regression models for continuous outcomes, analysis of variance (ANOVA) models (where all predictors are categorical), and analysis of covariance (ANCOVA) models (involving a mix of categorical and continuous predictors). ANOVA and ANCOVA models for normal data are generally discussed in the context of experimental designs (e.g., full factorial designs, randomized block designs) and experimental hypotheses are tested using F-statistics and multiple comparisons that are functions of expected mean squares. Historically, the distinction between ANOVA, ANCOVA, and linear regression analysis was reinforced in statistical software systems that included separate procedures adapted for ANOVA and standard linear regression modeling. Other procedures such as SAS PROC GLM integrated these two analyses in a linear modeling framework.

In fact, ANOVA- and ANCOVA-type analyses can be performed through proper specification of a linear regression model (e.g., main effects, interactions, nesting); see Neter et al. (1996). Analysts who wish to apply ANOVA-type procedures to complex sample survey data can do so using regression analysis procedures with indicator variable parameterization of categorical independent variables and interactions appropriate for the ANOVA-type model that they wish to fit to the data (e.g., indicator variables for levels of the main effects and interactions for a full factorial model).

Analysts should also exercise caution when interpreting intercept parameters (β_0) in linear regression models. The intercept, or the expected value of the response variable y when all of the predictor variables are fixed at value 0, is often not of much interest, because it may represent an expectation that is well outside the range of the observed data for a continuous predictor variable (e.g., what is the expected value of a continuous dependent variable when a continuous measure of height is equal to zero?). **Centering** of the continuous predictor variables in the model can make the intercept more interpretable:

$$y = \beta_0^* + \beta_1(x - \bar{x}) + \beta_2(z - \bar{z}) + \varepsilon \tag{7.44}$$

By subtracting the means of the predictor variables x and z from the observed values on each variable, and then using the "centered" predictors in the model, the reformulated intercept (denoted by an asterisk) now has the interpretation of the expected value on the response variable y when the predictors are equal to their *means*. One can think of this term as representing a *centercept*, or the overall average value on the response variable y.

We note that the choice of omitting the intercept in a regression model forces the expected value of y to be 0 when all of the predictor variables in the model are set to 0. This might be done when preliminary knowledge of the subject matter being studied dictates that the expected value of the response be 0 when all predictor variables are set to 0. Otherwise, omitting an intercept from a regression model is generally not recommended, and requires some form of theoretical or empirical support.

7.4.3 Standardization of the Dependent and Independent Variables

Analysts can also **standardize** all the variables being considered in a linear regression model (including indicator variables). One can standardize any variable by subtracting the mean for the variable, like centering, and then also dividing by the standard deviation of the variable:

$$x_{i,std.} = \frac{x_i - \bar{x}}{sd(x)} \tag{7.45}$$

This is often done in practice when predictor variables are measured on very different scales (e.g., income and grade point average), and re-scales all of the variables so that they are on the same scale (changes of one unit in standardized variables correspond to changes of one standard deviation in the unstandardized variables). The estimates of the regression parameters in a model where *all* variables have been standardized are often referred to as **standardized regression coefficients**, and these can be used to determine which predictor variable has the largest relative impact on the expected value of the response variable. Nothing changes about this process when analyzing complex sample survey data, but an analyst should note whether weighted estimates or unweighted sample estimates of the mean and standard deviation were used to standardize the variables (see Chapter 5).

7.4.4 Specification and Interpretation of Interactions and Non-Linear Relationships

Analysts should also exercise caution when interpreting regression parameters associated with predictor variables that define non-linear relationships between predictors and the outcome, or predictor variables that define interactions between two or more predictors. Consider the following linear regression model defining a non-linear (and specifically quadratic) relationship between x and y:

$$y = \beta_0 + \beta_1 x + \beta_2 x^2 + \varepsilon \tag{7.46}$$

In this model, the regression parameters β_1 and β_2 are *linked*, because they involve different transformations of the same predictor x, and β_1 alone in the presence of β_2 has no interpretation. Instead, the parameter β_2 measures the extent of the non-linearity in the relationship between x and y. If the estimate of β_2 suggests that the parameter is not different from 0, one can assume (unless higher order polynomial terms are significant) that the relationship between x and y is linear.

The same issue arises when considering linear regression models that involve interactions between predictors. Consider the following linear regression model:

$$y = \beta_0 + \beta_1 x + \beta_2 z + \beta_3 xz + \varepsilon \tag{7.47}$$

In this model, the three regression parameters β_1, β_2, and β_3 are once again *linked*, and one cannot simply interpret the regression parameter β_1 as being the relationship of the predictor variable x with the response variable y. Instead, this parameter represents the relationship of x with y when z is equal to zero. The relationship of x with y depends on the value of z (and therefore the value of β_3); put differently, different values of z will result in different relationships of x with y.

In general, interactions between two or more predictor variables are computed for entry into a regression model by saving the product of two or more variables as a new variable. Interactions can be easily computed for two or more continuous predictor variables, for two or more categorical predictor variables, or for combinations of both types of variables. When working with categorical predictor variables, relevant products of *all* dummy variables to be included in the model for a given categorical variable must be computed for all other predictor variables that are specified to interact with the categorical predictor. For example, to include the interaction of a three-category predictor with a continuous variable, two product terms must be computed and included in the regression model: the product of the continuous variable with each of the indicators for the two non-reference categories of the categorical predictor. Most software procedures will perform these tasks automatically, but analysts need to be extremely careful when including interaction terms in a regression model.

Plotting predicted values in linear regression models with significant interactions can be extremely helpful when attempting to interpret significant regression parameters (i.e., regression parameters statistically different from zero) associated with the interactions. Social scientists often think of one of the variables involved in an interaction (e.g., z in the example above) as a **moderator variable**, because that variable moderates the relationship of the other variable(s) involved in the interaction with the response variable (i.e., the relationship of x with the response variable depends on the value of z). The plot in Figure 7.2 illustrates the predicted values of a continuous response variable y based on the parameter estimates in two different regression models, showing two possible interactions between a continuous predictor variable x and a binary moderator variable z, taking on values of 1 and 0 for two different groups.

In the left panel of Figure 7.2, there appears to be an interaction of x with z; the relationship of x with the continuous outcome y clearly depends on the value of z. The right panel shows no interaction, because the relationship of x with y in both groups defined by z is essentially the same. Analysts often make the mistake of interpreting regression parameters associated with single predictors in models that include interactions between that predictor and other predictors as "main effects." For example, in the first model fitted in Figure 7.2 (left panel), an analyst may be tempted to interpret the regression parameter associated with the predictor x as being the "main effect" of x. In truth, the relationship of x depends on the value of z, even if the interaction between x and z *is not significant*, so there is no "main effect" of x. This problem can be eliminated by eliminating regression parameters representing interactions from the model if they are not significantly different from zero, which will be discussed in the next section. Nothing about the interpretation of interactions changes when fitting models to complex sample survey data.

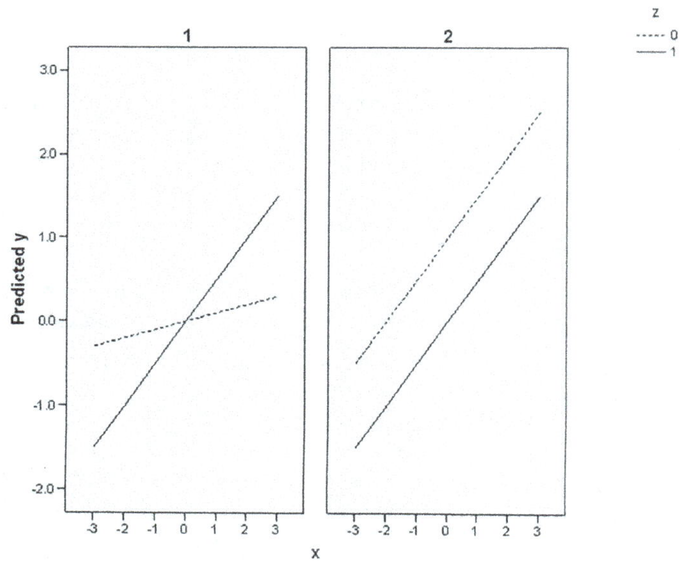

FIGURE 7.2
Hypothetical fits of two linear regression models, one with a significant interaction between x and z (left panel), and one without (right panel).

7.4.5 Model Building Strategies

A universal set of model-building steps that is widely acknowledged and will always lead to the best fitting linear regression model does not exist. Many statisticians have proposed practical guidelines for model fitting, keeping in mind the four steps in regression modeling discussed in Section 7.3. Out of many quality choices, we consider the model building steps proposed by Hosmer et al. (2013). These steps are summarized below:

1. Conduct exploratory bivariate analyses to get a sense of candidate predictors that appear to have a significant relationship with the response variable. Since the focus of this chapter is linear regression for continuous dependent variables, the initial bivariate analyses will be simple linear regression models of y on each of the individual candidate predictors. In later chapters that describe regression models for categorical and ordinal outcomes, other forms of bivariate analyses will be employed (e.g., chi-square tests).
2. Include those predictor variables that are *scientifically relevant* and have a bivariate relationship of significance $p < 0.25$ with the response variable in the initial multivariable model, and possibly consider variable selection techniques (e.g., backward selection), with discretion. Be wary of **multicollinearity**, which could arise from including predictors that are highly correlated with each other in the same model (see Section 7.3.3). Consider whether any initial results change when including *all* the predictors under consideration in the model (and not just those with $p < 0.25$ from the initial bivariate analyses); some predictors may emerge as significant only when adjusting for other predictors.
3. Verify the importance of the predictor variables retained in the model, using t-tests for individual coefficients and Wald tests for multiple coefficients (see Section 7.3.4), and assess whether the coefficients for *all* the predictor variables change substantially in the multivariable model, relative to the bivariate case; this represents the preliminary "main" model.
4. Examine the forms of the predictor variables: if they are categorical, are sample sizes in each category large enough to use the categories as they are in the model? If they are continuous, do they have linear relationships with the response variable? Or do the relationships appear to be non-linear? Residual diagnostics are useful as a part of this step.
5. Consider adding *scientifically relevant* interactions between the predictor variables to the model, one at a time, and do not retain them if they are not significant. Consider interactions involving predictors that may not have been significant in the preliminary "main" model, provided that they are scientifically relevant. Some predictors may not have significant relationships with the response variable by themselves, but may significantly moderate the relationships of other predictors with the response.
6. If any continuous or ordinal predictor variable has many zeroes, include an indicator variable that is equal to 1 for non-zero values and 0 for zero values in the model, in addition to the predictor variable in question, and see if the fit of the model has been improved.

The idea behind this last step (6) is that if we have a **semi-continuous predictor** with many zero values and continuous values, it is unlikely that we would have a true linear relationship passing through the mass of points at 0 and continuing through the range of non-zero values. If we retain the semi continuous predictor and also include the 0/1 indicator of a zero/positive value as a predictor variable in the model, this introduces a discontinuity, modeling the effect of being zero or non-zero and then for non-zero values the linear relationship of the predictor with the response variable.

We remind readers that there are many possible model-fitting strategies that one can follow; Harrell (2015) provides a comprehensive critique of alternative strategies. For example, another reasonable approach would be to start with a "full" multivariable model, including all predictors and interactions that are scientifically relevant, and then use backward selection (starting with the removal of non-significant interactions) with discretion, if variable selection and parsimony is critical. Otherwise, one can simply report inferences based on the full model.

7.5 Application: Modeling Diastolic Blood Pressure With the 2011–2012 NHANES Data

In this practical application of linear regression analysis for complex sample survey data, we consider building a predictive model of diastolic blood pressure (a continuous response variable) based on the probability sample of data collected from the U.S. adult population (ages ≥ 18) in the 2011–2012 National Health and Nutrition Examination Survey (NHANES). After exploring the bivariate relationships of the predictors of interest with diastolic blood pressure, we perform a naïve linear regression analysis that completely ignores the complex design features of the NHANES sample. Next, we perform a weighted regression analysis that ignores the stratification and cluster sampling of the NHANES sample design. Finally, we take all the important design features of the NHANES sample (stratification, cluster sampling, and weighting for unequal probability of selection, nonresponse, and post-stratification) into account at each step of the model building process. We also test whether using the final NHANES survey weights is necessary when estimating the model, explore design-adjusted model diagnostics, and illustrate an alternative Bayesian approach to fitting the final model.

Specifically, the design variables that the online documentation for the 2011–2012 NHANES data set states should be used for variance estimation[1] include SDMVPSU (which contains masked versions, or approximations, of the true primary sampling unit codes for each respondent for the purposes of variance estimation; see Section 4.3.1) and SDMVSTRA (which contains the "approximate" sampling stratum codes for each respondent, for variance estimation purposes). In addition, the appropriate survey weight to be used to generate finite population estimates of the regression parameters for the U.S. adult population for the years of 2011 and 2012 is WTMEC2YR. This survey weight variable was selected for analysis purposes instead of WTINT2YR because variables that will be used in the regression analyses were collected as a part of the physical examination, and the NHANES physical examination was performed on a *subsample* of all respondents (which required adjustments to the survey weights to account for the sub-sampling and nonresponse to the MEC follow-up phase of the NHANES data collection).

7.5.1 Exploring the Bivariate Relationships

In this application, we follow the regression modeling strategies recommended by Hosmer et al. (2013) to build a model for diastolic blood pressure (see Section 7.4.5). We will describe each of the steps explicitly as a part of the example. First, we consider a set of predictors of diastolic blood pressure that are scientifically relevant: age, gender, race/ethnicity, and marital status. We begin by identifying the relevant design variables for the 2011–2012 NHANES sample in Stata, requesting Taylor Series Linearization (TSL) for variance estimation, and indicating that identified strata with single PSUs should result in missing standard errors (see Chapters 3 and 4 for more on this issue):

```
svyset sdmvpsu [pweight = WTMEC2YR], strata(sdmvstra) ///
vce(linearized) singleunit(missing)
```

An initial descriptive summary of the diastolic blood pressure variable in the NHANES data set (BPXDI1) revealed several values of 0, and we set these values to missing in Stata before proceeding with the analysis:

```
gen bpxdi1 _ 1 = BPXDI1
replace bpxdi1 _ 1 = . if BPXDI1 == 0
```

We also generate an indicator variable for the subpopulation of adults (respondents with age greater than or equal to 18), for use in eventual unconditional subclass analyses:

```
gen age18p = 1 if age >= 18 & age != .
replace age18p = 0 if age < 18
```

With the subpopulation indicator defined, we now consider a series of simple bivariate regression analyses, to get an initial exploratory sense of the relationships of the candidate predictor variables with diastolic blood pressure. We make use of the `svy: regress` command to take the survey weights, stratum codes, and cluster codes into account when fitting these simple initial regression models, so that parameter estimates will be unbiased and variance estimates will reflect the complex design features of the NHANES sample. We first compute a weighted estimate of the mean age for the adult subpopulation, and then center the AGE variable at the weighted mean age based on the NHANES sample (46.36):

```
svy, subpop(age18p): mean age
gen agec = age - 46.36
```

Next, the continuous dependent variable, BPXDI1_1, is regressed separately on each of the candidate predictors. For the categorical predictor variables (i.e., race/ethnicity/RIDRETH1, gender/RIAGENDR, and marital status/MARCAT), we consider multi-parameter Wald tests in Stata (see Section 7.3.4) to assess the significance of the bivariate relationships. The Stata software allows users to perform these multi-parameter Wald tests by using `test` commands immediately after the models have been estimated:

```
svy, subpop(age18p): regress bpxdi1 _ 1 i.RIDRETH1
test 2.RIDRETH1 3.RIDRETH1 4.RIDRETH1 5.RIDRETH1

svy, subpop(age18p): regress bpxdi1 _ 1 i.marcat
test 2.marcat 3.marcat
```

```
svy, subpop(age18p): regress bpxdi1_1 i.riagendr
test 2.riagendr

svy, subpop(age18p): regress bpxdi1_1 agec
test agec
```

Note in the four Stata commands above how the indicator for the adult subpopulation (AGE18P) is explicitly specified for the analysis, via the use of the `subpop()` option. This ensures that Stata will perform an unconditional subclass analysis, treating the adult subpopulation sample size as a random variable and taking the full complex design of the NHANES sample into account.

We also note that after the dependent variable has been specified first following the `regress` command, the categorical predictor variables in the regression models are identified with the `i.` modifier, initiating dummy variable (or indicator) coding (referred to as **factor variable coding** in Stata). In the example commands above, we use the default factor variable coding, where the lowest alphanumeric value of a given categorical variable will be set as the reference category, with no corresponding indicator variable included in the model. The terms indicated in the `test` commands above (e.g., `2.RIDRETH1`) then refer to the specific regression parameters for each of the non-reference indicator variables temporarily generated by Stata to fit the model, and the labels for the parameters can be clearly identified in the model output. The `test` commands are used to perform a design-adjusted test of the null hypothesis that all of the listed regression parameters are equal to zero (or, in other words, the mean diastolic blood pressure does not vary as a function of a given predictor variable). Table 7.1 presents the results of these initial bivariate analyses.

Stata presents **adjusted Wald tests** for the parameters in each of these models by default, where the standard Wald F-statistic (Section 7.3.4) is multiplied by $(df - k + 1) / df$, with df = the design-based degrees of freedom, and k = the number of parameters being tested

TABLE 7.1

Initial design-based bivariate regression analysis results assessing potential predictors of diastolic blood pressure for the 2011–2012 NHANES adult sample

Predictor Variable	Parameter Estimate (Linearized SE)	Design-Adjusted Test Statistic	p-value
Race/Ethnicity (n = 5,112)		Wald $F(4,14) = 3.93$	0.02
Mexican American	—	—	—
Other Hispanic	−0.15 (1.46)	$t(17) = -0.11$	0.92
Non-Hispanic White	2.18 (0.74)	$t(17) = 2.94$	< 0.01
Non-Hispanic Black	2.29 (0.70)	$t(17) = 3.26$	< 0.01
Other Race	1.31 (0.70)	$t(17) = 1.85$	0.08
Age (Cent.) (n = 5,112)	0.04 (0.02)	$t(17) = 2.09$	0.05
Gender (n = 5,112)		Wald $F(1,17) = 15.01$	< 0.01
Male	—	—	—
Female	−2.20 (0.57)	$t(17) = -3.87$	< 0.01
Marital Status (n = 4,845)		Wald $F(2,16) = 0.85$	0.45
Married	—	—	—
Previously Married	−0.15 (0.70)	$t(17) = -0.21$	0.84
Never Married	−1.12 (0.84)	$t(17) = -1.33$	0.20

Note: — denotes reference category.

(Korn & Graubard, 1990). Under the null hypothesis, the resulting test statistic follows an F distribution with k and $df - k + 1$ degrees of freedom. For example, in the Wald test for the ethnicity predictor, there are $k = 4$ parameters being tested, and the design-based degrees of freedom are equal to 31 (ultimate clusters) minus 14 (strata), or 17. The denominator degrees of freedom for the adjusted test statistic are therefore $17–4 + 1 = 14$.

Note the different subpopulation sample sizes in Table 7.1; several of the adult cases appear to have missing data on the marital status variable. The design-based multi-parameter Wald tests and t-tests for the single parameters suggest that race/ethnicity, age, and gender each have potentially significant relationships with the response variable (diastolic blood pressure). Specifically, investigating the weighted parameter estimates in these simple models, males, non-Hispanic whites, non-Hispanic blacks, and elderly people appear to have the highest diastolic blood pressures at first glance, while marital status does not appear to be related to diastolic blood pressure. Following the guidelines of Hosmer et al., we therefore include the first three predictors in an initial model for the response variable measuring diastolic blood pressure, but we also consider whether marital status remains non-significant in the initial multivariable model.

7.5.2 Naïve Analysis: Ignoring Sample Design Features

In the first regression analysis, we *ignore* the final survey weights, stratification, and cluster sampling inherent to the NHANES sample design, we do not consider any interactions between the predictors, and we use standard OLS estimation to calculate the parameter estimates for the adult subpopulation:

```
regress bpxdi1_1 i.RIDRETH1 i.riagendr i.marcat agec if age18p == 1
```

When fitting regression models in Stata, the first variable listed after the main command is the response variable (BPXDI1_1), and the variables listed after the response variable represent the predictor variables in the model. The variable list is then generally followed by options (after a comma). In this example, we do not include any options; however, we do restrict the analysis conditionally to those subjects with age ≥ 18 by using the `if` modifier. We also once again use the `i.` modifiers to have Stata automatically generate indicator variables for selected levels of the categorical predictor variables (recall that Stata, by default, treats the lowest-valued level of a categorical predictor as the reference category; see Section 7.4.2 for syntax to manually choose the reference category). Table 7.2 presents OLS estimates of the regression parameters in this preliminary model, along with their standard errors and associated test statistics.

These initial parameter estimates suggest that age has a weak negative relationship with diastolic blood pressure (after adjusting for the other covariates), females tend to have significantly lower mean diastolic blood pressure, and Mexican American respondents tend to have the lowest mean blood pressures (significantly lower than non-Hispanic blacks and other ethnicities). In addition, we note that the coefficient for never married now appears to be significant, suggesting that those who were never married have a significantly lower mean diastolic blood pressure than those who are married. However, these parameter estimates may be biased, because the NHANES survey weights for respondents given a physical examination were not used to calculate nationally representative finite population estimates. In addition, the standard errors are likely understated, because the weights and the stratified cluster sample design of the NHANES sample were not considered. We therefore consider these results only for illustration purposes.

TABLE 7.2

Unweighted OLS estimates of the regression parameters in the initial diastolic blood pressure model

Predictor	Parameter Estimate	Standard Error	*t*-statistic (*df*)	*p*-value	95% CI
Intercept	72.163	0.568	127.13 (4836)	<0.001	(71.050, 73.276)
Ethnicity					
Other Hispanic	0.377	0.756	0.50 (4836)	0.618	(−1.105, 1.859)
Non-Hispanic White	0.942	0.610	1.54 (4836)	0.123	(−0.255, 2.139)
Non-Hispanic Black	2.509	0.640	3.92 (4836)	<0.001	(1.255, 3.763)
Other Race	1.776	0.677	2.62 (4836)	0.009	(0.448, 3.103)
Mexican American	—	—	—	—	—
Marital Status					
Married	—	—	—	—	—
Previously Married	0.697	0.444	−1.57 (4836)	0.116	(−1.567, 0.172)
Never Married	−2.440	0.462	−5.28 (4836)	<0.001	(−3.347, −1.534)
Gender					
Female	−2.532	0.339	−7.46 (4836)	<0.001	(−3.197, −1.867)
Male	—	—	—	—	—
Age (Centered)	−0.019	0.011	−1.77 (4836)	0.077	(−0.041, 0.002)

Notes: n = 4,845, R^2 = 0.021, F-test of null hypothesis that all parameters are 0: F(8, 4,836) = 12.78, p < 0.001
— denotes the reference category

7.5.3 Weighted Regression Analysis

Next, we consider weighted least squares (WLS) estimation for calculating the parameter estimates in the initial model. Note that we explicitly indicate in the Stata command (with the `pweight` option) that the NHANES survey weights for respondents given a physical examination (WTMEC2YR) should be included in the estimation to calculate unbiased estimates of the regression parameters:

```
regress bpxdi1_1 i.RIDRETH1 i.riagendr i.marcat agec ///
if age18p == 1 [pweight = WTMEC2YR]
```

Table 7.3 below presents weighted estimates of the regression parameters, in addition to **robust standard errors** automatically calculated by Stata's standard regression command (`regress`) when survey weights are explicitly specified with the `pweight` option. Recall from earlier chapters that this option generally refers to a "probability" weight; the full NHANES survey weights are the base "probability" weights, adjusted for nonresponse and calibrated to population control totals (see Section 2.7). The "robust" standard errors are "sandwich-type" standard errors (see Freedman, 2006, for an introduction) that are considered robust to possible misspecification of the correlation structure of the observations. In this part of the example, there is some misspecification involved: we have once again *ignored* the stratification and cluster sampling inherent to the 2011–2012 NHANES sample design when calculating the standard errors, meaning that they will likely be understated. Stata's automatic calculation of robust standard errors for the parameter estimates in the presence of survey weights is therefore an effective type of "safeguard" against this failure to incorporate the sample design features in the analysis (meaning that standard errors will not be understated), but we do not recommend following this approach in practice. Readers should be aware that not all software packages capable of

TABLE 7.3

Weighted Least Squares (WLS) estimates of the regression parameters in the initial diastolic blood pressure model

Predictor	Parameter Estimate	Standard Error	t-statistic (df)	p-value	95% CI
Intercept	71.806	0.587	122.33 (4836)	<0.001	(70.655, 72.957)
Ethnicity					
Other Hispanic	−0.095	0.741	−0.13 (4836)	0.898	(−1.547, 1.357)
Non-Hispanic White	1.913	0.625	3.06 (4836)	0.002	(0.688, 3.138)
Non-Hispanic Black	2.619	0.681	3.85 (4836)	<0.001	(1.284, 3.954)
Other Race	1.270	0.719	1.77 (4836)	0.077	(−0.140, 2.680)
Mexican American	—	—	—	—	—
Marital Status					
Married	—	—	—	—	—
Previously Married	0.243	0.610	0.40 (4836)	0.690	(−0.953, 1.440)
Never Married	−1.511	0.639	−2.36 (4836)	0.018	(−2.765, −0.258)
Gender					
Female	−2.510	0.447	−5.61 (4836)	<0.001	(−3.387, −1.633)
Male	—	—	—	—	—
Age (Centered)	−0.008	0.014	−0.54 (4836)	0.589	(−0.035, 0.020)

Notes: n = 4,845, R^2 = 0.018, F-test of null hypothesis that all parameters are 0: $F(8, 4{,}836) = 7.50$, $p < 0.001$
— denotes the reference category

survey data analysis perform this type of calculation automatically when standard regression commands are used with only survey weights specified. We only present this part of the analysis for illustrative purposes.

In Table 7.3, we note fairly large differences in the some of the parameter estimates relative to the OLS case (Table 7.2), especially in terms of the race/ethnicity parameters. When failing to incorporate the survey weights (Table 7.2), the differences between the ethnic groups were either being overstated (note that the difference in means between those with other ethnicities and Mexican Americans is smaller and no longer significant at the 0.05 level) or understated (the difference between Mexican Americans and whites is larger when accounting for the weights, with stronger evidence of a significant difference). The estimates in Table 7.3 represent nationally representative parameter estimates, and use of the OLS estimates in Table 7.2 would have painted an incorrect picture of the relationships of these variables with diastolic blood pressure; the weights appear to be informative about these relationships (see Theory Box 7.3 for more discussion of whether to use weights in regression modeling). We also note that the robust standard errors tend to be larger than the understated standard errors from Table 7.2, where no adjustments to the standard errors were made to account for the complex design features of the 2011–2012 NHANES sample. However, these standard errors are still not fully capturing the complex sample design effects in this case, as the effects of stratification and cluster sampling are still being ignored in the analysis.

7.5.4 Appropriate Analysis: Incorporating All Sample Design Features

We now use the `svy: regress` command in Stata to fit the initial finite population regression model to the adult subpopulation and take *all* the NHANES complex sample design features into account, calculating weighted estimates of the regression parameters and linearized estimates of the standard errors for the parameter estimates (incorporating the stratification and cluster sampling of the 2011–2012 NHANES sample). Note how an unconditional subpopulation analysis is also requested by specifying the binary AGE18P indicator in the `subpop()` option, similar to the bivariate analyses performed previously:

```
svyset sdmvpsu [pweight = WTMEC2YR], strata(sdmvstra)

svy, subpop(age18p): regress bpxdi1 _ 1 i.RIDRETH1 ///
i.riagendr i.marcat agec

estat effects, deff
```

We also use the post-estimation command `estat effects, deff` to request calculation of estimated design effects for the estimated regression parameters. Table 7.4 presents the estimated parameters in this initial "main" model.

First, we note in Table 7.4 that the weighted estimates of the parameters are identical to those reported in Table 7.3 from the previous analysis using WLS, confirming that these calculations are identical (assuming that the exact same data set is being analyzed). Second, the estimated parameters and tests of significance presented in Table 7.4 confirm most of the simple relationships observed in the initial design-based bivariate analyses (Table 7.1) and suggest that the relationships remain similar when taking other predictor variables into account in a multivariable analysis and continuing to account for *all* the sample design features (except for the linear relationship of age with diastolic blood

TABLE 7.4

Design-based estimates of the regression parameters in the initial "main" model for diastolic blood pressure, linearized standard errors for the estimates, design-adjusted test statistics and confidence intervals for the parameters, and design effects for the parameter estimates

Predictor	Est.	Linearized SE	*t*-statistic (*df*)	*p*-value	95% CI	DEFF
Intercept	71.806	0.578	124.3 (17)	< 0.001	(70.587, 73.025)	1.30
Ethnicity						
Other Hispanic	−0.095	1.272	−0.07 (17)	0.941	(−2.779, 2.589)	3.29
Non-Hispanic White	1.913	0.763	2.51 (17)	0.023	(0.304, 3.523)	2.18
Non-Hispanic Black	2.619	0.600	4.37 (17)	<0.001	(1.353, 3.885)	0.78
Other Race	1.270	0.604	2.10 (17)	0.051	(−0.005, 2.545)	0.79
Mexican American	—	—	—	—	—	—
Marital Status						
Married	—	—	—	—	—	—
Previously Married	0.243	0.672	0.36 (17)	0.722	(−1.175, 1.662)	3.00
Never Married	−1.511	0.899	−1.68 (17)	0.111	(−3.409, 0.386)	4.90
Gender						
Female	−2.510	0.539	−4.65 (17)	< 0.001	(−3.648, −1.372)	3.67
Male	—	—	—	—	—	—
Age (Centered)	−0.008	0.020	−0.38 (17)	0.712	(−0.050, 0.035)	4.23

Notes: Subclass $n = 4{,}845$, weighted $R^2 = 0.018$, adjusted Wald test for all parameters: $F(8,10) = 11.74$, $p < 0.001$
— denotes the reference category

pressure). When holding the other predictor variables in this model fixed, Non-Hispanic Whites and Blacks have significantly higher expected diastolic blood pressure values than Mexican Americans, females have significantly lower diastolic blood pressure than males, and interestingly, age does not appear to have a significant linear relationship with diastolic blood pressure. Age and marital status are, therefore, the only predictors that do not appear to be important, but we have only considered a linear relationship of age so far, and we have not considered any interactions. None of the sample sizes for the groups defined by the categorical variables appear to be extremely small, so we do not consider further recoding of these variables.

Compared to Table 7.3, we also note that after appropriately adjusting for all the complex sample design features, the standard errors have changed somewhat. The linearized standard errors now fully reflect the complex sample design effects due to the use of weights in estimation, the cluster sampling, and the stratification, and most of them have increased compared to the robust standard errors reported in Table 7.3. Indeed, seven of the nine estimated design effects presented in Table 7.4 (DEFF) are greater than 1, suggesting that the complex design features of the NHANES sample are generally resulting in a decrease in the precision of the parameter estimates relative to the precision that would have been achieved under a simple random sample of the same size (as would generally be expected when correctly accounting for these design features; see Section 2.4). The effects of the complex design on the standard errors are apparent.

We are willing to sacrifice some of the apparent losses of efficiency in these estimates due to the use of weights in estimation, given that our analyses thus far have suggested that the final survey weights are informative about these estimated regression parameters. We can formally test whether the weights are necessary for fitting this model using a procedure originally outlined by Fuller (1984):

1. Refit the model *without* using the weights for estimation but still accounting for the other complex sampling features (stratification and cluster sampling, or replicate weights if applicable);
2. Add the final survey weight variable to the unweighted model as an additional predictor, along with two-way interactions of the final survey weight variable with all other predictors in the working model of interest; and
3. Perform a design-adjusted Wald test of the null hypothesis that all of the coefficients associated with these new terms involving the final survey weights are equal to zero.

If this null hypothesis holds, the coefficients in the model of interest are not changing significantly as the values of the final survey weight variable change (i.e., the weights are not informative about these coefficients). We can perform this test using the following sequence of Stata commands:

```
* Specify all design features, but not the survey weights
svyset sdmvpsu, strata(sdmvstra)

* Add weights and all two-way interactions w/ the weights
svy, subpop(age18p): regress bpxdi1 _ 1 i.RIDRETH1 ///
i.riagendr i.marcat agec ///
WTMEC2YR c.WTMEC2YR#i.RIDRETH1 c.WTMEC2YR#i.riagendr ///
c.WTMEC2YR#i.marcat c.WTMEC2YR#c.agec
```

```
* Test whether the weights are informative
test WTMEC2YR 2.RIDRETH1#c.WTMEC2YR ///
3.RIDRETH1#c.WTMEC2YR 4.RIDRETH1#c.WTMEC2YR ///
5.RIDRETH1#c.WTMEC2YR 2.riagendr#c.WTMEC2YR ///
2.marcat#c.WTMEC2YR 3.marcat#c.WTMEC2YR ///
c.agec#c.WTMEC2YR
```

Note that continuous variables (including the weights) are specified using `c.` in the interaction terms. This design-adjusted Wald test produces the following output:

```
Adjusted Wald test

(1) WTMEC2YR = 0
(2) 2.RIDRETH1#c.WTMEC2YR = 0
(3) 3.RIDRETH1#c.WTMEC2YR = 0
(4) 4.RIDRETH1#c.WTMEC2YR = 0
(5) 5.RIDRETH1#c.WTMEC2YR = 0
(6) 2.riagendr#c.WTMEC2YR = 0
(7) 2.marcat#c.WTMEC2YR = 0
(8) 3.marcat#c.WTMEC2YR = 0
(9) c.WTMEC2YR#c.agec = 0

    F ( 9, 9) = 12.38
     Prob > F = 0.0005
```

We would therefore reject the null hypothesis that the 2011–2012 NHANES final survey weights are not informative about the estimated regression parameters in this model and proceed with the use of weights in estimation.

Returning to the results in Table 7.4, we make several important observations regarding the test statistics for the regression parameters. First, the degrees of freedom for the *t*-statistics based on the complex sample design of the NHANES (17) are calculated by subtracting the number of strata (14) from the number of sampling error computation units (31). These degrees of freedom are substantially different from those noted in Tables 7.2 and 7.3 ($df = 4{,}836$), where the complex sample design was not considered when performing the estimation. This shows how the primary sampling units (rather than the unique respondents) are providing the independent contributions to the estimation of distributional variance when one accounts for the complex sample design.

In addition, Stata presents an adjusted Wald test for all the parameters in the model (see the discussion of the Table 7.1 results). The numerator degrees of freedom for this adjusted statistic are equal to $k = p - 1$ (8 in this example, because eight parameters are being tested; the "null" or "reduced" model still contains the intercept parameter), and the denominator degrees of freedom are calculated as $df - k + 1$ (17–8 + 1 = 10 in this example). This adjusted Wald test suggests that a null hypothesis that all of the regression parameters are equal to 0 would be strongly rejected.

7.5.5 Initial Model Diagnostics

We now consider some initial model diagnostics to assess the fit of this preliminary model. After fitting the model that generated the results in Table 7.4, we save the residuals in a new

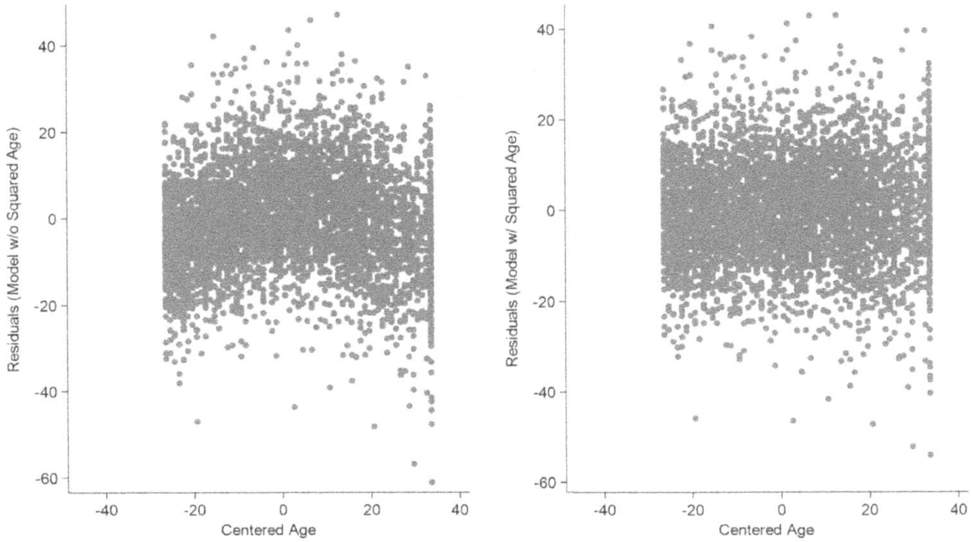

FIGURE 7.3
Plots of residuals vs. centered age for the diastolic blood pressure model, before and after the addition of the squared AGE variable to the model.

variable (RESIDS) in Stata, and then plot the residuals against the values of the continuous mean-centered age (AGEC) variable. The left-hand panel of Figure 7.3 presents this plot.

```
predict resids, resid
scatter resids agec
```

The first plot in Figure 7.3 indicates a well-defined curvilinear pattern of the residuals as a function of (centered) age, suggesting that the regression function has been misspecified. There is evidence that age has a quadratic relationship with diastolic blood pressure that has not been adequately captured by including a linear relationship of age with the response variable. We therefore add a squared version of age (AGECSQ) to the model to capture this relationship:

```
gen agecsq = agec * agec

svy, subpop(age18p): regress bpxdi1_1 i.RIDRETH1 ///
i.riagendr i.marcat agec agecsq

estat effects

predict resids2, resid
scatter resids2 agec
```

In the new model (see Table 7.5), the regression parameters for both the centered age predictor and the squared version of the age predictor are significantly different from 0 ($p < 0.01$), confirming that the relationship of age with diastolic blood pressure is in fact non-linear and quadratic in nature. The weighted R-squared of the new model becomes

TABLE 7.5

Estimates of the regression parameters in the intermediate model for the diastolic blood pressure response variable, prior to the inclusion of interaction terms

Predictor	Est.	Linearized SE	t-statistic (df)	p-value	95% CI	DEFF
Intercept	74.461	0.615	121.1 (17)	< 0.001	(73.164, 75.759)	1.50
Ethnicity						
Other Hispanic	0.015	1.114	0.01 (17)	0.990	(−2.335, 2.364)	2.77
Non-Hispanic White	1.822	0.793	2.30 (17)	0.035	(0.149, 3.495)	2.55
Non-Hispanic Black	2.064	0.612	3.37 (17)	0.004	(0.772, 3.355)	0.90
Other Race	1.098	0.584	1.88 (17)	0.077	(−0.134, 2.330)	0.80
Mexican American	—	—	—	—	—	—
Marital Status						
Married	—	—	—	—	—	—
Previously Married	0.556	0.608	0.91 (17)	0.373	(−0.726, 1.838)	2.69
Never Married	1.135	0.758	1.50 (17)	0.152	(−0.463, 2.734)	3.51
Gender						
Female	−2.214	0.509	−4.35 (17)	<0.001	(−3.289, −1.140)	3.56
Male	—	—	—	—	—	—
Age (Centered)	0.077	0.020	3.77 (17)	0.002	(0.034, 0.120)	4.54
Age (Centered) Squared	−0.012	0.001	−13.65 (17)	<0.001	(−0.013, −0.010)	2.88

Notes: Subclass n = 4,845, weighted R^2 = 0.098, adjusted Wald test for all parameters: $F(9,9)$ = 76.83, p < 0.001
— denotes the reference category

0.098, suggesting an improved fit by allowing the relationship of age with diastolic blood pressure to be non-linear. The right-hand panel of Figure 7.3 shows the improved distribution of the residuals as a function of age after adding the squared term, where there is no pattern evident in the residuals as a function of age.

7.5.6 Testing and Interpreting Interactions

Next, we consider testing specific interactions of interest, one at a time: the interactions between age (both predictors) and race/ethnicity, and the interactions between age (both predictors) and gender. This step essentially allows for testing whether the non-linear relationship of age with diastolic blood pressure tends to be moderated by these two demographic factors; for example, is the quadratic trend in diastolic blood pressure as a function of age flatter (i.e., more stable) for certain ethnic groups than others? We first add the interactions between age and race/ethnicity to the model, and then investigate a design-adjusted Wald test of the null hypothesis that all eight parameters associated with these two interactions are simultaneously equal to zero:

```
svy, subpop(age18p): regress bpxdi1_1 i.RIDRETH1 ///
i.riagendr i.marcat agec agecsq ///
i.RIDRETH1#c.agec i.RIDRETH1#c.agecsq

test 2.RIDRETH1#c.agec 3.RIDRETH1#c.agec ///
4.RIDRETH1#c.agec 5.RIDRETH1#c.agec 2.RIDRETH1#c.agecsq ///
3.RIDRETH1#c.agecsq 4.RIDRETH1#c.agecsq 5.RIDRETH1#c.agecsq
```

Note in the `svy: regress` command how the interactions are specified: the term `i.RIDRETH1#c.agec` indicates that an interaction (#) between the categorical predictor `RIDRETH1` (`i.`) and the *continuous* predictor AGEC (indicated with `c.`) should be included in the model. The subsequent `test` command then simply lists all eight parameters defining the two interaction terms, exactly as they are labeled in the output (including the `c.` notation for the continuous predictors).

We remind readers that when using the `test` commands in conjunction with survey regression commands, Stata performs an adjusted Wald test by default. The `nosvyadjust` option can be added to a `test` command if a user does not desire the additional adjustment to the test statistic. The multi-parameter Wald test for all the newly added interaction parameters essentially amounts to a design-based test of **change in R-squared** for comparing nested models (where in this case, one model includes the interactions, and one does not). The adjusted Wald test performed by Stata indicates that we can reject this null hypothesis ($F(8,10) = 5.92$, $p < 0.01$), which suggests that adding the interactions between both age terms and ethnicity is significantly improving the fit of the model (we will return to their interpretation shortly).

Next, we add the two-way interactions between the two age terms and gender (RIAGENDR) to the model, and again test the associated parameters using a design-adjusted Wald test:

```
svy, subpop(age18p): regress bpxdi1_1 i.RIDRETH1 ///
i.riagendr i.marcat agec agecsq ///
i.RIDRETH1#c.agec i.RIDRETH1#c.agecsq ///
i.riagendr#c.agec i.riagendr#c.agecsq

test 2.riagendr#c.agec 2.riagendr#c.agecsq
```

This Wald test once again suggests that at least one of the two regression parameters associated with the interactions between the age terms and gender is significantly different from zero ($F(2,16) = 4.09$, $p = 0.037$), so we have evidence in favor of the model including these interactions as well. Readers can use similar methods to test interactions between two (or more) categorical predictors.

After fitting this latest model including the new interaction terms, we can plot **marginal predicted values** for diastolic blood pressure based on the fitted model to assess variability in the relationship of age with diastolic blood pressure depending on race/ethnicity and gender. To do this, we first refit the model specifying an "interaction" between the original centered age variable and itself, so that Stata does not simply interpret the squared age term from the previous syntax as an arbitrary additional predictor, and instead knows that there is a curvilinear relationship between centered age and the outcome:

```
svy, subpop(age18p): regress bpxdi1_1 i.RIDRETH1 ///
i.riagendr i.marcat agec c.agec#c.agec ///
i.RIDRETH1#c.agec i.RIDRETH1#c.agec#c.agec ///
i.riagendr#c.agec i.riagendr#c.agec#c.agec
```

While this syntax may seem tedious, it is essential for correctly plotting curvilinear relationships involving continuous predictor variables. We generate the plots of marginal predicted values for diastolic blood pressure using the following syntax:

```
margins RIDRETH1, at(agec=(-30(5)30))
marginsplot

margins riagendr, at(agec=(-30(5)30))
marginsplot
```

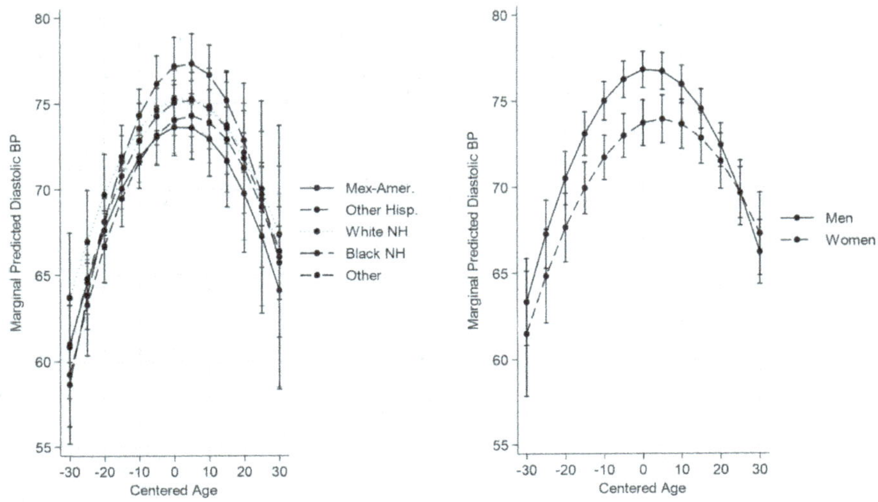

FIGURE 7.4
Plots of marginal predicted values based on the regression model including interactions between age and race/ethnicity as well as age and gender, illustrating differences in the curvilinear relationship of centered age with diastolic blood pressure depending on race/ethnicity and gender.

This `margins` syntax indicates that we wish to display marginal predicted values of diastolic blood pressure for each level of RIDRETH1 (plot 1) and RIAGENDR (plot 2), separately at all possible increments of five units in centered age (from −30 to 30; that is, −30, −25, −20, . . ., 25, 30). The marginal predicted values are computed by default in Stata by: 1) assuming that every case in the data set has the same particular values of RIDRETH1 (or RIAGENDR) and centered age, 2) computing the predicted values based on the fitted model for each case in the data set, and then 3) averaging the predictions. We then plot these marginal predicted values (including 95% confidence intervals for the predictions, where standard errors of the predictions are computed using the delta method) using the marginsplot command. The resulting plots are displayed in Figure 7.4.

Figure 7.4 shows that African Americans have a higher acceleration in diastolic blood pressure as a function of age, up until about middle age. The same can be said for males, where the gap between males and females in marginal predicted values widens approaching middle age, and then narrows in older ages. While these differences are subtle, they are significant (as shown earlier).

7.5.7 Design-Adjusted Diagnostics for the Full Model

We now carefully consider diagnostics for this latest model. We first refit the model, and then save variables containing residuals (EHAT1) and predicted values (YHAT1) based on the fitted model. Next, we generate a series of diagnostic plots to assess the assumptions underlying the model:

```
svy, subpop(age18p): regress bpxdi1 _ 1 i.RIDRETH1 ///
i.riagendr i.marcat agec c.agec#c.agec ///
i.RIDRETH1#c.agec i.RIDRETH1#c.agec#c.agec ///
i.riagendr#c.agec i.riagendr#c.agec#c.agec
```

```
predict ehat1, resid

symplot ehat1, name(sym_ehat1_1, replace) ///
title(Symplot of Residuals)

histogram ehat1, normal name(h_ehat1, replace) ///
title(Histogram of Residuals)

qnorm ehat1, name(qnorm_ehat1, replace) ///
title(Normal Q-Q Plot of Residuals)

predict yhat1, xb

scatter ehat1 yhat1, name(ehat1xyhat1, replace) ///
title(Residuals vs. Predicted Y)

graph combine sym_ehat1_1 h_ehat1 qnorm_ehat1 ///
ehat1xyhat1, rows(2)
```

We introduce one additional diagnostic plot that can be helpful for assessing model fit. The `symplot` command in Stata produces a **symmetry plot** of values on the residual variable based on the model, which is useful for determining whether the distribution of values on a given continuous variable appears to be symmetric in nature. Specifically, the distance below the median of the first ordered value in the distribution is plotted against the distance above the median of the last ordered value; the distance below the median of the second ordered value is plotted against the distance above the median of the second-to-last ordered value; and so forth. If the values in the symmetry plot lie on the straight diagonal line, there is evidence that the distribution is symmetric. The resulting plot in the upper-left panel of Figure 7.5 suggests that the distribution of the residuals based on the final fitted model is nearly symmetric around 0 (allowing for a handful of outliers), alleviating possible concerns about slight deviations from normality.

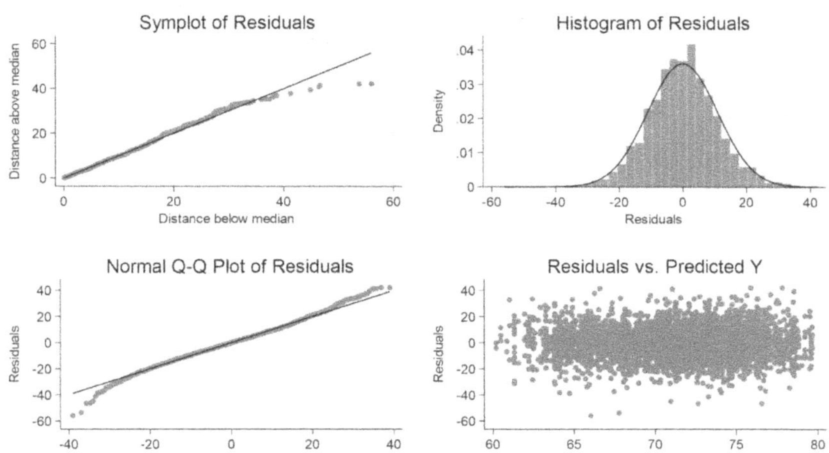

FIGURE 7.5
Diagnostic plots for the "final" regression model fitted to the diastolic blood pressure response variable in the 2011–2012 NHANES data set.

Collectively, the diagnostic plots presented in Figure 7.5 suggest that the residuals follow a symmetric distribution, and that assumptions of normality and constant variance for the residuals seem reasonable. Given any apparent violations of normality in the distribution of the residuals, the symmetry of the residuals around the expected value of 0 gives us confidence in the inferences that we are making. However, the diagnostic assessments that we have performed so far have neglected to account for the complex sampling features.

We now examine the residuals and check for the possibility of influential points and outliers in a manner that accounts for the complex sampling features. We do this using state-of-the-art diagnostic techniques for linear regression models fitted to complex sample survey data, which were discussed earlier in Section 7.3.3. These diagnostic tools have been implemented in the contributed R package `svydiags`. First, we load the contributed `survey` and `svydiags` packages in R, assuming that they have already been installed from a CRAN mirror:

```
library(survey)
library(svydiags)
```

We then load the 2011–2012 NHANES data, recode the dependent variable, compute the centered age variable, create a survey design object describing the complex sampling features, refit the final regression model discussed above using the `svyglm()` function (note that individual interactions are specific using a colon), and then generate the parameter estimates, standard errors, and tests of significance for the parameters, using the same approximate degrees of freedom method (number of clusters minus number of strata) employed by Stata:

```
# read in the full NHANES data set for the third edition
load("X:/. . ./nhanes1112.rdata")

nhanes1112$bpxdi1.1 <- nhanes1112$bpxdi1
nhanes1112$bpxdi1.1[nhanes1112$bpxdi1 == 0] <- NA
nhanes1112$agec <- nhanes1112$age - 46.36
nhanes1112$agec2 <- nhanes1112$agec ^ 2

dnhanes <- svydesign(id =~ sdmvpsu, strata =~ sdmvstra,
weights =~ wtmec2yr, nest = TRUE, data = nhanes1112)

finmod <- svyglm(bpxdi1.1 ~ as.factor(ridreth1) +
as.factor(riagendr) + agec + agec2 +
as.factor(ridreth1):agec + as.factor(ridreth1):agec2 +
as.factor(riagendr):agec + as.factor(riagendr):agec2,
subset = (age18p == 1), design = dnhanes)

summary(finmod, df.resid = degf(dnhanes))
```

Now that we have replicated the model fitting process in R and created an object named `finmod` containing all of the relevant information about the model, we can first employ the `svyCooksD()` function in the `svydiags` package to generate the modified Cook's D statistic for each case, measuring the influence of that case on the parameter estimates in the final model:

```
mcook <- svyCooksD(mobj = finmod, stvar = "sdmvstra",
clvar = "sdmvpsu", doplot = TRUE)
```

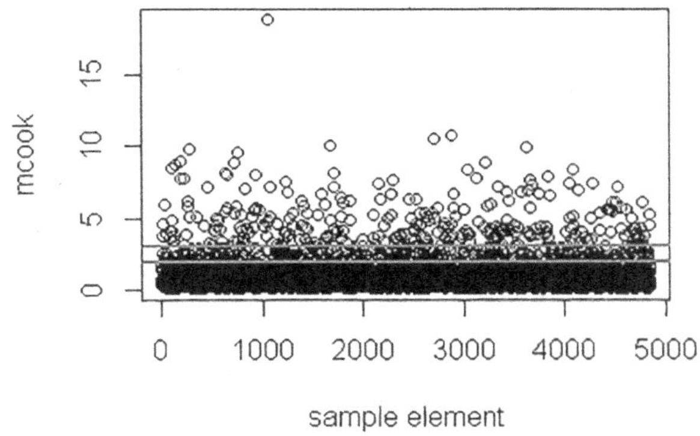

FIGURE 7.6
Plot of modified Cook's D statistics for the n = 4,845 cases analyzed in the final model, demonstrating the influence of each individual case and accounting for the complex sampling features of the 2011–2012 NHANES.

The plot of the modified Cook's D statistics for each case generated by the `doplot = TRUE` argument is shown in Figure 7.6. The two horizontal lines indicate rules of thumb for which cases have overly large influence on the parameter estimates in the model (values of 2 or 3). We see that several cases fall above the highest horizontal line, which warrants refitting the model excluding some of these extreme cases.

We specifically identify the four cases with the largest modified Cook's D statistics, and create a temporary data frame object excluding these specific rows:

```
mcook[mcook > 10]
    2168      3459      5543      5904
18.78218  10.01648  10.47560  10.80141

nhanes1112a <- nhanes1112[-c(2168,3459,5543,5904),]
```

We then create a new survey design object using this new data frame, and refit the model excluding these cases:

```
dnhanes2 <- svydesign(id =~ sdmvpsu, strata =~ sdmvstra,
weights =~ wtmec2yr, nest = TRUE, data = nhanes1112a)

finmod2 <- svyglm(bpxdi1.1 ~ as.factor(ridreth1) +
as.factor(riagendr) + as.factor(marcat) + agec + agec2 +
as.factor(ridreth1):agec + as.factor(ridreth1):agec2 +
as.factor(riagendr):agec + as.factor(riagendr):agec2,
subset = (age18p == 1), design = dnhanes2)

summary(finmod2, df.resid = degf(dnhanes2))
```

The results of this analysis (not shown) suggest slight shifts in the resulting parameter estimates, but none large enough to cause us to change our inferences. One could explore the impact of removing additional cases with large values of Cook's statistic at this point, but we do not consider these further. In practice, if the exclusion of a small number of influential points results in different inferences, a sound argument would be needed for why those individual cases should be removed from the analysis (e.g., extreme values for specific variables).

Next, we examine design-based calculations of **dfbetas**, which indicate the influence of individual observations on the parameter estimates in the final model. Using the original model fit object, we compute these values using the following function in the svydiags package:

```
dfbetas <- svydfbetas(mobj = finmod, stvar = "sdmvstra",
clvar = "sdmvpsu")
```

The resulting object shows, for each case used to fit the model, the changes in each of the parameter estimates that would occur if a given case was excluded from the analysis. We examine these changes for the case in row 2168, which was found above to have the largest modified Cook's D statistic:

```
b <- data.frame(dfbetas$Dfbetas)

b$X2168
```

```
[1]  -0.019095264  0.015358191  0.017390378  0.016738493  0.018028044  0.006663257 -0.005607790
[8]  -0.001839896  0.061285194  0.065315288 -0.043712952 -0.040747920 -0.058707437 -0.037383482
[15] -0.039585347 -0.075196788 -0.061617958 -0.058622136 -0.009626907 -0.009821412
```

The values above correspond to the changes in the estimates of the parameters that would occur from dropping this one case; some are noticeable. These examinations can therefore be used to quantify the impacts that individual cases are having on the parameter estimates, fully accounting for the complex sampling features in the calculations. We recommend performing these analyses alongside an examination of the Cook's D statistics to describe the impact that a given case is having on the parameter estimates if that case is being considered for exclusion from the analysis.

Next, we assess potential multicollinearity among the predictors in the final model. We start with the calculation of design-adjusted condition indices, using the approach outlined in Liao and Valliant (2012b). To implement this approach, we use the svycollinear() function in the svydiags package, which requires the user to work with the exact same data set that was used to fit the model of interest (i.e., all complete cases in that data set, including consideration of the complex sample design variables). We therefore create this complete data frame object first, and then use the following R code to set up this analysis:

```
nhanes.red <- nhanes1112[, c("bpxdi1.1", "ridreth1",
"riagendr", "marcat", "agec", "agec2", "age18p",
"wtmec2yr", "sdmvstra", "sdmvpsu")]
nhanes.red <- nhanes.red[complete.cases(nhanes.red),]

dnhanes <- svydesign(id =~ sdmvpsu, strata =~ sdmvstra,
weights =~ wtmec2yr, nest = TRUE, data = nhanes.red)
```

```
finmod <- svyglm(bpxdi1.1 ~ as.factor(ridreth1) +
as.factor(riagendr) + as.factor(marcat) + agec + agec2 +
as.factor(ridreth1):agec + as.factor(ridreth1):agec2 +
as.factor(riagendr):agec + as.factor(riagendr):agec2, design = dnhanes)

# need numeric version of design matrix for model
X <- model.matrix(~ as.factor(ridreth1) +
as.factor(riagendr) + as.factor(marcat) + agec + agec2 +
as.factor(ridreth1):agec + as.factor(ridreth1):agec2 +
as.factor(riagendr):agec + as.factor(riagendr):agec2, data = nhanes.red)

svycollinear(mod=X, w=nhanes.red$wtmec2yr,
Vcov=vcov(finmod), sc=TRUE, svyglm.obj=FALSE, rnd=3, fuzz=0.3)
```

We note that in this example, we create a numeric version of the **design matrix** used for the final model, including all the binary indicators for the factor variables. We cannot simply input the factor variables themselves for this function. As a result, we do not use the `svyglm` object created from the model fitting (`finmod`) directly for the mod argument in this analysis, which is why we input X instead and say that `svyglm.obj = FALSE`. The largest design-adjusted condition index that results is 24.225, which does not suggest an extreme problem with multicollinearity (where values greater than 30 generally indicate evidence of an issue; see Faraway, 2014). Given that this largest condition index is still approaching 30, we also investigate **variance inflation factors (VIFs)** associated with the individual coefficients.

We can compute design-adjusted VIFs for the coefficients in this final model using the approach described in Liao and Valliant (2012a). This can be accomplished using the design matrix and model fit objects created previously (removing the intercept column from the design matrix):

```
svyvif(mobj=finmod, X=X[,-1], w=nhanes.red$wtmec2yr, stvar = "sdmvstra",
clvar = "sdmvpsu")
```

We see some relatively large design-adjusted VIFs (labeled as `svy.vif` in the output), some of which are not surprising (e.g., those associated with squared age, which is a direct function of age). VIFs larger than 2.0 generally indicate potential concerns with variance inflation due to multicollinearity (Faraway, 2014). While we are confident that no sociodemographic predictors with large correlations have been included in this model, any unexpectedly large VIFs should be carefully investigated to potentially simplify the model and avoid unnecessary variance inflation in estimated coefficients due to high correlations of the predictor variables.

Finally, we compute standardized residuals based on the final fitted model, or residuals divided by the standard deviation of the residuals based on the fitted model (again accounting for the complex sampling features in the calculations). We can generate the standardized residuals by using the `svystdres()` function in the `svydiags` package:

```
st.resids <- svystdres(mobj = finmod, stvar = "sdmvstra",
clvar = "sdmvpsu", doplot = TRUE)
```

The plot of the standardized residuals that is automatically generated when using the `doplot = TRUE` argument (Figure 7.7) draws horizontal lines at values of –3 and 3 for the standardized residuals, where cases with residuals outside of this range would normally be considered outliers.

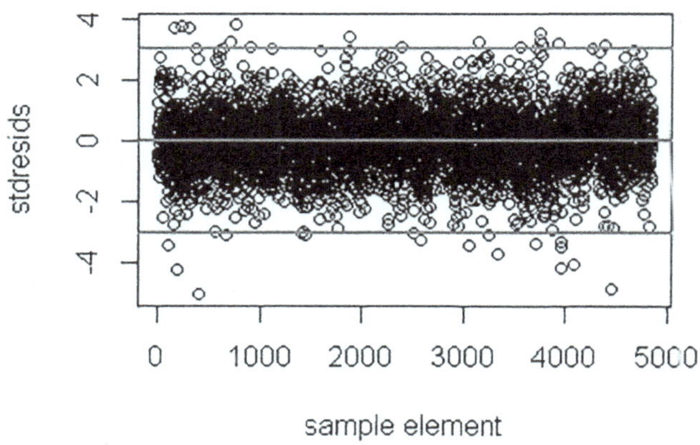

FIGURE 7.7
Plot of standardized residuals for the n = 4,845 cases analyzed in the final model, accounting for the complex sampling features of the 2011–2012 NHANES.

We identify those cases with standardized residuals less than −4:

```
st.resids$stdresids[st.resids$stdresids < -4]
     2735       3732       3138       1158       5815
-4.186686  -5.030542  -4.179745  -4.036181  -4.826700
```

These are cases that are not being fitted adequately by the final model. For example, the case in row 3732 was 80 years old and had a recorded diastolic blood pressure of 10 (!). Careful assessment of these outliers (accounting for the complex sampling features) can lead to the identification of situations like this, where there may have been data entry errors. If a value of 10 is unrealistic or not possible, this case could be dropped from the analysis moving forward.

7.5.8 Interpretation of Parameter Estimates in the Final Model

After carefully assessing the model diagnostics (accounting for the complex sampling features if possible) and possibly making changes to the model specification or dropping overly influential cases or extreme outliers (with discretion), the focus can turn to interpretation of the parameter estimates in the final model. Table 7.6 presents the parameter estimates and design-adjusted 95% confidence intervals for the parameters in the "final" model for this example.

The estimates in Table 7.6 provide strong evidence of the quadratic relationship of age with diastolic blood pressure when adjusting for other socio-demographic features, and show how there are still strong ethnicity and gender effects on blood pressure. We once again note the substantial improvement in the model R-squared values (from 0.018 to 0.105) due to the inclusion of the squared age term and the additional interactions.

If we wish to describe our "final" interpretation of the parameter estimates in this model, we need to be very careful with the estimated interaction coefficients. Importantly, the estimated coefficients for ethnicity, gender, and the two age predictors are *not* main effects for the entire population. Instead, because of the interactions included in the final model, these coefficients correspond to subgroups defined by the *reference categories* (or "0" values)

TABLE 7.6

Estimates of the regression parameters and 95% confidence/credible intervals for the parameters based on the final model for the diastolic blood pressure response variable (including significant interaction terms), following the standard design-based approach, the "q-weighted" approach outlined by Pfeffermann (2011), and the design-adjusted Bayesian approach.

	Design-Based Approach		q-weighted Approach		Bayesian Approach	
Predictor	**Est.**	**95% CI**	**Est.**	**95% CI**	**Posterior Mean**	**95% Credible Int.**
Intercept	74.93	(72.96, 76.90)	75.44	(73.40, 77.48)	74.90	(73.32, 76.64)
Ethnicity						
Other Hisp.	0.41	(−1.66, 2.47)	0.73	(−1.33, 2.79)	0.43	(−1.95, 2.64)
Non-Hisp. White	1.61	(−0.38, 3.61)	1.41	(−0.85, 3.66)	1.63	(−0.05, 3.28)
Non-Hisp. Black	3.52	(1.43, 5.61)	3.41	(1.19, 5.62)	3.55	(1.57, 5.51)
Other Race	1.42	(−0.40, 3.24)	1.17	(−0.94, 3.28)	1.45	(−0.75, 3.62)
Mexican American	—	—	—	—	—	—
Marital Status						
Married	—	—	—	—	—	—
Previously Married	0.29	(−1.06, 1.64)	—0.25	(−1.34, 0.84)	0.29	(−0.57, 1.13)
Never Married	1.11	(−0.53, 2.74)	0.55	(−0.70, 1.80)	1.12	(0.19, 1.95)
Gender						
Female	-3.10	(-4.70, −1.51)	-3.44	(-4.78, −2.10)	-3.09	(-3.98, −2.25)
Male	—	—	—	—	—	—
Age (Centered)	0.03	(−0.06, 0.11)	0.04	(−0.05, 0.13)	0.02	(−0.06, 0.10)
Age (Cent.) Squared	−0.01	(−0.02, −0.01)	−0.01	(−0.02, −0.01)	−0.01	(−0.02, −0.01)
Age × Ethnicity						
Age × Other Hisp.	0.06	(−0.05, 0.17)	0.02	(−0.11, 0.16)	0.07	(−0.04, 0.17)
Age × White	0.01	(−0.11, 0.13)	−0.002	(−0.13, 0.12)	0.01	(−0.07, 0.09)
Age × Black	0.07	(−0.02, 0.15)	0.07	(−0.01, 0.15)	0.07	(−0.02, 0.16)
Age × Other	0.04	(−0.09, 0.17)	0.04	(−0.09, 0.17)	0.04	(−0.06, 0.15)
Age Squared × Ethnicity						
Age Sq. × Other Hisp.	−0.001	(−0.008, 0.007)	−0.003	(−0.013, 0.007)	0.00	(−0.01, 0.01)
Age Sq. × White	0.002	(−0.003, 0.006)	0.002	(−0.003, 0.007)	0.00	(0.00, 0.01)
Age Sq. × Black	−0.004	(−0.009, 0.001)	−0.003	(−0.009, 0.003)	0.00	(−0.01, 0.00)
Age Sq. × Other	−0.001	(−0.006, 0.005)	0.001	(−0.005, 0.007)	0.00	(−0.01, 0.01)
Age × Gender						
Age × Female	0.05	(−0.01, 0.10)	0.03	(−0.03, 0.09)	0.05	(0.01, 0.09)
Age Sq. × Gender						
Age Sq. × Female	0.003	(−0.001, 0.007)	0.003	(−0.001, 0.007)	0.00	(0.00, 0.01)

Source: Subclass $n = 4{,}845$; weighted $R^2 = 0.105$

of the other variables involved in the interaction. So, the four estimated coefficients for the non-reference race/ethnicity categories correspond to race/ethnicity effects *specifically for individuals with a value of 0 on the two age variables* (or individuals with age equal to the population mean, given the centered age variable). It would therefore be erroneous to say, based on Table 7.6, that African Americans have a mean diastolic blood pressure that is higher by 3.52 units than Mexican Americans *in general*; this interpretation is only true *for individuals with age equal to the population mean.*

In the same manner, one would interpret the estimated coefficients for age and age-squared not as main effects for the entire population, but rather effects of age *for the reference categories of race/ethnicity and gender.* That is, for Mexican American males, the relationship of age with diastolic blood pressure is defined by a quadratic equation with intercept 74.93, linear age coefficient 0.028 (not significantly different from zero, based on the 95% confidence interval), and quadratic age coefficient −0.014 (significantly different from zero). The estimated interaction coefficients then represent *changes* in these age coefficients corresponding to the non-reference categories of race/ethnicity and gender. So why did we find in the Wald test that the overall interaction terms involving race/ethnicity were significant, given that none of these "change" coefficients seem to be different from zero in Table 7.6?

Keep in mind that we are looking at changes in relationships relative to *specific* reference categories. The difference in the quadratic age coefficients for whites and African Americans is largest, but we are not getting a test of this difference (given that Mexican American is the reference category). If we were to change the reference category to white, the change coefficient for African Americans would now be positive and significant (at the 5% level). In general, if one finds significant interactions in a linear regression model (or any regression model for that matter), very interesting insights are possible regarding subgroup differences in the relationships of interest. Great care is simply needed in interpreting these relationships.

7.5.9 The "q-weighted" Approach

As we noted in Table 7.5, design effects can be rather large for estimated regression coefficients when fully accounting for all the complex sample design features. Sometimes, these design effects (which we *should* be accounting for in the analysis) are driven by excess variability in the final survey weights. Are more efficient estimates possible, while still using the weights in estimation and accounting for the other design features as well?

Pfeffermann (2011) proposed a "q-weighted" approach for increasing the efficiency of weighted estimates of regression coefficients in design-based analyses. We now apply this approach to our "final" model and determine whether we can generate similar weighted estimates with increased efficiency. This approach proceeds as follows:

1. Fit a regression model to the final survey weights using the predictor variables in the regression model of substantive interest. [We note that under mild regularity conditions, this "q-weighted" approach will ultimately yield consistent estimates of the finite population regression parameters even if this initial regression model for the weights is misspecified in some way; see Pfeffermann (2011, p. 124).] In our example:

```
regress WTMEC2YR i.RIDRETH1 i.riagendr i.marcat ///
agec c.agec#c.agec i.RIDRETH1#c.agec ///
i.RIDRETH1#c.agec#c.agec i.riagendr#c.agec ///
i.riagendr#c.agec#c.agec if age18p == 1
```

2. Save the model-based predicted values of the survey weights for each case as a function of the predictor variables in the data set:

```
predict w_hat if age18p == 1, xb
```

3. Set any negative predictions to the smallest positive predicted value, and divide the final survey weights by their model-based predicted values:

```
summarize w _ hat, detail
replace w _ hat = 4809 if w _ hat < 0
gen q _ WTMEC2YR = WTMEC2YR / w _ hat
```

4. Use the new "q-weights" as the final survey weights in estimating the final regression model:

```
svyset sdmvpsu [pweight = q _ WTMEC2YR], ///
strata(sdmvstra)

svy, subpop(age18p): regress bpxdi1 _ 1 i.RIDRETH1 ///
i.riagendr i.marcat agec c.agec#c.agec ///
i.RIDRETH1#c.agec i.RIDRETH1#c.agec#c.agec ///
i.riagendr#c.agec i.riagendr#c.agec#c.agec
```

We compare weighted estimates and 95% confidence intervals based on this approach with those based on the original weights in Table 7.6. Some estimates (e.g., the gender coefficient) have improved efficiency, in terms of narrower confidence intervals, while others do not. In this example, there do not seem to be significant benefits of this approach. However, given the ease of implementing this approach and its demonstrated empirical benefits in terms of efficiency (Pfeffermann, 2011), we recommend that analysts fitting regression models to complex sample survey data consider applying it before presenting any final estimates.

7.5.10 Alternative Bayesian Approach

Finally, we consider an alternative design-adjusted Bayesian approach to fitting the final regression model identified following the steps outlined in this section. This Bayesian approach could be followed at each step of the model building process, but we illustrate it here for the "final" model estimated in this example.

Implementing the design-adjusted Bayesian approach to fitting a linear regression model is a straightforward extension of the simple intercept-only models that were fitted to the continuous variables of interest in Chapter 5, using the `csSampling` package in R. The only difference is that we need to add the predictor variables of interest to the specified models, and possibly specify prior distributions for the coefficients in the models (if prior information is available). We illustrate the Bayesian approach to fitting the final model in this example under the default assumption of a flat (non-informative) prior distribution for the vector of regression coefficients in the final model. We use the same reduced version of the NHANES data set for this example (with complete cases only) that we used to assess condition indices:

```
# need to normalize NHANES weights to match Stan
nhanes.red$wtsc <- nhanes.red$wtmec2yr / mean(nhanes.red$wtmec2yr)

# survey design object
nhanes.des <- svydesign(id = ~sdmvpsu, strata = ~sdmvstra,
weights = ~wtsc, nest = T, data = nhanes.red)
```

```
# Bayesian approach, flat prior
set.seed(41279)

model _ formula <- formula("bpxdi1.1|weights(wtsc) ~
as.factor(ridreth1) + as.factor(riagendr) +
as.factor(marcat) + agec + agec2 + as.factor(ridreth1):agec +
as.factor(ridreth1):agec2 + as.factor(riagendr):agec +
as.factor(riagendr):agec2")

mod.brms <- cs _ sampling _ brms(svydes = nhanes.des,
brmsmod = brmsformula(model _ formula), data = nhanes.red,
family = gaussian())

# Print results
mod.brms$stan _ fit
```

The means of the design-adjusted draws of each regression parameter from the survey-weighted posterior distribution for the regression parameters, in addition to 95% credible intervals for each regression parameter based on the design-adjusted draws, are included in Table 7.6. The output (not shown here; see the ASDA website for details) is a bit tricky to navigate, but each regression parameter is listed as `b[X]` in the same order as the output produced by `svyglm()`, and the intercept parameter is listed as `b _ Intercept`. Overall, we see posterior means quite like the point estimates produced by the design-based and q-weighted approaches, in addition to narrower 95% credible intervals, reflecting the gains in efficiency associated with this design-adjusted Bayesian approach that we have noted repeatedly in earlier chapters.

7.6 Quantile Regression

The models discussed in this chapter have exclusively focused on the analytic objective of modeling the *mean* of a continuous dependent variable as a function of one or more predictor variables of interest. The mean is only one feature of a distribution that a survey researcher might be interested in modeling; some researchers may be interested in modeling the **quantiles** of a given distribution as a function of predictor variables. For example, how does the median of a given continuous variable change as predictor variables change in value? We discussed descriptive approaches for quantiles in Chapter 5; here, we consider new developments in the regression modeling of quantiles in a way that accounts for complex sampling features.

There have been multiple methodological advances in design-based approaches to fitting these types of **quantile regression models** in recent years. Chen and Zhao (2019) described survey-weighted maximum likelihood estimation of quantile regression models using different types of smoothed weights to increase the efficiency of the estimated regression coefficients when the weights are informative, in combination with bootstrap methods of variance estimation. R code and examples of implementing their various approaches can be found at this website: https://github.com/yandzhao/Quantile-Regression-of-Survey-Data.

Wang (2021) derived a pseudo maximum likelihood (PML) estimation approach for fitting these models that has similar properties to those approaches described in Chapters 8 and 9 for generalized linear models and Chapters 11 and 13 for multilevel models. Wang also described an approach for variance estimation based on the rescaled Rao-Wu bootstrapping approach outlined in Chapter 3. Unfortunately, we are unaware of any software implementing this PML approach at the time of this writing. We will provide updates in this regard on the ASDA website.

Elliott and Xia (2021) developed an alternative Bayesian approach to fitting quantile regression models to complex sample survey data based on weighted Dirichlet process mixture models. These authors demonstrate the improved efficiency of this Bayesian approach relative to other design-based approaches, which has been noted repeatedly in earlier chapters. R code implementing this proposed Bayesian approach, with examples, can be found on GitHub.[2] These authors note that this approach can be computationally intensive and call for future implementation in Stan that may improve computational time. Even more recently, Nascimento and Gonçalves (2024) also proposed Bayesian approaches to fitting quantile regression models in the case of informative sampling, eliminating the need for bootstrap resampling as suggested by Chen and Zhao (2019). Like Elliott and Zia, these authors demonstrate improvements in the MSE of estimated parameters in quantile regression models relative to the approach of Chen and Zhao, and also provide R code implementing their approaches on GitHub.[3]

7.7 Exercises

7.1) Regression Model Building

a. Briefly outline and describe the four model building steps recommended in this chapter.
b. Briefly describe how each of the four steps changes when analyzing complex sample survey data.

7.2) Linear Regression Analysis using the ESS6 Russian Federation Data

This exercise asks you to perform a linear regression analysis using the ESS Round 6 Russian Federation data. Exercise 7.2 contains many parts and emphasizes model building and correct analysis of data collected from a complex sample survey. The data set, ess6_russia, is available for download from the ASDA website, in Stata, R, and SAS formats.

a. Explore the distribution of the Satisfaction with Government (STFGOV) variable using a weighted frequency table, a weighted bar chart, or a weighted histogram. This will be the dependent variable in the linear regression exercises to follow. Discuss the distribution of the variable and explain why it might be reasonable statistically to analyze this variable as a continuous dependent variable even though it consists of a small number of discrete levels.
b. Consider three possible scientifically relevant predictors of the dependent variable Satisfaction with the Government (STFGOV):

1. A left to right political scale collapsed into 3 categories, from the 0–10 scale LRSCALE (LR3CAT, 1=left to moderately left, 2=moderate, 3=moderately right to right);
2. Gender (GNDR, 1=Male, 2=Female); and
3. Self-rated health status (HEALTH, 1=Very good, 2=Good, 3=Fair, 4=Bad, 5=Very bad).

Describe how each predictor variable should be handled in the regression model (continuous or categorical).

c. Fit a set of bivariate linear regression models, where STFGOV is predicted by each possible predictor listed in part b. (one at a time), while incorporating the complex sample design features (STRATIFY and PSU) and the final survey weights (PSPWGHT). Based on these results, what are the F-test statistics, degrees of freedom, and p-values for each possible predictor variable? Which of the predictors seem to have significant associations with STFGOV?

d. Fit the "preliminary" model including all the predictors explored from part c., fully accounting for the complex sampling features. Request estimated design effects for the estimated coefficients. To initially assess model diagnostics, produce a residual vs. predicted scatter plot and a histogram of the residuals. Based on these results, answer the following questions:
   1. Do these plots indicate acceptable model fit?
   2. What other techniques might be used to assess model fit?
   3. What do the design effects tell you about the estimated coefficients for each predictor variable?
   4. How many degrees of freedom are used for the significance tests, and how are they calculated?
   5. What is the weighted R-Squared for this model, and what does this mean in plain English?

e. Perform a design-adjusted test of the two-way interaction of gender and left/right political orientation (the three-category variable) in the "preliminary" model. Write down the design-adjusted F-test statistic, its degrees of freedom, and its p-value under the null hypothesis that the coefficients associated with this two-way interaction are all equal to zero. Is the interaction term significant at the alpha = 0.05 level? Should the interaction be retained in your final model?

f. Fit your "final" model based on your answers in part e., and based on these results, prepare a table like Table 7.5. Be sure to either exclude or include your interaction terms depending on the decision you made in part e. Include variable labels, design effects, and descriptive columns in the table.

g. Is it necessary to use weights when estimating this "final" model? Perform a formal test of whether the weights are informative about the coefficients in this model and summarize the decision that you would make about using the final survey weights to fit this model because of this test.

h. Refit your final model using the alternative Bayesian approach discussed in the chapter. Comment on any changes in the inferences that you might make based on these results. Are the 95% credible intervals for the regression parameters narrower than seen in the design-based approach?

i. Write a brief paragraph (as you might for a research article) explaining the methods that you used to estimate the model parameters and estimate the standard errors of the estimated parameters in a way that accounts for the complex sample design of the ESS6 Russian Federation data. Also, discuss how your conclusions might have changed had you not performed a correct design-based analysis incorporating the complex sample design features and the final survey weights.

Notes

1 http://www.cdc.gov/nchs/data/nhanes/analytic_guidelines_11_12.pdf
2 https://github.com/mrelliot/WPDM
3 https://github.com/marcuslavagnole/BWQR_Informative_Sampling

8

Logistic Regression and Generalized Linear Models for Binary Survey Variables

8.1 Introduction

Perusal of the electronic codebook for any major survey data set quickly reveals that many key survey questions require only a simple "yes" or "no" answer. Consider these questions from three of the studies used for the examples in this book:

HRS	"Are you currently covered by Medicare health insurance?"
NHANES	"Have you ever been told by a doctor or health professional that you have diabetes or sugar diabetes?"
NCS-R	"Did you ever use drugs or alcohol so often that it interfered with your responsibilities at work, at school or at home?

The responses to such questions are coded as **binary variables** (or **dichotomous variables**). This chapter is devoted to **generalized linear models (GLMs)** for a binary survey variable—focusing primarily on the application of **logistic regression** analysis to complex sample survey data. There are many excellent texts on standard SRS methods for logistic regression analysis, including comprehensive introductions to the theory, methods and applications provided by Hosmer et al. (2013), Agresti (2012), and Allison (2012b). Readers wanting a full mathematical treatment of GLMs are referred to McCullagh and Nelder (1989) and McCulloch et al. (2008).

The general aims of this chapter are to: 1) introduce the fundamental concepts of regression for categorical data based on GLMs that are important to understand for effective application of the techniques to survey data; 2) provide a systematic review of the stages in fitting the logistic regression model to complex sample survey data; and 3) present example logistic regression analyses based on the 2012 HRS and NCS-R data that illustrate typical applications of the model building steps to actual survey data sets.

Section 8.2 will touch on the important underlying concepts of GLMs as they apply to **logistic regression** and **probit regression**, but will leave much of the underlying theory and detailed development to specialized texts on the topic. Sections 8.3 to 8.6 describe the theory and methods that are important in the four stages of building and interpreting a logistic regression model: specification, estimation, evaluation, and interpretation/inference for complex sample survey data. Section 8.7 takes the

DOI: 10.1201/9781003472094-9

reader through an example analysis in which a logistic regression model is fit to the binary indicator of having had a major depressive episode (MDE) at some point in the lifetime in the NCS-R data set; this example includes estimation and inference based on an alternative Bayesian approach (see Section 8.7.5). The chapter concludes in Section 8.8 with a comparative application of the logistic, probit, and complementary log-log regression techniques to the problem of modeling the probability of alcohol dependency in U.S. adults.

8.2 Generalized Linear Models (GLMs) for Binary Survey Responses

In principle, generalized linear models for a binary dependent variable and linear regression models for a continuous variable (Chapter 7) share a common aim—to estimate a regression equation that relates the expected value of the dependent variable y to one or more predictor variables, denoted by $\boldsymbol{x}$. In linear regression for a continuous dependent variable, the expected value of y is the conditional mean of y given a vector of covariates, $\boldsymbol{x}$, and is estimated by an equation that is linear in the regression parameters:

$$\hat{Y} = E(y \mid \boldsymbol{x}) = B_0 + B_1 x_1 + \cdots + B_p x_p \tag{8.1}$$

When y is a binary variable with possible values 0 and 1 ($y = \{0,1\}$), $E(y \mid \boldsymbol{x}) = \pi(\boldsymbol{x})$ is the conditional probability that $y = 1$ given the covariate vector $\boldsymbol{x}$. Throughout this chapter, the notation $\pi(\boldsymbol{x})$ will be used to represent this probability that $y = 1$ conditional on a vector of observed predictors, $\boldsymbol{x}$.

A naïve approach to regression analysis of a binary dependent variable is to model the $\pi(\boldsymbol{x})$ as a linear function of $\boldsymbol{x}$. There are several problems with this approach. First, as we will see in Section 8.4, the sample of observations on the dependent variable y is assumed to follow a binomial distribution—a severe violation of the normality and homogeneity of variance assumptions required for efficient least squares estimation of the coefficients of the linear regression model. Second, as shown in Figure 8.1, a naive linear regression model for $\pi(\boldsymbol{x})$ does not accurately capture the relationship between y and $\boldsymbol{x}$—and it may even produce predicted values for $\pi(\boldsymbol{x})$ that are outside the permissible range of 0 to 1.

The alternative would be to identify a non-linear function of $\pi(\boldsymbol{x})$, say $g(\pi(\boldsymbol{x}))$, that yields a fitted regression model that is linear in the coefficients for the model covariates, $\boldsymbol{x}$. Ideally, the estimated function, $g(\pi(\boldsymbol{x}))$, should also be chosen so that when it is transformed back [i.e., $\pi(\boldsymbol{x}) = g^{-1}(g(\pi(\boldsymbol{x})))$], the resulting values of the estimated $\pi(\boldsymbol{x})$ will fall in the range between 0 and 1.

In the terminology of generalized linear models, functions like $g(\pi(\boldsymbol{x}))$ are termed **link functions**. The two most common link functions used to model binary survey variables are the logit and the probit. The functional relationships of the logit and probit to the values of $\pi(\boldsymbol{x})$ are illustrated in Figure 8.2, where one can see the value of $\pi(\boldsymbol{x})$ that would result from a given value of the logit or probit. Theory Box 8.1 presents a latent variable interpretation of the logit and probit link functions and a mechanism for relating these linear functions in $\boldsymbol{x}$ to the $\pi(\boldsymbol{x})$ that are the quantities of true interest in the model estimation.

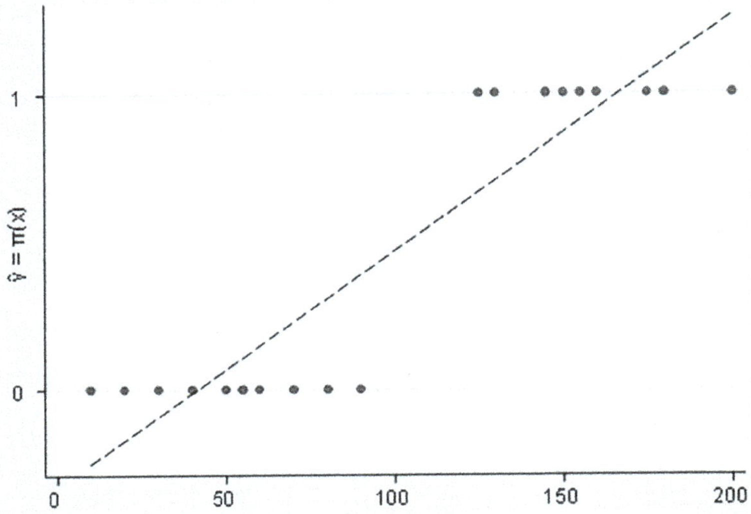

FIGURE 8.1
Naïve Use of Linear Regression for a Binary Dependent Variable.

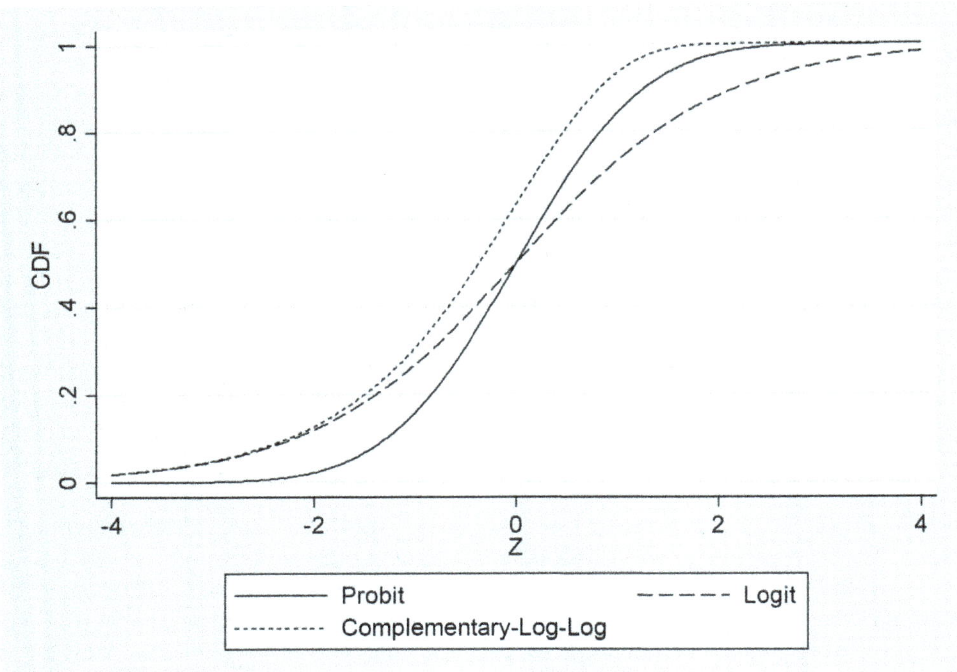

FIGURE 8.2
Cumulative Distribution Functions (CDFs) for the Logit, Probit, and C-L-L links.

8.2.1 The Logistic Regression Model

For a logistic regression model, the link function is the **logit**:

$$g(\pi(\boldsymbol{x})) = \text{logit}(\pi(\boldsymbol{x})) = z = \ln\left(\frac{\pi(\boldsymbol{x})}{1-\pi(\boldsymbol{x})}\right) = B_0 + B_1x_1 + \cdots + B_px_p. \tag{8.2}$$

The logit is non-linear in $\pi(x)$ but is presumed to be linear in the regression parameters, $\boldsymbol{B} = \{B_0, B_1, ..., B_p\}$. Based on the fitted regression model for the logit, the estimated value of $\pi(x)$ can be recovered by applying the inverse logit function:

$$\hat{\pi}(\boldsymbol{x}) = g^{-1}(g(\hat{\pi}(\boldsymbol{x}))) = \frac{\exp(\hat{B}_0 + \hat{B}_1x_1 + ... + \hat{B}_px_p)}{1+\exp(\hat{B}_0 + \hat{B}_1x_1 + ... + \hat{B}_px_p)} \tag{8.3}$$

The inverse logit function $g^{-1}(.)$ is the cumulative distribution function (CDF) for the logistic probability distribution. Under the logistic model (8.2), $\hat{\pi}(\boldsymbol{x})$ is the logistic CDF evaluated at the estimated logit corresponding to the covariate vector $\boldsymbol{x}$.

Unlike the linear regression model for a normally distributed variable y, there is no direct solution such as the method of least squares to estimate the regression coefficients in the logit model. Instead, an iterative estimation procedure such as the Newton-Raphson or Fisher Scoring algorithm (Agresti, 2012) is used to determine the values of the estimated coefficients that maximize the following weighted pseudo likelihood function:

$$PL(\boldsymbol{B} \mid X) = \prod_{i=1}^{n} \{\pi(\boldsymbol{x}_i)^{y_i} \cdot [1-\pi(\boldsymbol{x}_i)]^{1-y_i}\}^{w_i} \tag{8.4}$$

It is important to note that the link function is not an explicit transformation of the variable y but rather a transformation of the *expected value* of y, $E(y \mid \boldsymbol{x}) = \pi(\boldsymbol{x})$. Also, note that evaluation of the pseudo likelihood (8.4) requires both the original observations, y_i, the modeled values of $\hat{\pi}(\boldsymbol{x}_i)$, and in the case of complex sample survey data, the final survey weights w_i. We discuss this function in more detail in Section 8.4.

Each cycle in the iterative estimation of the logistic regression model requires four operations:

1) the value of the logit (8.2) is estimated for each survey respondent i based on the current iteration values of the estimated parameters: $z_i = \hat{B}_0 + \hat{B}_1x_{1i} + \cdots + \hat{B}_px_{pi}$.
2) the logit for each case is then transformed back to the probability scale by evaluating the logistic cumulative distribution function (CDF) at the value of z_i as illustrated in (8.3).
3) the value of the likelihood (8.4) is then evaluated at the $i = 1, \ldots, n$ values of the estimated $\hat{\pi}(\boldsymbol{x}_i)$ and observed y_i values.
4) the algorithm then adjusts the values of the individual parameter estimates $\hat{B}_j$ to maximize the likelihood function, and returns to repeat the cycle.

The iterative algorithm stops when the change in the estimates of the vector of $\boldsymbol{B}$ values no longer increases the value of the likelihood function. The values of the individual $\hat{B}_j$ at the final iteration are the final estimates.

THEORY BOX 8.1: A LATENT VARIABLE INTERPRETATION OF THE LOGIT MODEL

In specifying a logit (or probit) regression model for a binary outcome variable, the observed response on the binary outcome variable y can be viewed as determined by the value of a *latent* (or unobserved) continuous variable z. If the true value of the link function (or logit), z_i, for a survey respondent is less than or equal to some threshold value Z, the respondent answers with $y = 1$; if the value of $z_i > Z$, the respondent is expected to answer with $y = 0$. Since z_i is a latent variable and is never directly observed, its value for each respondent must be modeled as a function of observed survey variables.

The intercept parameter B_0 may be interpreted as a 'cutpoint' along the distribution of the logit (or probit) values. Changes in the predictor variables shift this 'cutpoint,' therefore changing the probability of a response of 1; thinking about the logistic regression model in this way will be useful for understanding ordinal logistic regression models in Section 9.3.

8.2.2 The Probit Regression Model

The probit regression model is an alternative to logistic regression for modeling a binary dependent variable. Probit regression models are also generalized linear models, and the procedures for estimating the models parallel those illustrated in Section 8.2.1 for logistic regression. The difference in the two approaches is that the link function changes to the **probit** (inverse Normal):

$$g(\pi(\boldsymbol{x})) = \Phi^{-1}(\pi(\boldsymbol{x})) = z = B_0 + B_1x_1 + \cdots + B_px_p \tag{8.5}$$

In contrast to logistic regression, where z is assumed to follow the logistic probability distribution, the probit is assumed to follow a standard normal distribution. Again, as in logistic regression, transforming back from the probit scale to the quantities of interest $\pi(\boldsymbol{x})$ requires an evaluation of the standard Normal CDF at the estimated value of the probit, $z_i = \hat{B}_0 + \hat{B}_1x_{1i} + \cdots + \hat{B}_px_{pi}$:

$$\hat{\pi}(\boldsymbol{x}_i)_{probit} = \Phi(z_i) = \text{Prob}(Z \le z_i) = \int_{-\infty}^{\hat{B}_0+\hat{B}_1x_{1i}+\cdots+\hat{B}_px_{pi}} \frac{1}{\sqrt{2\pi}} \cdot \exp\{Z^2\}dZ \tag{8.6}$$

With the probit link (8.5) replacing the logit link, and the normal probability transform (8.6) replacing the logistic CDF transform (8.3), the steps in estimating the probit regression model are identical to those outlined above for logistic regression. In general, inferences derived under logistic and probit regression models do not differ significantly.

8.2.3 The Complementary-Log-Log Model

A third less commonly used link function in regression analysis of a binary dependent variable is the **complementary-log-log** (C-L-L) link. Most software packages for survey data analysis that support logit and probit regression modeling also permit the user the option to choose the C-L-L link. The C-L-L link function is related to the Gompertz distribution, and the link may be referred to as the "gompit." The primary distinction for the C-L-L link is that unlike the logit or probit, the CDF is not required to be symmetric about the midpoint $\pi(x) = 0.5$ (see Figure 8.2). Its most common application is in situations where $\pi(x)$ is either very close to 0 or close to 1 (at one or the other extreme). Interested readers are referred to Allison (2012b) for applications of the C-L-L link function to both binary regression models and survival analysis.

Section 8.8 will provide a parallel comparison of a model fitted under all three links: the logit, the probit, and the C-L-L.

8.3 Building the Logistic Regression Model: Stage 1—Model Specification

The four stages in logistic regression modeling of survey data are identical to those presented in Chapter 7 for linear regression: 1) model specification; 2) estimation of model parameters and their standard errors; 3) model evaluation and diagnostics; and 4) interpretation of results and inference based on the final model. As in all statistical model building processes, stages 1–3 define an iterative process designed to sequentially refine and test the model. Several cycles of the model specification, estimation, and evaluation sequence are usually required before a final model is identified and inferences can be made concerning the modeled relationship in the larger survey population.

The logistic regression model is a flexible modeling tool that can simultaneously accommodate both categorical and continuous predictors as well as terms for the interactions among the predictors. As in all regression modeling, identification of a best logistic regression model for the survey data should follow a systematic, scientifically governed process whereby important candidate predictors are identified and evaluated, first individually and then in the multivariable context of other potentially important explanatory variables. Our recommendation is that survey analysts follow Hosmer et al.'s (2013) incremental process for specifying the initial model, refining the set of predictors, and then determining the final form of the logistic regression model:

- Perform initial bivariate analyses of the relationship of y to individual predictor variable candidates;
- Select the predictors that have a bivariate association with y at significance $p < 0.25$ as *candidates* for main effects in a multivariable logistic regression model (see also Harrell (2015), who argues that one should simply begin with *all* predictors of scientific interest in a multivariable model);
- Evaluate the contribution of each predictor to the multivariable model using the Wald test;
- Check the linearity assumption for continuous predictors;
- Check for scientifically justified interactions among the predictors.

Hosmer et al. (2013) advocate a final step involving the application of polynomial functions and smoothing splines to test that the logistic model is truly linear in the logit. Interested readers are referred to that text for a description of this final step in the evaluation of the fit of a logistic regression model.

8.4 Building the Logistic Regression Model: Stage 2—Estimation of Model Parameters and Standard Errors

Given the specification of a logistic regression model of the form $\text{logit}(\pi(x)) = B_0 + B_1x_1 + \cdots + B_px_p$, the second step in the model building process is to compute estimates of the regression parameters in the model along with their standard errors. For simple random samples, the logistic regression model parameters and standard errors can be estimated using the method of **maximum likelihood**. The likelihood function for a simple random sample (SRS) of n observations on a binary variable y with possible values 0 and 1 is based on the binomial distribution:

$$L(\boldsymbol{\beta} \mid \boldsymbol{x}) = \prod_{i=1}^{n} \pi(\boldsymbol{x}_i)^{y_i} \left[1 - \pi(\boldsymbol{x}_i)\right]^{1-y_i} \tag{8.7}$$

where $\pi(\boldsymbol{x}_i)$ is linked to the regression model coefficients through the logistic CDF:

$$\pi(\boldsymbol{x}_i) = \frac{\exp(\boldsymbol{x}_i\boldsymbol{\beta})}{1 + \exp(\boldsymbol{x}_i\boldsymbol{\beta})}. \tag{8.8}$$

When the survey data have been collected under a complex sample design, straightforward application of **maximum likelihood estimation (MLE)** procedures is no longer possible, for several reasons. First, the probabilities of selection (and responding) for the $i = 1, \ldots, n$ sample observations are generally no longer equal. Survey weights (see Section 2.7) are thus required to estimate the finite population values of the logistic regression model parameters. Second, the stratification and clustering of complex sample observations violates the assumption of independence of observations that is crucial to the standard MLE approach to estimating the sampling variances of the model parameters and choosing a reference distribution for the likelihood ratio test statistic (see Lumley and Scott (2014) and Lumley and Scott (2015) for more discussion of this testing issue).

Two general approaches have been developed to estimate the logistic regression model parameters and standard errors when analyzing complex sample survey data. Grizzle et al. (1969) first formulated an approach based on weighted least squares estimation. The WLS estimation method was originally programmed for logistic regression in the GENCAT software package (Landis et al., 1976) and still remains available as an option in programs such as SAS PROC CATMOD. Later, Binder (1981, 1983) presented a second general framework for fitting logistic regression and other generalized linear models to complex sample survey data. Binder proposed **pseudo maximum likelihood estimation (PMLE)** as a technique for estimating the model parameters. The PMLE approach to parameter estimation was combined with a linearized estimator of the variance-covariance matrix for the parameter estimates, taking complex sample design features into account (see also Kott, 2018, for what he terms a "design-sensitive" approach). Further development and evaluation of the PMLE approach was presented in Roberts et al. (1987), Morel (1989) and Skinner et al. (1989). The PMLE approach is now the standard method for logistic regression modeling in all of the major software packages that support analysis of complex sample survey data.

In theory, the finite population regression parameters for a generalized linear model of interest are those values that maximize a likelihood function for the $i = 1, \ldots, N$ elements in the survey population. For a binary dependent variable y (with possible values 0 or 1), the population likelihood can be defined as

$$L(\boldsymbol{B}\,|\,\boldsymbol{x}) = \prod_{i=1}^{N} \pi(\boldsymbol{x}_i)^{y_i}\left[1-\pi(\boldsymbol{x}_i)\right]^{1-y_i} \tag{8.9}$$

where under the logit link, $\pi(\boldsymbol{x}_i)$ is evaluated using the logistic CDF and the parameters in the specified logistic regression model: $\pi(\boldsymbol{x}_i) = \exp(\boldsymbol{x}_i\boldsymbol{B})/\left[1+\exp(\boldsymbol{x}_i\boldsymbol{B})\right]$. Note that as in Chapter 7, the finite population model parameters are denoted by the standard alphabetic $\boldsymbol{B}$ to distinguish them from the superpopulation model parameters denoted by $\boldsymbol{\beta}$. Here, as in linear regression, the distinction is primarily a theoretical one.

Estimates of these finite population regression parameters are obtained by maximizing the following unbiased estimate of the population likelihood, which is a survey-weighted function of the observed sample data and the $\pi(\boldsymbol{x}_i)$ values:

$$\begin{aligned} PL(\boldsymbol{B}\,|\,X) &= \prod_{i=1}^{n}\left\{\pi(\boldsymbol{x}_i)^{y_i}\cdot\left[1-\pi(\boldsymbol{x}_i)\right]^{1-y_i}\right\}^{w_i} \\ \textit{with: } \pi(\boldsymbol{x}_i) &= \exp(\boldsymbol{x}_i\boldsymbol{B})/\left[1+\exp(\boldsymbol{x}_i\boldsymbol{B})\right] \end{aligned} \tag{8.10}$$

Like the standard MLE procedure, this weighted pseudo-likelihood function can be maximized using the iterative Newton-Raphson method or related algorithms (see Theory Box 8.2 for more detail).

The next hurdle in fitting logistic regression models to complex sample survey data is to estimate the sampling variances and covariances of the parameter estimates. Binder (1983) proposed a solution to this problem that applied a multivariate version of Taylor Series Linearization. The result is a **sandwich-type variance estimator** of the form

$$var(\hat{\boldsymbol{B}}) = \left(\boldsymbol{J}^{-1}\right)var[S(\hat{\boldsymbol{B}})]\left(\boldsymbol{J}^{-1}\right) \tag{8.11}$$

where $\boldsymbol{J}$ is a matrix of second derivatives with respect to the $\hat{B}_j$ of the pseudo log-likelihood for the data (derived by applying the natural log function to the likelihood defined in (8.10)), and $var(S(\hat{\boldsymbol{B}}))$ is the variance-covariance matrix for the sample totals of the weighted "**score function**" for the individual observations used to fit the model. Interested readers can find a more mathematical treatment of this approach in Theory Boxes 8.2 and 8.3. Binder's linearized variance estimator in (8.11), $var(\hat{\boldsymbol{B}})$, is the default variance estimator in most major software packages for survey data analysis, and estimation of the components of this linearized variance is also critical to the design effect adjustments in the Bayesian analysis approach for complex sample survey data outlined in Section 3.8. Most software packages will also provide options to select a JRR, BRR, or bootstrap replication method to estimate the variance-covariance matrix, $var(\hat{\boldsymbol{B}})_{rep}$, for the estimated model coefficients.

THEORY BOX 8.2: PSEUDO MAXIMUM LIKELIHOOD ESTIMATION FOR COMPLEX SAMPLE SURVEY DATA

For a binary dependent variable and binomial data likelihood, the pseudo maximum likelihood approach to the estimation of the logistic regression model parameters and their variance-covariance matrix requires the solution to the following vector of estimating equations:

$$S(\boldsymbol{B}) = \sum_h \sum_\alpha \sum_i w_{h\alpha i} \boldsymbol{D}'_{h\alpha i} \left[(\pi_{h\alpha i}(\boldsymbol{B})) \cdot (1 - \pi_{h\alpha i}(\boldsymbol{B})) \right]^{-1} (y_{h\alpha i} - \pi_{h\alpha i}(\boldsymbol{B})) = \boldsymbol{0} \qquad (8.12)$$

where :

$\boldsymbol{D}_{h\alpha i}$ is the vector of partial derivatives, $\dfrac{\delta(\pi_{h\alpha i}(\boldsymbol{B}))}{\delta B_j}; j = 0, \ldots, p$

In (8.12), h is a stratum index, α is a cluster (or SECU) index within stratum h, and i is an index for individual observations within cluster α. The term $w_{h\alpha i}$ is the final survey weight for observation i. The term $\pi_{h\alpha i}(\boldsymbol{B})$ refers to the probability that the outcome variable is equal to 1 as a function of the parameter estimates and the observed data according to the specified logistic regression model. For a logistic regression model of a binary variable, this reduces to a system of $p + 1$ estimating equations (where p is the number of predictor variables, and there is one additional parameter corresponding to the intercept in the model):

$$S(\boldsymbol{B})_{\text{logisitic}} = \sum_h \sum_\alpha \sum_i w_{h\alpha i} (y_{h\alpha i} - \pi_{h\alpha i}(\boldsymbol{B})) \boldsymbol{x}'_{h\alpha i} = \boldsymbol{0} \qquad (8.13)$$

where: $\boldsymbol{x}'_{h\alpha i}$ *= a column vector of the p+1 design matrix elements for case* $i = \begin{bmatrix} 1 & x_{1,h\alpha i} & \cdots & x_{p,h\alpha i} \end{bmatrix}'$

For the probit regression model, the estimating equations reduce to:

$$S(\boldsymbol{B})_{\text{probit}} = \sum_h \sum_\alpha \sum_i w_{h\alpha i} \frac{(y_{h\alpha i} - \pi_{h\alpha i}(\boldsymbol{B})) \cdot \phi(\boldsymbol{x}'_{h\alpha i}\boldsymbol{B})}{\pi_{h\alpha i}(\boldsymbol{B}) \cdot (1 - \pi_{h\alpha i}(\boldsymbol{B}))} \boldsymbol{x}'_{h\alpha i} = \boldsymbol{0} \qquad (8.14)$$

where: $\phi(\boldsymbol{x}'_{h\alpha i}\boldsymbol{B})$ *is the standard normal probability density function evaluated at* $\boldsymbol{x}'_{h\alpha i}\boldsymbol{B}$

The weighted parameter estimates are computed by using the Newton-Raphson method to derive a solution for $S(\boldsymbol{B}) = \boldsymbol{0}$ (Agresti, 2012). Binder showed that the vector of weighted parameter estimates based on pseudo maximum likelihood estimation is *consistent* for $\boldsymbol{B}$ even when the sample design is complex; that is, the bias of this estimator is of order $1/n$, so that as the sample size gets larger (which is often the case with complex samples), the bias of the estimator approaches 0.

THEORY BOX 8.3: TAYLOR SERIES ESTIMATION OF $VAR(\hat{\boldsymbol{B}})$

The computation of variance estimators for the pseudo maximum likelihood estimates of the finite population parameters in the logistic regression model makes use of the $\boldsymbol{J}$ matrix of second derivatives:

$$\begin{aligned} \boldsymbol{J} &= -\left[\frac{\delta^2 \ln PL(\boldsymbol{B})}{\delta^2 \boldsymbol{B}} \right] \Bigg| \boldsymbol{B} = \hat{\boldsymbol{B}} \\ &= \sum_h \sum_\alpha \sum_i \boldsymbol{x}'_{h\alpha i} \boldsymbol{x}_{h\alpha i} w_{h\alpha i} \hat{\pi}_{h\alpha i}(\boldsymbol{B})(1 - \hat{\pi}_{h\alpha i}(\boldsymbol{B})) \end{aligned} \qquad (8.15)$$

Due to the weighting, stratification, and cluster sampling inherent to complex sample designs, J^{-1} is not equivalent to the variance-covariance matrix of the pseudo maximum likelihood parameter estimates, as is the case in the simple random sample setting (see Section 8.4). Instead, a **sandwich-type variance estimator** is used, incorporating the matrix J and the estimated variance-covariance matrix of the weighted score equations from (8.13):

$$V\hat{a}r(\hat{B}) = \left(J^{-1}\right) var[S(\hat{B})] \left(J^{-1}\right) \tag{8.16}$$

The symmetric matrix $var[S(\hat{B})]$ is the variance-covariance matrix for the $p + 1$ estimating equations in 8.12 above. Each of these $p + 1$ estimating equations is a summation over strata, clusters, and elements of the individual "scores" for the n survey respondents. Since each estimating equation is a sample total of respondents' scores, standard formulae for stratified sampling of ultimate clusters (Chapter 3) can be used to estimate the variances and covariances of the $p + 1$ sample totals. In vector notation:

$$var[S(\hat{B})] = \frac{n-1}{n-(p+1)} \sum_{h=1}^{H} \frac{a_h}{(a_h - 1)} \sum_{\alpha=1}^{a_h} (s_{h\alpha} - \bar{s}_h)'(s_{h\alpha} - \bar{s}_h)$$

which for n large is:

$$var[S(\hat{B})] \cong \sum_{h=1}^{H} \frac{a_h}{(a_h - 1)} \sum_{\alpha=1}^{a_h} (s_{h\alpha} - \bar{s}_h)'(s_{h\alpha} - \bar{s}_h) \tag{8.17}$$

where for the logistic link :

$$s_{h\alpha} = \sum_{i=1}^{n_\alpha} s_{h\alpha i} = \sum_{i=1}^{n_\alpha} w_{h\alpha i}(y_{h\alpha i} - \hat{\pi}_{h\alpha i}(B))\, x'_{h\alpha i} \text{ ; and}$$

$$\bar{s}_h = \frac{1}{a_h} \sum_{\alpha=1}^{a_h} s_{h\alpha}$$

The estimator of $var[S(\hat{B})]$ for the probit or C-L-L link is obtained by substituting the appropriate expressions for the individual score functions in the calculation of the $s_{h\alpha}$. For more details, interested readers should refer to Binder (1983).

8.5 Building the Logistic Regression Model: Stage 3—Evaluation of the Fitted Model

The next step in building the logistic regression model is to test the contributions of the individual model parameters and evaluate the overall goodness of fit of the model.

8.5.1 Wald Tests of Model Parameters

When fitting logistic regression models to data collected from simple random samples, the statistical significance of one or more logistic regression parameters can be evaluated using a **likelihood ratio test.** Under the null hypotheses H_0: $\beta_j = 0$ (a single parameter) or H_0: $\boldsymbol{\beta}_q = \mathbf{0}$ (a vector of q parameters), the following test statistic G follows a chi-square distribution with either 1 (for a single parameter) or q degrees of freedom:

$$G = -2\ln\left[\frac{L\left(\hat{\beta}_{\text{MLE}}\right)_{reduced}}{L\left(\hat{\beta}_{\text{MLE}}\right)_{full}}\right] \tag{8.18}$$

where: $L(\hat{\beta}_{\text{MLE}})$ = *the likelihood under the model evaluated at the mazimum likelihood estimates of* $\boldsymbol{\beta}$.

The *reduced* model in this case is the model excluding the q regression parameters to be tested, while the *full* model is the model including the q regression parameters. Both models should be fitted using *exactly* the same set of observations for this type of test to be valid.

As described in Section 7.3.4 for the linear regression model, complex sample designs invalidate the key assumptions that underlie the F-tests or likelihood ratio tests used to compare alternative models in simple random samples; we describe important developments in this area for generalized linear models fitted to data from complex samples below. Instead, Wald-type tests are often used to test hypotheses concerning the parameters of a specified logistic regression model. The default output from software procedures enabling analysts to fit logistic regression models to complex sample survey data provides a table of the estimated model coefficients, the estimated standard errors, and the test statistic and a "p-value" for the simple hypothesis test, H_0: $B_j = 0$ (see, for example, Table 8.1). Different software procedures may report the test statistic as a Student t-statistic, $t = \hat{B}_j \,/\, se(\hat{B}_j)$, or as a Wald X^2 statistic.

If the former of these two tests is output, the test statistic is referred to a Student t distribution with nominal design-based degrees of freedom (df = #clusters–#strata) to determine and report the p-value. If the output is in the form of the Wald X^2, the reference distribution for determining the p-value is χ_1^2, or a central chi-square distribution with 1 degree of freedom. The two tests are functionally equivalent. In fact, the absolute value of the Student t test statistic is simply the square root of the Wald X^2 test statistic for this single parameter hypothesis test.

More generally, software procedures enabling analysts to fit logistic regression models to complex sample survey data provide convenient syntax to specify design-adjusted Wald tests for a variety of hypotheses concerning the full vector of regression parameters. The general form of the null hypothesis for a Wald Test is defined by H_0: $\boldsymbol{CB} = \boldsymbol{0}$, where $\boldsymbol{C}$ is a matrix of constants that defines the hypothesis being tested (see Section 7.3.4.1 for examples). The Wald test statistic is computed as:

$$X^2_{Wald} = (\boldsymbol{C}\hat{\boldsymbol{B}})'\left[\boldsymbol{C}(var(\hat{\boldsymbol{B}}))\boldsymbol{C}'\right]^{-1}(\boldsymbol{C}\hat{\boldsymbol{B}}) \tag{8.19}$$

where $var(\hat{\boldsymbol{B}})$ is a design-consistent estimate of the variance-covariance matrix of the estimated logistic regression coefficients (see (8.11) for an example based on Taylor Series Linearization). Under the null hypothesis, this Wald test statistic follows a chi-square distribution with q degrees of freedom, where q is the rank, or number of independent rows, of the matrix **C**. This test statistic can also be converted into an approximate *F*-statistic by dividing the Wald X^2 test statistic by the degrees of freedom.

In Stata, the `test` post-estimation command is used to specify a multi-parameter hypothesis test after fitting a model. Section 7.5 has already illustrated the application of the `test` command for constructing hypothesis tests for parameters in the linear regression model. The syntax is identical for constructing hypothesis tests for the parameters of logistic regression models or other generalized linear models estimated in Stata. For example, after fitting a logistic regression model that includes as predictors two indicator variables, representing two of the three NHANES race/ethnicity categories (1 = White, 2 = Black, 3 = Other, with White considered as a reference category), the post-estimation command

```
test 2.race 3.race
```

tests the overall importance of race, that is, H_0: $\{B_{\text{Black}} = 0, B_{\text{Other}} = 0\}$, while the post-estimation command

```
test 2.race - 3.race
```

tests the hypothesis H_0: $\{B_{\text{Black}} - B_{\text{Other}} = 0\}$ or equivalently H_0: $\{B_{\text{Black}} = B_{\text{Other}}\}$.

We emphasize here once again that survey analysts must be cautious about interpreting single-parameter hypothesis tests when the estimated coefficient applies to an indicator variable for a multi-category predictor (e.g., levels of education) or when the model also includes an interaction term that incorporates the predictor of interest. In the former case, a significant result indicates that the expected value of the outcome for that category is significantly different from that of the *reference category*, but not necessarily other levels of the multi-category predictor. Tests of parameters for individual variables in a model with interactions involving that same variable are confounded and not easily interpreted (see Section 8.6 for an example of interpretation of interaction terms).

Lumley and Scott (2014, 2015) have developed design-adjusted likelihood ratio tests for the regression parameters in logistic regression models fitted to complex sample survey data. In addition, these authors have developed design-adjusted versions of the **Akaike Information Criterion (AIC)** and the **Bayesian Information Criterion (BIC)**, which can be used to compare the fits of competing logistic regression models that do not necessarily have a nested relationship (i.e., where one model contains a subset of the parameters in a larger model). These design-adjusted tools have been implemented in the contributed R package `survey`. We present examples of performing these tests and computing the design-adjusted information criteria in R in Section 8.7.

8.5.2 Goodness of Fit (GOF) and Logistic Regression Diagnostics

Summary statistics to measure overall **goodness of fit (GOF)** and methods for diagnosing the fit of a logistic regression model for individual cases or specific patterns of covariates have been developed for simple random samples. These GOF statistics and diagnostic tools have been included in the standard logistic regression procedures in most major statistical software packages. Chapter 5 in Hosmer et al. (2013) reviews these summary statistics and diagnostic methods in detail.

Included in these methods are: 1) two test statistics based on the Pearson and deviance residuals for the fitted model; 2) the **Hosmer-Lemeshow GOF test**; 3) classification tables comparing observed values of y to discrete classifications formed from the model's predicted values, $\hat{\pi}(x)$; and 4) the area under the ROC (receiver operating characteristic) curve. Several **pseudo-R^2 measures** have also been proposed as summary measures of the fit of a logistic regression model. However, since these measures tend to be incorrectly confused with the R^2 values (explained variability) in linear regression, we agree with Hosmer et al. (2013) that while they may be used by the analyst to compare the fits of alternative models, they should not be cited as a measure of fit in scientific papers or reports.

Archer and Lemeshow (2006) and Archer et al. (2007) extended the standard Hosmer-Lemeshow GOF test for application to complex sample survey data. The Archer-Lemeshow procedure is a modification of the standard Hosmer-Lemeshow test for GOF that takes the final survey weights and the stratification and cluster sampling features of the complex sample design into account when assessing the residuals, defined in this case by $y_i - \hat{\pi}(x_i)$, based on the fitted model. Archer and Lemeshow's paper should be consulted for more details, but Stata has implemented this design-adjusted approach for assessing GOF in the post-estimation command `estat gof`. We present examples of the use of this command in Section 8.7.

At the time that this third edition is being written, many survey analysis software procedures are still in a phase where summary measures of GOF are being translated or newly developed for application to logistic regression modeling of complex sample survey data. Furthermore, it may be some time before the major software packages routinely incorporate robust GOF evaluation procedures in the software procedures for complex sample survey data analysis. In lieu of simply bypassing the GOF evaluation entirely (outside of Stata), we recommend:

1) applying the Archer-Lemeshow GOF test and other available summary GOF measures when they are available in the software package; or
2) if the logistic regression procedure for complex sample survey data in the chosen software package does not provide capabilities to generate summary GOF measures, re-estimate the model using the final survey weights in the package's standard logistic regression procedure. The weighted estimates of parameters and predicted probabilities will be identical, and serious lack of fit should be quantifiable by exploring the residuals, even if the software procedure does not correctly reflect the variances and covariances of the parameter estimates given the complex sample design.

Summary measures such as the Archer-Lemeshow GOF test statistic have the advantage that they yield a single test of the overall suitability of the fitted model. But even when a summary GOF measure suggests that the model provides an acceptable fit to the data, a thorough evaluation of the fit of model may also include examination of the fit for specific patterns of covariates. Does the model fit well for some patterns of covariates x_j, but not for others? The number and statistical complexity of diagnostic tools that have been suggested in the literature preclude a detailed discussion here. We encourage survey analysts to consult Hosmer et al. (2013) for a description of the diagnostic options and guidance on how these computational and graphical methods may be applied using Stata, SAS, and other software packages.

Regarding model diagnostics for logistic regression models fit to complex sample survey data, we can offer the following recommendations at the time of this third edition:

1) use one or more of the techniques described in Chapter 5 of Hosmer et al. (2013) to evaluate the fit of the model for individual patterns of covariates. If the logistic regression modeling procedure for complex samples in your chosen software

package (e.g., SAS PROC SURVEYLOGISTIC) does not include the full set of diagnostic capabilities, use the standard procedures (e.g., SAS PROC LOGISTIC) with a weight specification (see Ryan et al., 2015, for details). As mentioned before, the weighted estimates of parameters and predicted probabilities will be identical and serious breakdowns in the model for specific covariate patterns should be identifiable through the residuals, even though the standard procedure does not correctly reflect the variances and covariances of the parameter estimates given the complex sample design; and

2) remember that regression diagnostics serve to inform the analyst of specific cases where the model provides a poor fit or cases that exert extreme influence on the overall fit of the model. They are useful in identifying improvements in the model or anomalies that warrant future investigation, but small numbers of "predictive failures" occur in almost all regression modeling and do not necessarily invalidate the entire model.

8.6 Building the Logistic Regression Model: Stage 4—Interpretation and Inference

In logistic regression modeling, one can make inferences concerning the significance and importance of individual predictor variables in several ways. As described above in Section 8.5.1, Wald X^2 tests or likelihood ratio tests may be employed to test the null hypothesis that a single coefficient is significantly different from zero, H_0: $B_j = 0$, or more complex hypotheses concerning multiple parameters in the fitted model. Confidence Intervals (CIs) for individual model coefficients may also be used to draw inferences concerning the significance of model predictors, and also provide information on the potential magnitude and uncertainty associated with the estimated relationships of individual predictor variables with a given binary outcome.

Recall from Section 6.4.5 that an estimated logistic regression coefficient is the natural logarithm of the odds ratio comparing the odds that $y = 1$ for a predictor with value $x + 1$ to the odds that $y = 1$ when that predictor has a value x. In addition, the estimated coefficient for an indicator variable associated with a level of a categorical predictor is the natural logarithm of the odds ratio comparing the odds that $y = 1$ for the level represented by the indicator to the odds that $y = 1$ for the reference level of the categorical variable. Consequently, the estimated coefficients are often referred to as being on the **log-odds scale**, as they indicate how the log-odds are expected to change as the corresponding predictor variable increases by one unit. In fact, the term "logit" is shorthand for "log-odds units". A design-based confidence interval for a logistic regression model parameter (for predictor variable j) is computed as:

$$CI_{1-\alpha}(B_j) = \hat{B}_j \pm t_{df,1-\alpha/2} \cdot se(\hat{B}_j) \tag{8.20}$$

Typically, $\alpha = 0.05$ is used (along with the design-based degrees of freedom, df) and the result is a 95% CI for the parameter. In theory, the correct inference to make is that over repeated probability sampling, 95 out of 100 confidence intervals computed in this way across the repeated samples are expected to include the true population value of β_j. If the estimated CI includes ln(1) = 0, analysts may choose to infer that H_0: $\beta_j = 0$ is accepted with a Type I error rate of $\alpha = 0.05$.

Inference concerning the significance/importance of predictors can be performed directly for the $\hat{B}_j s$ (on the log-odds scale). However, to quantify the magnitude of the effect of an individual predictor, it is more useful to transform the inference to a scale that is easily interpreted by scientists, policymakers, and the general public. As discussed in Section 6.4.5, in a logistic regression model with a single predictor, x_1, an estimate of the **odds ratio** corresponding to a one-unit increase in the value of x_1 can be obtained by exponentiating the estimated logistic regression coefficient:

$$\hat{\psi} = \exp\left(\hat{B}_1\right) \tag{8.21}$$

If the model contains only a single predictor, the result is an estimate of the **unadjusted odds ratio.** If the fitted logistic regression model includes multiple predictors, that is,

$$\text{logit}(\hat{\pi}(x)) = \ln\left[\frac{\hat{\pi}(x)}{1-\hat{\pi}(x)}\right] = \hat{B}_0 + \hat{B}_1 x_1 + \cdots + \hat{B}_p x_p, \tag{8.22}$$

the result $\hat{\psi}_j \left| \hat{\boldsymbol{B}}_{k \neq j} = \exp\left(\hat{B}_j\right)\right.$ is an **adjusted odds ratio.** In general, the adjusted odds ratio represents the *multiplicative* impact of a one-unit increase in the predictor variable x_j on the odds of the outcome variable being equal to 1, holding all other predictor variables constant. Confidence limits can also be computed for adjusted odds ratios:

$$CI\left(\psi_j\right) = \exp\left(\hat{B}_j \pm t_{df,1-\alpha/2} \cdot se\left(\hat{B}_j\right)\right) \tag{8.23}$$

Software procedures for logistic regression analysis of complex sample survey data generally offer the analyst the option to output parameter estimates and standard errors on the log-odds scale (the original $\hat{B}_j s$) or transformed estimates of the corresponding adjusted odds ratios and confidence intervals.

The adjusted odds ratios and confidence intervals can be estimated and reported for any form of a predictor variable, including categorical variables, ordinal variables, and continuous variables. To illustrate, consider a simple logistic regression model (based on the 2012 HRS) of the probability that a U.S. adult age 50+ has arthritis. The predictors in this model are gender, education level (with levels of less than high school, high school, some college, and more than college), and age:

$$\text{logit}(\pi(x)) = \hat{B}_0 + \hat{B}_1 I_{Male} + \hat{B}_2 I_{Educ,<HS} + \hat{B}_3 I_{Educ,HS} + \hat{B}_3 I_{Educ,SomeColl} + \hat{B}_5 X_{Age(yrs)}$$

where:

I_{Male} = indicator variable for male gender (female is reference);

$I_{Educ,<HS}, \ldots$ = indicators for education level (college or more is reference);

$X_{Age(yrs)}$ = repondent age in years.

The results from fitting this simple model in Stata (using the following commands) are summarized in Tables 8.1 and 8.2.

TABLE 8.1

Estimated logistic regression model for arthritis

Predictor *	Category	$\hat{B}$	$se(\hat{B})$	t	$P(t_{56} > t)$
INTERCEPT	Constant	−3.342	0.157	−21.25	< 0.01
GENDER	Male	−0.538	0.049	−10.90	< 0.01
EDCAT	<12 yrs	0.501	0.055	9.19	< 0.01
	12 yrs	0.441	0.047	9.29	< 0.01
	13–15 yrs	0.386	0.045	8.59	< 0.01
AGE (years)	Continuous	0.054	0.002	23.14	< 0.01

Source: 2012 HRS. $n = 20{,}438$.
Note: * Reference categories for categorical predictors are: GENDER (female); EDCAT (>15 yrs)

TABLE 8.2

Estimates of adjusted odds ratios in the arthritis model and 95% CIs for the adjusted odds ratios

Predictor *	Category	$\hat{\psi}$	$CI_{.95}(\psi)$
GENDER	Male	0.584	(0.529, 0.645)
EDCAT	<12 yrs	1.651	(1.480, 1.841)
	12 yrs	1.555	(1.413, 1.710)
	13–15 yrs	1.471	(1.344, 1.609)
AGE	Continuous	1.055	(1.051, 1.060)

Source: 2012 HRS. $n = 20{,}438$.
Note: * Reference categories for categorical predictors are: GENDER (female); EDCAT (>15 yrs).

```
svyset secu [pweight = nwgtr], strata(stratum)
svy: logit arthritis ib2.gender ib4.edcat nage
svy: logit arthritis ib2.gender ib4.edcat nage, or
```

Interpreting the output from this simple example, we can make the following statements:

- The estimated ratio of odds of arthritis for men relative to women is $\hat{\psi} = 0.584$;
- The estimated odds of arthritis for persons with less than a high school education are $\hat{\psi} = 1.651$ times the odds of arthritis for persons with a college education;
- The estimated odds of arthritis increase by a factor of $\hat{\psi} = 1.055$ for each additional year of age.

Note that for continuous predictors, the increment x to $x + 1$ can be a relatively small step on the full range of x. For this reason, analysts may choose to report odds ratios for continuous predictors that correspond to a greater increment in x. A common choice is to report the odds ratio for a one standard deviation increase in x. For example, the standard deviation of 2012 HRS respondents' age is approximately 10 years. The estimated odds ratio and the 95% confidence interval for the odds ratio associated with a one standard deviation increase in age are computed as follows:

$$\hat{\psi}_{10yrs} = e^{\hat{B}_5 \cdot 10} = e^{0.054\times 10} = \exp(0.054\times 10) = 1.72$$

$$CI_{.95}(\psi_{10yrs}) = \left(\exp(\hat{B}_5 \times 10 - t_{56,0.975} \times 10 \times se(\hat{B}_5)), \exp(\hat{B}_5 \times 10 + t_{56,0.975} \times 10 \times se(\hat{B}_5))\right)$$
$$= \exp(0.54 \pm 2.003 \times 10 \times 0.002) = (1.65, 1.79)$$

When interactions between predictor variables are included in the specified model, analysts need to carefully consider the interpretation of the parameter estimates. For example, consider an extension of the 2012 HRS model for the logit of the probability of arthritis that includes the two-way interaction between education level and gender. The estimated coefficients and standard errors for this extended model reported by Stata (using the command below) are shown in Table 8.3.

```
svy: logit arthritis ib2.gender ib4.edcat nage ///
ib2.gender#ib4.edcat
```

Note that when the interaction between gender and education is added to the model, the estimated parameter for less than high school education is substantially increased from $\hat{B}_2$= 0.501 to $\hat{B}_2$= 0.623. This is due to the fact that this parameter now represents the contrast in log-odds between less than high school education and college education for *females only* (the reference level for gender), and is combined with the parameter for the (<12 yrs × Male) product term to define the same contrast in log-odds for males. The parameter associated with the product term therefore represents a *change* in this contrast for males relative to females. The estimated change in log-odds for males with less than 12th grade education relative to males with college education is thus computed as 0.623–0.248 = 0.375. The change in this difference between females and males is apparent; there is a significantly larger less than high school vs. college education difference in the probability of arthritis for females.

To explore the impact of the interaction of GENDER and EDCAT on the estimated logits and odds ratios, assume that AGE is fixed at 65 years. Consider the possible patterns of covariates shown in columns (2)-(4) of Table 8.4:

TABLE 8.3

Estimated logistic regression model for arthritis, including the two-way interaction between education and gender

Predictor *	Category	$\hat{B}$	$se(\hat{B})$	t	$P(t_{56}>t)$
INTERCEPT	Constant	–3.342	0.161	–20.74	< 0.01
GENDER	Male	–0.570	0.087	–6.54	< 0.01
EDCAT	<12 yrs	0.623	0.079	7.92	< 0.01
	12 yrs	0.374	0.065	5.72	< 0.01
	13–15 yrs	0.337	0.069	4.88	< 0.01
AGE	Continuous	0.054	0.002	23.31	< 0.01
EDCAT × GENDER	<12yrs × Male	–0.248	0.098	–2.54	0.014
	12 yrs × Male	0.146	0.106	1.38	0.174
	13–15 yrs × Male	0.102	0.116	0.87	0.386

Source: 2012 HRS. n = 20,438.
Note: * Reference categories for categorical predictors are: GENDER (female); EDCAT (>15 yrs).

TABLE 8.4

Covariate Patterns, Logits, and Odds Ratios for the Arthritis Model

Pattern (*j*)	Gender	Education	Age	Logit: z_j	Odds ratio*
1	Male	<HS	65	−0.027	0.82
2	Male	HS	65	0.118	0.95
3	Male	Some College	65	0.037	0.88
4	Male	College	65	−0.402	0.57
5	Female	<HS	65	0.791	1.86
6	Female	HS	65	0.542	1.45
7	Female	Some College	65	0.505	1.40
8	Female	College	65	0.168	1.00

Note: * Relative to joint reference of Females with College Education.

To estimate the value of the logit for each covariate pattern, the estimated coefficients in Table 8.3 are applied to the corresponding values of the predictor variables:

$$\begin{aligned}\text{logit}(\pi(x)) =& -3.342 - 0.570 I_{Male} + 0.623 I_{Educ,<HS} + 0.374 I_{Educ,HS} \\ & + 0.337 I_{Educ,SomeColl} + 0.054 X_{Age(yrs)} - 0.248(I_{Male} \times I_{Educ,<HS}) \\ & + 0.146(I_{Male} \times I_{Educ,HS}) + 0.102(I_{Male} \times I_{Educ,SomeColl})\end{aligned}$$

Table 8.4 shows the estimated logits for the eight unique covariate patterns (with age fixed at 65). To evaluate the ratio of odds of arthritis for men and women of different education levels, a general technique to compare the estimated odds for two different patterns of covariates is used. Consider two "patterns" of covariate values x_1 and x_2. Using the example in Table 8.4, x_1 might be pattern 1, 65-year-old males with <HS education, and x_2 might be the reference category for the GENDER × EDCAT interaction, which would be 65-year-old women with college education. To obtain the estimated odds ratio that compares x_1 to x_2, the following five steps are required:

1) Based on the estimated model, compute the values of the logit function for the two sets of covariates: $\text{logit}_1 = \sum_{j=0}^{p} \hat{B}_j x_{1j}$; $\text{logit}_2 = \sum_{j=0}^{p} \hat{B}_j x_{2j}$. These are shown for the six example covariate patterns in Table 8.4. For example, logit_1 = −0.027 and logit_2 = 0.168 for the two patterns chosen for this illustration;
2) Compute the difference between the two logits, $\hat{\Delta}_{1:2} = \text{logit}_1 - \text{logit}_2$ = −0.027–0.168 = −0.195;
3) Exponentiate the difference in the two logits to estimate the odds ratio comparing x_1 to x_2, $\hat{\psi}_{1:2} = e^{\hat{\Delta}_{1:2}} = e^{-.195} = 0.82$.
4) To reflect the uncertainty in the estimated odds ratios, the estimates should be accompanied by an estimated confidence interval. The CI of an odds ratio comparing two arbitrary covariate patterns, x_1 and x_2, takes the general form of expression (8.23) above. The standard error estimate requires the derivation of the standard error of the difference in the two logits:

$$se(\hat{\Delta}_{1:2}) = \sqrt{\sum_{j=0}^{p} (x_{1j} - x_{2j}) var(\hat{B}_j) + 2\sum_{j<k} (x_{1j} - x_{2j})(x_{1k} - x_{2k}) cov(\hat{B}_j, \hat{B}_k)} \quad (8.24)$$

Note that any common values for x_j in logit_1 and logit_2 can be ignored in evaluating this standard error. The calculation of the standard error of the difference in logits requires the values of $var(\hat{B}_j)$ and $cov(\hat{B}_j, \hat{B}_k)$. These may be obtained in Stata by issuing the `estat vce` command after the model has been estimated.

5) Exponentiate the CI limits for the difference in logits to estimate the 95% CI for the odds ratio: $CI(\psi_{1:2}) = \exp[\hat{\Delta}_{1:2} \pm t_{df,.975} \cdot se(\hat{\Delta}_{1:2})]$.

Assuming that the sample design variables have already been specified using the `svyset` command, we include Stata commands below that facilitate the execution of these five steps, along with the Stata output generated:

```
* First, use svy: logit to estimate the model.
svy: logit arthritis ib2.gender ib4.edcat nage ///
ib2.gender#ib4.edcat
```

```
Survey: Logistic regression

Number of strata   =    56                 Number of obs     =      20,438
Number of PSUs     =   112                 Population size   =  90,193,806
                                           Design df         =          56
                                           F(8, 49)          =       96.61
                                           Prob > F          =      0.0000

--------------------------------------------------------------------------------------
              |             Linearized
   arthritis  | Coefficient  std. err.      t     P>|t|     [95% conf. interval]
--------------+-----------------------------------------------------------------------
    1.gender  |  -.5698561   .0871712     -6.54   0.000    -.7444809    -.3952312
       edcat  |
           1  |    6226258   .0786245      7.92   0.000      .465122     .7801297
           2  |     373729   .0653076      5.72   0.000     .2429022     .5045558
           3  |    3374743   .0691572      4.88   0.000     .1989357     .4760129
        nage  |   .0542511   .0023277     23.31   0.000     .0495882     .0589141
gender#edcat  |
         1 1  |  -.2484328   .0979783     -2.54   0.014    -.4447068    -.0521587
         1 2  |    1457494   .1059441      1.38   0.174    -.0664822     .3579809
         1 3  |   .1016149   .1161794      0.87   0.386    -.1311204     .3343503
       _cons  |  -3.342015   .1611268    -20.74   0.000    -3.664791    -3.019239
--------------------------------------------------------------------------------------
```

```
* men age 65 with < HS education vs. women age 65 with
* college education: odds ratio computation

nlcom exp((65* _ b[nage] + 1* _ b[1.gender] + 1* _ b[1.edcat] ///
+ 0* _ b[2.edcat] + 0* _ b[3.edcat] + 1* _ b[1.edcat#1.gender] ///
+ 0* _ b[2.edcat#1.gender] + 0* _ b[3.edcat#1.gender] ///
+ _ b[ _ cons]) - (65* _ b[nage] + 0* _ b[1.gender] ///
+ 0* _ b[1.edcat] + 0* _ b[2.edcat] + 0*[3.edcat] ///
+ 0* _ b[1.edcat#1.gender] + 0* _ b[2.edcat#1.gender] ///
+ 0* _ b[3.edcat#1.gender] + _ b[ _ cons]))
```

```
-------------------------------------------------------------------------------
  arthritis | Coefficient   Std. err.      z    P>|z|     [95% conf. interval]
------------+------------------------------------------------------------------
      _nl_1 |   .8222893    .0814669    10.09   0.000     .6626171    .9819615
-------------------------------------------------------------------------------
```

We see above how the `nlcom` command can be used to estimate the odds ratio of interest (referring to the parameter estimates saved in memory from the most recent fitted model) and compute a 95% confidence interval for the odds ratio. We would conclude (based on the 95% confidence interval for this odds ratio that does not include a value of 1) that men age 65 with less than high school education have significantly reduced odds of arthritis compared to women age 65 with college education (the estimated odds are about 18% lower).

The technique described above for estimating odds ratios applies generally to *any* observed patterns of covariates x_1 and x_2. The final column in Table 8.4 shows the estimated odds ratios comparing each of the eight covariate patterns based on the estimated logistic regression model including the interaction of gender and education level. Holding age constant at 65 years, the coding of gender (female reference) and education level (college reference) results in 65-year-old women with college education as the natural reference group.

A convenient way to visualize and interpret significant interactions like this in Stata is to plot **average marginal effects** of one covariate as a function of another covariate, as shown in Figure 8.3. These effects show changes in the **marginal predicted values** of the binary dependent variable for one category of a predictor of interest (e.g., education) relative to the reference category, separately for the levels of the other variable involved in the interaction (e.g., gender). Marginal predicted values are computed by default in Stata by 1) setting all cases in the analytic sample to have the same value on the covariate of interest (e.g., fixing all cases to have the same value of education), and then 2) computing predicted values for each case based on the fitted model and averaging the predictions. This can also be done separately for different subpopulations (e.g., males and females) when examining interactions.

In addition to computing average marginal effects based on these marginal predicted values, one can also compute design-adjusted standard errors of the average marginal effects by using the **delta method**, which is similar in principle to the TSL approximation method described in Chapter 3. Computing these marginal predicted values, computing average marginal effects, and then plotting the average marginal effects (in addition to confidence intervals for the effects based on their standard errors) is facilitated by the `margins` and `marginsplot` post-estimation commands in Stata.

We provide an example of using these commands below. First, after fitting the model above that includes the interaction between education and gender, we compute the average marginal effect of the first non-reference category of education (using the `dydx()` option), separately for males and females, and then plot the marginal effects using the `marginsplot` command. We then repeat this for the second non-reference category of education, and combine the plots:

```
* Fit model including interaction of gender and education.
svy: logit arthritis ib2.gender ib4.edcat nage ///
ib2.gender#ib4.edcat

* Compute marginal effects of less than high school
* and HS education (vs. college) on predicted
```

```
* probability of arthritis for males and females.
margins, dydx(1.edcat) by(gender)
marginsplot
graph save Graph "X:\. . .\fig83a.gph", replace
margins, dydx(2.edcat) by(gender)
marginsplot
graph save Graph "X:\. . .\fig83b.gph", replace
graph combine "X:\. . .\fig83a.gph" ///
"X:\. . .\fig83b.gph"
```

If one was looking at the average marginal effects corresponding to a one-unit change in the value of a continuous predictor, that variable would simply be included in the `dydx()` function. If one was looking to plot the effects of a categorical predictor as a function of selected values of a continuous predictor, one would use `at()` instead of `by()` in the commands above (see example below).

Figure 8.3 shows the plots that result from running the commands above, which include 95% confidence intervals for the average marginal effects by default. We see that the effect of less than high school education (vs. college education) on the predicted probability of arthritis is much stronger for females than for males, while this is not the case for the effect of high school education. We recommend visualizing significant interactions in this way, not just when fitting logistic regression models but when fitting any types of regression models where significant interactions emerge. Stata makes it especially easy to do this using the commands illustrated above.

Finally, we emphasize to readers that the relationships of predictor variables with the expected value of the outcome in regression models where the expected value is a

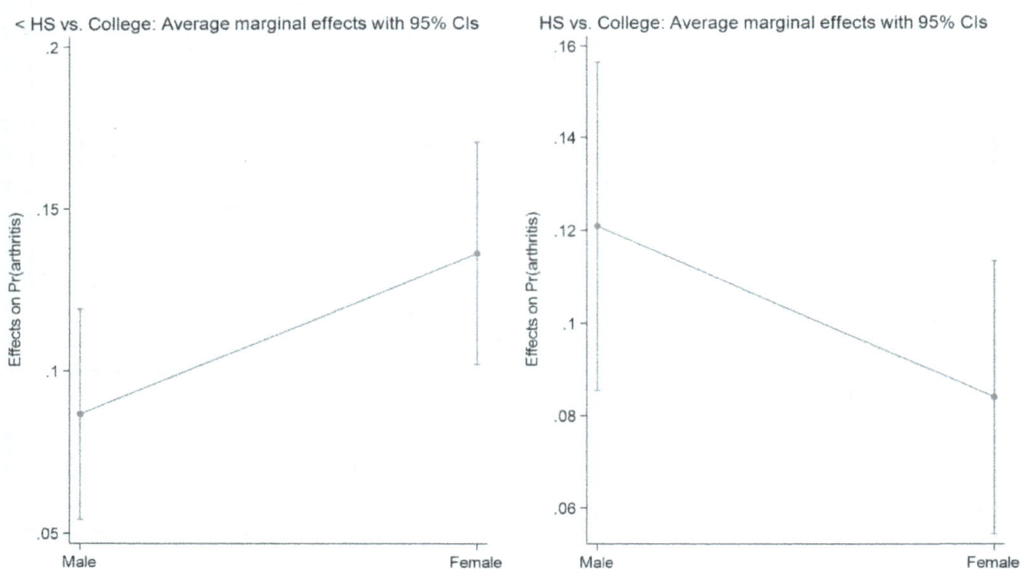

FIGURE 8.3
Plots of estimated marginal effects of less than high school and high school education (relative to college education) on the probability of arthritis for males and females separately.

Source: 2012 HRS.

non-linear function of the predictors (including logistic regression models) *depend on* the values of other covariates, *even if the model does not include interaction terms* (Ai and Norton, 2003). For this reason, when interpreting the relationships of a covariate with the predicted probability that a given survey variable is equal to 1, we recommend always plotting average marginal effects associated with that covariate (or with a given interaction) as a function of other covariates included in the model. This will allow the analyst to get a sense of whether the relationship of interest is robust to changing values of the other covariates included in the model. For example, given the significant interaction noted above, we can plot the average marginal effects of less than high school education (vs. college education) for males and females separately at different possible values of age, looking at 10-year increments:

```
margins, dydx(1.edcat) by(gender) at(nage=(50(10)90))
marginsplot
```

The resulting plot of average marginal effects (in addition to 95% confidence intervals for the effects) is displayed in Figure 8.4, and shows how the difference in this effect between males and females does change somewhat depending on a person's age, with the difference in this effect becoming smaller as a function of age. We will consider additional examples of the use of these plots for interpretation in the next section.

As mentioned in the Section 6.4.5 discussion of relative risk statistics, logistic regression and other generalized models are often used with complex sample survey data to estimate the "risk" of an outcome (e.g., risk of diabetes, risk of crime victimization, "risk" of successfully graduating from college) as a function of the model covariates. In this "risk estimation" context, the analysis of average marginal effects (expressed as the average

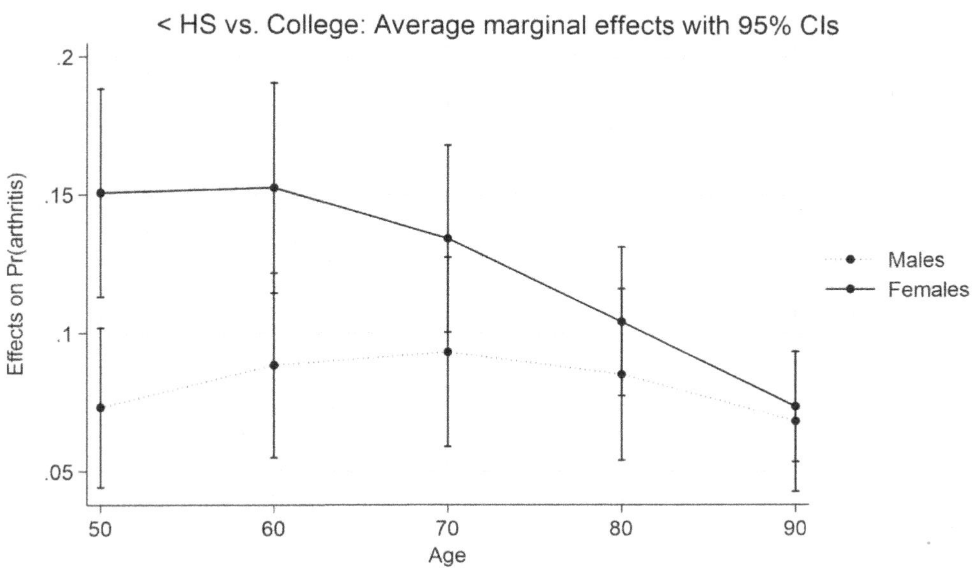

FIGURE 8.4
Plots of estimated marginal effects of less than high school education (relative to college education) on the probability of arthritis for males and females separately, considering different possible values of age.

Source: 2012 HRS.

probability of the outcome given the specified covariate levels) is closely tied to the estimation of the **attributable fraction (AF)** of the outcome that is associated with a specific exposure to a risk factor covariate (Graubard et al., 2007). For example, an epidemiologist may use a logistic or Poisson regression model to study the risk of emphysema as a function of lifetime smoking status. Based on the risk model, the Stata `margins` and `marginsplot` commands illustrated above could be used to compute and display the average marginal effect for lifetime smoking behavior, that is, the population average probability that if a person smokes for their lifetime they will develop emphysema. The attributable fraction for lifetime smoking is an estimate of the fraction of all incident cases of emphysema that would not occur if lifetime smoking behavior was eliminated from the population.

Heeringa et al. (2015a) provide an in-depth review of best practices for estimating attributable fractions from complex sample survey data and provide a general four-step algorithm for: 1) estimating the logistic regression or Poisson risk model; 2) computing weighted estimates of the AF statistics; 3) using JRR and bootstrap methods to estimate design-based standard errors; and 4) applying special methods for constructing confidence intervals for the AF population value (Greenland, 2004). Analysts interested in a detailed review of AF estimation based on complex sample survey data are referred to this 2015 publication. SAS code and an R function, along with documentation showing how to implement these four steps in design-based estimation and inference for the AF statistic, are available on the ASDA website.

8.7 Analysis Application

This section presents an example logistic regression analysis that follows the four general modeling stages described in Sections 8.3 to 8.6.

EXAMPLE 8.1: EXAMINING PREDICTORS OF A LIFETIME MAJOR DEPRESSIVE EPISODE IN THE NCS-R DATA

The aim of this example is to build a logistic regression model for the probability that a U.S. adult has been diagnosed (DSM-IV) with a major depressive episode (MDE) in their lifetime. The dependent variable is the NCS-R variable MDE, which takes a value of 1 for persons who meet lifetime criteria for major depression and 0 for all others. The following predictors are considered: AG4CAT (a categorical variable measuring age brackets, including 18–29, 30–44, 45–59, and 60+), SEX (1 = Male, 2 = Female), ALD (an indicator of any lifetime alcohol dependence), ED4CAT (a categorical variable measuring education brackets, including 0–11 years, 12 years, 13–15 years, and 16+ years), and MAR3CAT (a categorical variable measuring marital status, with values 1 = 'married', 2 = 'separated/widowed/divorced', and 3 = 'never married'). The primary research question of interest is whether MDE is related to alcohol dependence after adjusting for the effects of the demographic factors listed above.

8.7.1 Stage 1: Model Specification

The analysis session begins by specifying the complex design features of the NCS-R sample in the Stata `svyset` command. Note that we specify the "long" or Part 2 NCS-R survey weight (NCSRWTLG) in the `svyset` command. This is due to the use of the alcohol dependence variable in the analysis, which was measured in Part 2 of the NCS-R survey.

Recall that the NCS-R Part 2 questionnaire was only administered to a subsample of the NCS-R Part 1 respondents.

```
svyset seclustr [pweight = ncsrwtlg], strata(sestrat)
```

There are 42 sampling error strata and 84 sampling error computation units (two per stratum) in the NCS-R sampling error calculation model, resulting in 42 design-based degrees of freedom.

Following the recommendations of Hosmer and colleagues, the model building begins by examining the bivariate associations of MDE with each of the potential predictor variables. Since the candidate predictors are all categorical variables, the bivariate relationship of each predictor with MDE is analyzed in Stata by using the `svy: tab` command and requesting row percentages (as discussed in Chapter 6):

```
svy: tab ag4cat mde, row se
svy: tab sex mde, row se
svy: tab ald mde, row se
svy: tab ED4CAT mde, row se
svy: tab MAR3CAT mde, row se
```

Table 8.5 presents the results of these bivariate analyses, including the Rao-Scott F-tests of association. This table also presents estimates of the percentages of each predictor category that received the lifetime MDE diagnosis, along with the linearized standard errors of these weighted estimates.

TABLE 8.5

Initial bivariate design-based tests of association assessing potential predictors of lifetime major depressive episode (MDE) based on the NCS-R adult sample

Predictor	Rao-Scott F-test*	Category	% with MDE (SE)
AG4CAT	$F(2.76, 115.97) = 26.39$	18–29	18.4% (0.9)
	$p < 0.01$	30–44	22.9% (1.1)
		45–59	22.3% (1.3)
		60+	11.1% (1.0)
SEX	$F(1, 42) = 44.83$	Male	15.3% (0.9)
	$p < 0.01$	Female	22.6% (0.7)
ALD	$F(1, 42) = 120.03$	Yes	45.2% (0.3)
	$p < 0.01$	No	17.7% (0.7)
ED4CAT	$F(2.90, 121.93) = 4.30$	<12 yrs	16.3% (1.2)
	$p < 0.01$	12 yrs	18.6% (0.8)
		13–15 yrs	21.3% (1.0)
		16+ yrs	19.7% (1.1)
MAR3CAT	$F(1.90, 79.74) = 11.08$	Married	17.3% (0.7)
	$p < 0.01$	Previously	23.9% (1.5)
		Never	19.4% (1.2)

Source: NCS-R, $n = 5{,}692$.

Note: * See Chapter 6 for more details on the derivation of this test statistic, which can be used to test the null hypothesis of no association between the predictor variable and the outcome variable (MDE).

Based on these initial tests of association, all of the predictor variables appear to have significant bivariate associations with MDE, including ALD, the indicator of lifetime alcohol dependence, and all of the predictors appear to be good candidates for inclusion in the initial multivariable logistic regression model.

8.7.2 Stage 2: Model Estimation

The next step in the model building procedure is to fit the 'initial' multivariable model, examining the relationships of all five predictors with MDE when adjusting for the other predictors. Stata provides two commands for fitting the multivariable logistic regression model: `svy: logistic` and `svy: logit`. The default output for the `svy: logistic` command is estimates of adjusted odds ratios and 95% confidence intervals for the adjusted odds ratios. Note the use of the `i.` modifier, which indicates those predictor variables that should be considered categorical in the logistic regression model. When using this modifier, indicator variables will be created for each of the categories of the predictors, and the indicator for the lowest-valued category will be omitted from the model (by default). To change this default, one can use the `ib#.` modifier, which sets the category with value # as the reference category. We use this to set the reference category of SEX to 2 (females).

```
svy: logistic mde i.ag4cat ib2.sex ald i.ED4CAT ///
i.MAR3CAT
```

In `svy: logistic`, estimates of the parameters and standard errors for the logit model can be requested by using the `coef` option:

```
svy: logistic mde i.ag4cat ib2.sex ald i.ED4CAT ///
i.MAR3CAT, coef
```

Stata users who wish to see the estimated logistic regression coefficients and standard errors may also use the companion command, `svy: logit`:

```
svy: logit mde i.ag4cat ib2.sex ald i.ED4CAT i.MAR3CAT
```

Estimated odds ratios and 95% CIs can be generated in `svy: logit` by adding the `or` option:

```
svy: logit mde i.ag4cat ib2.sex ald i.ED4CAT i.MAR3CAT, or
```

Other software procedures (e.g., SAS PROC SURVEYLOGISTIC) will output both the estimated logistic regression coefficients (and standard errors) and the corresponding odds ratio estimates. (Readers should be aware that the `svy: logit` and `svy: logistic` commands differ slightly in the procedures for calculation of the standard errors for odds ratios.)

In each form of the Stata `svy: logistic` or `svy: logit` command, the dependent variable, MDE, is listed first, followed by the predictor variables (and their interactions, if applicable). Because the ALD variable is already coded as either 0 or 1, it does not require use of the `i.` modifier like the other categorical predictors.

Tables 8.6 and 8.7 summarize the output generated by fitting the initial MDE logistic regression model, which does not yet include any interactions between the predictors.

TABLE 8.6

Estimated logistic regression model for the lifetime MDE outcome (output generated by using the `svy: logit` command)

Predictor*	Category	$\hat{B}$	$se(\hat{B})$	t	$P(t_{42} > t)$
INTERCEPT		−1.583	0.121	−13.12	<0.001
AG4CAT	30–44	0.256	0.094	2.71	0.010
	45–59	0.206	0.092	2.26	0.029
	60+	−0.676	0.141	−4.78	< 0.001
SEX	Male	−0.577	0.077	−7.48	< 0.001
ALD	Yes	1.424	0.154	9.24	< 0.001
ED4CAT	12	0.079	0.097	0.82	0.418
	13–15	0.231	0.093	2.48	0.017
	16+	0.163	0.111	1.47	0.148
MAR3CAT	Previously	0.486	0.085	5.69	<0.001
	Never	0.116	0.108	1.07	0.290

Source: NCS-R, $n = 5{,}692$, adjusted Wald test for all parameters: $F(10, 33) = 28.07$, $p < 0.001$.
Note: * Reference categories for categorical predictors are: AG4CAT (18–29); SEX (Female); ALD (No); ED4CAT (<12 yrs); MAR3CAT (Married).

TABLE 8.7

Estimates of adjusted odds ratios for the lifetime MDE outcome

Predictor*	Category	$\hat{\psi}$	95% CI for ψ
AG4CAT	30–44	1.29	(1.067, 1.562)
	45–59	1.23	(1.022, 1.479)
	60+	0.51	(0.383, 0.677)
SEX	Male	0.56	(0.480, 0.656)
ALD	Yes	4.15	(3.042, 5.668)
ED4CAT	12	1.08	(0.890, 1.316)
	13–15	1.26	(1.044, 1.519)
	16+	1.18	(0.941, 1.471)
MAR3CAT	Previously	1.63	(1.369, 1.932)
	Never	1.12	(0.903, 1.396)

Source: NCS-R, $n = 5{,}692$, adjusted Wald test for all parameters: $F(10, 33) = 28.07$, $p < 0.001$.
Note: * Reference categories for categorical predictors are: AG4CAT (18–29); SEX (Female); ALD (No); ED4CAT (<12 yrs); MAR3CAT (Married).

8.7.3 Stage 3: Model Evaluation

The adjusted Wald tests in Stata for the AG4CAT, ED4CAT and MAR3CAT categorical predictors in this initial model are generated by using the `test` command:

```
test 2.ag4cat 3.ag4cat 4.ag4cat
test 2.MAR3CAT 3.MAR3CAT
test 2.ED4CAT 3.ED4CAT 4.ED4CAT
```

Note that we do not request these multi-parameter Wald tests for the SEX and ALD predictor variables, because they are represented by single indicator variables in the regression

TABLE 8.8

Design-adjusted *F*-Tests for the parameters associated with the categorical predictors in the initial MDE logistic regression model

Categorical Predictor	*F*-test Statistic	$P(\mathcal{F} > F)$
AG4CAT	$F_{(3,40)} = 19.03$	< 0.001
ED4CAT	$F_{(3,40)} = 2.13$	0.112
MAR3CAT	$F_{(2,41)} = 16.60$	< 0.001

model and the overall Wald test for each predictor is equivalent to the *t*-test reported for the single estimated parameter for that predictor. Further, note in each of the `test` statements that we include the *K* - 1 indicator variables generated by Stata for each of the categorical predictors (e.g., `2.ag4cat`) when the `i.` modifier is used to identify categorical predictor variables. Stata users can find the names of these indicators in the Stata Results window once the model has been fitted. Table 8.8 provides the design-adjusted *F* versions of the resulting Wald test statistics and associated *p*-values.

Two of the three design-adjusted Wald tests are significant at the 0.01 level. The exception is the Wald test for education [$F(3,40) = 2.13$, $p = 0.112$], which suggests that the parameters associated with education in this logistic regression model are not significantly different from zero and that education may not be an important predictor of lifetime MDE when adjusting for the relationships of the other predictor variables with the outcome. If the objective of the model-building process is the construction of a parsimonious model, education could probably be dropped as a predictor at this point. For the purposes of this illustration (and because of the marginal significance), we will retain education in the model moving forward.

8.7.4 Stage 4: Model Interpretation/Inference

Based on the results in Table 8.6 and Table 8.8, it appears that each of the predictors in the multivariable model has a significant (or marginally significant) relationship with the probability of MDE after adjusting for the relationships of the other predictors. Focusing on the primary predictor variable of interest, we see that the odds of having had a major depressive episode at some point in the lifetime are multiplied by 4.15 when a person has had a diagnosis of alcohol dependence at some point in their lifetime, when adjusting for the relationships of age, sex, education, and marital status. Of course, this model does not allow for any kind of causal inference, given that time ordering of the events is not available in the NCS-R data set; we can, however, conclude that there is strong evidence of an association between the two disorders in this finite population when adjusting for other demographic covariates.

We visualize the average marginal effects of alcohol dependence on the predicted probability of having a lifetime major depressive episode for different possible age categories by executing the post-estimation commands below and producing the plot shown in Figure 8.5.

```
margins, dydx(ald) by(ag4cat)
marginsplot
```

Considering the confidence interval limits, Figure 8.5 shows that alcohol dependence is expected to increase the probability of a lifetime major depressive episode by anywhere from 0.1 to 0.3, depending on the age category.

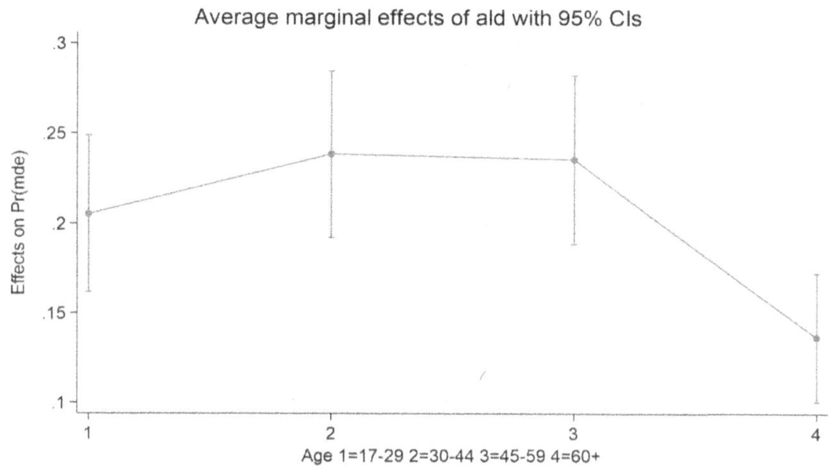

FIGURE 8.5
Plot of estimated average marginal effects of alcohol dependence on the probability of having a lifetime major depressive episode for different age categories.

We also note that relative to married respondents, respondents who were previously married have significantly higher (63% higher) odds of having had a major depressive episode in their lifetime when adjusting for the other covariates. Further, middle-age respondents have significantly higher odds of lifetime MDE (relative to younger respondents), while older respondents and males have significantly reduced odds of lifetime MDE (again relative to younger respondents and females). Each of these effects could be visualized as a function of other covariates included in the model by using the `margins` and `marginsplot` commands demonstrated above.

Respondent age is represented in the model as four grouped categories of age. Including grouped categories for age (or recoded categories of any continuous predictor, more generally) in a logistic regression model will result in estimates of the expected contrasts in log-odds for respondents in each of the defined categories, relative to the reference category. Since the model parameters are estimated separately for each defined age group (with age 18–29 as the reference), the model will capture any non-linearity of the relationship of age with MDE via the ordered age groupings.

Inspecting the estimated coefficients and odds ratios for the grouped age categories in Tables 8.6 and 8.7, it appears that there is indeed significant non-linearity in the relationship of age with the odds of MDE. Relative to the 18–29-year-old group, the odds of MDE increase by factors of 1.29 (age 30–44) and 1.23 (age 45–59) for the middle age ranges but decrease by a factor of 0.51 in the age 60 and older group. Such non-linear relationships of age are common in models of human disorders, and are possibly attributable to normal processes of aging and selective mortality. If the example model was estimated with age (in years) as a continuous predictor variable, at this stage in the model building process the analyst would re-estimate the model including both the linear and quadratic terms for age.

At this stage in the model building process, we have chosen to retain all of the candidate predictors. Next, we apply Archer and Lemeshow's (2006) design-adjusted test to assess the goodness of fit of this initial model:

```
estat gof
```

The design-adjusted *F*-statistic reported in the Stata Results window is $F_{A\text{-}L}(9,34) = 1.23$, with a *p*-value of 0.31. This suggests that the null hypothesis that the model fits the data well is *not* rejected. We therefore have confidence moving forward that the fit of this initial model is acceptable. (Note: the `estat gof` command currently does not work if a `subpop()` option has been used to fit the model. A work-around is to fit the model using the `if` modifier instead of the `subpop()` option, but we would only recommend doing this if the conditional analysis that results does not arbitrarily delete sampled clusters. See Chapter 4 for details.)

Next, we consider testing some scientifically relevant two-way interactions between the candidate predictor variables. For illustration purposes, we suppose that possible two-way interactions of sex with the other four covariates measuring age, lifetime alcohol dependence, education, and marital status are of interest, if sex is posited by an NCS-R analyst as being a possible moderator of the relationships of these other four covariates with lifetime MDE. We fit a model including these two-way interactions in Stata using the following command:

```
svy: logit mde i.ag4cat ib2.sex ald i.ED4CAT i.MAR3CAT ///
i.ag4cat#ib2.sex ib2.sex#c.ald i.ED4CAT#ib2.sex ///
i.MAR3CAT#ib2.sex
```

Note how the interactions are specified in this command: after all of the predictor variables are listed, a # sign is used to specify two-way interactions of the relevant variables, and any variables not indicated as categorical need to be specified as `c.` variables (or continuous variables) in the interaction terms (Stata will assume that variables included in the interaction terms are categorical by default). Table 8.9 presents the estimates of the regression parameters in this model generated by executing the command above in Stata.

At this point, the statistical question is whether these two-way interactions are making a significant additional contribution or improvement to the fit of this model to the NCS-R data. That is, are any of the parameters associated with the two-way interaction terms significantly different from 0? We can test this hypothesis by once again using design-adjusted Wald tests. The terms in the `test` commands below can be easily determined from the Stata Results window:

```
test 2.ag4cat#1.sex 3.ag4cat#1.sex 4.ag4cat#1.sex
test 1.sex#c.ald
test 2.ED4CAT#1.sex 3.ED4CAT#1.sex 4.ED4CAT#1.sex
test 2.MAR3CAT#1.sex 3.MAR3CAT#1.sex
```

The results of these four Wald F-tests are displayed in Table 8.10. Based on these results, we fail to reject the null hypotheses for all four of the tests, suggesting that these two-way interactions are actually not making a significant contribution to the fit of the model. We therefore do not consider these two-way interactions any further, and would proceed with making inferences based on the estimates from the model presented in Table 8.6.

We also illustrate calculation of the design-adjusted AIC criterion that has been implemented in the `survey` package of R (Lumley & Scott, 2015) for comparing the fits of competing models. First, we import the NCS-R data, and keep only those cases given the longer NCS-R interview (as no missing data are allowed on the design variables). Then, we load the newest version of the contributed `survey` package, declare an NCS-R design object containing the complex sampling features and the survey data, fit the MDE models without and with the interactions, and compute the design-adjusted AIC values for each of the models:

TABLE 8.9

Estimated logistic regression model for lifetime MDE, including two-way interactions of the other predictor variables with SEX

Predictor *	Category	$\hat{B}$	$se(\hat{B})$	t	$P(t_{42} > t)$
INTERCEPT	Constant	−1.600	0.134	−11.94	< 0.001
AG4CAT	30–44	0.220	0.114	1.94	0.059
	45–59	0.215	0.102	2.09	0.042
	60+	−0.646	0.175	−3.68	0.001
SEX	Male	−0.546	0.357	−1.53	0.134
ALD	Yes	1.553	0.211	7.36	< 0.001
ED4CAT	12	0.131	0.084	1.56	0.126
	13–15	0.297	0.117	2.54	0.015
	16+	0.242	0.152	1.59	0.118
MAR3CAT	Previously	0.418	0.111	3.78	<0.001
	Never	0.017	0.130	0.13	0.894
AG4CAT × SEX	30–44 × Male	0.097	0.201	0.48	0.633
	45–59 × Male	0.003	0.213	0.01	0.990
	60+ × Male	−0.038	0.302	−0.13	0.901
ALD × SEX	Yes × Male	−0.200	0.242	−0.83	0.413
ED4CAT × SEX	12 × Male	−0.138	0.271	−0.51	0.614
	13–15 × Male	−0.169	0.269	−0.63	0.534
	16+ × Male	−0.194	0.344	−0.56	0.576
MAR3CAT × SEX	Previously × Male	0.182	0.208	0.88	0.385
	Never × Male	0.232	0.212	1.09	0.280

Source: NCS-R. $n = 5{,}692$, adjusted Wald test for all parameters: $F(19,24) = 17.15, p < 0.001$.
Note: * Reference categories for categorical predictors are: AG4CAT (18–29 yrs); ALD (no); ED4CAT (<12 yrs); MAR3CAT (married); SEX(female).

TABLE 8.10

Design-adjusted Wald tests of first-order interactions of sex and other categorical predictors in the MDE logistic regression model

Interaction Term	F-test Statistic	$P(\mathcal{F} > F)$
AG4CAT × SEX	$F_{(3,40)} = 0.25$	0.863
ALD × SEX	$F_{(1,42)} = 0.68$	0.413
ED4CAT × SEX	$F_{(3,40)} = 0.13$	0.944
MAR3CAT × SEX	$F_{(2,41)} = 0.77$	0.472

```
load("X:/. . ./ncsr.rdata")

ncsr$marcatc <- factor(ncsr$mar3cat, levels = 1:3,
labels =c("Married", "Previously Married", "Never Married"))
ncsr$edcatc <-factor(ncsr$ed4cat, levels = 1:4,
labels =c("0-11", "12", "13-15","16+"))
ncsr$sexc <- factor(ncsr$sex, levels = 1:2,
labels=c("Male","Female"))
ncsr$sexc <- relevel(ncsr$sexc, ref="Female")
ncsr$agcatc <- factor(ncsr$ag4cat, levels = 1:4,
labels=c("18-29", "30-44", "45-59", "60+"))
```

```
ncsr.lg <- ncsr[!is.na(ncsr$ncsrwtlg),]

library(survey)

# design object
ncsrsvyp2 <- svydesign(strata=~sestrat, id=~seclustr,
weights=~ncsrwtlg, data=ncsr.lg, nest=T)

# fit first model
ex81 <- svyglm(mde ~ factor(agcatc) + factor(sexc) + ald +
factor(edcatc) + factor(marcatc), family=quasibinomial, design=ncsrsvyp2)

# compute design-adjusted AIC
AIC(ex81)

# fit model with interactions
ex81.1 <- svyglm(mde ~ factor(agcatc) + factor(sexc) + ald +
factor(edcatc) + factor(marcatc) +
factor(sex):factor(agcatc) + factor(sexc):ald +
factor(sexc):factor(edcatc) + factor(sexc):factor(marcatc),
family=quasibinomial, design=ncsrsvyp2)

# compute design-adjusted AIC
AIC(ex81.1)
```

The design-adjusted AIC for the first model is 5288.08, while the design-adjusted AIC for the second model is 5303.89. We would therefore select the first model as having a better fit given the number of parameters being estimated. We can also consider a design-adjusted likelihood ratio test comparing the fits of these two models (which is also only currently available in the contributed `survey` package for R):

```
# design-adjusted Likelihood Ratio Test
anova(ex81, ex81.1, test="Chisq", method="LRT")
```

```
Working (Rao-Scott) LRT for factor(agcatc):factor(sex)
factor(sexc):ald factor(sexc):factor(edcatc)
factor(sexc):factor(marcatc)
 in svyglm(formula = mde ~ factor(agcatc) + factor(sexc) +
ald + factor(edcatc) + factor(marcatc) +
factor(sex):factor(agcatc) +
    factor(sexc):ald + factor(sexc):factor(edcatc) +
factor(sexc):factor(marcatc),
    family = quasibinomial, design = ncsrsvyp2)
Working 2logLR = 3.412191 p = 0.90812
(scale factors: 3 1.5 1.1 0.84 0.78 0.65 0.57 0.34 0.23)
```

We see from the result of the design-adjusted likelihood ratio test (p = 0.90812) that we would fail to reject the (smaller) nested model that does not include the interactions as the preferable model. Using multiple approaches that account for the complex sampling features, we do not have any evidence of the interactions improving the fit of this model.

As a final part of our design-based analysis, we formally test whether the weights are necessary for estimation of the parameters in the "final" model, using the procedure that was outlined in Chapter 7. If the weights are not informative about the parameters of interest in this model, we may be able to increase the efficiency of the parameter estimates. We perform this test by specifying unweighted estimation in the `svyset` command (but still specifying the other design features), and performing a design-adjusted Wald test of the null hypothesis that the coefficients associated with the weight variable and all interactions of the weight variable with the other predictors in this model are equal to zero (Fuller, 1984):

```
# Test need for weighted estimation
svyset seclustr, strata(sestrat)

svy: logit mde i.ag4cat ib2.sex ald i.ED4CAT i.MAR3CAT ///
ncsrwtlg c.ncsrwtlg#i.ag4cat c.ncsrwtlg#ib2.sex ///
c.ncsrwtlg#c.ald c.ncsrwtlg#i.ED4CAT c.ncsrwtlg#i.MAR3CAT

test ncsrwtlg c.ncsrwtlg#2.ag4cat c.ncsrwtlg#3.ag4cat ///
c.ncsrwtlg#4.ag4cat c.ncsrwtlg#1.sex ///
c.ncsrwtlg#c.ald c.ncsrwtlg#2.ED4CAT ///
c.ncsrwtlg#3.ED4CAT c.ncsrwtlg#4.ED4CAT ///
c.ncsrwtlg#2.MAR3CAT c.ncsrwtlg#3.MAR3CAT
```

The resulting design-adjusted F-test statistic is $F(11, 32) = 7.17$, with $p < 0.001$. We therefore have strong evidence in favor of rejecting the null hypothesis and conclude that the final survey weights for Part II NCS-R respondents are informative about the coefficients in this final model. We would proceed with reporting the weighted estimates.

8.7.5 Alternative Bayesian Approach

We now consider an alternative design-adjusted Bayesian approach to fitting the "final" model determined from the analysis above. In previous chapters, we have repeatedly demonstrated that this model-based approach can increase the efficiency of the finite population inferences that we make from analyzing complex sample survey data; here, we see whether this is also the case for logistic regression models. We use functions in the contributed `csSampling` package in the R software to fit this model; see Section 3.8 for details.

First, we create a data frame object with only complete cases on the variables of interest, and then normalize the final survey weights for NCS-R Part II respondents:

```
# create complete data set with variables of interest
ncsr.red <- ncsr[, c("mde", "marcatc", "edcatc", "sexc",
"agcatc", "ald", "seclustr", "sestrat", "ncsrwtlg")]
ncsr.red <- ncsr.red[complete.cases(ncsr.red),]

# need to normalize NCS-R weights to match Stan
ncsr.red$wtsc <- ncsr.red$ncsrwtlg /
mean(ncsr.red$ncsrwtlg)
```

Next, we recreate the survey design object using the new normalized weights:

```
# survey design object
ncsr.des <- svydesign(id = ~seclustr, strata = ~sestrat,
weights = ~wtsc, nest = T, data = ncsr.red)
```

Now, we specify the logistic regression model of interest (note the use of `family = bernoulli()`), and initiate formation of the survey-weighted posterior distribution for the parameters in the model, along with design-adjusted draws of the parameters from that posterior (see Section 3.8):

```
# Bayesian approach, flat prior
set.seed(41279)
model_formula <- formula("mde|weights(wtsc) ~ agcatc + sexc
+ ald + edcatc + marcatc")

mod.brms <- cs_sampling_brms(svydes = ncsr.des, brmsmod =
brmsformula(model_formula), data = ncsr.red, family = bernoulli())

mod.brms$stan_fit
```

As we noted in Chapter 7, the output from this procedure (not shown here) can be a bit confusing to navigate, but the parameter estimates are labeled as `b[X]` and ordered in the same way that they would be when using `svyglm()` to fit the model, with the intercept labeled as `b_Intercept`. Table 8.11 compares the point estimates and 95% confidence/credible intervals between the design-based and Bayesian approaches to fitting the "final" logistic regression model from this application. We specifically extracted the posterior means and the 2.5 and 97.5 percentiles of the 1,000 design-adjusted draws of the parameters in this final model from the survey-weighted posterior distribution for the parameters. In Table 8.11, we see that the inferences are quite similar, and that in some cases, the credible intervals are wider than the design-based confidence intervals. We would arrive at quite similar conclusions based on the two approaches.

TABLE 8.11

Comparisons of inferences between the design-based and Bayesian approaches for the "final" logistic regression model

		Design-Based		Bayesian	
Predictor	**Category**	**Estimate**	**95% CI**	**Posterior Mean**	**95% Credible Interval**
INTERCEPT		−1.58	(−1.83, −1.34)	−1.58	(−1.87, −1.30)
AG4CAT	30–44	0.26	(0.07, 0.45)	0.26	(0.05, 0.48)
	45–59	0.21	(0.02, 0.39)	0.21	(−0.02, 0.42)
	60+	−0.68	(−0.96, −0.39)	−0.68	(−0.96, −0.41)
SEX	Male	−0.58	(−0.73, −0.42)	−0.58	(−0.72, −0.43)
ALD	Yes	1.42	(1.11, 1.73)	1.43	(1.21, 1.65)
ED4CAT	12	0.08	(−0.12, 0.27)	0.08	(−0.15, 0.30)
	13–15	0.23	(0.04, 0.42)	0.23	(−0.01, 0.46)
	16+	0.16	(−0.06, 0.39)	0.16	(−0.08, 0.41)
MAR3CAT	Previously	0.49	(0.31, 0.66)	0.48	(0.31, 0.66)
	Never	0.12	(−0.10, 0.33)	0.11	(−0.08, 0.32)

As we noted in earlier chapters, one could also specify prior distributions for specific coefficients in this model as well, if this information were available. For example, if a prior analysis suggested that the mean of the ALD coefficient was 1.0 and its posterior standard deviation (i.e., standard error) was 0.2, and we were to assume that this coefficient follows a Normal distribution, we could use the following code:

```
# With informative prior for ALD coefficient
mod.brms <- cs_sampling_brms(svydes = ncsr.des,
brmsmod = brmsformula(model_formula),
prior = c(set_prior("normal(1.0,0.2)", class = "b", coef="ald")),
data = ncsr.red, family = bernoulli())

mod.brms$stan_fit
```

In the resulting output, we see that the posterior mean of the design-adjusted draws of the ALD parameter is now 1.31 (lower than what was seen in Table 8.11, reflecting the effect of the prior information) and that the 95% credible interval is now (1.11, 1.50), which is narrower than the interval based on the default non-informative prior (see Table 8.11).

Recently, Salvatore et al. (2024) proposed methods for improving inferences based on logistic regression models by combining probability samples with non-probability samples and using Bayesian methods to analyze the combined data. In this setting, informative priors like the one considered above would be derived from the non-probability sample; we discuss methods for non-probability samples in more detail in Section 13.4. These authors not only show improvements in the MSE of parameter estimates in logistic regression models from using this method, but show that cost savings would be possible in data collection when using such an approach. We see this as a promising direction for future work, and the authors provide an R Shiny application implementing their methodology.[1]

8.8 Comparing the Logistic, Probit, and Complementary-Log-Log (C-L-L) GLMs for Binary Dependent Variables

This chapter has focused on logistic regression techniques for modeling $\pi(x)$ for a binary dependent variable. As discussed in Section 8.2, alternative generalized linear models for a binary dependent variable may be estimated using the probit or C-L-L link function. In discussing these alternative GLMs, we noted that inferences derived from logistic, probit, and C-L-L regression models should generally be consistent.

To illustrate this claim, consider the results in Table 8.12 for a side-by-side comparison of estimated logistic, probit, and C-L-L regression models. The example used for this comparison is a model of the probability that a U.S. adult is alcohol dependent (ALD). The data are from the NCS-R long interview (or Part II of the survey) and each model includes the same demographic main effects considered in Section 8.7 for the model of MDE: SEX, AG4CAT, ED4CAT, and MAR3CAT. The Stata commands for the estimation of the three models follow:

```
svyset seclustr [pweight = ncsrwtlg], strata(sestrat)
```

```
svy: logit ald i.ag4cat ib2.sex i.ED4CAT i.MAR3CAT
test 2.ag4cat 3.ag4cat 4.ag4cat
test 2.MAR3CAT 3.MAR3CAT
test 2.ED4CAT 3.ED4CAT 4.ED4CAT

svy: probit ald i.ag4cat ib2.sex i.ED4CAT i.MAR3CAT
test 2.ag4cat 3.ag4cat 4.ag4cat
test 2.MAR3CAT 3.MAR3CAT
test 2.ED4CAT 3.ED4CAT 4.ED4CAT

svy: cloglog ald i.ag4cat ib2.sex i.ED4CAT i.MAR3CAT
test 2.ag4cat 3.ag4cat 4.ag4cat
test 2.MAR3CAT 3.MAR3CAT
test 2.ED4CAT 3.ED4CAT 4.ED4CAT
```

Table 8.12 presents a summary of the estimated coefficients, standard errors, and p-values for simple hypothesis tests of the form H_0: $B_j = 0$. Table 8.13 presents the results for the design-adjusted Wald tests of the overall age, education, and marital status effects. Note that although the coefficients and standard errors for the probit model show the expected difference in scale from those of the logistic and C-L-L models, the three models produce

TABLE 8.12

Comparisons of Estimated Logistic, Probit, and C-L-L Models of Alcohol Dependency in U.S. adults

		Logistic			Probit			Complementary L-L		
Predictor*	**Category**	$\hat{B}$	$se(\hat{B})$	p	$\hat{B}$	$se(\hat{B})$	p	$\hat{B}$	$se(\hat{B})$	p
Intercept		−3.124	0.225	<0.01	−1.719	0.105	<0.01	−3.148	0.218	<0.01
SEX	Male	0.998	0.119	<0.01	0.471	0.056	<0.01	0.965	0.115	<0.01
AG4CAT	30–44	0.146	0.178	0.416	0.065	0.085	0.444	0.143	0.171	0.408
	45–59	−0.051	0.144	0.726	−0.034	0.067	0.609	−0.045	0.140	0.748
	60+	−1.120	0.212	<0.01	−0.531	0.093	<0.01	−1.083	0.209	<0.01
ED4CAT	12 yrs	−0.268	0.194	0.173	−0.124	0.095	0.200	−0.260	0.185	0.167
	13–15 yrs	−0.264	0.176	0.141	−0.124	0.085	0.152	−0.256	0.169	0.137
	16+ yrs	−0.736	0.197	<0.01	−0.340	0.092	<0.01	−0.713	0.190	<0.01
MAR3CAT	Previously	0.518	0.142	<0.01	0.255	0.070	<0.01	0.494	0.136	<0.01
	Never	0.065	0.169	0.701	0.039	0.077	0.616	0.060	0.164	0.713

Source: NCS-R. n = 5,692.
Note: * Reference categories for categorical predictors are: AG4CAT (60+ yrs); SEX (female); ED4CAT (<12 yrs); MAR3CAT (Married).

TABLE 8.13

Design-adjusted Wald tests of categorical predictors in the ALD models

	Wald F-test Statistic (p-value = $P(\mathcal{F} > F)$)		
Categorical Predictor	**Logistic**	**Probit**	**C-L-L**
AG4CAT	$F_{(3,40)}$ =12.06 (<0.001)	$F_{(3,40)}$ = 15.26 (<0.001)	$F_{(3,40)}$ = 11.52 (<0.001)
ED4CAT	$F_{(3,40)}$ = 4.80 (0.006)	$F_{(3,40)}$ = 4.79 (0.006)	$F_{(3,40)}$ = 4.77 (0.006)
MAR3CAT	$F_{(2,41)}$ = 6.54 (0.003)	$F_{(2,41)}$ = 6.66 (0.003)	$F_{(2,41)}$ = 6.50 (0.004)

similar *p*-values for the test of each parameter and would not lead to significant differences in inferences concerning the individual parameters. Given the similarity in these results, we recommend using the logit (or logistic) model for general applications. Faraway (2016) points out three advantages of this approach: simpler mathematical formulation of the models, ease of interpretation via odds ratios, and easier analysis of retrospectively sampled data.

8.9 Exercises

8.1) Consider what changes during logistic regression analysis when analyzing complex sample survey data, as compared to data assumed to be collected using a simple random sample. With that in mind, answer the following questions:

a. Why is standard Maximum Likelihood Estimation (MLE) not appropriate for logistic regression using data derived from a complex sample design? What method is used instead and why is this needed?

b. What are the model building steps used when estimating Logistic, Probit, or C-L-L models? Are they different than those covered in Chapter 7 (linear regression)?

c. List three methods used to evaluate the fit of a logistic regression model and explain how each helps in evaluation of model fit. What strategy might you employ if your software of choice does not include any design-adjusted diagnostic tests, such as ROC curves?

8.2) This question extends an analysis from Chapter 6 where we examined the relationship between gender and satisfaction with life, using the ESS6 Russian Federation data set (see the ASDA website to download the data). The exercise here adds to the previous analysis by predicting a binary variable representing high satisfaction with life (HIGH_SAT, 1=Yes, 0=No) with age, gender, an indicator of very good health, and marital status, using design-based logistic regression (with the TSL variance estimation method). Using a software package of your choosing, answer the following questions:

a. First, create an indicator variable of high satisfaction with life called HIGH_SAT using the STFLIFE variable, coded so that 1 = (STFLIFE = 9 OR 10), 0 = (0 <= STFLIFE <= 8), and missing values are set to system missing in your software.

b. Create an indicator of very good health called GOOD_HEALTH. Code this as GOOD_HEALTH = 1 (YES) if the variable HEALTH = 1, GOOD_HEALTH = 0 if HEALTH is equal to (2,3,4,5), and GOOD_HEALTH =. if HEALTH is any other value, including missing. Examine the distribution of the HEALTH variable. Based on this distribution, why is it important to collapse some of the cells in the HEALTH variable prior to performing a logistic regression analysis?

c. Using logistic regression models that account for the complex sample design features and the final survey weight, perform bivariate tests of the associations between HIGH_SAT and each predictor variable mentioned above: AG4CAT, GNDR, MAR3CAT, and GOOD_HEALTH. Prepare a table similar to Table 8.5 and identify which predictors are retained when using $p < 0.25$ as a cutoff.

d. Fit a "preliminary" model with the retained predictors from part c. Make sure to use the design variables and the final survey weight for design-based variance estimation. Based on these results, which predictors will you retain for the "final" model? Why would it be a good idea to retain all of the predictors regardless of these results?

e. Fit the "final" model with main effects and the two-way interactions of very good health (GOOD_HEALTH) and each additional predictor in your "final" model. Prepare a table similar to Table 8.9, and use these results to support your decision to either include or exclude the two-way interaction terms. What predictors will remain in your final model?

f. Test whether the use of the final survey weights is necessary when estimating this "final" model. What is your conclusion?

g. Using your "final" model from parts e. and f., prepare a summary table similar to Table 8.11 (just the design-based part) and write a short discussion (1–2 paragraphs) describing the results of your analyses. Make sure to interpret the estimated coefficient for each predictor, and discuss the overall approach used (method of estimation, method of variance estimation) to incorporate the complex sample design features.

h. To illustrate how one might use your estimated model in practice, include a detailed computation of the predicted probability of having high satisfaction in life for someone with a specified set of values on the covariates included in the final model. Clearly state the set of values and explain what the results mean.

8.3) Refit your final model from **Problem 8.2** using an alternative Bayesian approach (as implemented in the R functions that were demonstrated in this chapter). Would any of your inferences change compared to the design-based approach?

8.4) Fit a logistic regression model to the ESS6 Russian Federation data, where the probability of having voted in the last election is predicted by gender and marital status, among the subpopulation of those age 25 to 35 (young adult voters). Follow these steps:

a. Prior to fitting the model, create a subpopulation indicator called AGE25_35 that is equal to 1 if 25 <= AGEA <= 35, and 0 otherwise (if AGEA is not missing). Perform a cross-tabulation analysis to ensure that this variable is set to 1 only if age is 25 to 35, inclusive, and zero otherwise. What is the unweighted subpopulation sample size, and what percentage of respondents fall into this subpopulation?

b. Fit a logistic regression model with having voted in the last election as a binary outcome, using an unconditional (rather than conditional) approach with the subpopulation indicator created in part a., and examine the significance of the predictors gender (GNDR) and marital status (MARCAT, use MARCAT=1 or MARRIED as the reference group). Answer these questions:

1. Why is the unconditional approach important for correct subpopulation estimation (see Chapter 4)? What is the risk of using a conditional or "subsetting" approach?

2. Are both gender and marital status significant predictors of having voted within the subpopulation of interest? Provide statistical support for your conclusions.
3. Prepare a table similar to Table 8.6 and discuss/interpret your findings. Be sure to state that the results are for the subpopulation of interest and that a proper unconditional approach was used for subpopulation analysis.

8.5) This question focuses on the interpretation of logistic regression results when interactions are included in the model. The table below presents results from a logistic regression model where the probability of having alcohol dependence during the lifetime (1=Yes, 0=No) is regressed on gender (SEX: 1=M, 2=F) and an indicator of being age 45 and older, plus the interaction of these two terms. The NCS-R data are used in this example, and complex sample design features and the Part II weight are incorporated in the analysis through the use of the correct statements in SAS PROC SURVEYLOGISTIC:

```
proc surveylogistic data=ncsr ;
strata sestrat ;
cluster seclustr ;
weight ncsrwtlg ;
class sex (ref='F') / param=ref ;
model ald (event='1')= sex age45p sex*age45p ;
format sex sf. ;
run ;
```

Estimated logistic regression model for lifetime alcohol dependence

Parameter	Estimate	SE	t Value	Pr > |t|
Intercept	−3.0588	0.1063	−28.77	<.0001
Male	0.6948	0.1240	5.60	<.0001
Age 45+	−0.9572	0.1842	−5.20	<.0001
Age45+ × Male	0.7655	0.2358	3.25	0.0023

Source: NCS-R, n = 5,692, adjusted Wald test for all parameters: $F(3,42)=21.85$, p <.0001. Reference groups: Females and those Age 44 and younger. (Output generated by SAS PROC SURVEYLOGISTIC.)

Based on the results in the table above, prepare a short paragraph (as for publication) interpreting the results provided. Make sure to discuss what all of the estimated coefficients mean in the presence of the interaction term.

Note

1 https://bayesdataintegration.shinyapps.io/shiny_bayes_data_integration/

9

Generalized Linear Models for Multinomial, Ordinal, and Count Variables

9.1 Introduction

Chapter 8 covered generalized linear models for survey variables that are measured on a binary or dichotomous scale. The aim of this chapter is to introduce generalized linear modeling techniques for three other types of dependent variables that are common in survey data sets: **nominal categorical variables**, **ordinal categorical variables**, and **counts** of events or outcomes. Chapter 8 laid the foundation for generalized linear modeling, and this chapter will emphasize specific methods and software applications for three principal methods. Section 9.2 will introduce the "baseline" **multinomial logit regression model** for a survey variable with three or more nominal response categories. The **cumulative logit model** for dependent variables that are measured on an ordinal scale will be covered in Section 9.3. Regression methods for dependent variables that are counts (i.e., number of events, attributes, etc.), including **Poisson regression models** and **negative binomial regression models**, are presented in Section 9.4.

Procedures in the Stata and R software packages will be used to illustrate the applications of these methods, but the reader is encouraged to visit the companion website for this book to find each example replicated in the other major software packages that support these advanced modeling procedures.

9.2 Analyzing Survey Data Using Multinomial Logit Regression Models

9.2.1 The Multinomial Logit Regression Model

The **multinomial logit regression model** is the natural extension of the simple binary logistic regression model to survey responses that have three or more distinct categories. This technique is most appropriate for survey variables with nominal response categories; we present examples of these variables from the NHANES and the NCS-R data sets in Figure 9.1. It is common practice in surveys to use a detailed set of response categories to code the respondent's answer and then recode the multiple categories to a smaller but still scientifically useful set of nominal groupings. For example, the NCS-R public use data set contains a recoded labor force status variable, WKSTAT3, which combines the 11 questionnaire response codes for current work force status into three grouped categories: 1-Employed (EMP); 2-Unemployed (UN); and 3-Not in the labor force (NLF). The

DOI: 10.1201/9781003472094-10

(a) NHANES HUQ.040
What kind of place do you go to most often: is it a clinic, doctor's office, emergency room, or some other place?
1. CLINIC OR HEALTH CENTER.....................
2. DOCTOR'S OFFICE OR HMO......................
3. HOSPITAL EMERGENCY ROOM.................
4. HOSPITAL OUTPATIENT DEPARTMENT...
5. SOME OTHER PLACE.................................
6. REFUSED..
7. DON'T KNOW..

(b) NCS-R EM7.1
What about your current employment situation as of today -- are you?
1. EMPLOYED...
2. SELF-EMPLOYED...
3. LOOKING FOR WORK; UNEMPLOYED...........................
4. TEMPORARILY LAID OFF...
5. RETIRED..
6. HOMEMAKER...
7. STUDENT..
8. MATERNITY LEAVE..
9. ILLNESS/SICK LEAVE...
10. DISABLED...
11. OTHER (SPECIFY)...

FIGURE 9.1
Survey Questions with Multinomial Response Categories.

multinomial logit regression model is ideally suited for multivariable analysis of dependent variables like WKSTAT3.

Multinomial logit regression may also be applied to survey variables measured on Likert-type scales (e.g., 1 = Strongly agree, . . ., 5 = Strongly disagree) or other ordered categorical response scales (e.g., Self-rated health status: 1 = Excellent, . . ., 5 = Poor), but the cumulative logit regression model covered in Section 9.3 may be the more efficient technique for modeling such ordinal dependent variables.

To understand the multinomial logit regression model for a dependent variable y with K nominal categories, assume that category $y = 1$ is chosen as the **baseline category**. Multinomial logit regression is a method of simultaneously estimating a set of $K - 1$ logistic regression models that model the odds of being in category $y = 2, \ldots, K$ vs. the baseline category $y = 1$. Consider the example of the NCS-R recoded variable for labor force status, WKSTAT3, with three nominal categories: 1 = employed (EMP); 2 = unemployed (UN); 3 = not in the labor force (NLF). To fit the multinomial logit regression model to this "trinomial" dependent variable, two **generalized logits** are needed:

$$\begin{aligned}
\text{logit}(\pi(\text{"UN"} \mid \boldsymbol{x})) = \text{logit}(\pi_2) = \ln\left(\frac{\pi(y=2 \mid \boldsymbol{x})}{\pi(y=1 \mid \boldsymbol{x})}\right) = B_{2:0} + B_{2:1}x_1 + \cdots + B_{2:p}x_p \\
\text{logit}(\pi(\text{"NLF"} \mid \boldsymbol{x})) = \text{logit}(\pi_3) = \ln\left(\frac{\pi(y=3 \mid \boldsymbol{x})}{\pi(y=1 \mid \boldsymbol{x})}\right) = B_{3:0} + B_{3:1}x_1 + \cdots + B_{3:p}x_p
\end{aligned} \tag{9.1}$$

A natural question to ask at this point is, "Is it possible to simply estimate the multinomial logit regression model as a series of binary logistic regression models that consider only the response data for two categories at a time?" Strictly speaking, the answer is no. With the exception of the intercepts, the parameter estimates for the predictors ($\boldsymbol{x}$) in what Agresti (2012) labels the "separate-fitting" approach will be similar but not identical to those for simultaneous estimation of the multinomial logits. Standard errors for the former will be greater than those for the simultaneous estimation, and only the latter yields the full variance-covariance matrix that is needed to test hypotheses concerning the significance or equivalence of parameters across the estimated logits. Fortunately, almost all software packages that support analysis of complex sample survey data now include the capability for simultaneous estimation of the multinomial logit regression model.

9.2.2 Multinomial Logit Regression Model: Specification Stage

The specification stage of building a multinomial logit model parallels that described in detail in Section 8.3 for specifying a logistic regression model for a binary dependent variable. However, there are two aspects of the model specification that require special emphasis.

1. **Choice of the baseline category.** In the example model formulation for the two distinct logits in (9.1), category $y = 1$ is the selected baseline category. The survey analyst is free to choose which of the K categories they prefer to use as the baseline. This choice will not affect the overall fit of the multinomial logit model or overall tests of significance for the parameters associated with predictors included in the model. However, interpretation of the parameter estimates will depend on the selected baseline category, given how the generalized logits are defined. By default, Stata will use the lowest numbered category as the baseline category for estimating the logits and corresponding odds ratios. To choose a different category as the baseline for the multinomial logits, the Stata analyst can use the `baseoutcome(#)` option, where # represents the value of the desired baseline category. In general, when a choice of a baseline category is not clear based on research objectives, we recommend using the most common category (or mode) of the nominal dependent variable.
2. **Parsimony.** Because each of the $K - 1$ logits that form the multinomial logit model will include an identical vector of covariates, $\boldsymbol{x} = \{1, x_1, ..., x_p\}$, and each estimated logit will have $\boldsymbol{B}_k = \{B_{k:0}, B_{k:1}, ..., B_{k:p}\}$ parameters, the total number of parameter estimates will be $(K - 1) \times (p + 1)$. Consequently, to ensure efficiency in estimation and accuracy of interpretation, the final specification of the model should attempt to minimize the number of predictors that are either not significant or are highly collinear with other significant covariates. Analysts can use design-adjusted multiparameter Wald tests to determine the overall importance of predictors across the $K - 1$ logit functions, and we will consider an example of this in Section 9.2.6.

9.2.3 Multinomial Logit Regression Model: Estimation Stage

The principal difference in estimation for the multinomial logit model vs. the simple binary logit model of Chapter 8 is that the pseudo likelihood function for the data is based on the multinomial distribution (as opposed to the binomial) and the number of parameters and standard errors to be estimated increases from $p + 1$ for the binary logit model to $(K - 1) \times (p + 1)$ for the multinomial logit regression model. When survey data are collected from a sample with a complex design, the default in most current software procedures is to employ a multinomial version of Binder's Taylor Series Linearization estimator, $\hat{Var}(\hat{\boldsymbol{B}})_{TSL}$, to derive the estimated variance-covariance matrix of the model parameter estimates. Most software procedures will also provide BRR, JRR, or bootstrapping options to compute replication variance estimates, $\hat{Var}(\hat{\boldsymbol{B}})_{rep}$. Theory Box 9.1 provides a more mathematically oriented summary of the estimation of the multinomial logit regression parameters and their variance-covariance matrix when working with complex sample survey data.

In Stata, the `svy: mlogit` command is used to perform design-based estimation of the multinomial logit regression model coefficients and their standard errors. In SAS, analysts would employ the PROC SURVEYLOGISTIC procedure, with the LINK = GLOGIT option included for the MODEL statement. In R, analysts need to use the `svy_vglm()` function in the `svyVGAM` package. We will later introduce a Bayesian approach to fitting multinomial logit regression models that accounts for the complex sampling features via tools in the `csSampling` package for R as well. We introduce other software that can be used to fit multinomial logit models to complex sample survey data on the ASDA website.

THEORY BOX 9.1: ESTIMATION FOR THE MULTINOMIAL LOGIT REGRESSION MODEL

Estimation of the model parameters involves maximizing the following multinomial version of the pseudo likelihood function:

$$PL_{Mult}(\hat{\boldsymbol{B}} \mid \boldsymbol{X}) = \prod_{i=1}^{n} \left\{ \prod_{k=1}^{K} \hat{\pi}_k(\boldsymbol{x}_i)^{y_{i(k)}} \right\}^{w_i} \tag{9.2}$$

$$\begin{aligned} \textit{where}: & \; y_{i(k)} = 1 \text{ if } y = k \text{ for sampled unit } i, 0 \text{ otherwise;} \\ & \hat{\pi}_k(\boldsymbol{x}_i) \text{ is the estimated probability that } y_i = k \mid \boldsymbol{x}_i; \text{ and} \\ & w_i \text{ is the survey weight for sampled unit } i. \end{aligned}$$

The maximization involves application of the Newton-Raphson algorithm to solve the following set of $(K-1) \times (p+1)$ estimating equations, assuming a complex sample design with strata indexed by h and clusters within strata indexed by α:

$$S(\boldsymbol{B})_{Mult} = \sum_h \sum_\alpha \sum_i w_{h\alpha i}(y_{h\alpha i}^{(k)} - \pi_k(\boldsymbol{B}))\,\boldsymbol{x}'_{h\alpha i} = \boldsymbol{0} \tag{9.3}$$

$$\begin{aligned} & \textit{where}: \\ & \boldsymbol{x}'_{h\alpha i} = \text{ a column vector of the } p+1 \text{ design matrix elements for case } i \\ & \quad = \begin{bmatrix} 1 & x_{1,h\alpha i} & \cdots & x_{p,h\alpha i} \end{bmatrix}'; \\ & \boldsymbol{B} = \left\{ B_{2,0}, \ldots, B_{2,p}, \ldots, B_{K,0}, \ldots, B_{K,p} \right\} \text{ is a } (K-1) \times (p+1) \text{ vector of parameters;} \\ & \pi_k(\boldsymbol{B}) = \frac{\exp(\boldsymbol{x}'\boldsymbol{B}_k)}{1 + \sum_{k=2}^{K} \exp(\boldsymbol{x}'\boldsymbol{B}_k)}; \\ & \text{and } \pi_{1(base)}(\boldsymbol{B}) = 1 - \sum_{k=2}^{K} \pi_k(\boldsymbol{B}) \end{aligned}$$

The variance-covariance matrix of the estimated parameters takes the now familiar sandwich form, based on Binder's application of Taylor Series Linearization to estimates derived using pseudo maximum likelihood estimation:

$$V\hat{a}r(\hat{\boldsymbol{B}}) = (\boldsymbol{J}^{-1})\,\mathrm{var}[S(\hat{\boldsymbol{B}})](\boldsymbol{J}^{-1}) \tag{9.4}$$

The matrices J and var[$S(\hat{B})$] are derived as illustrated in Theory Box 8.3 for simple logistic regression, with the important change that both are now $(K - 1) \times (p + 1)$ symmetric matrices, reflecting the full dimension of the parameter vector for the multinomial logit regression model.

9.2.4 Multinomial Logit Regression Model: Evaluation Stage

Like simple logistic regression and all other forms of generalized linear models, the evaluation stage in building the multinomial logit regression model begins with Wald tests of hypotheses concerning the model parameters. With $(K - 1) \times (p + 1)$ parameter estimates, the number of possible hypothesis tests is almost limitless. However, there are a series of hypothesis tests that should be standard practice for evaluating these complex models. Standard t-tests for single parameters and Wald tests for multiple parameters should be used to evaluate whether specific coefficients are significantly different from zero in individual logit functions, that is, H_0: $B_{k:j} = 0$, or across all estimated logits, that is, H_0: $B_{2:j} = \ldots = B_{K:j} = 0$. Example questions that could drive hypothesis tests include: Is gender a significant predictor of the odds that a U.S. adult is unemployed vs. employed? Is gender a significant predictor in determining the labor force status of U.S. adults regardless of category?

Other multi-parameter Wald tests can be readily constructed to test custom hypotheses that are relevant for the interpretation of a given model. If gender significantly alters the odds that an adult is unemployed or not in the labor force (NLF) relative to employed, is the gender relationship equivalent for unemployment and NLF status? Examples of these general forms of hypothesis tests will be provided in the example in Section 9.2.6.

Fagerland and Hosmer (2012) developed a general-purpose goodness-of-fit test for assessing the consistency of predictions based on a fitted multinomial logit model with the observed nominal data for the dependent variable. While their article does not specifically address accounting for complex sampling features, they have developed an add-on Stata post-estimation command entitled `mlogitgof` that performs the goodness-of-fit test after using `svy: mlogit` to fit a model accounting for complex sampling features (including survey weights). We illustrate the use of this command in Stata for assessing the quality of fit for a multinomial logit model in our example below (see Section 9.2.6).

9.2.5 Multinomial Logit Regression Model: Interpretation Stage

The interpretation of the parameter estimates in a multinomial logit regression model is a natural extension of the interpretation of the estimates in the simple logistic regression model. Simply exponentiating a parameter estimate results in an adjusted odds ratio, corresponding to the multiplicative impact of a one-unit increase in the predictor variable, x_j, on the odds that the response is equal to k versus the baseline category:

$$\begin{aligned} \hat{\psi}_{k:j} &= \exp(\hat{B}_{k:j}) \\ CI(\hat{\psi}_{k:j}) &= \exp[\hat{B}_{k:j} \pm t_{df,1-\alpha/2} \cdot se(\hat{B}_{k:j})] \end{aligned} \quad (9.5)$$

$$\begin{aligned} &\textit{where}: \\ &\hat{B}_{k:j} = \text{ the parameter estimate corresponding to predictor } j \text{ in logit equation } k. \end{aligned}$$

If the survey analyst is interested in the impact of a one-unit increase in predictor x_j on the odds of belonging to one of two non-baseline categories, the following odds ratio estimates the multiplicative effect of a one-unit change in x_j on the odds of being in category k compared to category k':

$$\begin{aligned} \hat{\psi}_{k,k':j} &= \exp(\hat{B}_{k:j} - \hat{B}_{k':j}) \\ CI(\hat{\psi}_{k,k':j}) &= \exp[(\hat{B}_{k:j} - \hat{B}_{k':j}) \pm t_{df,1-\alpha/2} \cdot se(\hat{B}_{k:j} - \hat{B}_{k':j})] \end{aligned} \tag{9.6}$$

$$\begin{aligned} &\textit{where}: \\ &\hat{B}_{k:j}, \hat{B}_{k':j} = \text{ the parameter estimates corresponding to predictor } j \text{ in logit equations } k \text{ and } k'. \end{aligned}$$

More generally, the technique outlined in Section 8.6 can be applied to evaluate the relative odds of being in category k vs. the baseline category for any selected multivariate covariate patterns $\mathbf{x}_1$ and $\mathbf{x}_2$ or to compare the odds of being in category k vs. k' for any common covariate pattern, $\mathbf{x}$. We illustrate these techniques with examples in the following section.

9.2.6 Example: Fitting a Multinomial Logit Regression Model to Complex Sample Survey Data

In this example, we fit a multinomial logit regression model to the NCS-R response variable WKSTAT3C, which takes on three values: 1 = employed (EMP), 2 = unemployed (UN), and 3 = not in the labor force (NLF). Figure 9.2 is a simple bar graph that illustrates a weighted estimate of the population distribution for the WKSTAT3C variable, generated using the Stata code below; note that we use the NCSRWTLG variable, given our interest in using a variable from Part II of the survey, ALD, in the multivariable analysis.

```
tabulate WKSTAT3C, gen(wkstat)

* Vertical Bar Chart.
graph bar (mean) wkstat1 wkstat2 wkstat3 ///
[pweight=ncsrwtlg], percentages ///
bar(1,color(gs12)) bar(2,color(gs4)) ///
bar(3,color(gs8)) blabel(bar, format(%5.1f)) ///
bargap(7) scheme(s2mono) ///
legend (label(1 "Employed") ///
label(2 "Unemployed") label(3 "NLF")) ///
ytitle ("Percentage")
```

We consider as predictor variables AG4CAT, SEX, ALD (a lifetime diagnosis of alcohol dependence), MDE (a lifetime diagnosis of a major depressive episode), ED4CAT (1 = 0–11 yrs., 2 = 12 yrs., 3 = 13–15 yrs., and 4 = 16+ yrs.), and MAR3CAT (1 = married, 2 = previously

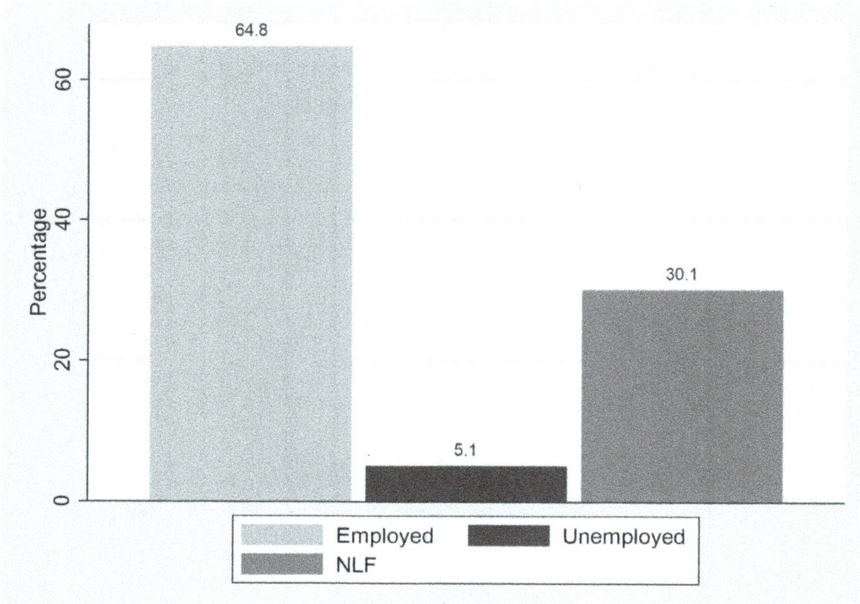

FIGURE 9.2
Weighted Distribution of WKSTAT3
(Source: NCS-R).

TABLE 9.1
Initial Bivariate Design-Based Rao-Scott Tests Assessing Potential Predictors of WKSTAT3C for the NCS-R Adult Sample

Categorical Predictor	*F*-test Statistic	$P(\mathcal{F} > F)$
AG4CAT	$F_{4.96,208.51} = 113.49$	< 0.001
SEX	$F_{1.87,78.75} = 27.33$	< 0.001
ALD	$F_{1.72,72.44} = 3.12$	0.057
MDE	$F_{1.73,72.86} = 4.67$	0.016
ED4CAT	$F_{5.15,216.12} = 27.64$	< 0.001
MAR3CAT	$F_{3.20,134.34} = 23.12$	< 0.001

married, 3 = never married). This example does not formally follow through with all the recommended model-fitting steps proposed by Hosmer et al. (2013). However, Table 9.1 does show the results from preliminary design-based analyses of the bivariate associations between WKSTAT3C and each of the six categorical predictors that are considered in the initial model specification. We show the code for examining one of the bivariate associations below.

```
svyset seclustr [pweight = ncsrwtlg], strata(sestrat)
svy: tab ag4cat WKSTAT3C, row se
```

The results for the Rao-Scott *F*-tests (see Chapter 6 for details) suggest that the four demographic predictors (AG4CAT, SEX, ED4CAT, and MAR3CAT) all have significant bivariate associations with work force status, and that the two diagnosis variables (ALD, MDE) have somewhat weaker associations.

To determine if these marginal associations remain significant when adjusting for the other predictors, the following Stata command sequence can be used to fit the multinomial logit regression model, taking the complex design of the NCS-R sample into account:

```
svyset seclustr [pweight = ncsrwtlg], strata(sestrat)

svy: mlogit WKSTAT3C ib2.sex ald mde i.ED4CAT i.ag4cat ///
i.MAR3CAT, rrr
```

Note that the `svy: mlogit` command is used to fit the model to the WKSTAT3C outcome. The nominal categorical dependent variable WKSTAT3C is listed first, followed by a list of the predictor variables in the model (which could be in any order), with categorical predictors indicated by the "`i.`" prefix. The `rrr` option is used in the `svy: mlogit` command to request output of the estimated odds ratios (which Stata interprets as *relative risk ratios*) and 95% confidence intervals. (This option can be omitted to generate the estimated coefficients on the logit scale.)

The default baseline category for the multinomial logit regression model in Stata will be the lowest-valued category, which in this example would be 1 = "Employed". Alternative baseline categories can be identified by using the `baseoutcome(#)` option in the `svy: mlogit` command, where # represents the value of the reference outcome category. For example, consider the following command:

```
svy: mlogit WKSTAT3C ib2.sex ald mde i.ED4CAT i.ag4cat ///
i.MAR3CAT, rrr baseoutcome(3)
```

This command would fit the multinomial logit model to WKSTAT3C with 3 = "Not in Labor Force" as the baseline category.

Table 9.2 provides the detailed Stata output for the estimated model, including coefficient estimates, linearized standard errors, *t*-statistics, and *p*-values for each of the (two)

TABLE 9.2

Estimated Multinomial Logit Regression Model for WKSTAT3C

		Logit 2: Unemployed vs. Employed			
Predictor*	**Category**	$\hat{B}_{2;j}$	$se(\hat{B}_{2;j})$	t	$P(t_{42} > t)$
INTERCEPT		−0.644	0.296	−2.17	0.035
AG4CAT	30–44	−0.852	0.295	−2.89	0.006
	45–59	−0.838	0.258	−3.25	0.002
	60+	1.828	0.295	6.20	< 0.001
SEX	Male	−1.393	0.198	−7.05	< 0.001
ALD	Yes	−0.164	0.357	−0.46	0.649
MDE	Yes	−0.140	0.157	−0.89	0.379
ED4CAT	12	−0.847	0.235	−3.60	0.001
	13–15	−1.365	0.258	−5.30	< 0.001
	16+	−1.731	0.310	−5.57	< 0.001
MAR3CAT	Previously	−0.590	0.225	−2.62	0.012
	Never	−2.785	0.380	−7.32	< 0.001

Predictor*	Category	$\hat{B}_{3:j}$	$se(\hat{B}_{3:j})$	t	$P(t_{42} > t)$
Logit 3: Not in Labor Force vs. Employed					
INTERCEPT		−0.379	0.173	−2.19	0.034
AG4CAT	30–44	−0.316	0.129	−2.46	0.018
	45–59	0.065	0.171	0.38	0.706
	60+	2.381	0.173	13.73	< 0.001
SEX	Male	−0.640	0.110	−5.82	< 0.001
ALD	Yes	0.333	0.130	2.56	0.014
MDE	Yes	0.099	0.088	1.12	0.269
ED4CAT	12	−0.651	0.141	−4.62	< 0.001
	13–15	−0.917	0.146	−6.26	< 0.001
	16+	−1.230	0.160	−7.70	< 0.001
MAR3CAT	Previously	−0.052	0.105	−0.50	0.621
	Never	0.553	0.132	4.18	< 0.001

Source: NCS-R, $n = 5{,}692$, adjusted Wald test for all parameters: $F(22,21) = 73.91$ $p < 0.001$.
Note: *Reference categories for categorical predictors are: 18–29 (AG4CAT); Female (SEX); No (ALD); No (MDE); <12 yrs (ED4CAT); Married (MAR3CAT)

TABLE 9.3

Estimates of Adjusted Odds Ratios for the Work Force Status Outcome (WKSTAT3C)

		Unemployed: Employed		NLF: Employed	
Predictor*	**Category**	$\hat{\psi}_{2:j}$	**95% CI for** $\psi_{2:j}$	$\hat{\psi}_{3:j}$	**95% CI for** $\psi_{3:j}$
AG4CAT	30–44	0.43	(0.24, 0.77)	0.73	(0.56, 0.95)
	45–59	0.43	(0.26, 0.73)	1.07	(0.76, 1.51)
	60+	6.22	(3.43, 11.28)	10.81	(7.62, 15.34)
SEX	Male	0.25	(0.17, 0.37)	0.53	(0.42, 0.66)
ALD	Yes	0.85	(0.41, 1.74)	1.40	(1.07, 1.82)
MDE	Yes	0.87	(0.63, 1.19)	1.10	(0.92, 1.32)
ED4CAT	12	0.43	(0.27, 0.69)	0.52	(0.39, 0.69)
	13–15	0.26	(0.15, 0.43)	0.40	(0.30, 0.54)
	16+	0.18	(0.10, 0.33)	0.29	(0.21, 0.40)
MAR3CAT	Previously	0.55	(0.35, 0.87)	0.95	(0.77, 1.17)
	Never	0.06	(0.03, 0.13)	1.74	(1.33, 2.27)

Note: *Reference categories for the categorical predictors are: 18–29 (AG4CAT); Female (SEX); No (ALD); No (MDE); <12 yrs (ED4CAT); Married (MAR3CAT)

generalized logits. Table 9.3 presents the estimated coefficients transformed into odds ratios and 95% confidence intervals for the odds ratios. We recommend the more concise display of the estimated odds ratios and confidence intervals in Table 9.3 for reporting the results of a multinomial logit regression model in scientific publications (e.g., Kavoussi et al., 2009).

To evaluate the fitted model, we perform multi-parameter Wald tests of the overall significance of each of the predictors (AG4CAT, SEX, ALD, MDE, MAR3CAT, and ED4CAT) across the logits using the `test` post-estimation command:

```
test 2.ag4cat 3.ag4cat 4.ag4cat
test 1.sex
test ald
test mde
test 2.MAR3CAT 3.MAR3CAT
test 2.ED4CAT 3.ED4CAT 4.ED4CAT
```

The Wald tests specified in these `test` statements are testing the null hypothesis that *all* parameters associated with each individual predictor (e.g., age, education level) in the two logits are not significantly different from zero. Table 9.4 provides the design-adjusted Wald test statistics and the associated *p*-values for these overall tests of the individual predictors.

Inspection of these overall test results shows that, as might be expected, the covariates AG4CAT, SEX, ED4CAT and MAR3CAT are all strongly significant determinants of the relative odds that an adult is employed, unemployed, or not in the labor force. Focusing on the relationships of alcohol dependence (ALD) and major depression (MDE), we note an interesting pattern. Adjusting for the socio-demographic variables, MDE (p = 0.330) does not appear to have a significant relationship, while ALD (p = 0.011) appears to have a significant relationship with work force status. Referring back to the *t*-tests of the individual logit parameters in Table 9.2, the ALD relationship becomes clearer. It appears that ALD is significantly related to the odds of being NLF relative to employed (where ALD increases these odds), but is *not* a significant predictor of unemployment status (relative to being employed).

Consider another example of a design-adjusted post-estimation Wald test: a test of the null hypothesis that the three education parameters in logit(2) are equivalent to those in logit(3) (i.e., the relationships of education in the two logits are equivalent). We would use the following command in Stata to test this null hypothesis:

```
test [2=3]: 2.ED4CAT 3.ED4CAT 4.ED4CAT
```

The *F*-test statistic and associated *p*-value for this Wald test are $F_{3,40} = 1.25$ and $P(\text{F} > F) = 0.30$. The design-adjusted Wald test statistic suggests a failure to reject the null hypothesis that these three pairs of coefficients are equal to each other. Note that we use numeric values for the outcome categories in square brackets after the `test` command.

As described in Section 9.2.5, the magnitudes and directions of significant relationships can be interpreted as odds ratios—odds for unemployed vs. employed for logit(2) and odds for NLF vs. employed for logit(3). Consider first the results for logit(2) comparing unemployed to employed work force status. From Table 9.3, the odds of being unemployed vs. employed

TABLE 9.4

Overall Wald Tests for the Predictors in the Multinomial Model for WKSTAT3C

Categorical Predictor	*F*-test Statistic	$P(\mathcal{F} > F)$
AG4CAT	$F_{(6,37)} = 83.59$	<0.001
SEX	$F_{(2,41)} = 35.75$	<0.001
ALD	$F_{(2,41)} = 5.05$	0.011
MDE	$F_{(2,41)} = 1.14$	0.330
ED4CAT	$F_{(6,37)} = 13.68$	<0.001
MAR3CAT	$F_{(4,39)} = 24.81$	<0.001

are significantly lower for middle-aged persons ($\hat{\psi}_{2:30-44} = \hat{\psi}_{2:45-59} = 0.43$) relative to younger persons, and men ($\hat{\psi}_{2:male} = 0.25$) relative to women. The odds of unemployment (vs. being employed) decrease with level of education ($\hat{\psi}_{2:12yrs} = 0.43$; $\hat{\psi}_{2:13-15yrs} = 0.26$; $\hat{\psi}_{2:16+yrs} = 0.18$), where less than high school education is the reference category, and are lower for previously married ($\hat{\psi}_{2:prev} = 0.55$) and never married persons ($\hat{\psi}_{2:never} = 0.06$) as compared to married persons. All of these interpretations are population estimates of these relationships when holding the other predictor variables in the model at fixed values.

Now, consider the results for the 'Not in Labor Force' (NLF) outcome. Relative to younger persons between the ages of 18 and 29, the odds of being NLF decrease significantly for 30–44-year-olds ($\hat{\psi}_{3:30-44} = 0.73$). There is a significant *increase* in the NLF odds for those in the age 60+ retirement years ($\hat{\psi}_{3:60+} = 10.81$), as one might expect. As observed for unemployment, the odds of being NLF vs. employed are significantly lower for men ($\hat{\psi}_{3:male} = 0.53$) and decrease with level of education ($\hat{\psi}_{3:12yrs} = 0.52$; $\hat{\psi}_{3:13-15yrs} = 0.40$; $\hat{\psi}_{3:16+yrs} = 0.29$). All else being equal, persons who were never married had higher odds of being NLF ($\hat{\psi}_{3:never} = 1.74$) than their married counterparts. Finally, alcohol dependency is associated with significantly increased odds ($\hat{\psi}_{3:ALD} = 1.40$) of being out of the labor force. For an example of reporting the results of a multinomial logit regression analysis of complex sample survey data in a scientific publication, we refer readers to Kavoussi et al. (2009).

Analysts can also display marginal predicted probabilities of having particular outcome categories as a function of selected covariates, as discussed in Chapter 8. For example, we can visualize the average marginal effects of alcohol dependency on the probability of not being in the labor force (outcome 3) for different age categories by using the following post-estimation commands in Stata:

```
margins, dydx(ald) by(ag4cat) predict(pr outcome(3))
marginsplot
```

The average marginal effects and 95% confidence intervals for the effects shown in Figure 9.3 illustrate a consistent tendency of alcohol dependence to increase the probability of not being in the labor force in a significant fashion (a value of 0 for the average marginal effect would mean that the probability did not change as the covariate changed from 0 to 1), regardless of age. We also note that the effects, while significant, do in fact depend on the values of other covariates, despite the absence of interaction terms in this model (as was discussed in Chapter 8). We can also plot marginal predicted probabilities of not being in the labor force as a function of alcohol dependence and age by using the following post-estimation commands:

```
margins, by(ald ag4cat) predict(pr outcome(3))
marginsplot
```

Figure 9.4 presents the resulting marginal predicted probabilities of not being in the labor force for each combination of values on these two variables. We feel that these types of visual displays enhance the presentation of multinomial logit regression results, and these post-estimation commands in Stata make it especially easy to generate these plots.

Finally, we assess the goodness-of-fit of this example multinomial logit model using the procedure described by Fagerland and Hosmer (2012). After installing the add-on command in web-aware Stata (users can type `findit mlogitgof`), an analyst simply needs to fit the model and then submit the `mlogitgof` command after fitting the model:

```
svy: mlogit WKSTAT3C ib2.sex ald mde i.ED4CAT i.ag4cat ///
i.MAR3CAT, rrr

mlogitgof
```

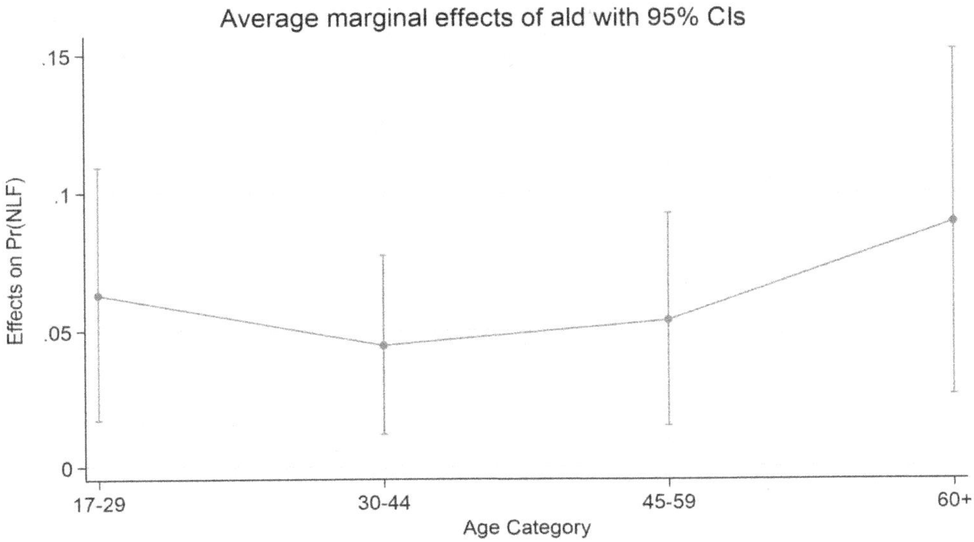

FIGURE 9.3
Average marginal effects of alcohol dependence on the probability of not being in the labor force (NLF), for different age categories.

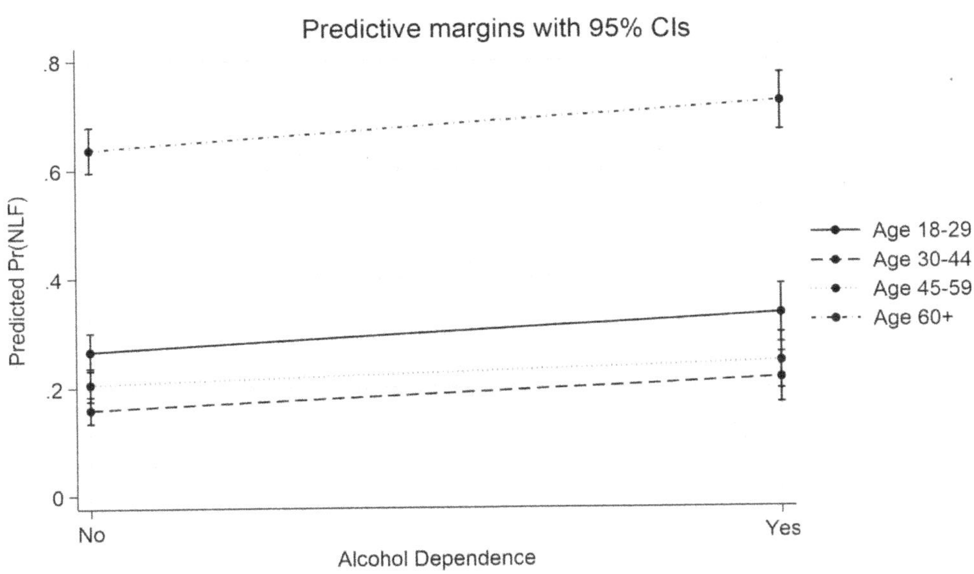

FIGURE 9.4
Marginal predicted probabilities of not being in the labor force, for different categories of alcohol dependence and age.

The resulting output suggests that we would reject the null hypothesis that the fitted model is the correct model:

```
Goodness-of-fit test for a multinomial logistic regression model
Dependent variable: WKSTAT3C

  number of observations  =   5679
number of outcome values  =      3
      base outcome value  =      1
        number of groups  =     10
    chi-squared statistic =     34.979
      degrees of freedom  =     16
        Prob > chi-squared =      0.004
```

Given the result of this test, which suggests that the model does not fit the data well, we would likely revisit the specification of the model and whether certain important predictor variables were omitted. In this simple illustration, we only considered the relationships of a small number of predictors. We could, for example, examine whether there were any interactions between the predictors that we did include (in the absence of any theory suggesting that other important predictors of employment status should have been included). We leave this as an exercise at the end of the chapter.

9.2.6.1 Alternative Bayesian Approach

We now illustrate the use of an alternative design-adjusted Bayesian approach to fitting this same multinomial logit model, using functions available in the contributed `csSampling` package for R. The approach is largely similar to that illustrated in Chapter 8, with the exception of the `family` argument required when specifying this type of model. First, we set up the complete-case version of the data frame object containing the variables of interest, in addition to the survey design object with the normalized versions of the Part II final survey weights:

```
# read in the full NCS-R data set for the third edition
load("X:/. . ./ncsr.rdata")

ncsr$marcatc <- factor(ncsr$mar3cat, levels = 1:3,
labels =c("Married", "Previously Married", "Never Married"))
ncsr$edcatc <-factor(ncsr$ed4cat, levels = 1:4,
labels =c("0-11", "12", "13-15","16+"))
ncsr$sexc <- factor(ncsr$sex, levels = 1:2,
labels=c("Male","Female"))
ncsr$sexc <- relevel(ncsr$sexc, ref="Female")
ncsr$agcatc <- factor(ncsr$ag4cat, levels = 1:4,
labels=c("18-29", "30-44", "45-59", "60+"))
ncsr$wkstatc <- factor(ncsr$wkstat3c, levels = 1:3,
labels=c("Emp", "Unemp", "NLF"))
```

```
ncsr.lg <- ncsr[!is.na(ncsr$ncsrwtlg),]

# create complete data set with variables of interest
ncsr.red <- ncsr.lg[, c("wkstatc", "marcatc", "edcatc",
"sexc", "agcatc", "ald", "mde", "seclustr", "sestrat", "ncsrwtlg")]
ncsr.red <- ncsr.red[complete.cases(ncsr.red),]

# need to normalize NCS-R weights to match Stan
ncsr.red$wtsc <- ncsr.red$ncsrwtlg /
mean(ncsr.red$ncsrwtlg)

# survey design object
ncsr.des <- svydesign(id = ~seclustr, strata = ~sestrat,
weights = ~wtsc, nest = T, data = ncsr.red)
```

Next, we carry out the Bayesian analysis with a non-informative (flat) prior distribution. Note the use of the argument `family = categorical(link = "logit")` to specify the multinomial logit model:

```
# Bayesian approach, flat prior
set.seed(41279)
model_formula <- formula("wkstatc|weights(wtsc) ~ sexc +
ald + mde + edcatc + agcatc + marcatc")
mod.brms <- cs_sampling_brms(svydes = ncsr.des,
brmsmod = brmsformula(model_formula), data = ncsr.red,
family = categorical(link = "logit"))

mod.brms$stan_fit
```

Given the number of parameters being estimated in this model, sampling from the posterior distribution will take some time, depending on the speed of the machine on which this code is being run. The resulting output displayed by `mod.brms$stan_fit` labels the coefficients by the category of the outcome defining the probability in the numerator of each generalized logit function, treating the lowest-valued category as the baseline category. This makes the factor coding used above critical for labeling and understanding the output. For example, each of the coefficients for the first generalized logit function, for the probability of unemployed vs. the probability of employed, are labeled as `b_muUnemp[X]` (consistent with the labels used above), and the coefficients are ordered as listed in the model formula, with the lowest numeric values set as the reference categories of the categorical predictors. The intercepts in each generalized logit function are labeled as `b_muUnemp_Intercept` and `b_muNLF_Intercept`, again referring to the labels above for the categorical dependent variable.

Table 9.5 presents the survey-weighted posterior means for each of the parameters defining this model, in addition to 95% credible intervals based on the design adjustments applied to each of the 1,000 draws from the survey-weighted posterior distribution for the parameters (see Section 3.8 for details). Comparing the results to Table 9.2, we see that the posterior means are quite similar to the survey-weighted point estimates from the design-based approach, and we would arrive at similar inferences regarding those parameters that are significantly different from zero. We also note that the credible intervals tend to be narrower than the corresponding 95% confidence intervals from the design-based approach (not shown in Table 9.2 for brevity). For example, when following the

TABLE 9.5

Estimated Posterior Means and 95% Credible Intervals for the Parameters in the Multinomial Logit Model for Labor Force Status

		Unemployed: Employed		NLF: Employed	
Predictor*	**Category**	**Posterior Mean**	**95% Credible Interval**	**Posterior Mean**	**95% Credible Interval**
Intercept		−0.67	(−1.16, −0.27)	−0.38	(−0.65, −0.15)
AG4CAT	30–44	−0.84	(−1.31, −0.41)	−0.32	(−0.53, −0.11)
	45–59	−0.83	(−1.27, −0.37)	0.07	(−0.15, 0.30)
	60+	1.85	(1.45, 2.28)	2.39	(2.16, 2.64)
SEX	Male	−1.39	(−1.69, −1.12)	−0.64	(−0.79, −0.51)
ALD	Yes	−0.21	(−1.02, 0.49)	0.33	(0.06, 0.62)
MDE	Yes	−0.14	(−0.50, 0.19)	0.10	(−0.07, 0.25)
ED4CAT	12	−0.85	(−1.18, −0.50)	−0.65	(−0.84, −0.48)
	13–15	−1.37	(−1.73, −0.96)	−0.92	(−1.12, −0.73)
	16+	−1.74	(−2.18, −1.31)	−1.24	(−1.45, −1.00)
MAR3CAT	Previously	−0.59	(−0.90, −0.30)	−0.05	(−0.22, 0.13)
	Never	−2.81	(−3.72, −1.96)	0.56	(0.36, 0.75)

(*Source:* NCS-R)

Note: *Reference categories for the categorical predictors are: 18–29 (AG4CAT); Female (SEX); No (ALD); No (MDE); <12 yrs (ED4CAT); Married (MAR3CAT)

design-based approach with TSL for variance estimation, the 95% confidence interval for the male coefficient in the generalized logit function for the probability of unemployed vs. the probability of employed is (−1.79, −0.99). The corresponding 95% credible interval when following the design-adjusted Bayesian approach is (−1.69, −1.12), reflecting more efficient inference that still incorporates the complex sample design effects.

9.3 Logistic Regression Models for Ordinal Survey Data

Survey questions with ordinal response options like the two examples in Figure 9.5 are common in survey practice. The HRS health rating question shown in Figure 9.5 part (a) exemplifies a question type in which the respondent is asked to assign discrete rankings to attributes—their own or those of others. A subtler form of ordinal response arises in questions like the NHANES activity question shown in part (b) of Figure 9.5. In that question, the ordinality of the response is not explicitly incorporated in a rating-type response scale but the response categories do implicitly capture ever increasing levels of physical activity.

As discussed in Chapter 5, survey analysts have historically treated descriptive analysis of ordinal response data in several different ways. The same is true in regression modeling for ordinal categorical data. At one extreme, the ordinal responses are treated as continuous variables. Many survey analysts do not hesitate to fit a standard linear regression model to ordinal response data. Others completely ignore the natural ordering of the response categories and analyze the ordinal data as though it were nominal in nature (i.e., apply multinomial logit regression as in Section 9.2).

(a) HRS
Would you say your health is excellent, very good, good, fair, or poor?

1. EXCELLENT
2. VERY GOOD
3. GOOD
4. FAIR
5. POOR

8. DK (Don't Know); NA (Not Ascertained)
9. RF (Refused)

(b) NHANES
Please tell me which of these four sentences **best** describes your usual daily activities:

1. You sit during the day and do not walk about very much
2. You stand or walk about quite a lot during the day, but do not have to carry or lift things very often
3. You lift or carry light loads, or have to climb stairs or hills often; or
4. You do heavy work or carry heavy loads.
7. Refused
9. Don't know

FIGURE 9.5
Ordinal Response Questions from the HRS and the NHANES.

In practical terms, neither approach is necessarily wrong. DeMaris (2004) identifies conditions under which he feels that a linear regression treatment leads to robust analysis—enough ordinal response options (5 or more), large n, and a response distribution that is not highly skewed across the ordinal range. Analysts who ignore the ordinality of such responses and apply the general multinomial logit regression technique of Section 9.2 are also certainly not wrong in their approach. However, such models require the estimation of many parameters [recall that one estimates $(K - 1) \times (p + 1)$ parameters when fitting a multinomial logit model] and therefore may not be the most efficient modeling option.

9.3.1 The Cumulative Logit Regression Model

A special class of logistic regression models has been developed for regression analysis of ordinal survey variables. These models differ from multinomial logit regression models for nominal categorical response variables in that they acknowledge the ordering of the response categories when estimating the relationships of the predictor variables with the probabilities of having certain responses. Ordinal models involve fewer parameters than multinomial logit regression models and are more parsimonious as a result. Standard statistical texts describe several approaches to specifying (parameterizing) an ordinal logistic regression model (Hosmer et al., 2013; Agresti, 2012). Here our focus will only be on the most common form, the **cumulative logit model**.

A **cumulative logit** is defined for the probability of having an ordinal response less than or equal to k, relative to the probability of having a response greater than k:

$$\begin{aligned}\text{logit}[P(y \le k)|x] &= \ln\left[\frac{P(y \le k)|x}{P(y > k|x)}\right] \\ &= \ln\left[\frac{P(y=1|x)+\ldots+P(y=k|x)}{P(y=k+1|x)+\ldots+P(y=K|x)}\right] \\ &= B_k - \left(B_1x_1 + B_2x_2 + \ldots + B_px_p\right)\end{aligned} \tag{9.7}$$

For an ordinal variable with K categories, $K - 1$ cumulative logit functions are defined. Each cumulative logit function includes a unique intercept or "cut point," B_k, but all share a *common* set of regression parameters for the p predictors, $\mathbf{B} = (B_1, \ldots, B_p)$. Consequently, a cumulative logit model for an ordinal response variable with K categories and $j = 1, \ldots, p$ predictors requires the estimation of $(K - 1) + p$ parameters—far fewer than the $(K-1) \times (p+1)$ parameters for a multinomial logit model.

By using logits defined by cumulative probability, the cumulative logit model captures trends across the adjacent categories of an ordinal response variable. By using a single set of regression parameters for the predictors, the model provides true parsimony in estimating the relationship between predictors and the profile of responses over the ordinal response categories. The concept of a "cumulative logit" is certainly more complex than that of a baseline category logit that is employed in the parameterization of simple logistic regression or even multinomial logit regression models. In actuality, though, it is simply an alternative way of parameterizing the model for estimating the probability that a response will fall in ordinal category $y = 1, \ldots, K$.

Consider the cumulative logit model in (9.7). If the cumulative logit is estimated for a covariate pattern $\boldsymbol{x} = \{x_1, x_2, \ldots, x_p\}$, then the transform

$$\hat{\varphi}(y \le k|\boldsymbol{x}) = \frac{\exp(\boldsymbol{x}\hat{\boldsymbol{B}})}{1+\exp(\boldsymbol{x}\hat{\boldsymbol{B}})} = \frac{\exp[\hat{B}_k - (\hat{B}_1x_1 + \hat{B}_1x_2 + \ldots + \hat{B}_px_p)]}{1+\exp[\hat{B}_k - (\hat{B}_1x_1 + \hat{B}_1x_2 + \ldots + \hat{B}_px_p)]} \tag{9.8}$$

estimates the cumulative probability (denoted here by $\hat{\varphi}(y \le k|\boldsymbol{x})$) that the response, y, is less than or equal to ordinal category k. To "recover" estimates of the individual response category probabilities, $\hat{\pi}_k(\boldsymbol{x})$, from the estimated cumulative probabilities simply requires taking the difference in the estimated cumulative probability through response categories k and $k-1$:

$$\hat{\pi}_k(\boldsymbol{x}) = \hat{\varphi}(y \le k|\boldsymbol{x}) - \hat{\varphi}(y \le k-1|\boldsymbol{x}) \tag{9.9}$$

where:

$$\hat{\varphi}(y \le 0|\boldsymbol{x}) = 0.$$

9.3.2 Cumulative Logit Regression Model: Specification Stage

The cumulative logit model is a specialized model that is applicable to true ordinal response variables such as a health satisfaction ratings and satisfaction scores, where the categorized response (excellent, very good, good, fair, poor) can be viewed as the observed representation of an underlying continuous measure of a latent attribute, belief, or attitude. The general steps in specifying and building a cumulative logit regression model—variable selection, evaluation of potential non-linearity for continuous predictors, tests of interactions, etc.—are identical to those applied in other regression models.

9.3.3 Cumulative Logit Regression Model: Estimation Stage

Several existing software procedures enable analysts to fit cumulative logit regression models to complex sample survey data (Liu, 2016). Procedures for pseudo maximum likelihood estimation of the parameters of the cumulative logit regression model are identical to those described in Section 9.2.3 for the multinomial logit regression model. The final parameter estimates are obtained by application of the Newton-Raphson algorithm to maximize the weighted multinomial likelihood given by (9.2) above. At each iteration, the values of $\hat{\pi}_k(\boldsymbol{x}_i)$ are obtained through the probability calculation sequence like that described in (9.8) and (9.9). Like multinomial logit regression, the default variance estimator in most current software systems is the multinomial version of Binder's Taylor Series Linearization estimator (9.4), with most major software packages also providing analysts the options to use BRR, JRR, or bootstrapping to compute replication variance estimates, $V\hat{a}r(\hat{\boldsymbol{B}})_{rep}$.

Analysts should be aware that different software packages use different parameterizations of the cumulative logit model. For example, the Stata `svy: ologit` procedure fits these models using the parameterization described in (9.7), in which the covariate adjustment, $\Delta_i = \sum_{j=1}^{p} B_j x_{ij}$, is *subtracted* from the cutpoint B_k for the *k*-th cumulative logit. Positive values for estimates of the regression parameters in the model will thus signify increases in the probability of higher-valued responses categories for higher values for the corresponding predictor variables. The SURVEYLOGISTIC procedure in the SAS software uses the following parameterization of the model by default:

$$\text{logit}[P(Y \leq k) \mid \boldsymbol{x}] = B_k + (B_1 x_1 + B_2 x_2 + \ldots + B_p x_p) \tag{9.10}$$

with the covariate adjustment *added* to the cut-point for cumulative logit *k*—reversing the sign on the regression parameters compared to the Stata default. Use of the DESCENDING option in the SAS SURVEYLOGISTIC procedure will yield results that are identical to the Stata default. Analysts need to carefully check how these models are being parameterized by the software procedures that they are using prior to interpreting the results from these types of analyses.

9.3.4 Cumulative Logit Regression Model: Evaluation Stage

Procedures (Student *t* tests, Wald tests) described previously for testing the significance of individual model parameters or multi-parameter predictors in the simple logistic model (Section 8.5.1) apply directly to the evaluation of fitted cumulative logit models. Unfortunately, goodness-of-fit measures and diagnostic tools like those available for binary logistic regression (Section 8.5.2) or multinomial logistic regression (Section 9.2.6) have not yet been developed for the more complex logistic regression models for ordinal response data and therefore are not yet available in statistical software procedures for complex sample survey data analysis.

Analysts can perform one type of diagnostic test for the cumulative logit model. The literature occasionally labels the cumulative logit model as the **proportional odds model**. This alternate label derives from a property of the model which results from the fact that each of the $K - 1$ cumulative logits are assumed to share a *common* set of regression coefficients, $\boldsymbol{B}$, for the model predictors. When analyzing complex sample survey data, tests of this assumption require the estimation of an alternative **generalized cumulative logit model** (Peterson & Harrell, 1990), where the regression parameters identified in (9.7) are allowed to vary depending on the choice of k (i.e., $B_{k:1}$ instead of B_1).

This "proportional odds", "equal slopes", or "parallel lines" assumption can be formally tested by fitting this alternative model (also using design-based methods) and performing a design-adjusted Wald test of the null hypothesis that all pairwise contrasts of the regression parameters for each predictor across the $K-1$ logit functions in the generalized model are equal to zero (e.g., $B_{1:1} - B_{2:1} = 0; B_{1:1} - B_{3:1} = 0$; etc.). This design-based test is currently implemented in the user-written Stata command `gologit2` (Stata users can submit the command `findit gologit2` for more details) and the CSORDINAL procedure of the SPSS Complex Samples module. Additional implementations will be noted on the companion website for this book when they become available.

In cases where this formal test rejects the null hypothesis of equal slopes, the survey analyst should first extend their model building investigation to evaluate whether the current predictor set should be modified (adding new predictors, interaction terms, or nonlinear effects) for the equal slopes assumption to hold. If this extended investigation fails to resolve the problem of apparent inequality in the slope parameters of the cumulative logit model, the analyst can ignore the ordinality of the response and revert to a generalized multinomial logit model for categorical response variables (Section 9.2) or consider more parsimonious generalized forms of the cumulative logit model (which are currently implemented in the Stata command `gologit2`). Often, despite the formal rejection of the proportional odds assumption, the interpretations of the results from the two models—cumulative and generalized logit—may be consistent and the analyst may still choose the cumulative logit model for its simplicity of form and interpretation.

9.3.5 Cumulative Logit Regression Model: Interpretation Stage

Like most regression models, the interpretation of results from a cumulative logit regression model can occur at two different levels. At the evaluation stage (see above), *t*-tests of single-parameter predictors or Wald tests of multi-parameter predictors will identify those predictors that have a significant relationship with the ordinal response variable. Examination of the estimated coefficients for the cumulative logits can inform the analyst about the directional nature of the relationships between the response variable and the predictors. For example, under the Stata default parameterization, positive values for estimates of the regression parameters for a continuous predictor correspond to increased probability of higher-valued response categories as the predictor value itself increases. Likewise, positive coefficients for parameters representing a level of a categorical predictor (e.g., males relative to females) suggest that relative to the reference category, the distribution of ordinal responses for the predictor category is shifted toward the higher values of the response distribution. In SAS, this interpretation would be reversed unless the DESCENDING option is used in PROC SURVEYLOGISTIC.

While estimates of the "cutpoint" intercept terms of the cumulative logit model are critical to model fit, they are typically of little further interest in the analysis and interpretation of the results. For interpretation and summarization of the model results, survey analysts may choose to go beyond simply establishing the significance and directionality of predictor effects. Following Agresti (2012), a more quantitative presentation of the direction and magnitude of significant relationships of predictors in cumulative logit regression models can be based on one of two related sets of statistics.

The first is the set of cumulative odds ratios that can be estimated directly from the fitted model. If the kth cumulative logit is estimated for two covariate patterns, $\boldsymbol{x}_1$ and $\boldsymbol{x}_2$, the following exponential function estimates a **cumulative odds ratio**:

$$\hat{\psi}_{y\leq k} = \exp[\hat{\boldsymbol{B}}'(\boldsymbol{x}_1 - \boldsymbol{x}_2)] = \exp[\hat{B}_1(x_{11} - x_{12}) + \cdots + \hat{B}_p(x_{p1} - x_{p2})] \tag{9.11}$$

The interpretation of the cumulative odds ratio statistic is slightly different from the standard odds ratio. For the given covariate patterns $\boldsymbol{x}_1$ and $\boldsymbol{x}_2$, the cumulative odds that the ordinal response, y, is less or equal to category k are $\hat{\psi}_{y\leq k}$ times greater for $\boldsymbol{x}_1$ than for $\boldsymbol{x}_2$. Cumulative odds ratios and confidence intervals are readily generated as output from software procedures for fitting cumulative logit models to complex sample survey data. In scientific reports and publications, they can be summarized in standard tables (see Table 9.7 below) or by using graphical displays of the type illustrated in Section 9.2.6.

The second (and related) set of statistics that can be used to quantify the relationships of individual predictors are estimates of cumulative probabilities of the form in (9.8). For example, holding all other predictors constant, expression (9.8) could be used to estimate men's and women's cumulative probability of response in category $y = 1, \ldots, K$. Estimated values and standard errors (or CIs) for $\hat{\varphi}(y \leq k \mid x)$, for men and women separately, could then be compared side-by-side in a tabular format or in a graph. To compare distributions of the estimated cumulative probabilities across the range of a continuous variable (e.g., age or blood pressure), Agresti (2012) recommends evaluating the cumulative logit for each category at the quartiles of that predictor's distribution (Q_{25}, Q_{50}, Q_{75}). We consider Stata commands for calculation and display of these types of marginal predicted probabilities in our example below.

Readers are encouraged to refer to Agresti (2012), Hosmer et al. (2013), Allison (2012b), and Long and Freese (2006) for additional tips on evaluating and interpreting the results of cumulative logit regression models.

9.3.6 Example: Fitting a Cumulative Logit Regression Model to Complex Sample Survey Data

This section considers an example of fitting a cumulative logit model to a recoded version of the satisfaction with life variable (STFLIFE) in the 2011 ESS Russia data set. The original variable was measured on a 0 to 10 scale, with 0 representing extremely dissatisfied and 10 representing very satisfied. For purposes of this example, this variable was recoded into a variable with five categories: 0–1, 2–4, 5, 6–8, and 9–10. We begin with a simple descriptive analysis, and run the Stata commands below to generate the *weighted* bar chart presented in Figure 9.6.

```
gen stflife2 =.
replace stflife2 = 1 if stflife >= 0 & stflife <= 1
replace stflife2 = 2 if stflife >= 2 & stflife <= 4
replace stflife2 = 3 if stflife == 5
replace stflife2 = 4 if stflife >= 6 & stflife <= 8
replace stflife2 = 5 if stflife >= 9 & stflife <= 10
```

```
tabulate stflife2, gen(sat)

graph bar sat* [pweight=pspwght], ///
ytitle("Weighted Proportion") legend(row(3) ///
lab(1 "0-1") lab(2 "2-4") lab(3 "5") lab(4 "6-8") ///
lab(5 "9-10"))
```

Before beginning the model building process, analysts should examine simple frequency distributions for ordinal response variables. If the majority of responses on a discrete ordinal outcome are grouped in single categories or highly skewed to the highest or lowest possible values, the cumulative logit model may not be the best choice, and simple binary or multinomial logit regression models would be more appropriate for a recoded version of the ordinal response variable. In this case, we see in Figure 9.6 that most of the individuals in Russia are estimated to respond in the 6–8 range on the original variable. This weighted population distribution across the life satisfaction categories and the implicit continuum of satisfaction that underlies this response scale suggest that this categorical variable would be a reasonable choice for a dependent variable in an ordinal regression model.

The objective in this simple example is to model the life satisfaction for Russian adults as a function of age, gender, and marital status. We remind readers that the model building steps discussed in Chapter 8 are always recommended when building a regression model. For purposes of brevity, we only consider fitting a single simple model using these three demographic variables and do not illustrate the full sequence of model building steps.

After specifying the complex sampling features of the ESS Russia data set using the `svyset` command (the cluster variable PSU, the final survey weight variable PSPWGHT,

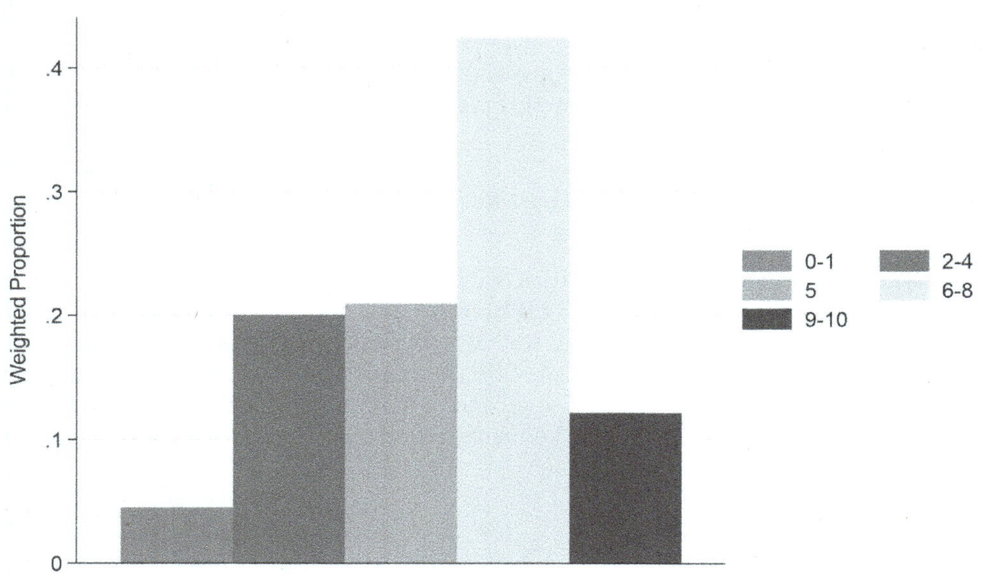

FIGURE 9.6
Bar Chart (Weighted) of STFLIFE2, the recoded version of the 2011 ESS Russia Variable STFLIFE (where 0 = extremely dissatisfied and 10 = very satisfied).

and the stratum variable STRATIFY), we use the `svy: ologit` command to fit the cumulative logit regression model:

```
svyset psu [pweight = pspwght], strata(stratify)

svy: ologit stflife2 i.agecat i.marcat male
svy: ologit stflife2 i.agecat i.marcat male, or
```

Similar to other regression commands in Stata, the response variable (STFLIFE2) is listed first, followed by a list of the predictor variables (no interactions are included in this model). The model is re-estimated with the `or` option to request that Stata provide output in the form of the estimated cumulative odds ratios and 95% confidence intervals.

Table 9.6 summarizes the estimated "cutpoint" parameters [or B_k values from (9.7)] for the logistic distribution in this model, along with the weighted estimates of the regression parameters for the non-reference categories of age, marital status, and gender. As noted above, these cut points are not of specific analytic interest but are required to compute predicted probabilities of the ordinal outcomes given input values on the predictor variables. Table 9.7 presents the corresponding estimates of the cumulative odds ratios and design-based 95% confidence intervals for the odds ratios (with standard errors computed using Taylor Series Linearization).

In Table 9.6, first note the estimates of the four intercept parameters. As described above, these parameter estimates are rarely of real analytic interest but are used by the analysis software to calculate predicted probabilities of being in one of the five ordered response categories for STFLIFE2.

Consider next the regression parameter estimates (Table 9.6) and estimated cumulative odds ratios (Table 9.7) for the non-reference categories of age, marital status, and gender. First, the t-test results for the three age coefficients suggest that the three oldest age categories (30–44, 45–59, and 60+) all have significantly different distributions on the ordinal outcome compared to the youngest age category (15–29). These three coefficients are all negative and significant at the 0.001 level. Given the parameterization that Stata is using

TABLE 9.6

Estimated Cumulative Logit Regression Model for STFLIFE2

Predictor	Category	$\hat{B}$	$se(\hat{B})$	t	$P(t_{176} > t)$
INTERCEPT	Cut 1	−3.711	0.214		
	Cut 2	−1.793	0.167		
	Cut 3	−0.835	0.159		
	Cut 4	1.384	0.154		
AGECAT*	2 = 30–44	−0.529	0.136	−3.89	< 0.001
	3 = 45–59	−0.746	0.143	−5.20	< 0.001
	4 = 60+	−0.808	0.166	−4.88	< 0.001
MARCAT*	2 = Previous	−0.209	0.105	−1.98	0.049
	3 = Never	−0.137	0.132	−1.04	0.300
GENDER*	1 = Male	−0.110	0.095	−1.15	0.251

Source: 2011 (Round 6) ESS Russian Federation data set, n = 2,415; Adjusted Wald Test for all parameters: $F(6,171) = 7.65, p < 0.001$;

Note: *Reference categories: AGECAT = 15–29, MARCAT = Married, GENDER = Female.

TABLE 9.7

Estimated Cumulative Odds Ratios in the Cumulative Logit Regression Model for STFLIFE2

		Cumulative Odds Ratio	
Predictor	**Category**	$\hat{\psi}_{y\leq kij}$	**95% CI for** $\psi_{y\leq kij}$
AGECAT*	2 = 30–44	0.589	(0.450, 0.771)
	3 = 45–59	0.474	(0.358, 0.630)
	4 = 60+	0.446	(0.321, 0.618)
MARCAT*	2 = Previous	0.811	(0.659, 0.999)
	3 = Never	0.872	(0.672, 1.131)
GENDER*	1 = Male	0.896	(0.743, 1.081)

Source: 2011 (Round 6) ESS Russian Federation data set, n = 2,415.
Note: *Reference categories: AGECAT = 15–29, MARCAT = Married, GENDER = Female.

(9.7), these estimates suggest that increasing age is negatively related to life satisfaction (where older age increases the probability of being in *lower* valued categories, indicating less satisfaction). The estimated cumulative odds ratios for these three age categories suggest that the odds of being in higher satisfaction categories are decreased by 41%, 53%, and 55%, respectively (adjusting for marital status and gender). From Table 9.7, we also see that the odds of being in higher satisfaction categories for previously married individuals are decreased by about 19% relative to married individuals. Finally, we note that males and females do not differ significantly in terms of distributions on life satisfaction when adjusting for age and marital status.

To test the proportional odds or "equal slopes" assumption that underlies the cumulative logit model, we first install the user-written gologit2 command in web-aware Stata, and then execute the following command:

```
gsvy: gologit2 stflife2 i.agecat i.marcat male, autofit
```

We note that the gologit2 command accommodates complex sample design features that have already been set using the svyset command via the gsvy prefix. This command fits the original cumulative logit model, and then (via the autofit option) fits a series of models that relax the proportional odds assumption (which requires estimation of more parameters). The fits of these models are compared automatically, and the command then yields results of these comparisons with explicit instructions for how to interpret these results:

```
Testing parallel lines assumption using the .05 level of
significance. . .

Step 1: Constraints for parallel lines imposed for
2.agecat (P Value = 0.7570)
Step 2: Constraints for parallel lines imposed for
3.agecat (P Value = 0.8772)
```

```
Step 3: Constraints for parallel lines imposed for
2.marcat (P Value = 0.4898)
Step 4: Constraints for parallel lines imposed for
3.marcat (P Value = 0.2593)
Step 5: Constraints for parallel lines imposed for male
(P Value = 0.1235)
Step 6: Constraints for parallel lines imposed for
4.agecat (P Value = 0.2091)
Step 7: All explanatory variables meet the pl assumption

Wald test of parallel lines assumption for the final model:

Adjusted Wald test

 ( 1)  [1]2.agecat - [2]2.agecat = 0
 ( 2)  [1]3.agecat - [2]3.agecat = 0
 ( 3)  [1]4.agecat - [2]4.agecat = 0
 ( 4)  [1]2.marcat - [2]2.marcat = 0
 ( 5)  [1]3.marcat - [2]3.marcat = 0
 ( 6)  [1]male - [2]male = 0
 ( 7)  [1]2.agecat - [3]2.agecat = 0
 ( 8)  [1]3.agecat - [3]3.agecat = 0
 ( 9)  [1]4.agecat - [3]4.agecat = 0
 (10)  [1]2.marcat - [3]2.marcat = 0
 (11)  [1]3.marcat - [3]3.marcat = 0
 (12)  [1]male - [3]male = 0
 (13)  [1]2.agecat - [4]2.agecat = 0
 (14)  [1]3.agecat - [4]3.agecat = 0
 (15)  [1]4.agecat - [4]4.agecat = 0
 (16)  [1]2.marcat - [4]2.marcat = 0
 (17)  [1]3.marcat - [4]3.marcat = 0
 (18)  [1]male - [4]male = 0

        F (18, 159) = 1.01
         Prob > F = 0.4566

An insignificant test statistic indicates that the final model
does not violate the proportional odds / parallel lines
assumption.
```

The F-test *does not* reject the null hypothesis, which indicates that the proportional odds assumption holds. We can therefore safely conclude that the assumption of proportional odds holds for our example model, and proceed with our interpretations as stated above.

If the null hypothesis of equal slopes/proportional odds/parallel lines had been rejected, we would be left with the difficult question of whether to ignore the apparent violation of this key model assumption in exchange for the simplicity of interpretation offered, or to opt instead for a generalized form of the model (Peterson & Harrell, 1990) or other modeling alternative that relaxes the equal slopes assumption but is far more complex and difficult to interpret. Agresti (2012) points out that for large sample sizes this test is extremely powerful against the null hypothesis that the proportional odds assumption is met. Agresti advises that even when the test rejects the assumption, the parsimony of the cumulative logit model parameterization may still make it a practical choice relative to the much more complex alternatives. We recommend comparing results from alternative models and using more flexible models if there are substantial differences in the resulting inferences.

Alternative Bayesian Approach. To fit the same cumulative logit regression model using a design-adjusted Bayesian approach, we use the same pattern of code for the functions in the `csSampling` package that we illustrated earlier for multinomial logistic regression (creating a complete data set with only the variables of interest, normalizing the weights, creating the survey design object, specifying the model, and then invoking the design-adjusted draws from the survey-weighted posterior). The only difference here is that the family is specified as `cumulative(link = "logit")`:

```
# read in the full ESS data set for the third edition
load("X:/. . ./ess6_russia.rdata")

ess6_russia$stflife2[ess6_russia$stflife >= 0 &
ess6_russia$stflife <= 1] <- 1
ess6_russia$stflife2[ess6_russia$stflife >= 2 &
ess6_russia$stflife <= 4] <- 2
ess6_russia$stflife2[ess6_russia$stflife == 5] <- 3
ess6_russia$stflife2[ess6_russia$stflife >= 6 &
ess6_russia$stflife <= 8] <- 4
ess6_russia$stflife2[ess6_russia$stflife >= 9 &
ess6_russia$stflife <= 10] <- 5

# create complete data set with variables of interest
ess.red <- ess6_russia[, c("stflife2", "agecat", "marcat",
"male", "psu", "stratify", "pspwght")]
ess.red <- ess.red[complete.cases(ess.red),]

# need to normalize NCS-R weights to match Stan
ess.red$wtsc <- ess.red$pspwght / mean(ess.red$pspwght)

# survey design object
ess.des <- svydesign(id = ~psu, strata = ~stratify, weights = ~wtsc,
nest = T, data = ess.red)

# Bayesian approach, flat prior
set.seed(41279)
model_formula <- formula("stflife2|weights(wtsc) ~
factor(agecat) + factor(marcat) + male")

mod.brms <- cs_sampling_brms(svydes = ess.des,
brmsmod = brmsformula(model_formula), data = ess.red,
family = cumulative(link = "logit"))

mod.brms$stan_fit
```

The results from this analysis are shown in Table 9.8. The parameters are labeled as `b[X]` in the output, in the same order that they are specified in the model formula (with the lowest-valued categories of each factor being the reference categories). The cutpoint parameters are

TABLE 9.8

Estimated Posterior Means and 95% Credible Intervals for the Parameters in the Cumulative Logit Model for STFLIFE2

Predictor	Category	Posterior Mean	95% Credible Interval
INTERCEPT	Cut 1	−3.72	(−4.01, −3.42)
	Cut 2	−1.80	(−2.05, −1.56)
	Cut 3	−0.84	(−1.09, −0.61)
	Cut 4	1.38	(1.13, 1.64)
AGECAT*	2 = 30–44	−0.53	(−0.75, −0.30)
	3 = 45–59	−0.76	(−1.00, −0.51)
	4 = 60+	−0.82	(−1.10, −0.55)
MARCAT*	2 = Previous	−0.21	(−0.39, 0.00)
	3 = Never	−0.14	(−0.38, 0.08)
GENDER*	1 = Male	−0.11	(−0.26, 0.04)

Source: 2011 (Round 6) ESS Russian Federation data set, n = 2,415.
Note: *Reference categories: AGECAT = 15–29, MARCAT = Married, GENDER = Female.

labeled as `b _ Intercept[1]`, `b _ Intercept[2]`, etc., and the same model parameterization is used here as was used in Stata. We note that the resulting inferences are very similar to those produced by the design-based approach in Table 9.6. We used a non-informative prior here, but informative priors could also be specified; see the example in Chapter 8.

9.4 Regression Models for Count Outcomes

9.4.1 Count Variables and Regression Modeling Alternatives

Regression models for dependent variables that are discrete counts of events or outcomes are also important in the analysis of survey data. Figure 9.7 provides an example of a survey question from the HRS that produces a count of falls that respondents age 65 and older have experienced in the past two-year period. The example question sequence illustrates a typical convention in survey measurement of counts. Respondents are initially queried as to whether any events occurred during the reference period. If the respondent answers "yes", a follow-up question captures a count of events (>0), which is recorded by the interviewer (or self-reported by the respondent) in the boxes. Respondents who answer "no" to the initial screening question are presumed to have a zero count of events.

It is clear that the question sequence encodes two pieces of information: a count of events (y) and the "exposure" time, t, during which the reported events occurred. The length of the observation period can be fixed (e.g., the last two years) or may vary from one sample respondent to another (e.g., since you last saw a doctor). Regression models for these types of counts aim to model the relationship between the count response y and predictor variables of theoretical interest $\boldsymbol{x}$, or equivalently, the relationship of $\boldsymbol{x}$ with the rate $\lambda = y / t$ at which the event occurred. The regression models described in this section model **rates**, or event counts per unit of time. Chapter 10 will introduce regression techniques for survival analysis or event history analysis that model the time to an event, where time itself is the dependent variable.

[If Age 65 or older]

Have you fallen down in the last 2 years?

1	Yes
2	No
8	Don't Know
9	Refused
Blank	Inapplicable, respondent is <65 years

How many times have you fallen in the last two years?

□□□ NUMBER OF TIMES
□ Don't Know
□ Refused

FIGURE 9.7
Example of a Survey Question Sequence Producing a Count Variable

(Source: 2012 HRS).

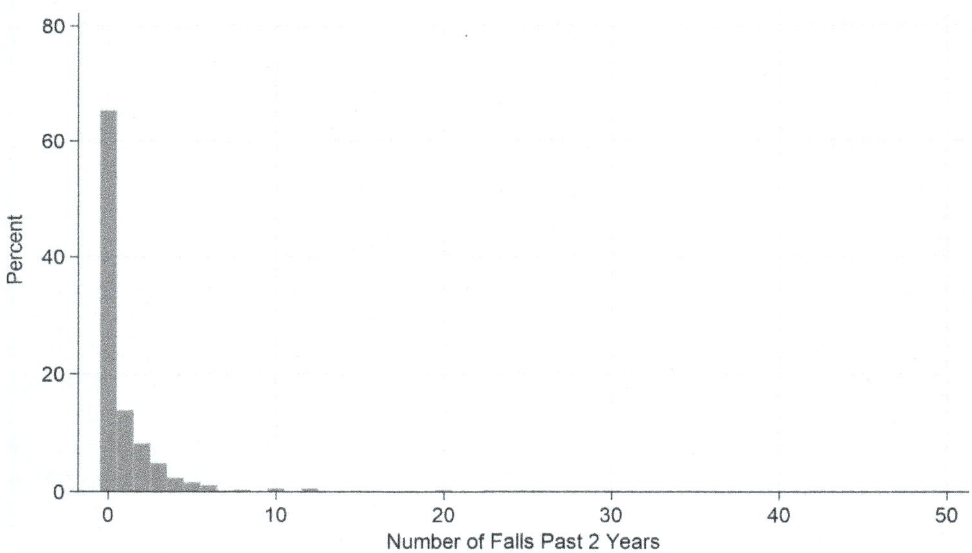

FIGURE 9.8
Histogram (weighted) of 2012 HRS Counts of Falls in the Past 24 Months (NUMFALLS24).

Figure 9.8 illustrates the estimated population distribution of counts that results from the 2012 HRS question on number of falls in the past two years, generated using the following command in Stata:

```
histogram numfalls24 [fweight = nwgtr], discrete percent
```

This figure illustrates two common properties of distributions of count-type survey variables. Often, the event of interest occurs rarely in the population and the distribution of counts is dominated by a very high proportion of zero values. The distribution is also highly skewed with declining frequencies for '1', '2', '3', etc. falls. The survey analyst can choose from a number of alternative approaches to regression modeling of these types of count variables.

Linear regression techniques are frequently used by survey analysts to model count data. Theory Box 9.2 discusses the pros and cons of trying to apply linear regression in the analysis of survey counts. Today, the preferred alternatives for regression modeling of count data are generalized linear models based in full or in part on the Poisson and negative binomial distributions.

THEORY BOX 9.2: LINEAR REGRESSION FOR COUNT VARIABLES

One alternative for modeling a count response might be to fit a standard linear regression model to a log-transformed version of the rate, denoted by $\lambda = y / t$. However, two problems can arise with this approach: the normality assumption for the residuals rarely holds for the counts or transformed rates, and the variance of $\log(\lambda_i)$ is by definition heterogeneous:

$$var[\log(\lambda_i)] \doteq \frac{var(\lambda_i)}{\lambda_i^2} = \frac{(1-\lambda_i)}{\lambda_i} \tag{9.12}$$

If the event of interest is fairly common and the distribution of counts is symmetrically (if not normally) distributed over the possible range, then a standard linear regression model of the form $y = B_0 + B_1 x_1 + \cdots + B_p x_p$ may be a practical approach. Initial graphical analyses can be very helpful in making this choice, and residual diagnostics also play an important role (see Chapter 7).

9.4.2 Generalized Linear Models for Count Variables

This section considers four related generalized linear models (GLMs) for regression modeling of count data: the Poisson regression model, the negative binomial regression model, and "zero-inflated" versions of both the Poisson and negative binomial models. The presentation here is deliberately simplified and will focus on the most important concepts for survey analysts, along with an application to the 2012 HRS data on falls. For a more in-depth mathematical treatment of these models, we encourage readers to see Chapter 8 of Long and Freese (2006) or Hilbe (2007).

9.4.2.1 The Poisson Regression Model

The simplest of the generalized linear models for count data is the **Poisson regression model**. The Poisson distribution, $y_i \sim Poisson(t_i \lambda_i)$, is a natural statistical distribution for describing counts of events, y_i, that randomly occur at some expected rate, λ_i, over a period of time, t_i. In the Poisson regression model, the natural **log link** function is employed to model this rate as a linear function of the predictors, $\boldsymbol{x}$:

$$\log(\lambda_i) = B_0 + B_1 x_{1i} + \cdots + B_p x_{pi} \tag{9.13}$$

A Poisson random variable y_i has the unique property that its mean is equal to its variance: $E(y_i \mid \boldsymbol{x}_i) = var(y_i \mid \boldsymbol{x}_i) = t_i \lambda_i > 0$, or $E(\text{count} \mid \boldsymbol{x})$ = time × $E(\text{rate} \mid \boldsymbol{x})$, where the expected count, conditional on the predictor variables, will be greater than zero. Hence, given the model for the log of the rate in (9.13), we have:

$$\begin{aligned} E(y_i \mid \boldsymbol{x}_i) &= t_i \lambda_i \\ &= t_i \exp\left(B_0 + B_1 x_{1i} + \cdots + B_p x_{pi}\right) \\ &= \exp\left[\log(t_i) + B_0 + B_1 x_{1i} + \cdots + B_p x_{pi}\right] > 0 \end{aligned} \tag{9.14}$$

When formulating a linear model for the expected value of the response variable, the natural log link transformation can be applied to the expected value for y_i defined in (9.14) to produce a convenient linear combination of predictor variables and regression parameters:

$$g\left[E(y_i \mid \boldsymbol{x}_i)\right] = \log\left[E(y_i \mid \boldsymbol{x}_i)\right] = \log(t_i) + B_0 + B_1 x_{1i} + \cdots + B_p x_{pi} \tag{9.15}$$

This is the standard Poisson regression model, incorporating an exposure time **offset variable**, $\log(t_i)$, which represents the observation period for individual i.

9.4.2.2 The Negative Binomial Regression Model

A key assumption of the Poisson regression model is that the mean and variance of the observed counts are equal: $E(y_i \mid \boldsymbol{x}_i) = var(y_i \mid \boldsymbol{x}_i) = t_i \lambda_i$. In practice, the variance of the count variable may differ from the mean. The **negative binomial regression model** is an extension of the Poisson regression model that relaxes this assumption by introducing a dispersion parameter, α, that allows the variance of the count to differ from the mean by a factor of $(1 + \alpha)$: $var_{NB}(y_i \mid \boldsymbol{x}_i) = E(y_i \mid \boldsymbol{x}_i) \cdot (1+\alpha) = t_i \lambda_i (1+\alpha)$. If the dispersion parameter is equal to 0, the negative binomial model reduces to the Poisson model.

Both the Poisson and negative binomial GLMs share the natural log link and "log rate" form of the generalized linear model, $\log(\lambda_i) = B_0 + B_1 x_{1i} + \cdots + B_p x_{pi}$, along with the expression for the expected count, $E(y_i \mid x_{1i}) = \exp\left[\log(t_i) + B_0 + B_1 x_{1i} + \cdots + B_p x_{pi}\right] > 0$. Pseudo maximum likelihood estimation of the Poisson regression model is based on a weighted Poisson likelihood function, while the extended negative binomial model employs a weighted version of the negative binomial likelihood in the estimation of the regression model parameters and the additional dispersion parameter, α (see Theory Box 9.3).

In general, the Poisson and negative binomial regression models should yield very similar estimates of the regression parameters and rate ratios. However, if the variance of the count data differs from the mean of the counts by a significant amount ($\alpha \neq 0$), fitting the regression model with the simpler Poisson likelihood will result in biased estimates of standard errors and test statistics.

9.4.2.3 Two-part Models: Zero-Inflated Poisson and Negative Binomial Regression Models

A **two-part model** is an appropriate alternative if the count variable of interest arises through a mixture of processes. Zero-inflated and zero-truncated versions of the Poisson and negative binomial regression models (Long & Freese, 2006) include a Part 1 logistic regression model, such as (8.2), of the probability that the count is zero and a Part 2 Poisson or negative binomial regression model (9.13) for the actual counts (See Theory Box 9.3). An example is the analysis of y = number of alcoholic drinks consumed per week, where the model that predicts abstinence ($y = 0$) is very likely to differ from the model that predicts frequency and consumption in that segment of the population that may drink alcohol either occasionally or often, as the case may be.

The two-part model allows the survey analyst to specify one set of predictors, z, for the logistic model that predicts the probability of a zero count and a second set of predictors, x, for the Poisson or negative binomial regression function for predicting positive counts, $y > 0$. Stata offers two commands for fitting these two-part models to count variables from complex sample designs: `svy: zip` (zero-inflated Poisson) and `svy: zinb` (zero-inflated negative binomial). The output from each of these programs will provide two sets of estimated coefficients. The first will be the estimated coefficients, standard errors, and test statistics for the Part 1 logistic model of the probability that $y = 0$:

$$\log\left[\frac{P(y=0)}{P(y>0)}\right] = A_0 + A_1 z_1 + \cdots + A_q z_q \tag{9.16}$$

Stata labels this part of the model the "inflation," because it will account for the excess of 0 counts that could not be properly modeled using only a one-part Poisson or negative binomial model. The second set of output produced by Stata for the zero-inflated models is the Part 2 estimates in the Poisson or negative binomial regression model for cases with positive counts, $y > 0$:

$$\log[E(y_i \mid x_i)] = \log(t_i) + B_0 + B_1 x_{1i} + \cdots + B_p x_{pi} \tag{9.17}$$

9.4.3 Regression Models for Count Data: Specification Stage

As outlined in Section 9.4.2, the model specification stage in regression analysis of count variables requires the analyst to apply scientific reasoning and quantitative evaluation in choosing between a one- and two-part model. The first question to ask is "Can the count variable be viewed as resulting from a mixture of processes that creates two groups, one that always responds with a zero and one that sometimes responds with a zero?" If the answer is "No: depending on x, events may occur at different rates, but the underlying process (e.g., accidents, bouts of flu, etc.) is common across the survey population", then a one-part Poisson (`svy: poisson`) or negative binomial regression model (`svy: nbreg`) is preferred. If the answer is "Yes," then either the zero-inflated Poisson model (`svy: zip`) or the zero-inflated negative binomial model (`svy: zinb`) may provide a better fit to the count data.

The choice between the simpler Poisson regression model and the extended negative binomial regression model will depend on whether the survey data satisfy the Poisson assumption, $E(y_i \mid x_i) = var(y_i \mid x_i) = t_i\lambda_i$, which is an assumption that can be formally evaluated by estimating the dispersion parameter α and its design-based 95% CI (see Section 9.4.4 below). Once the survey analyst has selected the appropriate GLM for their count data, the steps in variable selection, tests of non-linearity, and adding interaction terms can proceed as described previously for other forms of regression analysis.

9.4.4 Regression Models for Count Data: Estimation Stage

Estimation of the parameters in regression models for count data follows the general method of maximizing a weighted Poisson or negative binomial pseudo likelihood function (see Theory Box 9.3). Section 8.4 described the basic steps of the Newton-Raphson maximization algorithm for the example case of simple logistic regression.

In most current software procedures, the default variance estimator for estimates of the parameters in these models is once again a version of Binder's Taylor Series Linearization estimator (9.4) that is appropriate for the data likelihood (Poisson, negative binomial). As is true for all regression models covered in Chapters 7 through 10, most major software packages also provide analysts with the option to use BRR, JRR, or bootstrapping to compute replication variance estimates, $\hat{Var}(\hat{\boldsymbol{B}})_{rep}$.

In Stata, the following commands are available to estimate the various forms of regression models for count variables: `svy: poisson` (Poisson), `svy: nbreg` (negative binomial), `svy: zip` (zero-inflated Poisson), and `svy: zinb` (zero-inflated negative binomial). Outside of the Stata software package, programs for fitting the zero-inflated regression models to complex sample survey data are not widely available. However, Long and Freese (2006) outline how to fit these and other two-part models using separate programs for logistic regression and the standard Poisson or negative binomial regression.

THEORY BOX 9.3: PSEUDO LIKELIHOOD FOR THE POISSON AND NEGATIVE BINOMIAL REGRESSION MODELS

The weighted pseudo likelihood for estimating the parameters of the Poisson regression model (9.13) is:

$$
\begin{aligned}
PL(\boldsymbol{B} \mid y_i, \boldsymbol{x}_i) &= \prod_{i=1}^{n} \left\{ \frac{\lambda_i^{y_i} \exp(-\lambda_i)}{y_i!} \right\}^{w_i} \\
&= \prod_{i=1}^{n} \left\{ \frac{\left[t_i \exp(\boldsymbol{x}_i'\boldsymbol{B})\right]^{y_i} \exp[-t_i \exp(\boldsymbol{x}_i'\boldsymbol{B})]}{y_i!} \right\}^{w_i}
\end{aligned} \tag{9.18}
$$

The pseudo likelihood that is maximized to estimate the regression parameters, **B**, of the negative binomial regression model and the dispersion parameter, α, is:

$$
\begin{aligned}
PL(\boldsymbol{B}, \alpha \mid y_i, \boldsymbol{x}_i) &= \prod_{i=1}^{n} \left\{ \frac{\Gamma(y_i + \alpha^{-1})}{y_i!\Gamma(\alpha^{-1})} \cdot \left(\frac{\alpha^{-1}}{\alpha^{-1} + t_i\lambda_i} \right)^{\alpha^{-1}} \cdot \left(\frac{t_i\lambda_i}{\alpha^{-1} + t_i\lambda_i} \right)^{y_i} \right\}^{w_i} \\
&= \prod_{i=1}^{n} \left\{ \left(\frac{\Gamma(y_i + \alpha^{-1})}{y_i!\Gamma(\alpha^{-1})} \cdot \frac{\alpha^{-1}}{\alpha^{-1} + t_i \exp(\boldsymbol{x}_i\boldsymbol{B})} \right)^{\alpha^{-1}} \cdot \left(\frac{t_i \exp(\boldsymbol{x}_i\boldsymbol{B})}{\alpha^{-1} + t_i \exp(\boldsymbol{x}_i\boldsymbol{B})} \right)^{y_i} \right\}^{w_i},
\end{aligned} \tag{9.19}
$$

where $\Gamma(.)$ is the Gamma function.

9.4.5 Regression Models for Count Data: Evaluation Stage

The evaluation stage in building regression models for count data closely parallels that for other forms of generalized linear models. As described in Section 9.4.1, an important evaluation step in modeling count data is to ascertain whether the Poisson or the negative binomial form of the model best fits the data. When the two models are estimated for the same set of predictors, x, the Poisson model is "nested" in the extended negative binomial model. For complex sample survey data, Stata does not provide a likelihood ratio test of the null hypothesis that α is zero, but instead outputs a point estimate of α and the design-based 95% CI for the true value. If the 95% CI for α includes 0, one can use the simpler Poisson regression model.

Once the analyst has chosen the type of model, evaluation of the significance of parameters proceeds in standard fashion. Student t-tests for the significance of single parameters are routinely included along with the parameter estimates in the output from Poisson and negative binomial regression procedures in statistical software packages enabling survey data analysis. In Stata, design-adjusted Wald tests for multiple-parameter predictors or custom hypotheses for linear combinations of parameters can be obtained through use of the `test` post-estimation command.

Unfortunately, at the time of this writing, goodness of fit tests available for Poisson and negative binomial regression models fitted to simple random samples (e.g., the `estat gof` post-estimation command in Stata) have not yet been adapted for applications involving complex sample survey data. However, when analyzing complex sample survey data, graphical techniques that compare the modeled distribution of counts (0, 1, 2, etc.) to the observed distribution from the survey can be extremely useful in gauging the quality of the model fit over the range of responses. Here again, readers are referred to Long and Freese (2006) for examples and Stata command syntax. We aim to provide readers with any updates in this area on the ASDA website.

9.4.6 Regression Models for Count Data: Interpretation Stage

The interpretation of the results of regression models for counts is generally based on estimates and confidence intervals for a statistic termed the **rate ratio** (RR). Based on the identical log(rate) forms of the Poisson and negative binomial models in (9.15), some simple algebra yields:

$$\log\left[E(y_i / t_i \mid \boldsymbol{x}_i)\right] = B_0 + B_1 x_{1i} + \cdots + B_p x_{pi} \tag{9.20}$$

This formulation of the Poisson regression model enables exponentiated versions of the regression parameters associated with the predictor variables of interest to be interpreted as rate ratio statistics. Consider the ratio of two expected rates when one predictor, x_j, is increased by one unit:

$$\begin{aligned} E(y_i / t_i \mid x_{ji} + 1) &= \exp\left[B_0 + B_1 x_{1i} + B_2 x_{2i} + \cdots + B_j (x_{ji} + 1) + \cdots + B_p x_{pi}\right] \\ E(y_i / t_i \mid x_{ji}) &= \exp\left[B_0 + B_1 x_{1i} + B_1 x_{2i} \cdots + B_j x_{ji} + \cdots + B_p x_{pi}\right] \\ \Rightarrow \exp(B_j) &= \frac{E(y_i / t_i \mid x_{ji} + 1)}{E(y_i / t_i \mid x_{ji})} \\ &= \frac{\text{Expected rate at } (x_{ji} + 1)}{\text{Expected rate at } x_{ji}} = \hat{R}R_j \end{aligned} \tag{9.21}$$

Given estimates of the regression parameters in this model, we can therefore conclude that a one-unit increase in the predictor variable x_j will multiply the expected rate at which the event occurs by $\hat{R}R_j = \exp(\hat{\beta}_j)$, holding other predictors constant.

In a similar fashion to odds ratios from logistic regression models, one can estimate 100(1 − α)% confidence intervals for rate ratios as:

$$CI_{(1-\alpha)}(\hat{R}R_j) = \exp[\hat{B}_j \pm t_{df,1-\alpha/2} \cdot se(\hat{B}_j)] \tag{9.22}$$

9.4.7 Example: Fitting Poisson and Negative Binomial Regression Models to Complex Sample Survey Data

To illustrate the use of Stata commands for regression modeling of count survey data, Poisson and negative binomial regression models were fitted to a recoded variable, NUMFALLS24, derived from two variables measured in the 2012 HRS:

1. Have you fallen in the past two years? 1 = yes, 5 = no; and
2. If you have fallen in the past two years, how many times have you fallen?

Because only 2012 HRS respondents age 65 and older were asked about falls, we also generate a subpopulation indicator, AGE65P:

```
gen age65p = .
replace age65p = 1 if nage >= 65 & nage != .
replace age65p = 0 if nage < 65
```

We also generate an offset variable, OFFSET24 (measured in units of months, so that estimated rates represent falls per month):

```
gen offset24 = 24
```

Note that for the 2012 HRS falls measure, the reference period is fixed at 24 (months), enabling us to model the rate of falls per month. In other applications, the exposure time for individual respondents could vary. In such cases, the value of the exposure variable would be set equal to the exposure time recorded for each case.

Figure 9.8 from Section 9.4.1 presents the weighted distribution of the recoded variable NUMFALLS24. Table 9.9 below provides a simple (unweighted) statistical summary of the distribution of this variable—both for all cases and cases where the count is greater than 0.

```
sum numfalls24 if age65p == 1, detail
sum numfalls24 if numfalls24 >= 1 & age65p == 1, detail
```

TABLE 9.9

Distribution (unweighted) of the NUMFALLS24 Variable

Variable	*n*	Median	Mean	SD	Var	Min.	Max.
NUMFALLS24 (All)	10,594	0	1.2	3.3	11.0	0	50
NUMFALLS24 (>0)	3,835	2	3.3	4.8	23.5	1	50

(*Source:* 2012 HRS)

The output generated by submitting these initial commands in Stata (summarized in Table 9.9 above) gives us an initial sense of the distribution of this count outcome. First, we note that 10,594 adults provided a non-missing response to this question. We also note that the variance is substantially greater than the mean, which suggests that the Poisson distribution may not be the best choice of a distribution for this outcome measure. As a result, we will also fit a negative binomial regression model to this count response in Stata. Because nearly two-thirds of the observed counts are 0, a two-part zero-inflated version of the negative binomial regression model will also be estimated.

To keep the illustration simple, respondent reports of the number of falls in the past 24 months are modeled as a function of five predictors considered to be scientifically relevant: GENDER, AGE, ARTHRITIS, DIABETES, and BMI. We begin by centering AGE and BMI at the weighted estimates of their means for individuals age 65 and above, so that the estimated intercepts in our regression models can be meaningfully interpreted:

```
svyset secu [pweight = nwgtr], strata(stratum)

svy, subpop(age65p): mean nage
svy, subpop(age65p): mean R11BMI

gen nage_c = nage - 74.5
gen bmi_c = R11BMI - 27.7
```

For illustration purposes, we do not take the example analysis through all of the model building steps discussed in previous chapters. But we remind readers that the same steps of identifying key predictors, testing non-linearity of the relationships of continuous predictors, and investigating potential first-order interactions (see Section 8.3) would also apply in regression modeling of count data.

The three commands below, which follow the use of `svyset` to declare the essential complex sampling features of the 2012 HRS to Stata, are used to fit the Poisson, negative binomial, and zero-inflated negative binomial models to the generated count variable, NUMFALLS24:

```
svyset secu [pweight = nwgtr], strata(stratum)

svy, subpop(age65p): poisson numfalls24 ib2.gender ///
nage_c arthritis diabetes bmi_c, irr exposure(offset24)

svy, subpop(age65p): nbreg numfalls24 ib2.gender ///
nage_c arthritis diabetes bmi_c, irr exposure(offset24)

svy, subpop(age65p): zinb numfalls24 ib2.gender nage_c ///
arthritis diabetes bmi_c, inflate(ib2.gender nage_c ///
arthritis) irr exposure(offset24)
```

Note that in each model command, the Stata syntax includes two options: `irr` and `exposure()`. The `irr` option requests that Stata report the results in terms of estimated rate ratios associated with each of the predictors, along with confidence intervals for the rate ratios. The `exposure()` option specifies the exposure time offset for each sample case (fixed to 24 months in this case).

In this example, we elect to use a smaller set of predictors z for the part 1 logistic model of the zero-inflated negative binomial model. We provide the names of these variables (GENDER, AGE, and ARTHRITIS), which would normally be selected based on theoretical considerations of predictors that are relevant for never falling, as arguments for the `inflate()` option in the `svy: zinb` command. Another key feature of the commands is the use of `subpop(age65p)` for an appropriate unconditional analysis of the subpopulation of adults 65 years and older.

The remainder of each command is structured like all other regression commands for survey data in Stata. The recoded count response variable (NUMFALLS24) is listed first, followed by the predictor variables of interest. The `ib2.` coding used for GENDER indicates that we wish to treat females (GENDER = 2) as the reference group for GENDER when fitting the models.

Table 9.10 compares the estimates and 95% confidence intervals for the rate ratios for the Poisson and negative binomial regression models. The results in Table 9.10 suggest that the one-part Poisson and negative binomial regression models would lead to nearly identical conclusions concerning the significance and nature of the relationships of the chosen predictors with the rate of falls in the HRS subpopulation age 65 and above. The estimated confidence intervals for the rate ratios suggest that males, older adults, and those with arthritis or diabetes experience falls at a significantly higher rate. We would also infer that BMI does not predict the rate of falling when adjusting for all of the other predictors. The estimate of the dispersion parameter in the negative binomial model is $\hat{\alpha} = 3.8,\ CI_{.95}(\alpha) = (3.5,\ 4.2)$, indicating that the variance of the observed counts of falls is roughly 4.8 times the mean. This, in turn, implies that the negative binomial is the preferred model for this example.

Table 9.11 contains the Part 2 estimates of these same rate ratios from the zero-inflated negative binomial model, as well as the Part 1 estimates of the odds ratios in the logistic regression model for the probability that the reported count = 0. The results for the Part 1 logistic regression model confirm that older persons have significantly lower odds of zero falls in a two-year (24-month) period. After the Part 1 adjustment for the probability of zero falls, the estimated rate ratios from the Part 2 negative binomial regression model suggest that age does not have further significance in predicting the rate at which falls occur.

TABLE 9.10

Estimated Rate Ratios from the Poisson and Negative Binomial Regression Models for NUMFALLS24

		Poisson		Negative Binomial	
Predictor*	**Category**	$\hat{RR}_j$	$CI_{.95}(RR_j)$	$\hat{RR}_j$	$CI_{.95}(RR_j)$
SEX	**Male**	1.29	(1.101, 1.520)	1.27	(1.076, 1.492)
AGE (Cent.)	**Continuous**	1.01	(1.006, 1.024)	1.02	(1.008, 1.025)
ARTHRITIS	**Yes**	2.09	(1.788, 2.438)	2.10	(1.793, 2.459)
DIABETES	**Yes**	1.28	(1.113, 1.474)	1.30	(1.131, 1.488)
BMI (Cent.)	**Continuous**	1.00	(0.987, 1.021)	1.00	(0.988, 1.015)

(*Source:* 2012 HRS)

Note: Estimated dispersion parameter in the negative binomial model is $\hat{\alpha} = 3.8,\ CI_{.95}(\alpha) = (3.5,\ 4.2)$.
Source: 2012 HRS, n = 10,026. Adjusted Wald tests for all parameters:
Poisson: $F(5,51) = 30.51$, $p < 0.001$; Negative binomial: $F(5,51) = 26.91$, $p < 0.001$.
Note: * Reference categories for categorical predictors are: Female (SEX), No (ARTHRITIS), No (DIABETES).

TABLE 9.11

Estimated Rate Ratios and Odds Ratios from the Zero-Inflated Negative Binomial Regression Model for NUMFALLS24

		Part 2: NB Regression*		**Part 1: Logistic Zero Inflation#**	
Predictor	**Category**	$\hat{RR}_j$	$CI_{.95}(RR_j)$	$\hat{\psi}_j$	$CI_{.95}(\psi)$
SEX	**Male**	1.58	(1.31, 1.91)	4.53	(1.68, 12.30)
AGE (Cent.)	**Continuous**	1.00	(0.99, 1.01)	0.91	(0.89, 0.93)
ARTHRITIS	**Yes**	1.75	(1.41, 2.16)	0.40	(0.25, 0.64)
DIABETES	**Yes**	1.31	(1.13, 1.51)	—	—
BMI (Cent.)	**Continuous**	1.00	(0.99, 1.01)	—	—

(*Source:* 2012 HRS)
Note: * Estimate of dispersion parameter is $\hat{\alpha} = 3.0$, $CI_{.95}(\alpha) = (2.6, 3.4)$
Stata output for `svy: zinb` with the `irr` option is in the form of log-odds coefficients for the Part 1 logistic model. Estimates and CIs shown here are based on the standard odds ratio transformations in (8.21) and (8.23).

Interestingly, after adjusting for age and arthritis, men ($\hat{\psi}$ =4.53) have much higher odds of zero falls than women—a result that was not evident in the one-part Poisson and negative binomial regression models, where being male appeared to increase the rate at which falls occurred. Although women are more likely than men to experience at least one fall, men who are prone to experience falls appear to do so at a significantly higher rate than women of the same age and health status ($\hat{RR}$ =1.58).

The results in Table 9.11 also point out that all else being equal, persons with arthritis have significantly reduced odds of zero falls ($\hat{\psi}$ =0.40) and significantly higher adjusted overall rates of falls ($\hat{RR}$ =1.75). We also note that having diabetes continues to significantly increase the rate at which falls occur among those experiencing at least one fall. Adjusting for the other predictors, BMI is not significant in predicting the rate of falling.

In the complex sample survey context, there is currently no statistical test to aid in choosing between the standard negative binomial regression model and the zero-inflated alternative. Based on the results in Tables 9.10 and 9.11, the zero-inflated negative binomial regression model appears to offer scientific insights into the complexity of this simple count measure for falls among older adults that are not evident from the simpler one-part model.

9.4.7.1 Alternative Bayesian Approach

While we do not walk through a complete example of the design-adjusted Bayesian approach to fitting Poisson and negative binomial regression models with the contributed `csSampling` package in R here, we note that the same general approach illustrated for multinomial and ordinal logistic regression models earlier in this chapter can be applied to fit these models as well. The analyst simply needs to change the `family` argument to either `poisson`, `negbinomial`, `zero_inflated_poisson`, or `zero_inflated_negbinomial`. Each family will use the natural log link function by default.

For example, one could fit the zero-inflated negative binomial regression model with a natural log link for the non-zero counts to the HRS data analyzed above with the following functions in R:

```
model_formula <- formula("bf(numfalls24|weights(wtsc) ~
factor(gender) + nage_c + arthritis + diabetes + bmi_c,
zi ~ factor(gender) + nage_c + arthritis)")

mod.brms <- cs_sampling_brms(svydes = hrs.des,
brmsmod = brmsformula(model_formula), data = hrs.red,
family = zero_inflated_negbinomial)
```

Note the use of the additional `zi` formula in the overall model formula (captured in the `bf()` function, which allows for the specification of more complex functions when using the `brms` package), defining the logistic regression model for the zero inflation.

9.5 Exercises

9.1) Using the NCS-R data set and a software procedure of your choice, continue with model fitting for the multinomial logistic regression example presented in Section 9.2.6. Recall that the results of the goodness of fit test (from the Fagerland and Hosmer `mlogitgof` command) indicated a poor model fit. For this test, a p-value less than 0.05 indicates poor model fit for our multinomial logistic model predicting work status. This exercise extends the example by adding additional predictors to the model in an attempt to improve model fit.

a) Begin by setting the complex sample design variables (in Stata) or use an equivalent approach in your chosen software. Then, replicate the model results presented in Section 9.2.6, using the following model specification (Stata commands):

```
svy: mlogit WKSTAT3C ib2.sex ald mde ///
i.ED4CAT i.ag4cat i.MAR3CAT, rrr
```

If you are using Stata, download and use the `mlogitgof` command after fitting the above model. Make sure your results match those presented previously in this chapter.

b) Next, add two predictors to the above model: 1) an indicator of a social phobia diagnosis (DSM_SO, 1=Yes, 5=No), and 2) race/ethnicity in four categories (RACECAT, 1=Other/Asian, 2=Hispanic/Mexican, 3=Black, 4=White):

```
svy: mlogit WKSTAT3C ib2.sex ald mde i.ED4CAT ///
i.ag4cat i.MAR3CAT i.racecat i.DSM_SO, rrr
```

c) Perform post-estimation Wald tests for each predictor variable. Based on these results, which variables are significant in their contribution to the model? Should any variables be dropped from the model?

d) If you are using Stata, repeat the `mlogitgof` post-estimation command and examine the results. What is the p-value for the goodness of fit test? What does this indicate about the model in question?

e) Prepare a table of results similar to Table 9.3 and write a short paragraph describing the key findings. Make sure to discuss how you incorporated the complex sample design effects in the analysis.

f) **(Extra Credit, using Stata or other software that is able to produce plots of marginal predicted values):** Produce a plot of marginal predicted probabilities of not being in the labor force as a function of alcohol dependence and age, similar to Figure 9.4. How would you interpret the plot? Does the addition of race and social phobia as predictors in the model change the interpretation of the plot, and if so, how?

9.2) Generate a weighted bar chart examining the estimated distribution of the OBESE6CA outcome in the NCS-R population. Use the NCSRWTSH survey weight, or the Part 1 NCS-R weight. If an analyst were interested in predicting responses on this outcome for a member of this population, would ordinal (cumulative logit) regression be a reasonable analytic approach?

9.3) Fit an ordinal (cumulative logit) regression model to the OBESE6CA outcome in the NCS-R data (1 = BMI less than 18.5, 2 = 18.5–24.9, 3 = 25–29.9, 4 = 30–34.9, 5 = 35–39.9, 6 = 40+). Be careful to recognize the parameterization of the model given the software procedure that you choose to fit the model. Make sure to incorporate the complex features of the sample design and the Part 1 NCS-R weight (NCSRWTSH) in the analysis, and consider the following candidate predictors: age, gender, education, region, and race. Be careful with how you handle the categorical predictors. Complete the following exercises based on the estimated coefficients in this model:

a) Write the ordinal logit model that you are estimating in detail, including the regression parameters and the predictor variables, and using a general specification for the cutpoint *k*.

b) **(Extra Credit, Stata software):** Install the `gologit2` command and use this command post-estimation to test the proportional odds assumption. What would you conclude about the proportional odds assumption for this model? If the assumption is violated, why might you still choose to use ordinal logistic regression?

c) What is the interpretation of the relationship of age with the ordinal OBESE6CA outcome? Does age appear to have a relationship with obesity level? Justify your answer.

d) What is the interpretation of the relationship of gender with the OBESE6CA outcome? Does gender appear to have a relationship with obesity level? Justify your answer.

e) Which variables appear to have a significant relationship with obesity level? Are the directions of the relationships positive or negative (in terms of the actual labels for the values on the OBESE6CA outcome)?

f) Based on the estimated parameters in this model, compute the predicted probability that a 50-year-old white male in the East region with 0–11 years of education will report that their obesity level is < 18.5.

g) Does using a Bayesian approach to fitting this cumulative logit regression model change anything about your overall inferences? Show your R code, and compare the 95% credible intervals based on this approach to the 95% confidence intervals from the design-based approach.

9.4) This exercise uses the HRS 2012 data set and focuses on identifying predictors of a count of the number of days per week respondents drank alcohol (averaged over the previous three-month period), given that they have ever drank any alcohol. The analysis is performed for the subpopulation of those age 65+.

The outcome variable, DAYSDRKPERWK, was created from two variables from HRS 2012 Section C—Physical Health: NC128 and NC129. Question NC128 asks "Do you ever drink any alcoholic beverages such as beer, wine, or liquor?" If NC128=1 or "yes", the next question, NC129, asks "In the last three months, on average, how many days per week have you had any alcohol to drink? (For example, beer, wine, or any drink containing liquor.)". The coding for the outcome variable DAYSDRKPERWK ranges from 0 (never drank or drank < 1 day per week) to 7 (drank every day of an average week during the past 3 months). For purposes of this exercise, we will also assume an offset of 7 days (the total number of days in a week).

a) Using an appropriate software procedure, prepare a weighted bar chart of the response variable DAYSDRKPERWK, for the subpopulation of those age 65+ (AGE65P). Based on the resulting chart, provide a short description of the distribution.

b) Perform a weighted descriptive analysis of the response variable, within the subpopulation of interest. What is the weighted mean and variance of the response variable? What might this comparison suggest in terms of potential modeling approaches?

c) Fit a Poisson regression model to the DAYSDRKPERWK response variable, treating the number of days drank per week, on average, during the past three months as a count variable. Consider the predictor variables NAGE, GENDER, DIABETES, and MARCAT. Be sure to use the indicator AGE65P for a correct subpopulation analysis for those aged 65 and older, include an offset variable equal to 7 for all respondents (OFFSET) in the model, and use correct factor coding for categorical variables. Finally, account for the complex sample design of the HRS (stratum codes = STRATUM, sampling error computation units = SECU) when estimating standard errors for the parameter estimates, and compute unbiased estimates of the regression parameters using the final survey weights for respondents (NWGTR).

Generate a table similar to Table 9.10 (with just the Poisson regression section) presenting the results of the analysis, including weighted parameter estimates displayed as rate ratios (defined as the exponent of the parameter estimate), design-adjusted standard errors, and 95% confidence intervals for the parameters.

d) Use a post-estimation test statement (Stata command) or equivalent for each predictor in the model and prepare a table of results similar to Table 8.8 (from Chapter 8). Based on these results, what decisions would you make about the inclusion of these predictors in the model?

e) Refit the model using the same design-based approach, only assuming a negative binomial distribution for the count of days drank per week. What is the estimate of the dispersion parameter in this model? What does this suggest about the Poisson regression approach?

f) Compare the estimates of the parameters and the estimated risk ratios obtained using each approach, and discuss whether any of your inferences would differ depending on the assumed distribution for the counts.

10

Survival Analysis of Event History Survey Data

10.1 Introduction

Survival analysis, or **event history analysis** as it is often labeled in the social sciences, includes statistical methods for analyzing the time at which "failures" or events of interest occur. Common examples of **event times** of interest in the survival analysis literature include motor longevity (engineering), time of death, time to disease incidence or recovery (medicine and public health), time to unemployment, time to divorce, or time to retirement (social sciences).

Survival analysis is an important statistical topic that when treated in its full depth and breadth fills entire volumes. This chapter will only scratch the surface of this topic, focusing on basic theory (Section 10.2) and application for four survival analysis techniques that are commonly used with complex sample survey data. Section 10.3 will introduce nonparametric Kaplan-Meier analysis of the survivorship function. The Cox Proportional Hazards (CPH) model will be covered in Section 10.4, and Section 10.5 presents a description and application of logit and complementary log-log (C-L-L) models for discrete time event history data. Finally, Section 10.6 introduces parametric models for left- or interval-censored event history data.

Readers interested in the general theory and applications of survival analysis methods are referred to classic texts including Kalbfleisch and Prentice (2002), Lee (1992), and Miller (1981). Newer texts on applied methods (Hosmer et al., 2008) and several excellent user guides are also available that illustrate procedures for survival analysis in SAS (Allison, 2010) and Stata (Cleves et al., 2016).

10.2 Basic Theory of Survival Analysis

10.2.1 Survey Measurement of Event History Data

In the context of population-based survey research, survival analysis problems arise in three primary ways:

1) through longitudinal observations on individuals in a panel survey;
2) by administrative record follow-up of survey respondents; and
3) from retrospective survey measurement of events and times that events occurred.

DOI: 10.1201/9781003472094-11

The Health and Retirement Study (HRS) and other longitudinal panel studies prospectively follow and re-interview sample members over time. In this type of **prospective cohort design**, dates of events such as retirement, institutionalization, morbidity, and mortality are captured in the longitudinal data record. Administrative follow-up of a sample of survey participants using vital statistics, medical records, or other record-keeping systems also generates a prospective measurement of events of interest and the dates on which they occurred. The NHANES III linked mortality file (http://www.cdc.gov/nchs/data-linkage/index.htm) is an excellent example where U.S. Vital Statistics records were used to prospectively detect and record death of an NHANES III respondent for a period of years after the survey was conducted.

The third means of generating "survival data" in the survey context is by retrospective measurement of event histories. Figure 10.1 is an illustration of a question sequence from the NCS-R that uses retrospective recall to report and date events over the course of the respondent's lifetime (or other relevant observation period). Survey methodologists are aware of recall, telescoping, and non-random censoring errors that may be associated with such retrospective measurements (Groves et al., 2009). Techniques such as Event History Calendars that use key lifetime events and dates are often employed to anchor the retrospective recall measurement of events and times (Belli, 2000).

10.2.2 Data for Event History Models

Time itself is at the heart of survival analysis—time to occurrence of an event, *T*, or the time at which the observation on the survey subject is censored, *C*. Only one of these two

*D37. Think of the very first time in your life you had an episode lasting (several days or longer / two-weeks or longer) when most of the day nearly every day you felt (sad/or/discouraged/or/uninterested) and also had some of the other problems (you cited earlier). Can you remember your exact age?
YES..1
NO..5 **GO TO *D37b**
DON'T KNOW............................8 **GO TO *D37b**
REFUSED..................................9 **GO TO *D37b**

*D37a. (How old were you?)
__________ YEARS OLD **GO TO *D37b.1**
DON'T KNOW................998 **GO TO *D37b.1**
REFUSED........................999 **GO TO *D37b.1**

***D37b.** About how old were you (the first time you had an episode of this sort)?
IF "ALL MY LIFE" OR "AS LONG AS I CAN REMEMBER,"
PROBE: Was it before you first started school?
IF NOT YES, PROBE: Was it before you were a teenager?
__________ YEARS OLD
BEFORE STARTED SCHOOL 4
BEFORE TEENAGER 12
NOT BEFORE TEENAGER............................. 13
DON'T KNOW.. 998
REFUSED... 999

FIGURE 10.1
Example of NCS-R Retrospective Event Recall Questions.

times is directly measured. *T* is measured if the event of interest occurs during the survey observation or follow-up period. The censoring time, *C*, is measured in cases where the respondent is **lost to follow-up** (due to noncontact, refusal, or other reasons; a type of **right-censoring**), if the survey observation period has ended and no event has yet occurred (also a type of right-censoring), the event occurred prior to the onset of survey measurement (known as **left-censoring**), or the event occurred at some point in-between two points of survey measurement (known as **interval censoring**). Estimation of the survival analysis model requires assumptions concerning the relationship of *T* and *C*—the most common being that time to event and censoring time distributions are independent of each other after adjusting for any covariates in the statistical model used to analyze the times of interest.

The majority of survival analysis problems deal with time to a single event or the first occurrence of the event. Often the transition is to an **absorbing state** (e.g., mortality). This is the class of applications that we will primarily consider in this chapter. However, readers should be aware that event history methods have been extended to include models of repeated events (e.g., heart attacks, marriages) or **spells** of time between events (e.g., unemployment episodes, or events that may be due to multiple causes or **competing risks**; see Yamaguchi, 1991 and Allison, 2010).

In the literature and in applications, survival models are often referred to as either **continuous time** or **discrete time** models. The dependent variables *T* and *C* may be measured on a continuous scale (e.g., number of milliseconds until failure) or a discrete scale (e.g., number of years of marriage until divorce, on an integer scale). Discrete survival times often arise in longitudinal survey designs where survey measurement occurs on a periodic basis (e.g., every two years for the HRS) and it is difficult to precisely pinpoint the date and time when the event occurred.

Figure 10.2 illustrates the general nature of the observational data that are available for a survival analysis of survey data. The figure illustrates the **observational window** spanning the time period at which survey observations begin to the point in time when the survey observations end. In the figure, the first horizontal "time" arrow illustrates the case where an event occurs and the time to the event, *T*, is recorded. This tends to be the type of measurement recorded for the majority, if not vast majority, of survey respondents, depending on the type of event.

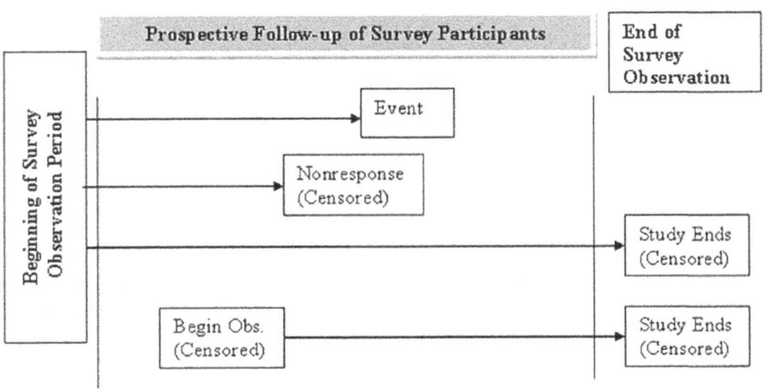

FIGURE 10.2
Prospective View of Event History Survey Data.

The second arrow in Figure 10.2 represents a case that is followed from the beginning of the survey observation window but is lost to follow-up (right-censored) before the study is complete. This often happens in panel surveys that are unfortunately subject to the attrition of survey participants over time. A second form of censored observation is represented by the third arrow in the figure. This case is successfully monitored from the beginning to the end of the observation window, but the event of interest does not occur. In cases 2 and 3, the time that is recorded in the data set is the censoring time, *C*; an event time is never observed. The censoring mechanism that is illustrated by cases 2 and 3 in Figure 10.2 is termed **right censoring**. Right-censored observations such as these are important in survival analysis since they remain representative of the **population at risk** up to the point that further observation is censored. In the analysis examples presented in this chapter, we will assume **Type 1 Censoring**, where any censoring occurs at random over the fixed period of observation.

Survey observations may also be **left-censored**, as illustrated by the fourth and final horizontal arrow in Figure 10.2. Left censoring of survey observations can occur for several reasons, including ineligibility in the initial period of survey observation, initial nonresponse, or having the event of interest occur prior to the onset of event measurement (e.g., someone started drinking at age 12, but a survey question only asks about initiation of drinking after age 15).

10.2.3 Important Notation and Definitions

Prior to examining survival analysis models in more detail, several important statistical definitions are required. The **probability density function** for the event time is denoted by $f(t)$, and is defined as the probability of the event at time t (for continuous time), or by π_m, denoting the probability of failure in the interval $(m, m + 1)$ for discrete time. The corresponding cumulative distribution functions (CDFs) are defined in the standard fashion:

$$\begin{aligned}F(t) &= \int_0^t f(t)dt \text{ for continuous } t;\text{ or}\\ F(m) &= \sum_{k\le m}^{m} \pi(k) \text{ for } t \text{ measured in discrete intervals of time.}\end{aligned} \tag{10.1}$$

The CDFs for survival time measure the probability that the event occurs at or before time t (continuous) or before the close of time period m (for discrete time).

The **survivor function** or **survivorship function**, $S(t)$, is the complement to the CDF and is defined as follows:

$$\begin{aligned}S(t) &= 1 - P(T \le t) = 1 - F(t) \text{ for continuous time; or}\\ S(m) &= 1 - F(m) \text{ for discrete time.}\end{aligned} \tag{10.2}$$

The value of the survivor function for an individual is the probability that the event has *not yet occurred* at time t (continuous) or prior to the close of observation period m (discrete time).

The concept of a **hazard** or **hazard function** plays an important role in the interpretation of survival analysis models. A hazard is essentially a conditional probability. For continuous time models, the hazard is $h(t) = f(t)/S(t)$, or the conditional probability that the event will occur at time *t* given that it has not occurred prior to time *t*. In discrete time models, this same conditional probability takes the form $h(m) = \pi(m) / S(m-1)$.

10.2.4 Models for Survival Analysis

Survival analysis models are classified into four major types, based on the assumptions that are made concerning the probability distributions of survival times. We briefly discuss these four types of survival analysis models in this section.

Parametric survival models assume a parametric distribution for the probability density function $f(t)$, and allow analysts to estimate parameters that describe the relationships of covariates with the hazard function. A very common example is the exponential distribution with rate parameter 1. In this case, $f(t) = e^{-t}$, $F(t) = 1 - e^{-t}$, and $S(t) = e^{-t}$. This distribution is characterized by a constant hazard function, $h(t) = f(t) / S(t) = e^{-t} / e^{-t} = 1$. Additional common examples of parametric models for survival times include the Weibull distribution (often used for human mortality), the log-normal distribution, and the gamma distribution. For more details on parametric models for survival data, we refer readers to Kalbfleisch and Prentice (2002) or Miller (1981).

Software procedures for fitting parametric survival models to complex sample survey data are not yet widely available. Stata currently provides the `svy: streg` command for users interested in fitting these models to survey data, but similar procedures have not yet been widely implemented in other statistical packages. As a result, our focus in this chapter will be on nonparametric and semi-parametric survival models that can currently be fitted in several major software packages while taking complex sample designs into account. We revisit parametric survival models for the cases of left- and interval-censored data in Section 10.6.

Nonparametric survival models make no assumptions concerning the probability density functions of survival times, and include approaches like **Kaplan-Meier estimation** and **life table methods**. These empirically-based methods estimate survivor functions and hazards solely using the observed survey data, and do not rely on any parametric assumptions (i.e., assumed probability models). We present examples of applying these methods to survey data in Section 10.3.

The popular **Cox Proportional Hazards (CPH) model** is labeled as a **semi-parametric survival model**. The CPH model makes no strong assumptions concerning the underlying probability distribution for event times, but does assume proportionality in the regression parameterization of the individual hazard associated with the covariate vector, $\boldsymbol{x}$. We consider an example of fitting a CPH model to survey data in Section 10.4.

Finally, a special class of survival models exists for **discrete time event history data**. The models and methods introduced above are largely applicable for survival times that are roughly continuous in nature, taking on many possible values on a continuum. Some survival times may be measured in terms of discrete units of time only; for example, in a panel survey covering five years, a researcher might wish to analyze the number of years until a person loses their job (1, 2, 3, 4, or 5) as a function of selected covariates. Because these failure times are measured in terms of a small number of discrete values, alternative methods are needed. Section 10.5 introduces **discrete time logit models** and **complementary log-log models** for discrete time event history data.

In each of the following sections, the example applications will be based on a survival analysis of the age of onset for a major depression episode (MDE) question from the NCS-R (see Figure 10.1). The event of interest for the survival analysis is thus the first MDE that the respondent experiences. The event time is the age that they report experiencing their first MDE. Individuals who have never in their life prior to the NCS-R interview had an MDE are assumed to provide right-censored data; that is, we analyze their ages at interview, but the ages are considered to be right-censored for the purposes of the survival analysis. In Sections 10.4 and 10.5, we illustrate the fitting of Cox proportional hazards and discrete time logit regression models to the NCS-R data in Stata, considering gender (SEX), age at interview (INTWAGE), marital status (MAR3CAT), education (ED4CAT), and race/ethnicity (RACECAT) as possible predictors of the hazard of having a first MDE at a given age. Note that in this analysis, each NCS-R respondent's observation window (Figure 10.2) begins at birth and continues until: 1) the age at which they first experience a major depression episode (MDE), or 2) the age at which they are interviewed for NCS-R.

10.3 Kaplan-Meier Estimation of the Survivor Function

The **Kaplan-Meier (K-M) estimator** or **product limit estimator** is a nonparametric estimator of the survivorship function, $S(t)$. It can be applied to event times that are measured on a continuous time (t) basis or as counts of events measured over discrete time periods, $m = 1, \ldots, M$.

10.3.1 K-M Model Specification and Estimation

A general form of the K-M estimator for complex sample survey data follows:

$$\hat{S}(t) = \prod_{t(e) \le t} (1 - \hat{D}_{t(e)} / \hat{N}_{t(e)}) \tag{10.3}$$

where:

$t(e)$ = times at which unique events $e = 1, \ldots, E$ are observed;

$\hat{D}_{t(e)} = \sum_{i=1}^{n} \mathrm{I}[t_i = t(e)] \cdot \delta_i \cdot w_i$; $\hat{N}_{t(e)} = \sum_{i=1}^{n} \mathrm{I}[t_i \ge t(e)] \cdot w_i$;

$\mathrm{I}[\cdot] = \textit{indicator} = 1$ if [expression] is true, 0 otherwise;

$\delta_i = 1$ if the observed time for case i is a true event, 0 if censoring time;

w_i = the final survey weight for observation i.

The survivorship at any point in time t is estimated as a product over unique event times, where $t(e) \le t$, of the estimated conditional survival rate at each event time t, $(1 - \hat{D}_{t(e)} / \hat{N}_{t(e)})$. Note how at each unique event time the rate of events is a ratio of the weighted estimate of total events at $t(e)$ to the estimated population "at risk" at time $t(e)$. Note also that the final survey weights (possibly incorporating adjustments for nonresponse and post-stratification/calibration) are incorporated into the estimation of the survivorship function.

Unlike the regression-based methods described later in this chapter, the Kaplan-Meier estimator of survivorship does not permit adjustment for individual covariates. However, the Kaplan-Meier estimates of survivorship can be computed separately for subgroups, $h = 1, \ldots, H$, of the survey population (e.g., separately for men and women). The `svykm()` function in the R `survey` package, PROC KAPMEIER in SUDAAN, and the SPSS

Complex Samples Cox Regression program (product limit estimator option) provide the capability to conduct a full K-M analysis for complex sample survey data, including estimation for subpopulations. Stata provides the capability to compute and plot weighted estimates of the survivor function, but currently does not permit design-based estimation of the standard errors of the estimates or confidence intervals. At present, PROC PHREG in SAS permits survey-weighted Kaplan-Meier estimation and display of the survivor function for different subgroups (in conjunction with PROC SGPLOT; see Gardiner, 2015), but this procedure does not support design-adjusted estimates of standard errors and confidence intervals, nor does it allow for the use of non-integer weights. Any changes in terms of the capabilities of these software procedures will be reported on the ASDA website.

Under complex sample designs, SUDAAN employs a Taylor Series Linearization (TSL), BRR, or JRR estimator of the standard error of $\hat{S}(t)$, and the `svykm()` function in the `survey` package in R employs TSL. The TSL estimates of $se(\hat{S}(t))$ require the generation of a large matrix of derivative functions that may lead to computational difficulties. SUDAAN allows users to circumvent this problem by estimating the K-M survivorship and standard errors for a subset of points (e.g., deciles) of the full survivorship distribution, while the `svykm()` function allows users to compute standard errors as an option. Another alternative that we recommend is to employ the JRR option in SUDAAN to estimate the full survivorship distribution and confidence intervals for the individual $\hat{S}(t)$. Using the TSL or JRR estimates of standard errors, confidence limits for individual values of $\hat{S}(t)$ are then derived using a special two-step transformation/inverse-transformation procedure (see Kalbfleisch & Prentice, 2002 and Theory Box 10.1 for details).

THEORY BOX 10.1: CONFIDENCE INTERVALS USING KAPLAN-MEIER ESTIMATES OF $\hat{S}(t)$.

A two-step (transformation, inverse-transformation) procedure can be used to estimate confidence intervals for the survivorship at time t, $S(t)$.

In the first step, the 100(1 – α)% CI is estimated on the log[-log($\hat{S}(t)$)] scale:

$$CI_{g(S(t))} = g(\hat{S}(t)) \pm t_{df,1-\alpha/2} \cdot se(g(\hat{S}(t))) \tag{10.4}$$

where:

$$g(\hat{S}(t)) = \log(-\log(\hat{S}(t)));\ \textit{and}$$

$$se(g(\hat{S}(t))) = \sqrt{\frac{\text{var}(\hat{S}(t))}{\left[\hat{S}(t)\cdot\log(\hat{S}(t))\right]^2}}$$

For simple random sample data, $var(\hat{S}(t))$ is based on Greenwood's (1926) estimator. For complex samples, $var(\hat{S}(t))$ is estimated by Taylor Series linearization (RTI, 2012) or through JRR methods.

The inverse transformation, $\exp(-\exp(\cdot))$, is then applied to the lower and upper limits of $CI_{1-\alpha}(g(S(t)))$ to construct the 100(1 – α)% CI for $S(t)$.

10.3.2 K-M Estimator—Evaluation and Interpretation

Since the K-M estimator is essentially "model-free," the evaluation phase of the analysis is limited to careful inspection and display of the estimates of $\hat{S}(t)$ and the associated standard errors. This is best done using plots of the estimated survivorship functions against time. Such plots are an effective way to display the overall survivorship curves and to compare the survival outcomes for the distinct groups (see Figure 10.3 below).

Under SRS assumptions, two quantitative tests (a Mantel-Haenszel test and a Wilcoxon test) comparing observed and expected "failures" over time can be used to evaluate whether the survivorship curves for two groups are equivalent (i.e., test a null hypothesis defined as $H_0 : S_1(t) = S_2(t)$). Analogous Wald-type X^2 test statistics could be developed for complex sample applications, but to the best of our knowledge these have not been implemented in the current software procedures that support Kaplan-Meier estimation for complex sample survey data. In this chapter, we will focus on the graphical display of the K-M estimates of the survivorship function, using these plots to inform decisions about predictors to use in Cox proportional hazards and discrete time regression models of survival.

10.3.3 K-M Survival Analysis Example

We begin our analysis example for this chapter by computing and plotting weighted **Kaplan-Meier estimates** of the survivor functions (10.3) for four NCS-R subgroups defined by the race variable (RACECAT). Even if a full K-M analysis is not the goal, this initial step is useful to visually examine the survivor functions for different subgroups and determine whether: a) the functions appear to differ substantially, and b) the survival functions appear to be roughly parallel (one of the assumptions underlying the Cox proportional hazards model, discussed in Section 10.4 below).

We first define a new variable AGEONSETMDE as being equal to the age of first MDE (MDE_OND) if a person has had an MDE in their lifetime, or the age of the respondent at the interview (AGE) if a person has not had an MDE in their lifetime:

```
gen ageonsetmde = age
replace ageonsetmde = MDE _ OND if mde == 1
```

The recoded variable, AGEONSETMDE, will serve as the time-to-event variable in this example. Next, we declare the structure of the NCS-R time-to-event data in Stata using the `stset` command. This is a required specification step that must be run before invoking most Stata survival analysis commands:

```
stset ageonsetmde [pweight = ncsrwtsh], failure(mde==1)
```

With the initial `stset` command, we define AGEONSETMDE as the variable containing the survival times (ages of first MDE onset, or right-censored ages at NCS-R interview for those individuals who had not yet experienced an MDE in their lifetime), and the indicator of whether an event actually occurred or whether the data should be analyzed as right-censored. This indicator is simply the indicator of lifetime MDE in the NCS-R data set (MDE), equal to 1 if the individual has ever had an MDE, and 0 otherwise (the right-censored cases). Note that we also define a "pweight" (or final survey weight) for the time-to-event data with the `[pweight = ncsrwtsh]` addition. Once the `stset` command has been submitted, Stata displays the following output:

```
      failure event: mde == 1
obs. time interval: (0, ageonsetmde]
 exit on or before: failure
           weight: [pweight=ncsrwtsh]
------------------------------------------------------------------------------
     9282   total obs.
        0   exclusions
------------------------------------------------------------------------------
     9282   obs.remaining, representing
     1829   failures in single record/single failure data
   385696   total analysis time at risk, at risk from t =      0
                                 earliest observed entry t =      0
                                      last observed exit t =     99
```

We see that there are 1,829 recorded "failure events" or individuals having experienced an MDE in their lifetime (with MDE = 1). Stata also processes the total number of observations at risk at each age starting with age 0, and the last observed exit from the risk set (an individual was interviewed at age 99 and reported never having an episode of MDE in their lifetime). Stata saves these variables in memory for any subsequent survival analysis commands, similar to how `svyset` works for the identification of complex sample design variables (strata, clusters, weights).

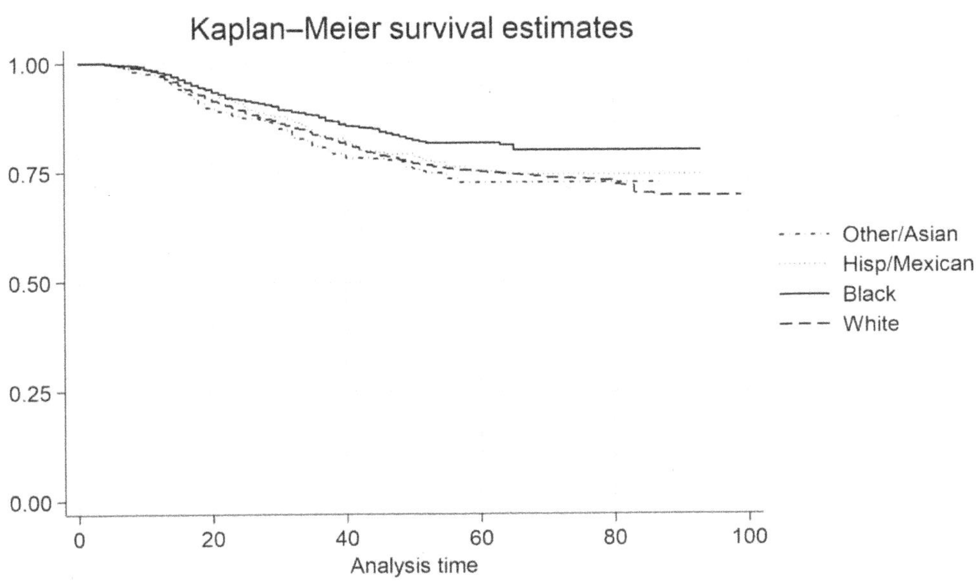

FIGURE 10.3
Weighted Kaplan-Meier Estimates of the Survivor Functions for Lifetime Major Depressive Episode for Four Ethnic Groups

Source: NCS-R.

Stata allows users to plot weighted population estimates of the survivor functions (provided that the final survey weights were specified in the stset command) for different groups by using the sts graph command with a by() option. We now apply this command to the NCS-R data:

```
sts graph, by(racecat) legend(lab(1 "Other/Asian") ///
lab(2 "Hisp/Mexican") lab(3 "Black") lab(4 "White"))
```

This command generates the plot in Figure 10.3.

The resulting plot in Figure 10.3 is very informative. First, by age 90, only about 25% of the persons in each race group will have had a major depressive episode. Further, the black race group consistently has a smaller percent of persons that have experienced a major depressive episode by each age considered. We also note that there is no strong evidence of distinct crossing of the estimated survival functions across ages; an assumption of parallel lines would seem fairly reasonable for these four subgroups.

Stata users can also use the sts list command to display estimates of the overall survivor function at specified time points (or ages in this example). The command below requests estimates of the survivor function at seven unique time points:

```
sts list, at(10 20 30 40 50 60 70)
```

Submitting this command results in the following Stata output:

```
          failure _d: mde == 1
    analysis time _t: ageonsetmde
              weight: [pweight=ncsrwtsh]

                Beg.                        Survivor
    Time       Total          Fail          Function
----------------------------------------------------------
     10      9171.84       139.004          0.9850
     20      8169.95       632.016          0.9165
     30      6390.66       401.574          0.8659
     40      4736.49       327.113          0.8155
     50      2997.27       186.745          0.7760
     60      1779.61       59.0606          0.7568
     70      949.309       22.2811          0.7438
----------------------------------------------------------
Note: survivor function is calculated over full data and
      evaluated at indicated times; it is not calculated from
      aggregates shown at left.
```

We note that design-adjusted standard errors for these estimates of the survivor function are not available in Stata. At present, PROC KAPMEIER in SUDAAN does provide the capability to compute weighted estimates of the survivor function along with design-based standard errors and 95% confidence intervals for the survivor function. Example SUDAAN syntax required to conduct this descriptive analysis for the NCS-R MDE data is as follows:

TABLE 10.1

Selected Results from the Kaplan-Meier Analysis of the NCS-R Age of Onset for MDE Data in SUDAAN

Age of Onset	$\hat{S}(t)$	SUDAAN $se(\hat{S}(t))$	SUDAAN $CI_{.95}(S(t))$
10	0.985	0.001	(0.982. 0.988)
20	0.917	0.004	(0.908. 0.924)
30	0.866	0.005	(0.855, 0.876)
40	0.816	0.005	(0.804, 0.826)
50	0.776	0.006	(0.764, 0.787)
60	0.757	0.006	(0.745, 0.769)
70	0.744	0.007	(0.730, 0.757)

```
proc kapmeier;
nest sestrat seclustr;
weight ncsrwtsh;
event mde;
class racecat;
strhaz racecat;
time ageonsetmde;
setenv decwidth=4;
output / kapmeier=all filename="c10 _ km _ out" filetype=sas;
replace;
run;
```

Table 10.1 contains an extract from the complete set of results produced by this analysis. The table displays the total population K-M estimates of the survivor function, along with the standard errors and 95% CIs for the values of the survivor function at ages t = 10, 20, 30, . . ., 70. Note that the weighted estimates of $\hat{S}(t)$ at these ages match those obtained in the weighted analysis using Stata's `sts list` command. The full output includes age-specific estimates of the survivor function for the total population and separately for each of the four race/ethnic groups defined by RACECAT. In PROC KAPMEIER, the stratified K-M analysis is invoked by the use of the STRHAZ *varname*; statement. SUDAAN does not provide the capability to plot the estimated curves and confidence intervals for the survivorship at each age of onset; however, the analysis output can be saved using SUDAAN's OUTPUT statement and exported to a software package that has a general graphics capability to produce the desired plots.

10.4 Cox Proportional Hazards Model

The Cox Proportional Hazards (CPH) model (Cox, 1972) is one of the most widely-used methods for survival analysis of continuous time event history data. CPH regression provides analysts with an easy-to-use and easy-to-interpret multivariable tool for examining the relationships of selected covariates with expected hazard functions. Despite its widespread use and the semi-parametric nature of the underlying model, survey analysts must be careful to examine the critical "proportional hazards" assumption and to ensure that the model is fully and correctly identified. Focusing largely on clinical trials, Freedman

(2008) cautions against blind use of Cox regression models without performing some initial descriptive analyses of the survival data.

10.4.1 Cox Proportional Hazards Model: Specification

The Cox model for proportional hazards is specified as follows:

$$h(t \mid \boldsymbol{x}_i) = h_0(t)\exp(\sum_{j=1}^{P} B_j x_{ij}) \tag{10.5}$$

In (10.5), $h(t \mid \boldsymbol{x}_i)$ is the expected hazard function for an individual i with vector of covariates $\boldsymbol{x}_i$ at time t. The model includes a **baseline hazard function** for time t, $h_0(t)$, that is common to all population members. The individual hazard at time t is the product of the baseline hazard and an individual-specific factor, $\exp\left(\sum_{j=1}^{P} B_j x_{ij}\right)$, that is a function of the regression parameters, $\boldsymbol{B} = \{B_1, \ldots, B_p\}$ and the individual covariate vector, $\boldsymbol{x}_i = \{x_{i1}, \ldots, x_{ip}\}$. Individual hazards are therefore a proportional scaling of the baseline hazard: $h(t \mid \boldsymbol{x}_i) \propto h_0(t)$. There is no separate intercept parameter in the CPH model specification; the baseline expectation (i.e., the hazard when all covariates are equal to 0) is absorbed into $h_0(t)$.

The CPH model is most applicable to continuous or nearly continuous measurement of event and censoring times. Available software for estimating the CPH model anticipates the possibility that multiple events/censoring may occur at any time point t and incorporates rules for handling such ties. For example, Stata assumes that events occur before censoring if there are ties—censored cases at time t remain in the risk set for events at time t. If the time scale for the survey observations is very coarse (e.g., years in a study of children age 5–18) or explicitly discrete (e.g., the two-year period between consecutive HRS survey interviews), many events and censorings may occur at time period t. If the measurement of time is coarse and ties are common, analysts might consider the complementary-log-log discrete time model of Section 10.5 as an alternative to CPH regression.

The values of covariates in the CPH model may be fixed (e.g., gender, race) or they may be **time-varying covariates**, meaning that the values of the covariates can change depending on the time at which the set of sample units "at risk" is being evaluated. Stata allows users to set up time-varying covariates for use in the `svy: stcox` command for estimating the CPH model. If the desired model includes a large number of time-varying covariates, an alternative to the CPH model is to recode the event and censoring times into discrete time units and apply the discrete time logistic or complementary log-log models described in Section 10.5.

10.4.2 Cox Proportional Hazards Model: Estimation Stage

The CPH regression parameters B_j are estimated using a **partial likelihood** procedure, based on the conditional probability that an event occurred at time t [see Freedman (2008) for a nice primer, or Lee (1992) for more details]. For estimation, the E observed event times are ordered such that $t_{(1)} < t_{(2)} < \ldots < t_{(E)}$. The **risk set** at a given time t, $t = 1, \ldots, E$, is the set of respondents who: 1) have not experienced the event prior to time t, and 2) were not randomly censored prior to time t. The probability that an event occurs at time t for the i-th respondent, conditional on the risk set R_t, is defined as follows:

$$P(t_i = t \mid x_i, R_t) = \frac{\exp\left(\sum_{j=1}^{P} B_j x_{ij,t}\right)}{\sum_{l \in R_t} \exp\left(\sum_{j=1}^{P} B_j x_{lj,t}\right)} \tag{10.6}$$

where R_t= {set of cases still "at risk" at time t}.

This conditional probability in (10.6) is computed for every event that occurs. The partial likelihood function for the survey observations of event times is then defined as the product of these probabilities (see Theory Box 10.2). An iterative mathematical algorithm is then applied to derive estimates of the regression parameters that maximize the partial likelihood function.

Binder (1992) describes the Taylor Series Linearization approach for estimating $var(\hat{\mathbf{B}})_{TSL}$ that accounts for the complex sample design. BRR, JRR, or bootstrap methods can also be used to compute replicated estimates of the variances and covariances of the model coefficients, $var(\hat{\mathbf{B}})_{Rep}$. Routines capable of fitting CPH models and computing design-based estimates of standard errors for the parameter estimates are currently implemented in Stata, SAS, the SPSS Complex Samples module, SUDAAN, the `survey` package in R, Mplus, and IVEware (see Appendix A).

THEORY BOX 10.2: PARTIAL LIKELIHOOD FOR THE COX PROPORTIONAL HAZARDS MODEL

The partial likelihood that forms the basis for estimating the regression parameters of the CPH model is the product of conditional probabilities—one for each distinct event, $e = 1, \ldots, E \le n$ that is recorded in the survey period. Each conditional probability component of this partial likelihood is the ratio of the time $t(e)$ hazard for the case experiencing the event (case i) to the sum of hazards for all sample cases remaining in the risk set at $t(e)$:

$$L(B) = \prod_{i=1}^{n} \left[\frac{h(t(e)_i \mid x_i)}{\sum_{j=1}^{n} I\left[t(0) < t(e)_i \le t(e)_j\right] \cdot h(t(e)_j \mid x_j)} \right]^{\delta_i \cdot w_i}$$

where:

$t(0)$ is the time that observations begin;

$t(e)_i$ and $t(e)_j$ are the respective event (censoring) times for respondents i and j;

$I\left[t(0) < t(e)_i \le t(e)_j\right]$ is the 0,1 indicator that respondent i is in the risk set when unit j experiences an event;

$\delta_i = 1$ if unit i experiences the event, or 0 if unit i is censored;

w_i = the final survey weight for unit i.

In the partial likelihood for complex sample survey applications, the contribution of the *i*-th case to the partial likelihood is raised to the power of its final survey weight, w_i. Censored cases that remain in the risk set contribute to the denominator of the conditional probabilities that events occur for other cases; however, the direct contributions of censored cases to the product likelihood are factors of 1 (due to the $\delta_i = 0$ exponent).

10.4.3 Cox Proportional Hazards Model: Evaluation and Diagnostics

Procedures for constructing confidence intervals and testing hypotheses for the estimated parameters of the CPH model parallel those described in previous chapters for the parameters of other linear and generalized linear regression models (e.g., Section 7.5).

Diagnostic tools for CPH models fitted to complex sample survey data sets are still in the early stages of development. Inspection of the Kaplan-Meier estimates of the survivor functions for different groups can be helpful as an initial step, to see if the survival functions are approximately parallel. A better graphical check of the proportional hazards assumption that is currently implemented in Stata is a plot of $-\ln(-\ln(\hat{S}_h(t)))$ against $\ln(t)$, where $\hat{S}_h(t)$ is a weighted estimate of the survival function for group h (see Figure 10.4). Based on the specified CPH model, the transformed versions of the survival functions should be parallel as a function of $\ln(t)$.

At the time of the writing of this third edition, only selected software procedures have incorporated standard residual diagnostics for CPH models into the CPH modeling procedures for complex sample survey data. For example, it is possible to generate **partial Martingale residuals** (e.g., in SUDAAN) for the purpose of checking the functional forms of continuous covariates in the Cox model. These residuals are computed as

$$M_i = \delta_i - \hat{H}_i(t), \tag{10.7}$$

where δ_i represents an indicator variable for whether or not a given sample case i had the event of interest occur (1 = event, 0 = censored), and $\hat{H}_i(t)$represents a weighted estimate of the **cumulative hazard function** based on the fitted model (or the cumulative sum of all instantaneous probabilities of failure up until time t; one can also think of this function as the "total rate" of experiencing an event up to time t). These residuals can be plotted against the values of individual continuous covariates to check for the possibility of nonlinear relationships, which would indicate that the functional forms of continuous covariates may have been misspecified.

Any additional developments allowing users to generate these residuals and further evaluate the goodness of fit of CPH models will be highlighted on the ASDA website. For additional discussion of fitting CPH models to complex sample survey data sets, we refer interested readers to Cleves et al. (2016, Section 9.5).

10.4.4 Cox Proportional Hazards Model: Interpretation and Presentation of Results

Interpretation of results and finite population inference from the CPH model is typically based on comparisons of **hazards**—the conditional probabilities that the event will occur at time t given that it has not occurred prior to time t. Given estimates of the regression parameters in the model, consider the ratio of estimated hazards if predictor variable x_j is incremented by one unit and all other covariates remain fixed:

$$\begin{aligned} \hat{HR}_j &= \frac{h_0(t)\exp(\hat{B}_1 x_1 + \cdots + \hat{B}_j(x_j+1) + \cdots + \hat{B}_p(x_p))}{h_0(t)\exp(\hat{B}_1 x_1 + \cdots + \hat{B}_j(x_j) + \cdots + \hat{B}_p(x_p))} \\ &= \exp(\hat{B}_j) \end{aligned} \quad (10.8)$$

The one-unit change in x_j will multiply the expected hazard by $\hat{HR}_j = \exp(\hat{B}_j)$. This multiplicative change in the hazard function is termed the **hazard ratio**. If x_j is an indicator variable for a level of a categorical predictor, then $\hat{HR}_j = \exp(\hat{B}_j)$ is the relative hazard compared to the hazard for the reference category of the categorical predictor.

The procedure for developing a 100(1 – α)% CI for the population hazard ratio parallels that used to build a CI for the odds ratio statistic in simple logistic regression:

$$CI\left(HR_j\right) = \exp\left(\hat{B}_j \pm t_{df,1-\alpha/2} \cdot se\left(\hat{B}_j\right)\right) \quad (10.9)$$

10.4.5 Example: Fitting a CPH Model to Complex Sample Survey Data

We now turn to fitting the CPH model to the NCS-R age of major depression episode onset data. The appropriate data structure for this type of analysis features one row per survey respondent, with that row containing measures of the survival time, a censoring indicator, covariates of interest, and complex sample design features (stratum and cluster codes and survey weights). Prior to fitting the CPH model to the NCS-R data, we once again declare the variables containing the relevant NCS-R sampling error codes to Stata [note that we use the survey weight variable NCSRWTSH and request a Taylor Series Linearization approach (Binder, 1992) for variance estimation]:

```
svyset seclustr [pweight=ncsrwtsh], strata(sestrat)
```

Next, we submit the `stset` data preparation command required by Stata for survival analysis. Note that a common `pweight` variable must be specified in the `svyset` and `stset` commands:

```
stset ageonsetmde [pweight = ncsrwtsh], failure(mde==1)
```

After defining the complex sample design variables and the `stset` coding of the dependent variable AGEONSETMDE, the CPH model is fit using the `svy: stcox` command [note that value 2 (female) for the predictor variable SEX is explicitly set as a reference category]:

```
svy: stcox age ib2.sex i.MAR3CAT i.ED4CAT i.racecat
```

Note that we use the "`i.`" prefix to declare certain predictor variables (SEX, MAR3CAT, ED4CAT, and RACECAT) as categorical, so that Stata includes the appropriate dummy variables in the model. The syntax of the `svy: stcox` command is fairly simple and similar to other regression analysis commands for survey data within Stata, provided that the two previous steps (declaring the time-to-event data and the complex design variables) have been followed carefully. What makes the command unique is the fact that a dependent variable (the time variable AGEONSETMDE in this case) is *not* specified; the first variable listed in the command, AGE, is a predictor variable in the CPH model. The initial use of the `stset` command directs Stata to set up the appropriate CPH model, and Stata reads the list of variables in the `svy: stcox` command as the predictors to be included in the model. Submitting this command produces the estimated hazard ratios in Table 10.2.

The design-adjusted *F*-test of the null hypothesis that all of the regression parameters in the CPH model are equal to zero suggests that some of the predictors have a significant relationship with the hazard of MDE onset at any given age, and the design-based 95% confidence intervals for the hazard ratios also support this finding. Specifically, Hispanic and Black respondents have marginally ($p < 0.10$) and significantly ($p < 0.01$) reduced hazards of onset of MDE at any given age, holding the other covariates fixed, relative to the Asian/Other ethnic group. This agrees with the initial evidence provided by the Kaplan-Meier estimates of the survival functions for the four ethnic groups.

Further, with every one-year increase in age at the time of interview (a birth cohort predictor), the hazard of MDE onset is multiplied by 0.95, or decreased by about 5%, holding the other covariates fixed. This suggests that younger individuals at the time of the interview have a higher hazard of MDE onset at any given age than older individuals. Finally, previously married individuals have a hazard of MDE onset that is about 65% higher at any given age compared to currently married individuals, and males have a hazard of MDE onset that is about 36% lower at any given age than females, holding the other covariates fixed.

We now consider the graphical check of the proportional hazards assumption described in Section 10.4.3 above. Recall that this graphical check involves a plot of $-\ln(-\ln(\hat{S}_h(t)))$ against $\ln(t)$, where $\hat{S}_h(t)$ is a weighted estimate of the survival function for group h and

TABLE 10.2

Estimated Hazard Ratios for the Cox Proportional Hazards Model of Age of Onset of a Major Depression Episode

Predictor*	Estimated Hazard Ratio	Estimated SE	*t*-statistic (df)	*p*-value	95% CI
Education					
12 Years	0.94	0.06	−0.85 (42)	0.40	(0.83, 1.08)
13–15 Years	1.05	0.06	0.79 (42)	0.44	(0.93, 1.18)
16+ Years	0.91	0.06	−1.42 (42)	0.16	(0.80, 1.04)
Race/Ethnicity					
Hispanic	0.78	0.10	−1.86 (42)	0.07	(0.59, 1.02)
Black	0.62	0.09	−3.22 (42)	0.003	(0.46, 0.84)
White	1.08	0.13	0.66 (42)	0.51	(0.85, 1.37)
Marital Status					
Previously Married	1.65	0.099	8.37 (42)	<0.001	(1.46, 1.87)
Never Married	1.08	0.096	0.91 (42)	0.37	(0.91, 1.30)
Gender					
Male	0.64	0.040	−7.28 (42)	<0.001	(0.56, 0.72)
Age	0.95	0.002	−20.80 (42)	<0.001	(0.947, 0.956)

(*Source:* NCS-R)

Note: n = 9,282, Design-adjusted F-test of null hypothesis that all parameters are 0: $F(10, 33) = 53.02$, $p < 0.001$

Note: * Reference categories for categorical predictors are: <12 yrs (Education); Asian/Other (Ethnicity); Married (Marital Status); Female (Gender).

t represents the time variable. Based on the specified CPH model, the transformed versions of the survival functions should be parallel as a function of $\ln(t)$. Remember that the `stset` command specification of the dependent variable and the `pweight` variable remain in effect. The `stphplot` command is used to generate this specific diagnostic plot:

```
stphplot, by(racecat) legend(lab(1 "Other/Asian") ///
lab(2 "Hisp/Mexican") lab(3 "Black") lab(4 "White"))
```

This command generates the plot in Figure 10.4.

From Figure 10.4, we see minimal evidence of the transformed survival curves crossing during the ages when the majority of the data were collected, suggesting that an assumption of proportional hazards would be reasonable for the four race/ethnicity groups. If the transformed survival curves were not approximately parallel (for example, there is evidence at the highest ages of a possible cross between the White ethnic group and the Hispanic/Mexican and Other/Asian groups), one could fit a model allowing for the effects of ethnicity to vary depending on the age at which a risk set is being formed for estimation. This could be done by including interactions of race with age in this model, which would be time-varying covariates. For more on how to do this when using `svy: stcox`, Stata users can see `help tvc _ note`.

Optionally, Stata users could also plot $\ln(-\ln(\hat{S}_h(t)))$ against t (a variation of the parallel lines check) and/or adjust the estimates for mean values of covariates using the following options (the `adj()` option includes dummy variables as covariates, such as SEXF as an indicator for females, that are equivalent to the dummy variables generated by Stata for the model fitting):

```
stphplot, by(racecat) adj(age sexf prevmarr nevmarr ///
ed12 ed1315 ed16) nolntime ///
legend(lab(1 "Other/Asian") lab(2 "Hisp/Mexican") ///
lab(3 "Black") lab(4 "White"))
```

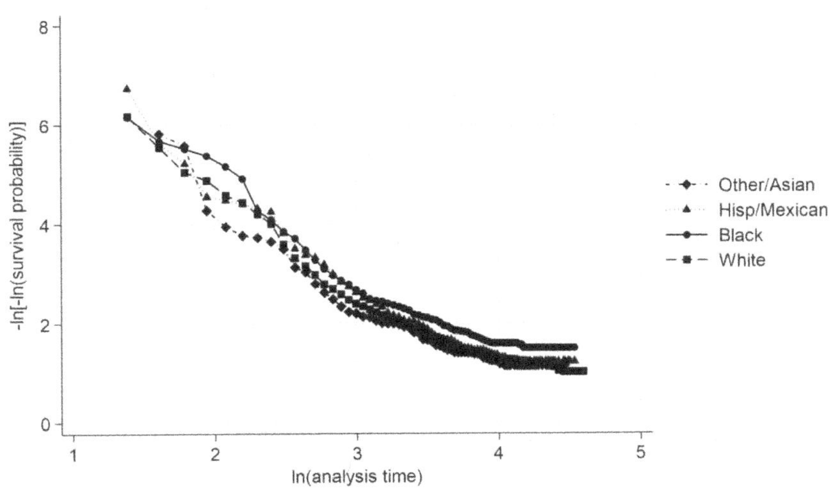

FIGURE 10.4
Graphically Examining the Proportional Hazards Assumption for the Fitted Cox Model of Onset of Major Depression Episode (MDE), with No Adjustment for Covariates.

Source: NCS-R.

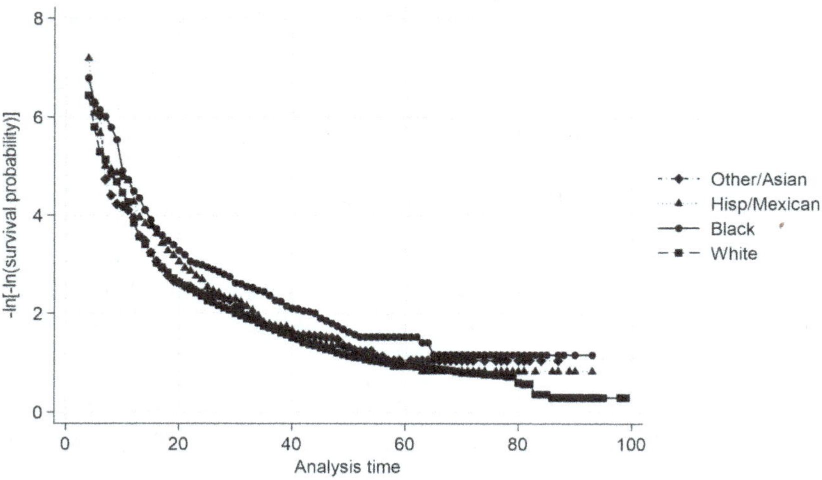

FIGURE 10.5
Graphically Examining the Proportional Hazards Assumption for the Fitted Cox Model of Onset of Major Depression Episode (MDE), with Adjustment for Covariates.

Source: NCS-R.

The resulting plot in Figure 10.5 shows that when adjusting for the means of the covariates (and using the alternative specification of the plot enabled by the nolntime option), we still do not have evidence against an assumption of proportional hazards.

10.5 Discrete Time Survival Models

Survival models for events measured in discrete time are needed only when events are reported in discrete time units. For a dependent variable representing a discrete survival time, there are $m = 1, \ldots, M$ observed intervals, such as time periods between waves of a longitudinal survey, or time measured in years (year 1, year 2, etc.). In some cases, the "discrete" nature of the survival data may be due to **coarsening** in the dating of events (within the past four weeks, did you . . .?) or even deliberate grouping of times to provide disclosure protection for individual survey respondents.

10.5.1 The Discrete Time Logistic Model

A popular choice for modeling hazard functions when working with discrete survival outcomes is the **discrete time logit model** (Allison, 2010; Singer & Willett, 1993; Yamaguchi, 1991). This model is defined as follows:

$$\begin{aligned}\ln\left(\frac{h_{i,m}}{1-h_{i,m}}\right) &= B_{0,m} + \boldsymbol{x}_{i,m}\boldsymbol{B} \\ &= B_{0,m} + B_1 x_{1,m} + \cdots + B_p x_{p,m}\end{aligned} \tag{10.10}$$

In (10.10), $h_{i,m}$ refers to the hazard of failure at discrete time m for respondent i, $B_{0,m}$ is a time-specific intercept term that applies to all individuals at time m, $\boldsymbol{x}_{i,m}$ is a row vector of values of covariates (possibly time-varying) for respondent i, and $\boldsymbol{B}$ is the column vector of regression parameters. It is important to note that (10.10) models the logit of the individual hazard, $h_{i,m}$. To recover the estimated hazard from the estimated logit model requires the inverse logit transformation:

$$\hat{h}_{i,m} = \frac{\exp(\hat{B}_{0,m} + \boldsymbol{x}_{i,m}\hat{\boldsymbol{B}})}{1+\exp(\hat{B}_{0,m} + \boldsymbol{x}_{i,m}\hat{\boldsymbol{B}})} \tag{10.11}$$

Once the discrete time logit model has been estimated, estimates of the individual hazards, standard errors, and CIs can be readily generated using Stata's `predict` post-estimation command.

Allison (2010) describes an alternative discrete time model that is based on the complementary-log-log (C-L-L) link:

$$\log[-\log(1-h_{i,m})] = \alpha_m + B_1 x_{1,m} + \cdots + B_p x_{p,m} \tag{10.12}$$

The C-L-L version of the discrete time model is most applicable when the true model for the underlying continuous time process that generates the discrete time measures is the CPH model of Section 10.4. An advantage of the C-L-L discrete time model is that $H\hat{R}_j = \exp(\hat{B}_j)$. Exponentiating the estimated $\hat{B}_j$ yields an estimate of the **hazard ratio** corresponding to a one unit change in x_j, which is identical to the interpretation for the CPH model.

The regression parameters from the two discrete time models do have slightly different interpretations—log(odds) for the logit version, log(hazard) for the C-L-L—however, both models share the following features:

- Identical **person-time** data set formats are required to fit the two models;
- Both models permit direct estimation of the relationship of time with the hazard, through the $\hat{B}_{0,m}$ in the logit form and the $\hat{\alpha}_m$ in the C-L-L model;
- Both models permit the use of time-varying covariates, $\boldsymbol{x}_{i,m}$;
- For complex sample survey data, pseudo maximum likelihood estimation is used to estimate the regression parameters in each model (see Section 8.2).

In most cases, the choice of the logit or the C-L-L form of the discrete time survival model will not affect the survey analyst's interpretation of the significance and nature of the relationship of the model covariates to the hazard function.

10.5.2 Data Preparation for Discrete Time Survival Models

Discrete time survival models require a special data input structure that is labeled a **person-time** data format. The person-time data input format contains multiple records for each respondent: one record to represent each discrete time interval during which the respondent was observed, *up to and including* the time interval when the event of interest actually occurred or the observation was censored. A respondent who experiences the event at the third (m = 3) discrete time point will have three records in the data set.

Another respondent whose observation period is censored at time $m = 5$ will have five person-time data records in the data set.

Each person-time record will contain a value for a binary dependent variable, Y_{im}, taking on value 1 if the event occurred for respondent i at time period m, or value 0 if no event occurred or the respondent was censored at time m. The sequence of person-time records extends from the first discrete time point, $m = 1$, until the time period when an event actually occurred (i.e., until the dependent variable is equal to 1), or until time M (the last possible discrete time period) if no event occurred (all records for this type of censoring will have the dependent variable equal to 0). If a respondent drops out of the sample or leaves the study for some other reason prior to the end of the observation period, they will have a person-time record for each time point at which they were observed, plus one additional record for the time point at which they were lost. All of their records will have the dependent variable $Y_{im} = 0$.

The discrete time models offer an advantage in that they directly model the relationship of time with the hazard of the event. Each person-time record therefore requires one or more variables to indicate the time period. There are several ways to parameterize the relationship of time in the discrete time model. The method that we recommend is to create a set of indicator variables—one for each time period, $\boldsymbol{I} = \{I_1, \ldots, I_M\}$. Each person-time record will contain all M indicator variables, but only the indicator corresponding to the time period m will have $I_m = 1$. All other time indicators (for times $m' \neq m$) for that record will be set to $I_{m' \neq m} = 0$. When this parameterization is used, the logit or C-L-L model should be estimated without the overall intercept term. (If the overall intercept is modeled, one time indicator variable should be dropped.) We illustrate the implications of this coding for the specification of a discrete time logit model in (10.13) below:

$$\begin{aligned} \ln\left(\frac{h_{i,m}}{1-h_{i,m}}\right) &= \boldsymbol{B}_0 \cdot \boldsymbol{I} + B_1 x_{1,m} + \cdots + B_p x_{p,m} \\ &= B_{0,1} \cdot 0 + \cdots + B_{0,m} \cdot 1 + \cdots + B_{0,M} \cdot 0 + B_1 x_{1,m} + \cdots + B_p x_{p,m} \\ &= B_{0,m} + B_1 x_{1,m} + \cdots + B_p x_{p,m} \end{aligned} \tag{10.13}$$

There are two advantages to this "time indicator" parameterization of the discrete time models. First, no functional relationship (e.g., linear, quadratic) is imposed on the relationship between the hazard and time. Second, if time periods are not of equal length or must be collapsed (if event times are sparse), the individual time period intercepts will adjust for the variability in the size of the time units and the regression parameters for the covariates will not be biased.

An alternative parameterization is to model the sequential discrete time unit indices ($m = 1, 2, 3, \ldots, M$) as linear or quadratic predictors. For the discrete time logit model:

$$\ln\left(\frac{h_{i,m}}{1-h_{i,m}}\right) = B_0 + \gamma_1 m + \gamma_2 m^2 + B_1 x_{1,m} + \cdots + B_p x_{p,m} \tag{10.14}$$

This is a much more parsimonious parameterization of time and may be preferred if the total number of discrete times is large (say, $M > 15$). (For simplicity of presentation and comparability to the CPH model results in Table 10.2, this parameterization is used in the analysis example in Section 10.5.5.) One practical analysis strategy is to first fit the discrete time model using the unconstrained time indicator parameterization (10.13). If the estimated coefficients show a linear or quadratic trend, the model could be re-estimated more parsimoniously using the model in (10.14).

In addition to the binary dependent variable and the indicators representing the time period, each person-time record in the input data set will include the vector of p covariates, $x_m = \{x_{1,m}, \ldots, x_{p,m}\}$. Individual variables in the covariate vector may be time-invariant (e.g., sex assigned at birth) or may be time-varying (e.g., body weight in kilograms).

Finally, in complex sample survey data sets, the final survey weight and the sampling error stratum and cluster codes are assigned to each person-time record. In most cases, the weight value assigned to each time point will be the final survey weight assigned to each survey respondent at the start of the observation period ($m = 1$). However, in cases such as longitudinal surveys where there may be substantial panel attrition due to nonresponse over the multiple waves in the observation period, the adjusted time-specific weight at time t for each observed case may provide a better weighted representation of the population risk set at time t. The person-time data structure for discrete time models allows the user this flexibility to adjust the time-specific weight for prior nonresponse.

The following examples should help to clarify the required person-time data structure. Assume that a sample of individuals is observed for five waves of survey data collection, and a discrete survival time T is recorded for each individual, equal to 1 for events that occur in a fixed interval of time prior to Wave 1, 2 for events that occur between Wave 1 and Wave 2, and so forth ($T = 1, 2, 3, 4, 5$). Table 10.3 illustrates the person-time data inputs for two hypothetical respondents who experience the event of interest: the first at the second wave ($m = 2$), and the second at the fourth wave ($m = 4$).

In Table 10.3, note the presence of a time-invariant survey weight (which may be time-varying depending on the design of the sample), and stratum and cluster codes for sampling error estimation (also time-invariant). In this hypothetical example, time is represented in the model as a single integer-value covariate, making this input consistent with model parameterization (10.14) above. If the time-period indicator parameterization of the model (10.13) was chosen, each person-time record would include five indicator variables (one for each of the $M = 5$ possible observation times in this hypothetical example). Since the two example cases in Table 10.3 experienced the event of interest, the value of the dependent variable is coded as $Y_{im} = 1$ for the event time, and $Y_{im} = 0$ for all person-year records preceding the event. The X_{im} variable (e.g., body mass index) is a time-varying covariate, and the Z_i variable (e.g., indicator of female sex assigned at birth) is a time-invariant covariate.

TABLE 10.3

Example Data Structure for Discrete Time Hazard Models, for Two Individuals Experiencing an Event of Interest

ID	Weight	Stratum	Cluster	Time (t)	Y_{im}	X_{im}	Z_i
1	1.56	1	1	1	0	23	1
1	1.56	1	1	2	1	25	1
2	0.82	1	2	1	0	14	0
2	0.82	1	2	2	0	16	0
2	0.82	1	2	3	0	17	0
2	0.82	1	2	4	1	17	0

TABLE 10.4

Example Data Structure for Discrete Time Hazard Models, for Censored Individuals

ID	Weight	Stratum	Cluster	Time (t)	Y_{im}	X_{im}	Z_i
3	1.25	2	1	1	0	19	1
3	1.25	2	1	2	0	19	1
3	1.25	2	1	3	0	18	1
3	1.25	2	1	4	0	21	1
3	1.25	2	1	5	0	22	1

Table 10.4 illustrates an example of the person-time data structure for a censored respondent who did not experience the event of interest by the fifth and final wave of data collection.

Note in Table 10.4 that the person-time records for this right-censored case include the same input variables as the event cases in Table 10.3; however, the values of the dependent variable, including the final time point, are all coded $Y_{im} = 0$.

10.5.3 Discrete Time Models: Estimation Stage

Mathematically, estimation of the regression parameters for the discrete time logit and C-L-L models for complex sample data is a direct application of the pseudo maximum likelihood methods described in Section 8.3 for the simple logistic and C-L-L models for a binary dependent variable. Likewise, procedures for estimation of $var(\hat{\boldsymbol{B}})_{TSL}$ or alternatively $var(\hat{\boldsymbol{B}})_{Rep}$ are identical to those for the standard logit and C-L-L generalized linear models. Provided that the data are structured correctly in the person-time input format described above, any statistical software procedure capable of fitting logistic or C-L-L regression models to complex sample survey data can be used to fit these models. As pointed out by Allison (2010), some analysts' natural intuition leads them to believe that because the required person-time data structure involves multiple observations for the same person, an extra level of clustering may be introduced in the data that is similar to the clustering that occurs in repeated measures data (see Chapter 11). Fortunately, as explained in Theory Box 10.3, this is actually not a concern in the estimation of sampling variances for the parameters in discrete time models. In modeling event history data, the issue of correlation among the person-time records arises only when working with discrete event time models that permit individuals to have multiple events.

THEORY BOX 10.3: THE LACK OF A CLUSTERING EFFECT IN PERSON-TIME DATA

The explanation for the absence of a "clustering effect" in the person-time data is that the pseudo likelihood for the complete set of observed event and censoring times can be factored into separate pseudo-likelihoods for the individual event times, $m = 1, \ldots, M$. Consequently, estimation for each component of the factored likelihood can be treated as if its set of person-time records were independent of those used to estimate the hazard probabilities for other discrete time points. See Allison (1982) for a mathematical proof that the full likelihood for the discrete time model can be factored into time-period specific likelihoods.

10.5.4 Discrete Time Models: Evaluation and Interpretation

As described above, procedures for confidence interval construction, hypothesis testing, and general model evaluation for the discrete time models follow directly from those described in Chapter 8 for the logit and C-L-L models for a binary dependent variable.

10.5.4.1 Discrete Time Logit Model

The regression coefficients in the discrete time logit model are estimated on the log-odds scale, where the odds in question pertain to the conditional hazard probabilities. Exponentiating the estimated regression coefficients yields the estimated odds ratios: $\hat{\psi}_j = \exp(\hat{B}_j)$. Confidence intervals for these estimated odds ratios are constructed in the now familiar fashion: $CI(\psi_j) = \exp\left(\hat{B}_j \pm t_{df,1-\alpha/2} \cdot se(\hat{B}_j)\right)$. Given pseudo maximum likelihood estimates of the regression parameters and a vector of covariates, $x_{i,m}$, the estimated hazard at time m can be computed as follows:

$$\hat{h}_{i,m} = \frac{\exp(\hat{B}_{0,m} + \boldsymbol{x}_{i,m}\hat{\boldsymbol{B}})}{1+\exp(\hat{B}_{0,m} + \boldsymbol{x}_{i,m}\hat{\boldsymbol{B}})} \tag{10.15}$$

Note that this is mathematically identical to the computation of a predicted probability based on a fitted logistic regression model.

10.5.4.2 Discrete Time Complementary Log-Log (C-L-L) Model

An advantage of the C-L-L model for discrete time data is that population inferences and discussion of results use the **hazard ratio** statistic: $\hat{HR}_j = \exp(\hat{B}_j)$. Procedures for constructing confidence intervals and interpreting the hazard ratio statistics generated from the C-L-L model for discrete time data are identical to those presented for the CPH model in Section 10.4.4.

With this background, we now consider fitting the discrete time logit and discrete time C-L-L models to the NCS-R data on age of onset for major depression.

10.5.5 Fitting a Discrete Time Hazard Model to Complex Sample Survey Data

An important first step in fitting the discrete time hazard model is to transform the NCS-R input data into the appropriate person-time format. The following Stata commands "expand" the NCS-R data set from a one record per person file to a multiple record per person file, with records from year = 1 to year at age of interview (AGE). The use of the "`by caseid: gen pyr = _n`" command below creates a new variable called PYR which represents person years of life ranging from 1 to year at age of interview (AGE):

```
expand age
sort caseid
by caseid: gen pyr = _n
```

The NCS-R data set now includes one "person-year" time record for each year of life for a given individual (e.g., 45 rows for a 45-year-old). Next, the indicator variable MDETV is generated for each person-year record. The variable MDETV = 1 if the value of time

(PYR) corresponding to the person-year record is equal to the age of onset of MDE, and MDETV = 0 otherwise:

```
gen mdetv = 1 if pyr == MDE _ OND
replace mdetv = 0 if pyr != MDE _ OND
```

The variable MDETV will be the dependent variable when fitting the logit and C-L-L regression models to these data. The following example Stata list output for the *first respondent* in the NCS-R data set shows the person-year data for this discrete time modeling example:

```
list caseid age ncsrwtsh sestrat seclustr pyr mdetv ///
ageonsetmde if caseid == 1
```

```
     +---------------------------------------------------------------------------+
     | caseid   age   ncsrwtsh   sestrat   seclustr   pyr   mdetv   ageons~e |
     |---------------------------------------------------------------------------|
  1. |      1    41    2.02426         1          2     1       0         34 |
  2. |      1    41    2.02426         1          2     2       0         34 |
  3. |      1    41    2.02426         1          2     3       0         34 |
  4. |      1    41    2.02426         1          2     4       0         34 |
  5. |      1    41    2.02426         1          2     5       0         34 |
     |---------------------------------------------------------------------------|
  6. |      1    41    2.02426         1          2     6       0         34 |
  7. |      1    41    2.02426         1          2     7       0         34 |
  8. |      1    41    2.02426         1          2     8       0         34 |
  9. |      1    41    2.02426         1          2     9       0         34 |
 10. |      1    41    2.02426         1          2    10       0         34 |
     |---------------------------------------------------------------------------|
 11. |      1    41    2.02426         1          2    11       0         34 |
 12. |      1    41    2.02426         1          2    12       0         34 |
 13. |      1    41    2.02426         1          2    13       0         34 |
 14. |      1    41    2.02426         1          2    14       0         34 |
 15. |      1    41    2.02426         1          2    15       0         34 |
     |---------------------------------------------------------------------------|
 16. |      1    41    2.02426         1          2    16       0         34 |
 17. |      1    41    2.02426         1          2    17       0         34 |
 18. |      1    41    2.02426         1          2    18       0         34 |
 19. |      1    41    2.02426         1          2    19       0         34 |
 20. |      1    41    2.02426         1          2    20       0         34 |
     |---------------------------------------------------------------------------|
 21. |      1    41    2.02426         1          2    21       0         34 |
 22. |      1    41    2.02426         1          2    22       0         34 |
 23. |      1    41    2.02426         1          2    23       0         34 |
 24. |      1    41    2.02426         1          2    24       0         34 |
 25. |      1    41    2.02426         1          2    25       0         34 |
     |---------------------------------------------------------------------------|
 26. |      1    41    2.02426         1          2    26       0         34 |
 27. |      1    41    2.02426         1          2    27       0         34 |
 28. |      1    41    2.02426         1          2    28       0         34 |
 29. |      1    41    2.02426         1          2    29       0         34 |
 30. |      1    41    2.02426         1          2    30       0         34 |
     |---------------------------------------------------------------------------|
 31. |      1    41    2.02426         1          2    31       0         34 |
 32. |      1    41    2.02426         1          2    32       0         34 |
 33. |      1    41    2.02426         1          2    33       0         34 |
 34. |      1    41    2.02426         1          2    34       1         34 |
```

Note that the sampling error stratum and cluster codes and the final survey weight value are included on each person-year record for the respondent. This respondent was 41 years old when interviewed and first had an MDE at age 34. Although there are more data records for this individual (ages 35–41), only the records from ages 1 to 34 (the event time) should be analyzed when we fit the model.

Next, we fit the standard design-based logit regression model (Chapter 8) to the person-year data. The `svy: logistic` command assumes that the complex sample design features of the NCS-R data set have already been declared to Stata (see the `svyset` command in Section 10.4.5). However, unlike the CPH model, Stata's `stset` command is *not* required. The following predictors of the event are included in the discrete time logit model: PYR (person-year), AGE (age at interview), SEX, ED4CAT, RACECAT, and MAR3CAT. We note that including AGE in the model serves to introduce a cohort effect in the model; in other words, do persons of different ages at the time of the interview have different hazards?

```
svy: logistic mdetv pyr age ib2.sex i.ED4CAT ///
i.racecat i.MAR3CAT if pyr <= ageonsetmde
```

The use of the `if` modifier is especially important in the model commands; we only want to analyze those person-years for times equal to or less than the age of MDE onset (or age of interview for right-censored individuals never having experienced an MDE). Fitting the logit version of the model results in the estimates of odds ratios presented in Table 10.5.

Focusing on the results for the discrete time logit model, we see that the estimated odds ratios in this discrete time hazard model are remarkably similar to the estimated

TABLE 10.5

Estimated Odds Ratios for the NCS-R Discrete Time Logit Model of Age to Onset of a Major Depression Episode

Predictor*	Estimated Odds Ratio	Estimated SE	*t*-statistic (df)	*p*-value	95% CI
Education					
12 Years	0.98	0.06	−0.30 (42)	0.76	(0.86, 1.12)
13–15 Years	1.10	0.06	1.62 (42)	0.11	(0.98, 1.23)
16+ Years	0.98	0.06	−0.31 (42)	0.76	(0.86, 1.11)
Race/Ethnicity					
Hispanic	0.78	0.11	−1.84 (42)	0.07	(0.59, 1.02)
Black	0.63	0.09	−3.05 (42)	<0.01	(0.47, 0.86)
White	1.08	0.13	0.63 (42)	0.54	(0.85, 1.37)
Marital Status					
Previously Married	1.64	0.10	8.10 (42)	<0.01	(1.45, 1.85)
Never Married	0.97	0.08	−0.40 (42)	0.69	(0.81, 1.15)
Gender					
Male	0.64	0.04	−7.14 (42)	<0.01	(0.57, 0.73)
Person-year	1.03	0.002	15.82 (42)	<0.01	(1.03, 1.04)
Age	0.94	0.002	−23.82 (42)	<0.01	(0.94, 0.95)

Source: $n = 385{,}696$, Design-adjusted F-test of null hypothesis that all parameters are 0: $F(11, 32)=53.63$, $p<0.001$.
Note: * Reference categories for categorical predictors are: <12 yrs (Education); Asian/Other (Ethnicity); Married (Marital Status); Female (Gender).

hazard ratios found for the CPH model fitted in Section 10.4.5, which suggests that the two approaches result in similar estimates of the relationships of these socio-demographic predictors with the risk of MDE onset at any given age. For example, the discrete time hazard model suggests that holding all other covariates fixed (including person-year, or the age at which a person might be at risk), males have 36% lower odds of having an MDE for the first time than females.

Given that the discrete time hazard modeling approach can be implemented using standard design-based logistic regression modeling techniques, any of the diagnostics discussed in Chapter 8 could be applied when evaluating the fit of this model. We also note that this makes Bayesian approaches to fitting discrete time hazard models possible, using the exact same approach that was illustrated for logistic regression models in Chapter 8. Analysts just need to make sure that the data set has been structured appropriately for the discrete time modeling first; after this, the exact same alternative Bayesian approach to fitting design-adjusted logit models can be applied.

Finally, the comparable discrete time complementary-log-log (C-L-L) model is fit to the person-time data using Stata's `svy: cloglog` command:

```
svy: cloglog mdetv pyr age ib2.sex i.ED4CAT ///
i.racecat i.MAR3CAT if pyr <= ageonsetmde, eform
```

Table 10.6 presents the results for the C-L-L model in the form of estimates and confidence intervals for the hazard ratios (note the use of the `eform` option in the command above).

TABLE 10.6

Estimated Hazard Ratios for the NCS-R Discrete Time C-L-L model of Age to Onset of a Major Depression Episode

Predictor*	Estimated Hazard Ratio	Estimated SE	*t*-statistic (df)	*p*-value	95% CI
Education					
12 Years	0.98	0.06	−0.30 (42)	0.77	(0.86, 1.12)
13–15 Years	1.10	0.06	1.61 (42)	0.11	(0.98, 1.23)
16+ Years	0.98	0.62	−0.30 (42)	0.76	(0.86, 1.11)
Race/Ethnicity					
Hispanic	0.78	0.10	−1.84 (42)	0.07	(0.60, 1.02)
Black	0.63	0.09	−3.04 (42)	<0.01	(0.47, 0.86)
White	1.08	0.13	0.63 (42)	0.54	(0.85, 1.37)
Marital Status					
Previously Married	1.64	0.10	8.11 (42)	<0.01	(1.45, 1.85)
Never Married	0.97	0.08	−0.41 (42)	0.69	(0.81, 1.15)
Gender					
Male	0.64	0.04	−7.14 (42)	<0.01	(0.57, 0.73)
Person-year (time)	1.03	0.002	15.82 (42)	<0.01	(1.03, 1.04)
Interview Age	0.94	0.002	−23.83 (42)	<0.01	(0.94, 0.95)

Sourec: Design-adjusted F-test of null hypothesis that all parameters are 0: $F(11, 32) = 53.65$, $p < 0.001$.

Note: * Reference categories for categorical predictors are: <12 yrs (Education); Asian/Other (Ethnicity); Married (Marital Status); Female (Gender).

Compared to the results in Table 10.4 for the CPH model (hazard ratios) and the discrete time logit model results in Table 10.5 (odds ratios), we again see a similar pattern of results in the estimated hazard ratios for the C-L-L discrete time model for age of onset of major depression. Adjusting for the other predictors and a linear time relationship, education level does not appear to have a significant relationship with the hazard of MDE. Black individuals have a significantly lower hazard for MDE relative to persons of Asian and Other race/ethnicities. Adjusting for the other predictors in the model, previously married persons have a greater hazard of MDE than married adults, and the estimated hazard for men is about 64% of that for women (all else being equal).

10.6 Methods for Left- and Interval-Censored Data

The methods and examples presented thus far in this chapter have primarily focused on the case of right-censored time-to-event data collected in surveys. As mentioned earlier, left censoring can also occur (where, for example, an event may occur prior to the onset of measurement), in addition to **interval censoring** (where, for example, an event occurs sometime between measurement occasions, but not at either occasion exactly). Parametric time-to-event models are available for such cases, and Stata has fortunately adapted these types of models to the design-based analysis setting.

The command available in Stata for this type of design-based time-to-event modeling is `svy: stintreg`. Unlike the earlier approaches demonstrated in this chapter, this command does not rely on the use of `stset` to define the time-to-event data structure. Instead, this command requires the specification of two `interval` variables that represent measures of time, as follows:

```
svyset . . .
svy: stintreg x1 x2, interval(ltime rtime) ///
distribution(weibull)
```

In this example, x1 and x2 represent covariates in the model. Note the required `interval()` option, with the required variables `ltime` and `rtime`. If a given survey respondent does not have censored data, both variables should be equal to the time point at which the event occurred. If a given survey respondent has left-censored data, the `ltime` variable should be set to missing or zero, and the `rtime` variable should be equal to the first time point of measurement (before which the event occurred). If a given survey respondent has interval-censored data, the `ltime` variable should be equal to the lower endpoint of the interval during which the event occurred, and the `rtime` variable should be equal to the upper endpoint of the interval during which the event occurred. Finally, if a given survey respondent has right-censored data, the `ltime` variable should be equal to the last time point at which a measurement was collected (without the event occurring), and the `rtime` variable should be set to missing.

We also note that this command requires the specification of a parametric distribution for the failure times. We simply used Weibull in this case to illustrate the command. Other popular options for parametric survival models (such as exponential and Gompertz) are also available. Finally, as with all other `svy`: commands, the complex sample design features would need to be declared first in `svyset` to implement the design-based analysis approach. Stata provides a nice tutorial on the interpretation of these types of parametric models at https://www.stata.com/stata-news/news33-1/spotlight-stintreg/.

10.7 Exercises

10.1) This exercise considers the NCS-R data. The objective of this analysis is to model the gender-specific hazard of first onset of a social phobia (DSM_SO, 1=Yes, 5=No); in other words, do males and females in the NCS-R population have different hazards of onset of a social phobia at any given age?

Begin the analysis by generating an age of first onset of social phobia variable, equal to the age of onset of social phobia (SO_OND) if DSM_OND is equal to 1 (or Yes), or equal to age at interview (AGE) if DSM_OND equals 5 (or No), or missing. Note that cases not qualifying for social phobia are *right-censored* cases in terms of age of social phobia onset, so you should also compute an indicator variable for the censored cases. Present the code that you used to compute these variables, which will serve as the "time-to-event" variable and the right-censoring indicator in this analysis.

10.2) Generate weighted (using the NCS-R Part II sampling weight = NCSRWTLG) Kaplan-Meier estimates of the survival curves for each gender by using the indicator variable SEXM (1 = Male, 0 = Female). What do the survival curves suggest about the hazards of initial social phobia diagnosis as a function of gender?

10.2a) **(Extra Credit)** Use a software procedure capable of producing design-based standard errors and confidence limits (for example, the `svykm()` function in the R `survey` package, or SUDAAN) for Kaplan-Meier estimates. Generate weighted (using NCSRWTLG) survival curves with correct standard errors and confidence limits for the two groups defined by the variable SEXM. Again, what do the survival curves suggest about the hazards of initial social phobia diagnosis as a function of respondent gender?

10.3) Fit a Cox Proportional Hazards (CPH) model to the age of onset of social phobia data, recognizing the right censoring and the complex sample design features of the NCS-R (stratum codes = SESTRAT, cluster codes = SECLUSTR, NCS-R Part II sampling weight = NCSRWTLG). Consider as predictors of the hazard function the categorical education variable representing highest education level attained (ED4CAT: 1=0–11 Years, 2=12 Years, 3=13–15 Years, and 4=16+ Years) and the indicator variable for males (SEXM). Use the education variable as though it is time-invariant in the model. Based on the weighted estimates of the regression parameters in this model and their estimated standard errors:

a. Is there evidence of significant differences between gender categories and the education categories in terms of the hazard functions? If so, what are the differences?

b. Compute the estimated hazard ratio of initial social phobia onset for females compared to males, holding the education variables constant, and compute a 95% confidence interval for this hazard ratio. How would you interpret these results in plain English?

10.4) **(Using Stata or other software capable of design-based visualization)** Test the assumption of proportional hazards for the two gender groups using the graphical methods discussed in this chapter. Does this assumption seem justified? What would be an alternative analytic approach if this assumption seems violated?

10.5) Using the data management methods discussed in this chapter, construct an expanded version of the NCS-R data set that could be used to fit a discrete time logit model to the age of onset of social phobia data. Generate a table showing the data structure for the first two respondents in the NCS-R data set.

10.6) Fit a discrete time logit model to the expanded data set considering the same predictors of first onset of social phobia used to fit the CPH model, ED4CAT and SEXM. Make sure to recognize the complex sample design features when fitting the model, and don't forget to include both person-year and interview age (AGE) as predictors in the model (see Section 10.5.5). Also, make sure to limit the records analyzed to include the records up to and including the age of onset or censoring. Based on the data structure and output from this model, answer the following questions:

a. Why do we include the interview age as a predictor in this model?
b. What mistake would occur if you skipped the limiting of observations to up to and including the age of onset/censoring?
c. Would we make the same inferences as from the CPH model about the differences in the hazards for the four education groups and for males compared to females?
d. Prepare a table comparing the estimated odds ratios from the discrete time logit approach to the estimated hazard ratios from the CPH modeling approach, and write a short paragraph interpreting these results.
e. Does anything change about your inferences when using a Bayesian approach to fit the discrete time logit model? See Chapter 8 for details.

11

Analysis of Longitudinal Complex Sample Survey Data

11.1 Introduction

In this chapter, we provide an overview of the newest literature on appropriate analytic techniques for **longitudinal data** collected in a **panel survey**, where the initial sample design was complex in nature (possibly featuring stratification, cluster sampling, and weighting for unequal probability of selection, as discussed in prior chapters), and attempts were made to repeatedly measure the initially sampled units at multiple follow-up waves. Many different patterns of missing data can emerge in panel surveys featuring repeated measurement of the same units over time; for example, some cases may provide complete data at every measurement occasion, others may choose not to participate at particular waves, and others may drop out of the panel survey permanently at a particular time point. This introduces the possibility of several alternative design-based and model-based approaches that simultaneously account for: 1) the correlation of the repeated measurements on a given individual; 2) the complex sampling features; and 3) the different patterns of nonresponse (Thompson, 2015). With this chapter, we attempt to describe and illustrate those methods that are advocated in the current literature as being most appropriate for a particular analytic objective.

A variety of survey designs can be used to study changes in selected population phenomena over time. **Repeated cross-sectional surveys** such as the NHANES, which are defined by the cross-sectional measurement of repeated independent samples from a given population over time, enable researchers to study marginal changes in the features of a population over time. **Rotating panel designs** such as the Current Population Survey and true **panel survey** designs such as the Panel Study of Income Dynamics [PSID] (Hill, 1992) and the Health and Retirement Study [HRS] (Juster & Suzman, 1995) all play an important role in understanding both societal and *individual-level* change over time. Binder (1998) and Kalton and Citro (1993) provide sound overviews of the key issues associated with longitudinal sample designs, both at the design and analysis stages. Menard (2008) and Lynn (2009, 2021) review state-of-the-art research on design, measurement, and analysis techniques for longitudinal surveys. In this chapter, we focus on appropriate techniques for the analysis of data collected from panel surveys, where the same individuals are measured repeatedly over time and estimates of expected trajectories in measures of interest for individuals in selected population subgroups are the scientific focus.

We first consider the possible analytic objectives that analysts of panel survey data may have in mind in Section 11.2. For each of these objectives, we consider potential analytic techniques proposed in the literature that one might employ and describe their

DOI: 10.1201/9781003472094-12

implementation in detail. Finally, we consider several examples of analyzing real longitudinal survey data from the Health and Retirement Study (HRS) in Section 11.3, illustrating the use of the alternative methods with existing software and comparing the results across the competing methods that have been proposed in the extant literature. We conclude with some thoughts on directions for future research in Section 11.4.

11.2 Alternative Analytic Objectives with Longitudinal Survey Data

This section reviews possible analytic objectives that those working with longitudinal survey data may have in mind. A clear analytic objective is essential for choosing an appropriate method for analyzing the longitudinal survey data.

11.2.1 Objective 1: Descriptive Estimation at a Single Time Point

Analysts with this objective in mind are interested in making descriptive inferences about a specific target population at a particular point in time. For example, an analyst of the HRS data may be interested in describing the mean total household assets of the HRS target population in 2012, or estimating the proportion of individuals in the HRS target population who chose to pursue part-time employment in 2012. These objectives involve a desire to make descriptive inference in a *single wave* from the overall panel survey, and therefore rely on methods quite like those discussed in Chapter 5 and Chapter 6.

The key complication of these approaches is how to account for the fact that many of the individual elements initially sampled for the panel survey may not respond in the specific wave of interest. For this type of **wave nonresponse**, three primary approaches are generally advocated in the literature (Kalton, 1986; West & McCabe, 2017):

1. **Weight adjustment.** Adjust the final survey weights from the initial wave of the panel survey, which likely reflect a) unequal probabilities of selection, b) differential unit nonresponse in the initial wave, and c) calibration adjustments given population controls from the initial wave, for the conditional probability of responding at the follow-up wave in question (given that a sampled case responded in the initial wave). This approach assumes that nonresponse in the current wave is occurring at random *conditional on the observed measures recorded in earlier waves* and that wave nonresponse does *not* depend on the measures being recorded in the current wave of interest. It is also crucial when using this approach that *all* cases who responded in the wave of interest be analyzed, so that relevant data are not unnecessarily deleted (e.g., when only analyzing cases that may have responded in *all* waves of the longitudinal study; see Si et al., 2022).
2. **Multiple imputation.** Impute the values of the measures of interest in the current wave multiple times using measures recorded in previous waves, including the initial wave, and then apply multiple imputation (MI) combining rules (Little & Rubin, 2019) to the results of analyses of the imputed data sets to make overall inferences (see Chapter 12 for more details regarding multiple imputation). The analyses of the multiple imputed data sets would account for the final survey weights from the initial wave of the panel survey, in addition to any available

stratification codes, cluster codes, or replicate weights for variance estimation. Like the weight adjustment method, this method assumes that wave nonresponse is occurring at random *conditional on the predictors used in the imputation models.* See Kalaycioglu et al. (2016) for a discussion of the effectiveness of alternative MI approaches for handling wave nonresponse depending on the types of outcome variables being analyzed.

3. **Methods for non-ignorable missing data.** This class of methods does *not* assume that wave nonresponse is occurring at random conditional on measures observed in previous waves, but rather allows the wave nonresponse to depend on some arbitrary function of the measures of interest in the current wave. Popular approaches to making descriptive inferences in this case include **pattern-mixture models** and **selection models** (Little & Rubin, 2019; West & McCabe, 2017; Andridge et al., 2019; Little et al., 2020), but these approaches all rely on assumptions regarding the relationship of wave nonresponse with the measures of interest in the current wave. For this reason, **sensitivity analyses** define a critical component of analyses assuming that the wave nonresponse is not ignorable, where the sensitivity of inferences to different assumptions about the wave nonresponse mechanism is carefully examined (Andridge et al., 2019; Little et al., 2020).

Simple complete-case analyses of the data in a current wave that only account for complex sampling features from the initial wave of data collection assume that the wave nonresponse is occurring *completely at random*, independent of measures collected in earlier waves or the measures collected in the current wave. This is an extremely strong assumption, and one that often proves to be incorrect. We encourage analysts of longitudinal survey data with these descriptive objectives in mind to employ each of the approaches outlined above and examine the sensitivity of their inferences to the assumptions underlying each approach. We illustrate this approach in Section 11.3.

11.2.2 Objective 2: Estimation of Change across Two Waves

Analysts with this objective in mind are interested in making inference about the change in a descriptive parameter of interest over a period defined by two waves of the larger panel survey. For example, an HRS analyst may be interested in making inference about the mean change in total household assets in the HRS target population from 2016 to 2018. If all panel survey respondents initially sampled were to provide complete information for the measures of interest at each of the two waves, this analysis would be straightforward: one would compute the change measure of interest for each respondent, save the change measure in a new variable, and apply the standard descriptive techniques discussed in Chapter 5 and Chapter 6, accounting for the complex sampling features from the initial wave of data collection.

More specifically, the simplest type of analysis is the analysis of individual change over two points in time denoted by t and $t-k$, for example,

$$\bar{\delta}_{t-k,t} = \frac{\sum_i w_i \cdot (y_{t,i} - y_{t-k,i})}{\sum_i w_i} \tag{11.1}$$

where w_i is a survey weight for sample elements i;

or,

$\hat{y}_{t,i} = \hat{\beta}_0 + \hat{\gamma} y_{t-k,i} + \hat{\beta}_1 x$; where $\hat{\gamma}, \hat{\beta}_1$ are weighted estimates of regression parameters

We considered an example of analyzing the mean change between two points in time in Example 5.16 from Chapter 5.

In real panel surveys, this ideal scenario is seldom realized, and analysts need to account for the differential attrition that produced the ultimate survey respondents in each of the two waves of interest. Analysts will likely encounter one of two scenarios in this case:

1. **If the measure of change is relative to the initial wave of data collection**, one of the three methods outlined for Objective 1 in Section 11.2.1 can be employed, where the weights for cases with complete data at both time points are adjusted for wave nonresponse at the follow-up wave, or missing values at the follow-up wave are first imputed assuming that wave nonresponse is either occurring at random or not at random, and then the resulting complete cases are analyzed further accounting for the complex sampling features at the initial wave. These approaches are illustrated in detail by West and McCabe (2017), and considered further in Section 11.3.
2. **If the measure of change is based on two waves aside from the initial wave of data collection**, where differential wave nonresponse in each of the two waves of interest is a real possibility, approaches based on imputation are generally recommended (Kalton, 1986). In these cases, one can impute missing values in both waves of interest multiple times using measures recorded in earlier waves, and then analyze the measures of change in the imputed data sets accounting for the complex sampling features from the initial wave of data collection.

We once again reiterate that analyses of change across two waves that do not account for the possibility of differential attrition over time run the risk of being substantially biased, as respondents providing complete data over time may represent a selected subset of the full initial sample of elements with unique properties. We encourage analysts to use one of the methods outlined above when estimating measures of change across two waves, and we illustrate these methods in Section 11.3.

11.2.3 Objective 3: Trajectory Estimation Based on Three or More Waves

In this third case, analysts have explicit interest in estimating expected within-person trajectories for outcomes over time, using three or more waves of information to estimate the trajectories (where some individuals may only be able to contribute fewer than three waves). This is the most complicated scenario, where different individuals may respond in different combinations of waves, and there may be many different patterns of attrition to account for. We now consider five general classes of approaches that have been proposed in the literature for fitting models to these types of longitudinal survey data.

11.2.3.1 Approach 1: Weighted Multilevel Modeling

Following this first approach, one would begin by creating a vertical data set, containing multiple measurements (or rows) per survey respondent (across the waves) and multiple respondents per sampling cluster (if applicable), and then employ approaches for weighted estimation of multilevel models (Pfeffermann et al., 1998) to estimate both a) the overall expected trajectories of interest, and b) between-subject variance in the trajectories of interest (Skinner & Holmes, 2003). The complication in this case is how the appropriate weights should be computed for weighted estimation of the multilevel models. This approach would proceed as follows:

a. The Level 1 weights are generally the wave-specific weights associated with each individual responding in a given wave, incorporating different adjustments for attrition in each wave. The calculations of these wave-specific weights for each responding individual need to reflect appropriate adjustments to the initial survey weights for each individual for differential attrition patterns at each wave, so that they effectively recover the original probability sample in each wave. These wave-specific weights also need to be appropriately scaled so that their sum is equal to the available number of data points within each individual; a failure to do this can introduce bias in the weighted estimates of multilevel model parameters (Pfeffermann et al., 1998; Rabe-Hesketh & Skrondal, 2006; Carle, 2009).
b. The baseline survey weight represents the subject-specific survey weight, where subjects would generally define Level 2 of the multilevel model. If there is no nesting of individuals within sampled clusters, these weights do not need to be computed conditionally or scaled. If individuals *are* nested within sampling clusters (which would define Level 3 of the multilevel model), the weights should be computed *conditionally*, reflecting the conditional probabilities of selection for individuals within a randomly sampled cluster. They would also need to be scaled within each higher-level cluster; again, one option would be to normalize the conditional weights so that they sum to the sample size of individuals within a given cluster (Carle, 2009).
c. To model sample design clustering, primary sampling units (if available) can define the third level of the multilevel model. The Level 3 weights for each PSU will be the reciprocal of the primary stage probability of selection for the PSU. If PSU-level weights are provided by the data producer, they need to be accounted for in the estimation. When PSUs are treated as Level 3 of the multilevel model, the subject-specific weights at Level 2 need to be *conditional* weights, as noted above (see also Theory Box 11.1), reflecting probabilities of selection conditional on a given PSU being selected. If PSU-level weights are *not* provided by the data producer, one can assume that a simple random sample of PSUs was selected (where each PSU would receive a weight of 1), but this may introduce bias in the estimates of the between-cluster variance based on the multilevel model. Primary stage stratification effects are generally accounted for via the inclusion of fixed stratum effects in the third level of the multilevel model (Pfeffermann et al., 1998).

Multilevel models can extend the simple "two-point" change models discussed in Section 11.2.2 to examine trends in repeated measures or growth curves based on multiple observations collected from the individual sampled persons (or units of analysis more generally). In this section, we present a synthesis of the existing literature on suggested approaches to using multilevel models for longitudinal analysis of complex sample survey data (see Hsu et al., 2018, for a more in-depth review). We then consider examples of these approaches using available software in Section 11.3.

We first describe general notation for multilevel models that can be fitted to longitudinal survey data arising from complex sample designs. Let y_{it} be the response for individual i at time t, which may vary randomly over time. We assume that the randomly sampled individual i has some 'permanent' status that partly defines the observed responses, which can be denoted by b_i. These values for the randomly sampled individuals are assumed to arise from a normal distribution with a fixed mean (for the population) of 0 and a variance of τ^2 (see McCulloch & Neuhaus, 2011, for more discussion about this assumption). Then, observed values for y_{it}, conditional on a randomly sampled value of b_i, are assumed to arise from a normal distribution with mean $\beta_t + b_i$ (where β_t is the population mean of the response at time t) and variance σ^2. More succinctly,

$$\begin{aligned} b_i &\sim N(0,\tau^2) \\ y_{it} \mid b_i &\sim N(\beta_t + b_i, \sigma^2) \end{aligned} \tag{11.2}$$

Thinking in a multilevel modeling context, τ^2 represents between-individual variance, while σ^2 represents within-individual variance over time. The variance of y_{it} is thus partitioned into variance across individuals and transitory variance, and can be written as $\tau^2 + \sigma^2$.

A simple model for an outcome y measured over time can then be written as follows:

$$y_{it} = \beta_t + b_i + e_{it} \tag{11.3}$$

In this model, $t = 1, \ldots, T$, where T represents the total number of waves, $i = 1, \ldots, n$ (persons), β_t is the population mean of the outcome at time t, b_i is a long-term difference from β_t for person i (a random variable corresponding to a random individual effect), and e_{it} is a transitory effect at time t for person i (each individual will randomly vary around their 'permanent' state at time t).

It is important to note that the e_{it} values may be correlated over time (Veiga et al., 2014). One possible model expressing this correlation could be the **first-order autoregressive [AR(1)] model**:

$$e_{it} = \rho e_{i(t-1)} + \varepsilon_{it} \tag{11.4}$$

The model in (11.3) can thus be rewritten as follows, using the AR(1) model for the transitory effects:

$$y_{it} = \beta_t + b_i + \rho e_{i(t-1)} + \varepsilon_{it} \tag{11.5}$$

In this specification, ε_{it} represents random measurement error, and the b_i and ε_{it} are assumed to be independent. We also assume that $E(b_i) = E(\varepsilon_{it}) = 0$, $var(b_i) = \tau^2$, $var(\varepsilon_{it}) = \sigma^2_{\varepsilon}$, and that the e_{it} and ε_{it} are mutually independent and stationary. We can then write:

$$var(e_{it}) = \frac{\sigma^2_{\varepsilon}}{(1-\rho^2)} \tag{11.6}$$

In this section, we focus on a multilevel modeling approach that has been developed to fit models of this form to longitudinal data from complex sample surveys (Skinner & Holmes, 2003). To fit this model, we must first address the question of how the longitudinal survey weights should be incorporated in the model estimation. Pfeffermann et al. (1998) laid the groundwork for appropriate methods of incorporating sampling weights into the estimation of parameters in multilevel models for complex sample survey data. They also proposed appropriate variance estimators for the estimated parameters (based on the delta method, which is like linearization). The methods proposed by these authors can be used to estimate two-level multilevel models for longitudinal survey data given survey weights computed for *both* the Level 2 units (individuals) and the Level 1 units (repeated observations). Theory Box 11.1 describes the procedure for constructing the weights for the individual levels in detail.

Unfortunately, the multilevel modeling approach proposed by Pfeffermann et al. (1998) does not allow for auto-correlation of transitory effects within persons, as described in (11.5) above. The original approach was described for the case where Level 2 units were PSUs and Level 1 units were sampled individuals, and auto-correlated errors from different individuals within the same cluster would not be expected or realistic. Skinner and Holmes (2003) developed a method to circumvent this problem and enable the methods proposed by Pfeffermann et al. (1998) to be applied. Veiga et al. (2014) further expanded on this approach and have developed both Stata and R software implementing it. We include examples of fitting these types of models on the ASDA website.

THEORY BOX 11.1: WEIGHTED ESTIMATION IN TWO-LEVEL MULTILEVEL MODELS.

Following Pfeffermann et al. (1998), the final survey weights for individuals at level 2 and repeated observations at level 1 are computed as follows. In general, given a longitudinal study design with these two levels, the appropriate weight for the data from individual i at time t, w_{it}, is computed as

$$w_{it} = \frac{1}{\pi_i \pi_{t|i}}, \quad (11.7)$$

where π_i represents individual i's original probability of inclusion in the sample, and $\pi_{t|i}$ represents the probability that individual i responds at time t. The $\pi_{t|i}$ values might be estimated using a propensity modeling approach (e.g., Lepkowski & Couper, 2002), where the probability of responding at time t is predicted using a logistic regression model with predictors from previous waves. We assume that some reasonable methodology has been used for estimation of these response probabilities in this discussion.

For purposes of the multilevel modeling approach discussed in this section, define the **level 2 weight** as

$$w_i = \frac{1}{\pi_i}, \quad (11.8)$$

and define the **level 1 weight** as

$$w_{t|i} = \frac{1}{\pi_{t|i}}. \quad (11.9)$$

Then, we have

$$w_{it} = w_i w_{t|i} \tag{11.10}$$

and

$$w_{t|i} = \frac{w_{it}}{w_i}. \tag{11.11}$$

A reasonable approach suggested by Skinner and Holmes (2003) is to let $w_i = w_{i1}$, or to let the level 2 weight for individual i be the inverse of the probability of being selected for the sample *and* responding at the baseline wave. Then, $w_{1|i} = w_{i1} \ / \ w_{i1} = 1$. This procedure will result in less variance in the level 1 weights denoted by $w_{t|i}$.

The computed level 1 weights can then be scaled, as recommended by Pfeffermann et al. (1998) and Skinner and Holmes (2003) to minimize small-sample estimation bias. Scaling weights (e.g., normalizing weights; see Chapter 4) generally does not have an impact on the estimates of parameters in *multivariable* models, but this is not the case when fitting *multilevel* models to complex sample survey data. Skinner and Holmes (2003, p. 212) suggest rescaling the level 1 weights described above as follows:

$$w^*_{t|i} = \frac{t^*(i) w_{t|i}}{\sum_{t=1}^{t^*(i)} w_{t|i}} \tag{11.12}$$

In this notation, $t^*(i)$ represents the last wave at which individual i provides a response. Note that this method ensures that the average rescaled weight for a given individual i is equal to 1. Rabe-Hesketh and Skrondal (2006) outline additional options for scaling the level 1 weights, one of which we consider in the example in this section. Carle (2009) examines the performance of these various scaling methods via empirical simulations, and arrives at the general conclusion that the choice of the weight scaling method does not have a significant impact on inferences; it simply needs to be done in one way or another. The main point is that some form of scaling needs to be employed to reduce potential biases in the estimates of the multilevel model parameters.

Rabe-Hesketh and Skrondal (2006) further investigated methods of fitting multilevel models to complex sample survey data (and specifically binary outcomes). Based on their research, these authors stressed the importance in pseudo maximum likelihood (PML) estimation (see Chapter 8) of incorporating appropriately scaled weights at *all* levels of a multilevel study (e.g., school-level, or level 2, weights and student-level, or level 1, weights, and not simply student-level weights). Addressing a limitation of the approaches proposed by Pfeffermann et al. (1998) and Skinner and Holmes (2003), these authors also discussed the computation of "sandwich" estimates of standard errors (based on a Linearization approach) to incorporate stratification and primary stage clustering of a complex sample into the calculation of design-based standard errors for the parameter estimates derived using the PML approach.

Their proposed method is implemented in the **`gllamm` (Generalized Linear Latent and Mixed Models) command** that these authors developed for Stata (visit http://www.gllamm.org), in addition to the mixed-effects modeling procedures in Stata (e.g., the `mixed` command). The underlying assumption of this approach is that the primary stage clusters represent the highest-level units in the multilevel design (e.g., students at Level 1 nested within schools at Level 2, and schools at Level 2 nested with primary stage clusters at Level 3). Via simulation, these authors found good coverage for confidence intervals based on the robust standard errors. Thinking about multilevel models for longitudinal survey data, we can apply the same models, where the repeated observations across waves define the Level 1 observations, the sampled individuals define Level 2, and the primary sampling units define level 3 of the data hierarchy. We consider an example of this type later in Section 11.3.

Skinner and Vieira (2007) specifically focused on the impacts of cluster sampling on variance estimation when analyzing longitudinal survey data with multilevel models. They concluded that the simple method of including a random cluster effect in a three-level multilevel model for longitudinal survey data (where again, primary sampling units define Level 3 of the multilevel study design, individuals define Level 2, and the repeated measurements define Level 1) has the potential to seriously underestimate effects of clustering on the standard errors of parameter estimates. In a practical application, model-based standard errors for key parameter estimates in the three-level model were nearly identical to those found in a two-level model *excluding* the random cluster effects, clearly suggesting that attempting to incorporate the effects of cluster sampling via random cluster effects and relying on model-based standard errors would be erroneous. However, these authors did not consider the effects of informative sampling weights at each level on parameter estimates in the models, where if the weights are informative (especially about the variance components), one may see different results.

Skinner and Vieira showed that calculation of 'robust' standard errors (Goldstein, 2003, p. 80) in a three-level multilevel model including random PSU/cluster effects at Level 3 resulted in standard errors that most closely resembled those computed by applying the classical linearization approach to the computation of design-based standard errors (incorporating clustering, and considered by the authors to be a 'gold standard') for parameter estimates in a *two-level* model for the longitudinal data, including random individual effects only (a method similar to that proposed by Rabe-Hesketh and Skrondal). The example provided by these authors, however, did *not* consider the effects of weighting or stratification, and the authors acknowledged that additional work was needed (and ongoing) in this area.

Additional work by Vieira and Skinner (2008) found that point estimators for the regression parameters and variance-covariance parameters in a model for longitudinal survey data based on pseudo maximum likelihood (PML) estimation, when combined with a variance estimator based on Taylor Series Linearization (allowing for clustering of the sample), had the best performance overall, providing more support for the work of Rabe-Hesketh and Skrondal (2006). However, their simulations only considered respondents at all possible waves of a panel survey ("complete" respondents, meaning respondents with any wave nonresponse were dropped), and once again ignored survey weights that might be adjusted for other types of attrition or wave nonresponse. We note that the general PML estimation method and linearization variance estimation method proposed by these authors can accommodate single sampling weights for the individuals, possibly adjusted for other forms of nonresponse, which varies from the multilevel modeling approach accommodating weights at both level 2 (sampled individuals) and level 1 (repeated observations) of the longitudinal hierarchy.

Critical to the success of the weighted multilevel modeling approach is the calculation of appropriate adjustments for attrition in each of the waves. This approach makes use of all available information collected in the panel survey, which is an attractive alternative to only analyzing cases with complete data and has been shown empirically to improve estimates of trajectories in a *marginal* (and not multilevel) modeling context (Si et al., 2022; West et al., 2024; see also Section 11.2.3.3). In the monotone missing data case, if one were to analyze only cases with *complete* data in all of the waves of interest, the risk of bias is again strong if those individuals represent a selected subset of the full sample in terms of the outcomes of interest. Some type of adjustment would seemingly be needed for the initial survey weights when following this type of complete case approach and throwing out the information for cases without complete data (e.g., Schmidt & Woll, 2017), but the existing literature on such adjustments is sparse. See Lynn and Watson (2021) for a recent overview of possible weighting strategies in this context.

The existing literature has examined how to account for stratified cluster sampling in variance estimation when fitting multilevel models to complete cases in longitudinal survey data sets (Feder et al., 2000), but the literature on computation of appropriate time-varying nonresponse adjustments for survey weights in this context is also relatively sparse. Da Silva and Skinner (2011) and Earp et al. (2018) describe potential approaches for estimating response propensity over time. Recently, West et al. (2024) studied this problem in more detail via simulation and demonstrated empirically that using *all available observations* from each respondent, with appropriate time-varying weighting adjustments for attrition, tends to produce the best estimates of trajectories. However, this work was performed in a marginal, and not multilevel, modeling context (see Section 11.2.3.3). We explore these different possible approaches in Section 11.3, but encourage more work in this area, especially with respect to optimal weighting adjustments for attrition when multilevel modeling is the analytic objective.

11.2.3.2 Approach 2: Covariance Structure Modeling

This second approach employs marginal covariance structure modeling to fit models to longitudinal data collected from complex sample surveys. Recall the basic model specified in (11.5), using the AR(1) model for the transitory effects:

$$y_{it} = \beta_t + u_i + \rho v_{i(t-1)} + \varepsilon_{it} \quad (11.13)$$

Now, let $y_i = (y_{i1}, \ldots, y_{iT})'$ be a vector of observations on the dependent variable of interest collected on the i-th individual, assuming no nonresponse. Then, we have

$$E(y_i) = \beta = (\beta_1, \ldots, \beta_T)' \quad (11.14)$$

and

$$var(y_i) = \tau^2 J_T + \sigma^2 V_T(\rho) \quad (11.15)$$

where J_T is a $T{\times}T$ matrix of 1s, and $V_T(\rho)$ is a $T{\times}T$ matrix with the (t,t') and (t',t) elements defined by $\rho^{(t'-t)}$, for $1 \le t \le t' \le T$.

This model therefore has T parameters for the mean of the outcome, but only three parameters for the variance of the outcome (σ^2_ε, σ^2_u, and ρ). Denote this set of three parameters by φ, and define the sample estimator of the variance-covariance matrix of the vector of observations on a given individual i as:

$$\hat{S} = \sum_{i \in s} w_i (y_i - \bar{y})(y_i - \bar{y})' / \sum_{i \in s} w_i \tag{11.16}$$

The mean in (11.16) is actually a vector of weighted estimates of means at the T time points. Then, let $\hat{A} = vech(\hat{S})$, or a vector containing the $T(T + 1)/2$ distinct elements of $\hat{S}$, and let $A(\varphi) = vech(var(y_i))$, or a vector containing the $T(T + 1)/2$ distinct elements of the variance-covariance matrix for y_i defined in (11.15) above. Estimation of the parameters defining the variance-covariance matrix then proceeds by obtaining a generalized least squares (GLS) estimator of φ, denoted by $\hat{\varphi}_{GLS}$, as the iterative solution that minimizes

$$\left[\hat{A} - A(\varphi)\right]' \hat{V}^{-1}(\hat{A})\left[\hat{A} - A(\varphi)\right], \tag{11.17}$$

where $\hat{V}(\hat{A})$ is a design-consistent estimator of the variance-covariance matrix of $\hat{A}$, possibly obtained using linearization or replication methods for variance estimation.

Given the generalized least squares estimate of the set of variance-covariance parameters denoted by φ, a goodness of fit statistic is readily obtained as

$$X_W^2 = \left[\hat{A} - A(\hat{\varphi}_{GLS})\right]' \hat{V}^{-1}(\hat{A})\left[\hat{A} - A(\hat{\varphi}_{GLS})\right], \tag{11.18}$$

where provided that the model is correct and the sample is large enough for $\hat{V}(\hat{A})$ to be a good approximation to the variance-covariance matrix of $\hat{A}$, then the goodness of fit statistic in (11.18) will be distributed as a chi-squared statistic with $k - 3$ degrees of freedom, with $k = T(T + 1) / 2$. It is important to note that $\hat{V}(\hat{A})$ may be unstable with a small number of primary sampling units in the design, and Skinner and Holmes (2003) present alternatives.

Skinner and Holmes (2003, pp. 209–210) also discuss two possible approaches to dealing with nonresponse due to attrition when using the covariance structure modeling approach. One approach involves only considering the attrition sample who responded at all points up to time T, and developing adjusted weights denoted by w_{iT} that incorporate both the probability of selection and the probability of responding until time T. The procedure described above can then be applied using the adjusted weights. These authors also describe a more sophisticated approach that we do not describe in detail here; interested readers can refer to the text for more details.

The covariance structure modeling approach offers the advantage of easily handling serial correlation and provides users with a goodness of fit test. Unfortunately, the covariance structure approach described here has yet to be readily implemented in any general-purpose statistical software packages. Further, Skinner and Holmes (2003) report evidence of bias in estimates of variance components when using the GLS approach described above with the V matrix estimated from the data.

Veiga et al. (2014) presented an alternative approach that combined the ideas of multilevel modeling and covariance structure modeling. Following this approach, we once again recognize longitudinal survey data collected from complex samples as having a three-level structure. At Level 3, we have the primary sampling units (PSUs, or ultimate clusters/SECUs; see Chapter 4), potentially having associated sampling weights. At Level 2, we have survey respondents nested within clusters, each with their own *conditional* survey weights. At Level 1, we have the repeated measures across waves, nested within the survey respondents. Veiga et al. (2014) propose fitting a multilevel model to data with this structure that includes random PSU effects (to capture within-PSU correlations), and *wave-specific* random effects of individuals (essentially, the random errors at each time point associated with the survey responses from individuals are allowed to have unique variances and covariances).

To ensure that the proposed model is identified and all parameters are estimable, the model does not include a fixed overall intercept or fixed effects of time-varying covariates (outside of time itself). This approach therefore enables estimation of changes in the means of a particular outcome variable over time for individuals in a target population, along with the fixed effects of time-invariant covariates and changes in the variances (and covariances) of the outcome measures over time, but prevents estimation of the fixed effects of other time-varying covariates.

Veiga et al. (2014) present an example of fitting this model (accounting for PSU-specific weights and conditional survey weights at the individual level) using an *unstructured* variance-covariance matrix for the random errors, allowing for unique variances of the errors at each of eight waves, and unique covariances of the errors for each pair of waves. They note that other more constrained covariance structures are possible, and find that the *Toeplitz* structure resulted in the most parsimonious model. They use the method of **probability-weighted iterative generalized least squares (PWIGLS)** to fit the models and employ a linearization-based variance estimation approach that is both design- and model-consistent, similar to Pfeffermann et al. (1998). However, they only consider the weights of respondents who have provided *complete data* at all of the waves (where these base sampling weights might be adjusted for differential probabilities of providing complete data to account for those respondents who did not provide complete data). Extensions of this approach that incorporate all available information for cases that have responded in at least one wave are still needed. These authors have implemented this model-fitting procedure in the Stata programming language Mata and the contributed package `pwlmm` for the R software. We include examples of using both of these tools on the ASDA website.

11.2.3.3 Approach 3: Weighted GEE Estimation

In this third approach, extensions of the generalized estimating equation (GEE) technique originally proposed by Liang and Zeger (1986) that use either case-specific or observation-specific weights to account for complex sampling features and different patterns of attrition over time (Robins et al., 1995; Fitzmaurice et al., 1995; Sutradhar & Kovacevic, 2000; Roberts et al., 2009; Carrillo et al., 2010) are used to estimate the parameters in a model of interest (which presumably would include, at a minimum, time as a predictor of interest). In this approach, which enables an analyst to model changes in the marginal mean of a dependent variable over time while also accounting for one of several possible correlation structures for the repeated measurements, baseline sampling weights (possibly adjusted for unit nonresponse in the initial wave) are adjusted by weights that account for the probability of having a particular *response pattern* over time, and individuals are treated as clusters for variance estimation purposes.

The general case-specific weighted GEE approach assumes that the initial wave includes the entire baseline sample of individuals (with corresponding base sampling weights that may be adjusted for the probability of responding at the initial wave), and proceeds as follows (Fitzmaurice et al., 1995; Raghunathan, 2016):

a. Create a vertical data set (with multiple rows per sampled case), including a response indicator for each wave following the initial wave, the response pattern in previous waves, time-invariant auxiliary variables for each case, and the values of any available lagged survey information from previous waves (see Table 11.1).
b. Use a well-specified logistic regression model to estimate the probability of responding for case i at wave t, denoted by p_{it}, conditional on the response pattern in previous waves (i.e., based on a model fitted to only those cases that have the same pattern of response indicators in previous waves as case i). Use the auxiliary variables and the lagged survey responses available in the vertical data set (not shown in Table 11.1) as predictors, and save the predicted probabilities. A failure to carefully specify the response propensity model for a given wave conditioning on prior wave information (under the assumption that missing responses are **missing at random**, conditional on the prior wave information) will lead to bias in the weighted GEE estimates.
c. Compute the probability of having a particular response pattern as the product of the estimated probabilities of responding (or not responding) in each wave, up until the final wave. For cases that do not respond in all waves, some of the estimated probabilities used to compute this product will be $(1 - p_{it})$, or the probability of *not responding* in a particular wave conditional on information from previous waves. (We note that the probability of dropping out permanently in a particular wave that is *not* the final wave can be represented as the product of all conditional probabilities of *not responding* from that wave forward.)
d. Invert the final case-specific probabilities of having a particular response pattern, and adjust the base sampling weights by these inverted predictions. Proceed to estimate the parameters in the model of interest using GEE with each case's contribution to the estimating equation multiplied by the adjusted case-specific weight.

We note that cases 1, 2, and 3 in the hypothetical example in Table 11.1 have a **monotone pattern of missing data**, while case 4 has a **non-monotone pattern of missing data**. In longitudinal surveys with several waves, many non-monotone patterns could be possible. Raghunathan (2016, pp. 138–139) argues that a multiple imputation approach may be more effective in these non-monotone scenarios, given the difficulty of estimating the probabilities of responding in particular waves conditional on unique patterns of response outcomes in previous waves that are necessary for this case-specific weighting approach. The difficulty arises from the number of unique patterns of response outcomes in previous waves that would arise in these non-monotone scenarios, meaning that some of the logistic regression models would be fitted to small subsets of cases. We discuss this recommended type of multiple imputation approach for non-monotone patterns of missing data in Section 11.2.3.4.

Notably, weighted GEE is a **marginal modeling approach**, enabling inference about the average trajectories of interest in the target population, but not inference about subject-specific variability in the trajectories of interest. If an analyst of longitudinal survey data with a monotone missing data pattern is interested in estimating between-subject variability

TABLE 11.1

An illustration of the data preparation required to compute case-specific weights for the weighted GEE approach, considering a longitudinal study with four waves.

Case ID	Wave	Response Indicator (Outcome)	Predicted Probability of Outcome	Cumulative Probability	Lagged Cumulative Probability	Response Pattern in Prior Waves*	Final Case-Specific Weight Adjustment
1	2	1	p_{12}	p_{12}	—	—	—
1	3	1	p_{13}	$p_{12} \times p_{13}$	p_{12}	1	—
1	4	1	p_{14}	$p_{12} \times p_{13} \times p_{14}$	$p_{12} \times p_{13}$	1, 1	$1 / (p_{12} \times p_{13} \times p_{14})$
2	2	1	p_{22}	p_{22}	—	—	—
2	3	0	$1 - p_{23}$	$p_{22} \times (1 - p_{23})$	p_{22}	1	—
2	4	0	$1 - p_{24}$	$p_{22} \times (1 - p_{23}) \times (1 - p_{24})$	$p_{22} \times (1 - p_{23})$	1, 0	$1 / [p_{22} \times (1 - p_{23}) \times (1 - p_{24})]$
3	2	1	p_{32}	p_{32}	—	—	—
3	3	1	p_{33}	$p_{32} \times p_{33}$	p_{32}	1	—
3	4	0	$1 - p_{34}$	$p_{32} \times p_{33} \times (1 - p_{34})$	$p_{32} \times p_{33}$	1, 1	$1 / [p_{32} \times p_{33} \times (1 - p_{34})]$
4	2	1	p_{42}	p_{42}	—	—	—
4	3	0	$(1 - p_{43})$	$p_{42} \times (1 - p_{43})$	p_{42}	1	—
4	4	1	p_{44}	$p_{42} \times (1 - p_{43}) \times p_{44}$	$p_{42} \times (1 - p_{43})$	1, 0	$1 / [p_{42} \times (1 - p_{43}) \times p_{44}]$

Note: * The response pattern in prior waves is used to define the subset of cases that will be used to fit the logistic regression model predicting the probability of responding in the current wave for this particular case. In Wave 2, all cases will be used to estimate the probability of responding in Wave 2.

in trajectories of interest, the weighted multilevel modeling approaches discussed earlier in Section 11.2.3.1 need to be employed. This approach also assumes that the missing data are arising at random, conditional on the covariates included in the response propensity models underlying the weighted GEE method. We consider examples of the weighted GEE approach using case-specific weights later in Section 11.3.

Extensions of the weighted GEE approach allowing for time-varying weights (rather than weights assumed to be constant for a given individual over time) have also been developed and are currently implemented in the SAS and SUDAAN software (Carrillo & Karr, 2013; Thompson, 2015). SAS implements this approach in the GEE procedure, available in releases 9.4 and above, which allows users to model the conditional probabilities of response using the MISSMODEL keyword. SUDAAN uses GEE when fitting regression models with weights by default and requires users to compute their own time-varying weights as input (per Table 11.1). This approach is especially convenient for accommodating different patterns of attrition and makes maximum use of the available information collected over time from respondents, which has repeatedly been demonstrated to be optimal for marginal estimation of trajectories (Si et al., 2022; West et al., 2024).

Following this approach, the time-varying weights for *cases responding at a particular wave* would simply be the base sampling weights from the initial wave, multiplied by the inverted value of the cumulative product of the conditional probabilities of responding (or not responding) up to a particular time point where a case responded (see the "Cumulative Probability" column in Table 11.1, where only cases with the response indicator equal to 1 would be analyzed). The weighted estimating equations then simply allow each individual contribution from a responding sample case at each time point to be weighted by this wave-specific weight. In addition to the implementations in SAS and SUDAAN, this type of weighted GEE approach with time-varying weights can also be implemented using standard design-based regression modeling procedures in other software, treating the individual as the sampling cluster and using TSL for variance estimation (Si et al., 2022; West et al., 2024). Doing so only allows for an exchangeable working correlation structure for the repeated observations within a respondent.

11.2.3.4 Approach 4: Multiple Imputation Analysis

This fourth approach begins with all available longitudinal survey data in a wide format, with one row per panel survey respondent, and uses sequential regression imputation techniques (Raghunathan et al., 2001; Raghunathan, 2016; see Chapter 12 for details) to impute missing values for respondents that did not provide information in particular waves. Missing values on key variables are imputed multiple times based on well-specified models for the variables with missing data, which are generally fitted using the respondents providing complete data in all of the waves. These imputation models take advantage of the fact that the same variables are measured in different waves, where an imputation model for a key outcome in a later wave would likely benefit from having a measure of that same outcome in a previous wave as a predictor.

After this imputation process has been repeated multiple times, resulting in multiple complete wide-format data sets, one would fit the model of interest to each imputed data set (possibly a weighted GEE model, using the final Wave 1 survey weights for each individual, or a multilevel model, with the baseline survey weights serving as Level-2 weights and the Level-1 weights for each individual set to 1.0, given the "complete" data), and then combine the estimates from each imputed data file into a single set of estimates and design-adjusted standard errors (Little & Rubin, 2019; Sikkel et al., 2009).

This approach clearly relies on well-specified imputation models, and may be limited if the individuals providing complete data in all of the waves represent a very specific subpopulation (where the imputation models being used may not apply to everyone in the population of interest). Careful assessment of the quality and fit of the imputation models is key to the performance of this approach, and existing software tools now allow analysts to impute missing values based on assumed non-ignorable missing data mechanisms (e.g., West & McCabe, 2017; West & Little, 2013) to examine the sensitivity of multiple imputation estimates to assumptions made about the mechanism underlying the wave nonresponse. We consider an example of this multiple imputation approach in Section 11.3.

11.2.3.5 Approach 5: Calibration Adjustment for Respondents with Complete Data

Following this fifth approach (Sikkel et al., 2009), we would use the ideas of post-stratification adjustment discussed in Chapter 2 to calibrate the final baseline survey weights of individuals providing *complete data* in all of the waves of interest to reflect the original population represented by the *full sample* of respondents in the first wave. Suppose, for example, that the sample in the first wave of a five-wave panel survey represents a population that is 50% male and 50% female (after applying the Wave 1 survey weights in estimation). In subsequent waves, males respond at lower rates. When examining the individuals who provided complete data in the first five waves of the panel survey and the population that they represent, we might find that the population represented is 35% male and 65% female (after applying the baseline survey weights). Clearly some adjustment of the final baseline survey weights is needed (presumably to give more weight to those males who responded in all five waves), and this is where one can consider this calibration approach.

The calibration approach proceeds as follows:

a. Identify auxiliary variables in the first wave that are measured for the full sample of respondents and are generally predictive of both 1) key survey measures and 2) the probability of responding in future waves.
b. Compute weighted estimates of population distributions across subgroups defined by cross-classifications of these auxiliary variables, based on the respondents in the first wave. These weighted estimates serve as the benchmark population features for the calibration approach.
c. Compute weighted estimates of population distributions across the same subgroups based on those respondents that provided complete data in all of the waves of interest.
d. Based on the ratios of population percentages in the different subgroups defined by b) and c) above, compute calibration adjustments for the Wave 1 sampling weights of the respondents who provided complete data in all of the waves of interest. For example, if 5% of the population appears in a certain subgroup (based on the benchmark estimate) and 2.5% of the population is estimated to come from that same subgroup based on the respondents with complete data across the waves, those respondents would have their weights adjusted by a factor of 2.

This approach clearly relies on the measurement of good auxiliary variables in the first wave, which is a definite advantage of panel surveys where the same measures are collected over time. This approach also discards potentially useful information that would otherwise be analyzed when following some of the approaches described in earlier subsections. We evaluate the quality of estimates based on this approach relative to estimates found when using the other approaches in Section 11.3.

11.3 Alternative Longitudinal Analyses of the HRS Data

We now consider illustrations of each of the approaches outlined in Section 11.2 using current software procedures. In each of these illustrations, we analyze data from the 2006, 2008, 2010, and 2012 waves of the HRS, treating the sample of responding financial reporters with non-zero weights in 2006 as the "full sample" in the initial wave (meaning that there is no nonresponse in the first wave, for purposes of these illustrations). Our variable of interest measured for financial reporters in each of these waves is the total annual household income, which was computed based on the income of the respondent and their spouse only (if applicable). Given a handful of extreme reports (including zeroes) and general evidence of a right skew in the distribution of the reported annual income values in each of these four waves, we consider a natural log transformation of the reported income value (plus $1, to enable analysis of the zero reports) in each wave as our outcome variable of interest.

We start with a merged HRS data set in the "wide" (or "horizontal") format, with one row per responding financial reporter. This data file (available on the ASDA website) contains the sample design information in the baseline wave (2006), including the final survey weight, KWGTR, the stratum variable, STRATUM, and the sampling error computation unit variable, SECU. Also available for responding financial reporters in each of the four waves are measures of age (in years), arthritis (yes/no), diabetes (yes/no), years of education (0–11, 12, 13–15, 16+), race/ethnicity (Hispanic, Non-Hispanic White, Non-Hispanic Black, Non-Hispanic Other), marital status (married/previously married/never married), number of falls in the past two years, and self-rated health (1 = Excellent, 2 = Very Good, 3 = Good, 4 = Fair, 5 = Poor). We consider these variables as possible covariates when illustrating the different types of analyses discussed above.

We first create the four log-transformed income variables using the available responses in each wave, and assess histograms for these four transformed measures. All of these histograms (not shown) suggest that the log-transformed income values are roughly normally distributed, with the exception of some outliers:

```
use "X:\. . .\hrswide.dta"

gen ln_inc06 = ln(H8ITOT + 1)
gen ln_inc08 = ln(H9ITOT + 1)
gen ln_inc10 = ln(H10ITOT + 1)
gen ln_inc12 = ln(H11ITOT + 1)

hist ln_inc06
hist ln_inc08
hist ln_inc10
hist ln_inc12
```

11.3.1 Example: Descriptive Estimation at a Single Wave

Our analytic objective in these illustrations of descriptive estimation at a single wave is to estimate the mean total household income in 2008 in the HRS target population. Of the baseline sample of 11,789 participating financial reporters with non-zero weights in 2006 who also had a valid value for log-transformed annual household income, 10,574 (89.7%) also provided valid income reports in 2008.[1] We wish to estimate the mean total household income in 2008 in a manner that repairs possible nonresponse bias in the estimates due to financial reporters who did not participate in the 2008 wave. We consider a simple complete case analysis, in addition to the three approaches outlined above in Section 11.2.1: weighting adjustment, multiple imputation, and adjustment that allows for non-ignorable missing data.

11.3.1.1 Complete Case Analysis

Following this approach, we simply perform a design-based descriptive analysis of the available log-transformed household income values in 2008, and back-transform the weighted estimate and the 95% confidence limits to the original dollar scale. The results from this initial analysis are reported in Table 11.2.

```
* Complete case analysis for 2008 log-income.
svyset secu [pweight = kwgtr], strata(stratum)
svy: mean ln_inc08
matrix list r(table)
matrix a = r(table)
di exp(a[1,1])
di exp(a[5,1])
di exp(a[6,1])
```

11.3.1.2 Weight Adjustment

Following this second approach, we first fit a response propensity model, predicting response in the 2008 wave with selected covariates in 2006, including the total household income in 2006. We only include these covariates in the model (after imputing modal values for small counts of cases with missing data on the covariates) and do not consider any interactions for the sake of illustration. In a real analysis, we may want to consider interactions between covariates, provided that they are predictive of both response propensity and the key variable(s) of interest in the later wave.

TABLE 11.2

Descriptive estimates of mean total household income in 2008 for the HRS target population following alternative estimation approaches

Approach	*n*	Estimated Mean	95% CI
Complete Case Analysis	10,574	34,223.97	32,467.83, 36,075.10
Response Propensity Adjustment	10,574	33,309.15	31,580.11, 35,132.86
Multiple Imputation	11,789*	33,281.21	31,568.93, 35,086.35

Note: * After imputation (total number of complete cases in 2006).

We also note that this approach relies on being a respondent in 2006 as well, since cases without values for the covariates in 2006 would have missing values for estimates of response propensity in 2008. While that is not the case here, since we are working with all 2006 respondents, this would be the case in HRS more generally. As an alternative, one could recalibrate the most recent HRS survey weight for each 2008 respondent (from the most recent wave in which the respondent participated) to known population control totals in 2008, which would make the most use of the available respondent information.

We use the final 2006 survey weights to estimate the response propensity model parameters (Wun et al., 2007; see also the discussion of this issue in Chapter 2), and then compute predicted probabilities of response based on the weighted parameter estimates. The final 2006 survey weights are then multiplied by the inverses of the mean estimated response propensities within deciles of the response propensities, which is a technique that is used to minimize variability in the ultimate adjusted sampling weights that result from these nonresponse adjustments (Valliant et al., 2018). Here is the code that one can use in Stata to do this:

```
* Weight adjustment approach for 2008 log-income.

* Compute response indicator for 2008.
gen resp08 = 1 if ln_inc08 != .
replace resp08 = 0 if ln_inc08 == .
tab resp08, miss

* Modal imputation of missing covariate values.
replace selfrhealth_06 = 3 if selfrhealth_06 == .
replace marcat_06 = 2 if marcat_06 == .
replace diabetes_06 = 0 if diabetes_06 == .
replace arthritis_06 = 1 if arthritis_06 == .
replace racecat = 2 if racecat == .
replace edcat = 2 if edcat == .

* Response propensity model, with predictions.
svyset secu [pweight = kwgtr], strata(stratum)
svy: logit resp08 ln_inc06 i.selfrhealth_06 age_06 ///
 i.marcat_06 diabetes_06 arthritis_06 i.racecat i.edcat
predict phat, p

* Form deciles of response propensity, and compute means
* within each decile for weight adjustment.
xtile dec = phat, nq(10)
egen mean_phat = mean(phat), by(dec)
gen adj_kwgtr = kwgtr * (1 / mean_phat)

* Estimation using adjusted weights.
svyset secu [pweight = adj_kwgtr], strata(stratum)
svy: mean ln_inc08
matrix a = r(table)
di exp(a[1,1])
di exp(a[5,1])
di exp(a[6,1])
```

Of the covariates used in the response propensity model, only self-rated health, age, and marital status are significantly associated with *both* the response indicator and log-transformed income in 2008 (results not shown), meaning that some of the covariates have the desired properties for reduction of nonresponse bias (see Chapter 2 for discussion). Some of the covariates (e.g., log-transformed income and education category) are strong predictors of log-transformed income in 2008 only, which should help to reduce the variance in the weighted estimates. We note in Table 11.2 that the adjusted estimate of the mean total household income in 2008 has been reduced by nearly $1,000 after the adjustment; respondents in 2008 tended to be those with higher incomes.

11.3.1.3 Multiple Imputation

Following this third approach, which like the weight adjustment approach above assumes that the missing data in 2008 are **missing at random** (or, in other words, the missing data mechanism is ignorable conditional on the covariates used in the imputation model), we first predict missing values for 2008 log-transformed income multiple (5) times, reflecting the variability in the predictive distribution for this variable, and then we analyze each data set and combine the results into a single set of estimates and standard errors (see Chapter 12 for more discussion of the multiple imputation analysis approach).

Here is the code that we can use in Stata to do this, employing the same covariates used in the weighting adjustments. Per recommendations of Reiter et al. (2006), we take the relatively simple approach of including deciles of the final 2006 survey weights and the sampling strata as predictors in the imputation models, to account for the complex sampling features when imputing the missing income values. We consider alternative methods of accounting for complex sampling features in imputation in Chapter 12.

```
* Create deciles of the 2006 survey weights.
xtile wgt_dec = kwgtr, nq(10)

* Set data structure and register both complete "regular"
* variables and variables with missing data to be imputed.
mi set flong
mi register imputed ln_inc08
mi register regular ln_inc06 selfrhealth_06 age_06 ///
   marcat_06 diabetes_06 arthritis_06 racecat edcat ///
   stratum wgt_dec
mi describe

* Implement regression imputation after setting a seed.
set seed 41279

mi impute chained (regress) ln_inc08 = ln_inc06 ///
   i.selfrhealth_06 age_06 i.marcat_06 diabetes_06 ///
   arthritis_06 i.racecat i.edcat i.wgt_dec i.stratum, ///
   noisily augment add(5) burnin(5)

* Set complex sampling features in MI framework.
mi svyset secu [pweight = kwgtr], strata(stratum)
```

```
* Multiple Imputation estimation.
mi estimate, vartable: svy: mean ln_inc08
matrix list r(table)
matrix a = r(table)
di exp(a[1,1])
di exp(a[5,1])
di exp(a[6,1])
```

We note in Table 11.2 that the overall inference that we would make regarding mean income in 2008 is very similar to that produced by the weighting adjustment for nonresponse. The estimated fraction of missing information (FMI) value for the MI estimate (produced by the `vartable` option above) is 3.5%, which is well below the unweighted item-missing data rate of 10.3% for this income variable. We were therefore able to effectively recover missing information on income in 2008 using the covariates in the imputation model, and we do not see immediate evidence of a non-ignorable missing data mechanism for income (which would be suggested if the FMI value was larger than the item-missing data rate; Nishimura et al., 2016).

11.3.1.4 Imputation Using a Selection Model

Finally, we fit a probit selection model, which enables us to initially assess the correlation of the residuals in a substantive model for income with the residuals in a probit model for an indicator of providing a valid income measure in 2008. If these residuals are positively correlated, it could mean, for example, that someone providing a higher-than-expected income value (a positive residual unexplained by the covariates in the substantive model) has a higher-than-expected value on the 2008 response indicator (a positive residual, once again unexplained by the covariates in the selection model). Such a mechanism would suggest that missingness on income is non-ignorable, given that those with larger income in 2008 are more likely to respond.

To fit this type of selection model, we do need to identify good **instrumental variables**, which are variables that are predictive of the response indicator in 2008 but not the substantive variable of interest (log-transformed income in 2008). From the analysis above, we know that diabetes and arthritis serve this role well, as they were significant predictors in the response propensity model but not the regression model for income in 2008. We therefore include these predictors in the selection (response) equation of the probit selection model, but not in the substantive equation. See West and McCabe (2017) for more discussion of this issue.

We first fit the selection model to the original data set (prior to imputation, and using the original HRS sample design variables), and then examine the estimated correlation of the residuals in the substantive and selection equations:

```
svyset secu [pweight = kwgtr], strata(stratum)

svy: heckman ln_inc08 ln_inc06 selfrhealth_06 age_06 ///
     i.marcat_06 i.edcat, ///
     select(resp08 = diabetes_06 arthritis_06 ///
     selfrhealth_06 age_06 i.marcat_06)
```

The estimated residual correlation parameter is denoted as `rho` in the Stata output. This weighted estimate is 0.017, with a design-based 95% confidence interval of (–0.020, 0.053).

The fact that 0 is included in this interval suggests that there is a negligible correlation between the unique residuals associated with the income reports in 2008 (given the predictors in the substantive model) and the unique residuals associated with the probability of responding (again given the predictors in the selection model). This means that the weighting and imputation methods above (which assumed ignorable missing data mechanisms) should provide sufficient bias correction. The imputation model used for income would therefore apply for both respondents and non-respondents in 2008, meaning that the imputation model fitted to respondents would not be biased in any way (where the selection model approach would provide estimates of a substantive model corrected for non-ignorable selection bias). For more details on imputation approaches using fitted selection models if this correlation were to be non-negligible, see West and McCabe (2017).

11.3.2 Example: Change across Two Waves

We now consider examples of possible approaches when the analytic objective involves estimation of the mean change in a variable of interest across two waves for a given individual in the target population. We consider measures of total household income provided by financial reporters in 2006 and 2010, and our objective is to estimate the mean change in household income (in dollars) for an individual from the HRS target population across these four years. In this example, the variable of interest that we would ideally like to compute for each financial reporter in the sample is the difference in actual income between 2010 and 2006. However, we don't have income reports from 2010 for financial reporters who didn't respond in both waves; only 9,402 (79.75%) of the 11,789 responding financial reporters in 2006 also provided income information in 2010. We therefore consider the same weighting and imputation approaches discussed in Section 11.3.1 to address potential bias in estimates of this mean change due to wave nonresponse in 2010.

11.3.2.1 Complete Case Analysis

Following this approach, we first compute the change in income from 2006 to 2010 for each financial reporter who provided values of income in both waves. We then perform a simple design-based descriptive analysis of the changes in income, only for those "complete case" individuals who responded in both waves. The results from this initial complete case analysis, generated using the Stata code below, are reported in Table 11.3.

```
* Complete case analysis: mean change in income,
* 2006 to 2010.
gen incdiff_06_10 = H10ITOT - H8ITOT
svyset secu [pweight = kwgtr], strata(stratum)
svy: mean incdiff_06_10
```

The 95% confidence interval for the mean change clearly does not include 0, and the estimated mean change is negative. This suggests that on average, individuals in the HRS target population experienced a significant decline in total household income from 2006 to 2010, a decline that is not unexpected given the impact of the global financial recession of 2008.

TABLE 11.3

Descriptive estimates of mean change in total household income from 2006 to 2010 for those individuals in the HRS target population, following alternative estimation approaches

Approach	*n*	Estimated Mean Change	95% CI
Complete Case Analysis	9,402	−6,551.40	−10,289.72, −2,813.09
Response Propensity Adjustment	9,402	−6,115.54	−9,526.65, −2,704.43
Multiple Imputation	11,789*	−3,455.83	−9,656.79, 2,745.13
Calibration	9,402	−6,341.66	−9,950.37, −2,732.95

Note: * After imputation (total number of complete cases in 2006).

11.3.2.2 Weight Adjustment

For this second approach, we adjust the final 2006 survey weights of individuals providing valid income reports at each of these two waves to account for those individuals who reported income in 2006 but not 2010 (who would be dropped from the complete case analysis, potentially introducing bias in the estimate of mean change). We compute the adjusted weights exactly as described in Section 11.3.1 above, only using an indicator of response in 2010, and then estimate the mean change in income from 2006 to 2010 using the adjusted weights for each of the financial reporters who responded in both waves:

```
* Compute response indicator for 2010.
gen resp10 = 1 if ln_inc10 != .
replace resp10 = 0 if ln_inc10 == .
tab resp10, miss

* Modal imputation of missing covariate values.
replace selfrhealth_06 = 3 if selfrhealth_06 == .
replace marcat_06 = 2 if marcat_06 == .
replace diabetes_06 = 0 if diabetes_06 == .
replace arthritis_06 = 1 if arthritis_06 == .
replace racecat = 2 if racecat == .
replace edcat = 2 if edcat == .

* Response propensity model.
svyset secu [pweight = kwgtr], strata(stratum)
svy: logit resp10 ln_inc06 i.selfrhealth_06 age_06 ///
 i.marcat_06 diabetes_06 arthritis_06 i.racecat i.edcat
predict phat, p

* Form deciles of response propensity, and compute means
* within each decile for weight adjustment.
xtile dec = phat, nq(10)
egen mean_phat = mean(phat), by(dec)
gen adj_kwgtr = kwgtr * (1 / mean_phat)

* Estimation of mean change using adjusted weights.
svyset secu [pweight = adj_kwgtr], strata(stratum)
svy: mean incdiff_06_10
```

We note in Table 11.3 that our inferences would hardly be affected at all relative to the complete case analysis, and that the 95% confidence interval for the mean change once again does not include 0. This confidence interval is also slightly narrower than that for the complete case analysis, which reflects some gains in efficiency due to the use of covariates in the response propensity model that are strongly predictive of income in 2010.

11.3.2.3 Multiple Imputation

We now apply the same regression imputation approach from Section 11.3.1 to impute missing values on reported income in 2010 multiple (5) times. Given that the imputed values are on the natural-log scale, we first back-transform the imputed values to the original dollar scale (by exponentiating the imputed values), and then compute the difference in income between 2006 and 2010 for each responding financial reporter in each of the five imputed data sets. We also place bounds on the imputed values of these changes in income from 2006 to 2010 for non-respondents in 2010, ensuring that the imputed change values fall within the range of the original observed data:

```
* Multiple imputation of 2010 log-income.

* Create deciles of the 2006 sampling weights.
xtile wgt_dec = kwgtr, nq(10)

* Set data structure and register both complete "regular"

* variables and variables with missing data to be imputed.
mi set flong
mi register imputed ln_inc10
mi register regular ln_inc06 selfrhealth_06 age_06 ///
   marcat_06 diabetes_06 arthritis_06 racecat edcat ///
   stratum wgt_dec
mi describe

* Implement regression imputation
set seed 41279
mi impute chained (regress) ln_inc10 = ln_inc06
i.selfrhealth_06 age_06 i.marcat_06 diabetes_06 ///
arthritis_06 i.racecat i.edcat ///
   i.wgt_dec i.stratum, ///
   noisily augment add(5) burnin(5)

* Compute bounded change scores in each imputed data set.
sum incdiff_06_10 if _mi_m == 0
sum ln_inc10 if _mi_m == 0
replace ln_inc10=14.92 if ln_inc10 > 14.92 & ln_inc10 != .
gen new_chg0610 = exp(ln_inc10) - exp(ln_inc06)
replace new_chg0610 = -12300000 if new_chg0610 < -12300000
replace new_chg0610 = 2062968 if new_chg0610 > 2062968 & new_chg0610 !=
.

* Set complex sampling features for MI analysis.
mi svyset secu [pweight = kwgtr], strata(stratum)
```

```
* Multiple Imputation estimation.
mi estimate, vartable noisily: svy: mean new_chg0610
```

We note in Table 11.3 that our estimate of the mean change in income has been reduced and would no longer be considered significant after multiple imputation of the missing income values. Closer inspection of the design-based estimates from each imputed data set reveals smaller (negative) estimates of mean changes and larger within-imputation variance than was observed for the complete cases, meaning that nonrespondents likely had smaller changes in general.

This provides an example where multiple imputation, which imputes missing values for specific variables of interest rather than relying on global nonresponse adjustments that are "variable-agnostic", is picking up additional information for the missing cases that the weighting adjustments may not have been fully capturing. We also note that the FMI value for the mean change in income is 9.5%, which once again is much smaller than the unweighted nonresponse rate of 20.25% for income in 2010. We are therefore recovering much of the missing information using the available covariates, and we no longer have as strong evidence of a significant negative change in total household income.

After fitting a selection model to these data using the same approach described in Section 11.3.1, we once again find evidence of a negligible correlation between the residuals in the substantive model and the residuals in the selection model (estimated value of `rho` = 0.033, 95% CI = 0.003, 0.063), suggesting that there is minimal evidence of a non-ignorable missing data mechanism at work here, and that the multiple imputation estimates assuming an ignorable missing-at-random mechanism should be considered reasonable.

We note that the multiple imputation approach illustrated here could be extended to more complex models and more waves, using the same basic syntax. In some cases, imputed data sets may need to be restructured into vertical formats before being analyzed (for example, when fitting a multilevel model to the imputed data), but this process is straightforward (see Section 11.3.3) and facilitated by the `mi reshape` command in Stata. In general, when imputing missing values in longitudinal survey data sets, we recommend that users first impute the missing values in a wide-format data set (using information from adjacent waves in the imputation models, as we have done here), and then fit the models of interest, which may require restructuring the data. We provide additional examples of multiple imputation analyses in Chapter 12.

11.3.2.4 Calibration

Finally, we apply the calibration approach outlined in Section 11.2.3.5 to estimate the mean change in income from 2006 to 2010. Starting with the wide-format data set and the full sample in 2006, we first compute the sums of the final 2006 survey weights (i.e., estimates of the population sizes) within each of the 32 cross-classes defined by gender, race, and education category (selected simply for illustration purposes as significant predictors of income), and we then save these sums in an independent data set:

```
use "X:\. . .\hrswide.dta", clear

* Modal imputation of missing covariate values.
replace racecat = 2 if racecat == .
replace edcat = 2 if edcat == .

* Compute sums of 2006 weights in cross-classes
```

```
* defined by sex, race, and education.
egen cal_class = group(racecat edcat gender)
collapse (sum) kwgtr, by(cal_class)
rename kwgtr popsize

save "X:\. . .\cal_pop_sizes.dta", replace
```

Next, we repeat this process, but only for those cases with complete data in 2006 and 2010. We then merge the two independent data sets by calibration class, and compute the calibration adjustments for each class as the ratios of the estimated population sizes based on the 2006 (baseline) sampling weights for the full sample to the estimated population sizes based on the cases with complete data in 2006 and 2010:

```
* Repeat process for cases with complete data.
use "X:\. . .\hrswide.dta", clear

* Modal imputation of missing covariate values.
replace racecat = 2 if racecat == .
replace edcat = 2 if edcat == .

gen ln_inc10 = ln(H10ITOT + 1)

* Compute response indicator for 2010.
gen resp10 = 1 if ln_inc10 != .
replace resp10 = 0 if ln_inc10 == .
tab resp10, miss

* Only keep cases with complete data.
keep if resp10 == 1

* Compute sums of 2006 weights in cross-classes
* defined by sex, race, and education.
egen cal_class = group(racecat edcat gender)
collapse (sum) kwgtr, by(cal_class)
rename kwgtr sumrespwgts

save "X:\. . .\cal_resp_sizes.dta", replace

* Merge the two data sets of estimated population sizes.
merge 1:1 cal_class using "X:\. . .\cal_pop_sizes.dta"

* Compute calibration adjustments for each cross-class.
gen cal_adj = popsize / sumrespwgts
drop _merge

save "X:\. . .\cal_resp_sizes.dta", replace
```

Finally, we re-open the original wide data set, merge the calibration adjustments into the data set (matching by cross-class), multiply the final 2006 survey weights by the calibration adjustments, and verify that the sums of the calibrated weights in each cross-class for respondents with complete data are equal to the sums of the baseline weights in each cross-class for the full sample:

```
* Open original data set and merge calibration adjustments.

use "X:\. . .\hrswide.dta", clear

* Modal imputation of missing covariate values.
replace racecat = 2 if racecat == .
replace edcat = 2 if edcat == .

egen cal_class = group(racecat edcat gender)
sort cal_class
gen ln_inc06 = ln(H8ITOT + 1)
gen ln_inc10 = ln(H10ITOT + 1)

merge m:1 cal_class using "X:\. . .\cal_resp_sizes.dta"

* Compute response indicator for 2010.
gen resp10 = 1 if ln_inc10 != .
replace resp10 = 0 if ln_inc10 == .
tab resp10, miss

* Compute calibrated weights for cases with complete data.
gen kwgtr_cal = kwgtr * cal_adj if resp10 == 1

* Verify that sums of calibrated weights for cases
* with complete data are equal to sums of base weights
* for full sample.
total kwgtr, over(cal_class)
total kwgtr_cal if resp10 == 1, over(cal_class)
```

After this verification, we estimate the mean change based on the cases with complete data in 2006 and 2010, using the calibrated weights for estimation and estimating variance accounting for the other complex sampling features (including the post-stratification information):

```
* Estimate mean change using complete cases.
gen incdiff_06_10 = H10ITOT - H8ITOT

svyset secu [pweight = kwgtr_cal], strata(stratum) ///
poststrata(cal_class) postweight(popsize)

svy: mean incdiff_06_10
```

We note in Table 11.3 that our inference based on the calibration approach would largely be similar to that based on the complete case analysis and the response propensity adjustment: there is a significant reduction in mean income from 2006 to 2010. The above approach could be repeated with other possible cross-classes (e.g., based on 2006 income) to assess the sensitivity of the inference to the choice of the cross-classes for calibration.

11.3.2.5 Accounting for Refreshment Samples When Estimating Mean Change

In many panel surveys that employ rotating panel designs (e.g., the U.S. Current Population Survey), estimates of mean changes in particular measures from one wave to the next are not based entirely on panel respondents, but include refreshment samples that are specific to particular waves. In these cases, imputation of missing values for the refreshment sample cases in waves where they are not included would not make logical sense. In these settings, one should employ the approaches described for making inference about differences in estimates between subgroups in Chapter 5, treating the two waves being compared as separate subgroups and making sure to use the final survey weights computed for respondents at each wave in the analysis. The covariance of these two estimates would then need to account for the fact that individuals from the same sampling cluster may appear in both waves. However, should the fact that the same *individual* might also appear in both waves also be accounted for when estimating this covariance? In our experience with several empirical studies, treating the individual as a lower stage of multistage cluster sampling when estimating the variance of this type of difference in means across waves of a panel study does not make a substantial difference in the covariance of the estimates.

Consider this hypothetical example using a stacked version of the 2006 and 2008 HRS data sets, where individuals responding in both waves would have two rows in the resulting data set and a common household ID + person number code (HHIDPN, nested within the HRS sampling clusters), and the final survey weight variables would be specific to each wave. We consider the approach outlined in Chapter 5 for comparing estimated mean total income from the two waves, using two different specifications of the sample design: one that only accounts for common sampling clusters across the two waves when estimating the covariance of the two estimated means, and one that accounts for the common sampling clusters as first-stage sampling units (with a small sampling fraction within strata of 0.01), along with individuals as second-stage sampling units within the first-stage sampling clusters (again with a small sampling fraction of 0.01):

```
* Load subset of HRS 2006-2008 data.
use "X:\. . .\hrssub0608.dta", clear

* Ignore individuals as second-stage clusters.
svyset secu [pweight = weight], strata(stratum)
svy: mean income, over(wave)
lincom c.income@2008.wave - c.income@2006.wave

* Account for individuals (HHID) as second-stage clusters,
* incorporating small FPCs.
gen samprate = 0.01
svyset secu [pweight = weight], strata(stratum) ///
    fpc(samprate) || hhidpn, fpc(samprate)
svy: mean income, over(wave)
lincom c.income@2008.wave - c.income@2006.wave
```

Both approaches produce nearly identical inferences, where the standard error of the weighted estimate of the difference in the means accounts for the single-stage or multistage cluster sampling. The linearized standard error of the weighted estimate of the mean difference ($3,749.24) in the first case is $8,431.68, while the linearized standard error of the same weighted estimate in the second case (accounting for individuals as second-stage sampling clusters, given the repeated measurements on some individuals) is $8,432.28. Assuming slightly larger FPCs (e.g., 0.05 at either level) does not make a difference. When sampling fractions are generally small (either for the ultimate clusters at the first stage or individuals at the second stage), the contributions of lower-stage clusters (individuals in this case) to the covariance of estimates from different waves in a panel study are generally quite small, meaning that the simple approach of accounting for the common sampling clusters across both waves would be sufficient in practice.

11.3.3 Example: Weighted Multilevel Modeling

We now consider an example of fitting a multilevel model to the 2006, 2008, 2010, and 2012 waves of the HRS data, using both wave-specific survey weights at Level 1 and respondent-specific survey weights at Level 2 in the estimation. We have two objectives in this example: 1) using all available information from the respondents across the four waves, estimate trajectories in annual income for both males and females, and compare the trajectories statistically; and 2) estimate any remaining between-respondent variance in the intercepts and trajectories among individuals in the HRS target population during this time period. Per the HRS documentation (Heeringa & Connor, 1995), survey weights representing inverses of the products of the probability of being sampled and the probability of responding at a given wave were computed for responding households at each wave, to account for wave nonrespondents.

We first illustrate the necessary data management syntax to go from a "wide" format data set with one row per respondent to the required "vertical" format with multiple rows per respondent. We start by computing log-transformed versions of all available annual income reports from the four waves, select the variables of interest, and set the 2006 survey weight as the respondent-specific (baseline) sampling weight at Level 2. We then use the `reshape` command in Stata to transform the HRS data set from wide to long format and scale the wave-specific sampling weights per the recommended "normalizing" method discussed by Chambers and Skinner (2003).

```
* Reshape the original wide data.
use "X:\. . .\hrswide.dta", clear
gen ln_inc1 = ln(H8ITOT + 1)
gen ln_inc2 = ln(H9ITOT + 1)
gen ln_inc3 = ln(H10ITOT + 1)
gen ln_inc4 = ln(H11ITOT + 1)

* Initial wave-specific survey weights.
gen wgt1 = kwgtr
gen wgt2 = lwgtr
gen wgt3 = mwgtr
gen wgt4 = nwgtr

* Baseline sampling weight for respondents from 2006.
gen basewgt = kwgtr

* Keep variables of interest.
keep hhid pn gender ln_inc1-ln_inc4 wgt1-wgt4 ///
    secu stratum basewgt

* Reshape data set from wide to vertical format.
reshape long ln_inc wgt, i(hhid pn) j(year)
```

```
* Drop missing income values (wave nonresponse).
drop if ln_inc == .

* Form a unique ID variable.
egen newid = concat(hhid pn)
destring newid, gen(newid_num)

* Compute conditional Level 1 weights and rescale weights.
gen level1wgt = wgt / basewgt
gen level2wgt = basewgt

* "Method 2" rescaling (normalizing weights).
egen sumw = sum(level1wgt), by(newid)
egen nj = count(ln_inc), by(newid)
gen level1wgt_r = level1wgt * nj/sumw
```

The data now have the appropriate vertical structure for the weighted multilevel modeling analysis, and we illustrate this structure below for the first 20 rows in the newly restructured vertical data set:

	newid	year	level1wgt_r	level2wgt	ln_inc	gender
1.	000003010	1	.9055811	4093	10.59636	1
2.	000003010	2	.9485039	4093	10.69197	1
3.	000003010	3	1.063333	4093	11.01484	1
4.	000003010	4	1.082582	4093	10.4913	1
5.	010001010	1	1.181266	7434	9.229064	1
6.	010001010	2	1.320145	7434	9.236106	1
7.	010001010	3	.7501688	7434	9.575053	1
8.	010001010	4	.7484209	7434	9.392745	1
9.	010004010	1	.9737449	5217	11.34865	1
10.	010004010	2	.9731849	5217	10.98191	1
11.	010004010	3	1.05307	5217	12.1168	1
12.	010013010	1	.9750476	5373	9.569203	1
13.	010013010	2	.9623446	5373	17.9101	1
14.	010013010	3	1.010979	5373	9.675646	1
15.	010013010	4	1.051629	5373	9.585827	1
16.	010013040	1	.9386594	5440	10.91874	2
17.	010013040	2	1.063756	5440	9.785154	2
18.	010013040	3	.9560866	5440	5.70711	2
19.	010013040	4	1.041498	5440	9.72466	2
20.	010038010	1	.9557571	5217	12.21305	1

Note that there are up to four rows per respondent (financial reporter) in this restructured data set, containing a unique respondent identifier, a year identifier (1 = 2006, 2 = 2008, etc.), the scaled wave-specific weights (with mean 1.0 for each respondent), the respondent-specific 2006 weights, the wave-specific log-transformed income values, and the subject-specific gender codes.

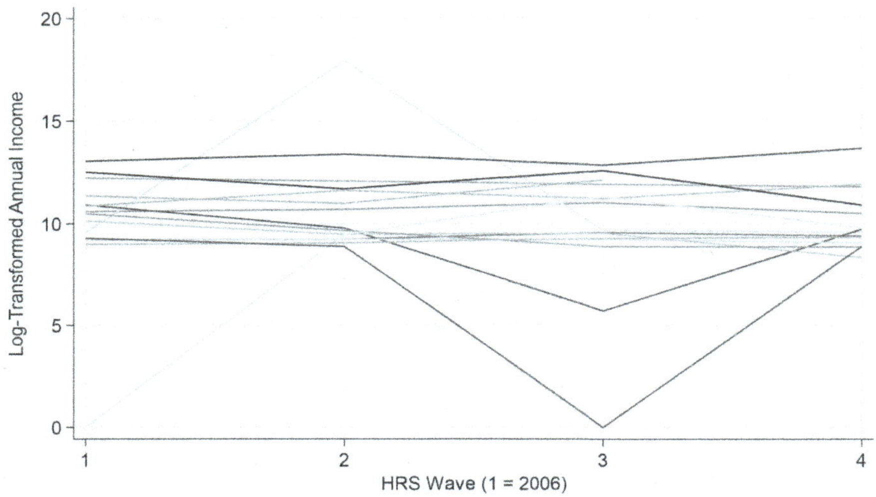

FIGURE 11.1
Multiple line plot showing trends in log-transformed annual income from 2006–2012 for a small subsample of HRS financial reporters.

For analysis purposes, we recode the YEAR variable to represent number of years since 2006 (YRSSINCE06), taking on values 0, 2, 4, and 6:

```
gen yrssince06 = 0 if year == 1
replace yrssince06 = 2 if year == 2
replace yrssince06 = 4 if year == 3
replace yrssince06 = 6 if year == 4
```

After this initial data management, we can make use of Stata's graphical capabilities for panel data to examine an initial **multiple line plot** for a small subsample of HRS respondents. We generate the plot in Figure 11.1 using the following two commands (additional editing was performed using Stata's graph editor):

```
* Plot data for small subsample of subjects.
xtset newid_num year
xtline ln_inc if newid_num <= 10200000, overlay ///
legend(off) ytitle(Log-Transformed Annual Income) ///
ttitle(HRS Wave (1 = 2006))
```

From this small subsample of respondents, we see substantial between-respondent variance in the mean values of log-transformed annual income (suggesting the inclusion of a random intercept in a multilevel model for the data). We do not see evidence of substantial variability among the respondents in their trajectories, although there are clearly some unusual respondents experiencing substantial variation over time. We also look at a plot of unweighted mean values for males and females (one should save the restructured vertical data file first under a new name before collapsing the data set for purposes of this plot):

```
* Plot means across waves for males and females.
collapse (mean) ln_inc, by(gender year)
```

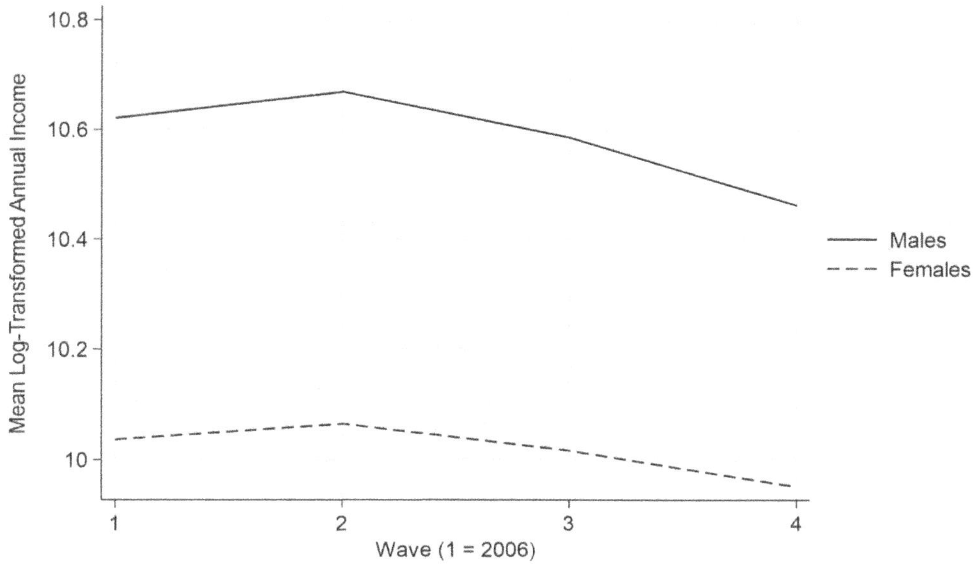

FIGURE 11.2
Multiple line plot showing trends in mean log-transformed annual income from 2006 (HRS Wave 1) to 2012 (HRS Wave 4) for males and females.

```
xtset gender year
xtline ln_inc, overlay ///
ytitle(Mean Log-Transformed Annual Income) ///
ttitle(Wave (1 = 2006))
```

In Figure 11.2, we note that males tend to have a higher (unweighted) mean than females in each wave, and we also detect a possible curvilinear trend in mean log-transformed annual income. Our models will allow for this possibility, but we don't see strong evidence of different trends for males and females.

We are now ready to fit the weighted multilevel model of interest using the `mixed` command in Stata. After re-opening the saved data set with the vertical (or long) format, we first compute a squared version of the new centered year variable, enabling us to fit a model including fixed effects of years since 2006, years since 2006 squared, gender, and interactions between gender and each of the year variables. Next, we compute a variable containing a unique code for each of the HRS sampling error computation units (which are coded 1 or 2 for each of the HRS sampling strata), for the purpose of computing robust standard errors reflecting the intra-cluster correlations arising from the cluster sampling (Rabe-Hesketh & Skrondal, 2006). The effects of the sampling strata will be accounted for by including fixed stratum effects in the model. Finally, we also include random intercepts, random effects of years since 2006, and random effects of years since 2006 squared for each respondent, allowing these three random effects to have a general variance-covariance structure:

```
* Fit multilevel model of interest.
use "X:\. . .\hrslong.dta", clear
gen yrs06sq = yrssince06*yrssince06
```

```
egen unique _ secu = group(stratum secu)

mixed ln _ inc c.yrssince06##i.gender c.yrs06sq##i.gender ///
      i.stratum [pweight = level1wgt _ r] ///
      || newid _ num: yrssince06 yrs06sq, ///
      variance cov(unstruct) ///
      pweight(level2wgt) pwscale(size) ///
      vce(cluster unique _ secu)
```

We carefully note important options used in this syntax to fit the multilevel model accounting for the respondent-specific and (scaled) wave-specific survey weights. The `[pweight = level1wgt _ r]` option specifies the conditional, scaled wave-specific weights at Level 1, while the `pweight(level2wgt)` option specifies the respondent-specific weights in 2006 at Level 2. Although this is unnecessary given the earlier data management, we also include the `pwscale(size)` option, which would automatically normalize the Level 1 weights within each Level 2 respondent. The `cov(unstruct)` option allows the three random effects to have unique variances and covariances. Finally, the `vce(cluster unique _ secu)` option requests that robust, sandwich-type standard errors be computed reflecting the intra-cluster correlations.

Given the complexity of these models and the resulting Stata output, we include the actual Stata output from fitting this model below:

```
Obtaining starting values by EM:

Performing gradient-based optimization:

Iteration 0: log pseudolikelihood = -3.022e+08
Iteration 1: log pseudolikelihood = -3.020e+08
Iteration 2: log pseudolikelihood = -3.019e+08
Iteration 3: log pseudolikelihood = -3.019e+08
Iteration 4: log pseudolikelihood = -3.019e+08

Computing standard errors:

Mixed-effects regression                       Number of obs     =    40,325
Group variable: newid_num                      Number of groups  =    11,789

                                               Obs per group:
                                                             min =         1
                                                             avg =       3.4
                                                             max =         4

                                               Wald chi2(60)     =   5134.14
Log pseudolikelihood = -3.019e+08              Prob > chi2       =    0.0000

                           (Std. Err. adjusted for 112 clusters in unique_secu)
---------------------------------------------------------------------------------
                   |              Robust
            ln_inc |      Coef.   Std. Err.      z    P>|z|     [95% Conf. Interval]
-------------------+-------------------------------------------------------------
        yrssince06 |   .0240185   .0150499     1.60   0.111    -.0054787    .0535158
          2.gender |  -.5619222   .0354198   -15.86   0.000    -.6313439   -.4925006
                   |
gender#c.yrssince06 |
```

```
              2 |  .0053894   .0220435     0.24   0.807    -.0378151     .048594
                |
        yrs06sq | -.0094561   .0028249    -3.35   0.001    -.0149928   -.0039193
                |
gender#c.yrs06sq |
              2 |  .0012008   .0036437     0.33   0.742    -.0059407    .0083423
                |
        stratum |
              2 |  .1688801   .1066258     1.58   0.113    -.0401026    .3778629
              3 |  .3205789   .0595461     5.38   0.000     .2038708    .4372871
              4 |  .4617349   .0456192    10.12   0.000     .3723229    .5511469
              5 |  .2687635   .0929799     2.89   0.004     .0865264    .4510007
              6 |  .2377024   .1813596     1.31   0.190    -.1177559    .5931606
              7 |  .7435081   .0893011     8.33   0.000     .5684811    .9185352
              8 |  .3798766   .0447116     8.50   0.000     .2922436    .4675097
              9 |  .5331056   .1310302     4.07   0.000     .2762911      .78992
             10 |  .8097707   .1514721     5.35   0.000     .5128909    1.106651
             11 |  .2496312   .2342885     1.07   0.287    -.2095659    .7088282
             12 |  .2918107    .122146     2.39   0.017     .0524089    .5312125
             13 |  .6212048   .0965329     6.44   0.000     .4320038    .8104057
             14 |  .4835137   .0452381    10.69   0.000     .3948487    .5721786
             15 |  .5270103   .0804145     6.55   0.000     .3694008    .6846199
             16 |  .2257717   .0735858     3.07   0.002     .0815462    .3699972
             17 |  .5952817   .0961691     6.19   0.000     .4067938    .7837696
             18 |  .7651221   .1399946     5.47   0.000     .4907378    1.039506
             19 |  .6316019   .1115861     5.66   0.000     .4128972    .8503067
             20 | -.1454788   .1675017    -0.87   0.385    -.4737762    .1828186
             21 |  .4390896   .1013933     4.33   0.000     .2403624    .6378168
             22 |  .8055653   .1827756     4.41   0.000     .4473317    1.163799
             23 |  .6225997   .0432323    14.40   0.000      .537866    .7073334
             24 |  .4333461   .0686031     6.32   0.000     .2988864    .5678057
             25 |  .5283591   .0798621     6.62   0.000     .3718321     .684886
             26 |   .482608   .0969039     4.98   0.000     .2926797    .6725362
             27 |  .5894716   .0470894    12.52   0.000      .497178    .6817651
             28 |  .4051973   .0730532     5.55   0.000     .2620156    .5483789
             29 |  .5262572   .0470189    11.19   0.000     .4341018    .6184125
             30 |  .4670825   .0444934    10.50   0.000     .3798771     .554288
             31 |   .464193   .0497094     9.34   0.000     .3667643    .5616216
             32 |  .2108956   .1077358     1.96   0.050    -.0002627    .4220538
             33 |  .1931331   .1164993     1.66   0.097    -.0352014    .4214676
             34 | -.0059096    .269963    -0.02   0.983    -.5350274    .5232082
             35 |  .3609932   .1165093     3.10   0.002     .1326391    .5893473
             36 |  .0170487   .1874262     0.09   0.928       -.3503    .3843974
             37 |   .168937   .0842175     2.01   0.045     .0038738    .3340003
             38 |  .4454747   .0547952     8.13   0.000      .338078    .5528713
             39 |  .4070119   .0440516     9.24   0.000     .3206725    .4933514
             40 |  .6901032   .0896625     7.70   0.000     .5143679    .8658385
             41 |  .5353654   .0989309     5.41   0.000     .3414644    .7292664
             42 |  .3910658   .0430351     9.09   0.000     .3067186     .475413
             43 |  .3842676   .1073781     3.58   0.000     .1738104    .5947247
             44 |  .3769977   .0595536     6.33   0.000     .2602749    .4937205
             45 |  .5021343   .0583944     8.60   0.000     .3876834    .6165852
             46 |  .4560482    .049509     9.21   0.000     .3590122    .5530841
             47 |  .1898394   .0471173     4.03   0.000     .0974913    .2821875
             48 |  .3858483   .0685417     5.63   0.000      .251509    .5201877
             49 |  .1357135   .1524849     0.89   0.373    -.1631514    .4345784
             50 |   .197195   .0916388     2.15   0.031     .0175862    .3768037
             51 |   .272965   .1546447     1.77   0.078     -.030133     .576063
             52 | -.7738671   .1884314    -4.11   0.000    -1.143186   -.4045483
             53 |  .4938157   .0796057     6.20   0.000     .3377914      .64984
             54 | -.5024009   .3015392    -1.67   0.096    -1.093407    .0886051
             55 |  .0630471   .1286935     0.49   0.624    -.1891876    .3152818
             56 |  .7020623   .2148519     3.27   0.001     .2809604    1.123164
                |
          _cons |   10.3206   .0478293   215.78   0.000     10.22685    10.41434
------------------------------------------------------------------------------
```

Random-effects parameters	Estimate	Robust std. err.	[95% conf.	interval]
newid_num: Unstructured				
var(yrssince06)	.1110225	.0402469	.0545558	.2259338
var(yrs06sq)	.0041059	.0010209	.0025221	.0066842
var(_cons)	1.069344	.1142976	.8672335	1.318556
cov(yrssince06,yrs06sq)	-.019302	.0059776	-.031018	-.0075861
cov(yrssince06,_cons)	-.0279811	.0583628	-.1423701	.086408
cov(yrs06sq,_cons)	-.003295	.008148	-.0192647	.0126747
var(Residual)	.8532461	.0755888	.7172435	1.015037

Interpreting this output, we note the following:

1. Stata computes robust "sandwich-type" standard errors for the parameter estimates, adjusting for the clustering of observations by sampling error computation unit (of which there are 112) and also accounting for the survey weights at each level. These standard errors therefore reflect the complex sampling features and the model being fitted (i.e., variances are estimated with respect to the model and the sample design; see Rabe-Hesketh and Skrondal (2006) for details).
2. Stata is attempting to maximize a pseudo-likelihood function, incorporating the survey weights at each level in the estimation.
3. Interpreting the weighted estimates of the fixed effects, we see clear evidence of a substantial gender gap, with males tending to have a significantly higher mean. We also see strong evidence of an "upside-down U-shaped" curvilinear pattern in log-transformed annual income for both males and females, given the significant negative fixed effect of years since 2006 squared and the non-significant fixed effect of years since 2006; this is entirely consistent with Figure 11.2 (and the downturn in income levels in 2008 coincides with the advent of the major recession in that year). The interaction terms are both non-significant, suggesting similar trajectories for males and females.
4. Based on the weighted estimates of the variances of the random effects and the approximate confidence intervals for these variance components, we see substantial evidence of variance in the intercepts among respondents, and less evidence of variance among respondents in the trajectories. The unexplained variance in the intercepts might be explained by including fixed effects of other subject-level covariates (e.g., education).

Finally, we note that software implementing Bayesian approaches to fitting weighted multilevel models is in initial stages of development (primarily for the Stan platform) but is not yet widely available. For recent methodological advances in this area, please see do Silva and Moura (2022) and Savitsky and Williams (2022), each of whom provide Stan code implementing the proposed approaches as supplementary material.

11.3.4 Example: Weighted GEE Analysis

We now consider an example of the weighted GEE approach discussed earlier, using case-specific weights accounting for the different patterns of wave nonresponse in the HRS. Preparation for a weighted GEE analysis of this type requires fitting several logistic regression models to wave-specific response indicators for an outcome variable of interest, each time conditioning on the response pattern in the prior waves. We therefore fit a series of design-based logistic regression models, predicting response to the income question in 2008, 2010, and 2012, each time using respondents with different patterns of responses in the prior waves as the subpopulations of interest (note that for purposes of this illustration, all 11,789 panel cases were once again considered respondents in the baseline wave of 2006). As covariates in these logistic regression models, we consider the reported income value (log-transformed) in 2006, along with several non-missing baseline covariates from 2006. We fit these models to the original HRS data set in the wide format:

```
* Open original wide-format HRS data set.
use "X:\. . .\hrswide.dta", clear

* Modal imputation of missing covariate values.
replace selfrhealth_06 = 3 if selfrhealth_06 == .
replace marcat_06 = 2 if marcat_06 == .
replace diabetes_06 = 0 if diabetes_06 == .
replace arthritis_06 = 1 if arthritis_06 == .
replace racecat = 2 if racecat == .
replace edcat = 2 if edcat == .

gen ln_inc06 = ln(H8ITOT + 1)
gen ln_inc08 = ln(H9ITOT + 1)
gen ln_inc10 = ln(H10ITOT + 1)
gen ln_inc12 = ln(H11ITOT + 1)

* Compute response indicator for 2008.
gen resp08 = 1 if ln_inc08 != .
replace resp08 = 0 if ln_inc08 == .
tab resp08, miss

* Response propensity model (2008).
svyset secu [pweight = kwgtr], strata(stratum)
svy: logit resp08 ln_inc06 i.selfrhealth_06 age_06 ///
   i.marcat_06 diabetes_06 arthritis_06 i.racecat i.edcat
predict phat08, p

* Compute response indicator for 2010.
gen resp10 = 1 if ln_inc10 != .
replace resp10 = 0 if ln_inc10 == .
tab resp10, miss

* Response propensity model (2010), respondents in 2008.
svyset secu [pweight = kwgtr], strata(stratum)
svy, subpop(if resp08 == 1): logit resp10 ln_inc06 ///
   i.selfrhealth_06 age_06 ///
   i.marcat_06 diabetes_06 arthritis_06 i.racecat i.edcat
predict phat10_11, p
```

```
* Response propensity model (2010), nonrespondents in 2008.
svyset secu [pweight = kwgtr], strata(stratum)
svy, subpop(if resp08 == 0): logit resp10 ln_inc06 ///
   i.selfrhealth_06 age_06 ///
   i.marcat_06 diabetes_06 arthritis_06 i.racecat i.edcat
predict phat10_10, p

* Compute response indicator for 2012.
gen resp12 = 1 if ln_inc12 != .
replace resp12 = 0 if ln_inc12 == .
tab resp12, miss

* Response propensity model (2012), 111 pattern.
svyset secu [pweight = kwgtr], strata(stratum)
svy, subpop(if resp08 == 1 & resp10 == 1): logit resp12 ///
   ln_inc06 i.selfrhealth_06 age_06 ///
   i.marcat_06 diabetes_06 arthritis_06 i.racecat i.edcat
predict phat12_111, p

* Response propensity model (2012), 110 pattern.
svyset secu [pweight = kwgtr], strata(stratum)
svy, subpop(if resp08 == 1 & resp10 == 0): logit resp12 ///
   ln_inc06 i.selfrhealth_06 age_06 ///
   i.marcat_06 diabetes_06 arthritis_06 i.racecat i.edcat
predict phat12_110, p

* Response propensity model (2012), 101 pattern.
svyset secu [pweight = kwgtr], strata(stratum)
svy, subpop(if resp08 == 0 & resp10 == 1): logit resp12 ///
   ln_inc06 i.selfrhealth_06 age_06 ///
   i.marcat_06 diabetes_06 arthritis_06 i.racecat i.edcat
predict phat12_101, p

* Response propensity model (2012), 100 pattern.
svyset secu [pweight = kwgtr], strata(stratum)
svy, subpop(if resp08 == 0 & resp10 == 0): logit resp12 ///
   ln_inc06 i.selfrhealth_06 age_06 ///
   i.marcat_06 diabetes_06 arthritis_06 i.racecat i.edcat
predict phat12_100, p
```

Next, we compute the cumulative probabilities of responding at each wave for each individual case:

```
* Compute cumulative response probabilities.

gen cumprob1 = 1

gen cumprob2 = phat08 if resp08 == 1
replace cumprob2 = 1-phat08 if resp08 == 0

gen cumprob3 = phat08 * phat10_11 ///
   if resp08 == 1 & resp10 == 1
```

```
replace cumprob3 = phat08 * (1-phat10_11) ///
    if resp08 == 1 & resp10 == 0
replace cumprob3 = (1-phat08) * phat10_10 ///
    if resp08 == 0 & resp10 == 1
replace cumprob3 = (1-phat08) * (1-phat10_10) ///
    if resp08 == 0 & resp10 == 0

gen cumprob4 = phat08 * phat10_11 * phat12_111 ///
    if resp08 == 1 & resp10 == 1 & resp12 == 1
replace cumprob4 = phat08 * phat10_11 * (1-phat12_111) ///
    if resp08 == 1 & resp10 == 1 & resp12 == 0
replace cumprob4 = phat08 * (1-phat10_11) * phat12_110 ///
    if resp08 == 1 & resp10 == 0 & resp12 == 1
replace cumprob4 = phat08 *(1-phat10_11)*(1-phat12_110) ///
    if resp08 == 1 & resp10 == 0 & resp12 == 0
replace cumprob4 = (1-phat08)*phat10_10 *phat12_101 ///
    if resp08 == 0 & resp10 == 1 & resp12 == 1
replace cumprob4 = (1-phat08)*phat10_10*(1-phat12_101) ///
    if resp08 == 0 & resp10 == 1 & resp12 == 0
replace cumprob4 = (1-phat08)*(1-phat10_10)*phat12_100 ///
    if resp08 == 0 & resp10 == 0 & resp12 == 1
replace cumprob4 = (1-phat08)*(1-phat10_10)*(1-phat12_100) ///
    if resp08 == 0 & resp10 == 0 & resp12 == 0
```

Next, we compute the overall cumulative probability of having a particular response pattern for each case (CUMPROB4), reshape the data set into the required vertical format, and multiply the base sampling weight for each panel respondent in 2006 by the inverse of the overall cumulative probability of having a particular response pattern:

```
* Compute overall probability of response pattern.
gen cumprob_case = cumprob4

* Reshape data set, keeping variables of interest.
gen ln_inc1 = ln(H8ITOT + 1)
gen ln_inc2 = ln(H9ITOT + 1)
gen ln_inc3 = ln(H10ITOT + 1)
gen ln_inc4 = ln(H11ITOT + 1)

* Baseline sampling weight for individual from 2006.
gen basewgt = kwgtr

keep hhid pn gender marcat_06 diabetes_06 arthritis_06 ///
    racecat edcat ln_inc1-ln_inc4 secu stratum basewgt ///
    cumprob_case

reshape long ln_inc wgt, i(hhid pn) j(year)

* Form a unique ID variable.
egen newid = concat(hhid pn)
destring newid, gen(newid_num)

* Compute the case-specific weights for weighted GEE.
gen casewt = basewgt * (1 / cumprob_case)
```

```
* Compute measure of years since 2006, and squared version.
gen yrssince06 = 0 if year == 1
replace yrssince06 = 2 if year == 2
replace yrssince06 = 4 if year == 3
replace yrssince06 = 6 if year == 4

gen yrs06sq = yrssince06*yrssince06
```

We then employ the weighted GEE approach to examine population estimates of the marginal change over time in log-transformed income for males and females. We treat individuals as clusters, include fixed stratum effects, set the correlation structure to be unstructured (given the unbalanced data set, where some cases do not have complete responses), and test the interactions between both years since 2006 and years since 2006 squared with gender:

```
* Fit weighted GEE model.
xtset newid_num year
xtgee ln_inc c.yrssince06##i.gender c.yrs06sq##i.gender ///
   i.stratum [pweight = casewt], corr(uns)
estat wcorrelation
```

Examining the output, we no longer see strong evidence of a non-linear trend over time when using this marginal modeling approach (unlike the multilevel modeling approach in Section 11.3.3, where we were explicitly conditioning on random subject effects):

```
Iteration 1: tolerance = .24071225
Iteration 2: tolerance = .00511372
Iteration 3: tolerance = .00025702
Iteration 4: tolerance = .00001743
Iteration 5: tolerance = 1.379e-06
Iteration 6: tolerance = 1.222e-07

GEE population-averaged model                  Number of obs      =    40,325
Group and time vars:         newid_num year    Number of groups   =    11,789
Link:                              identity    Obs per group:
Family:                            Gaussian                  min =         1
Correlation:                   unstructured                  avg =       2.5
                                                             max =         4
                                               Wald chi2(60)      =    247.93
Scale parameter:                    2.28706    Prob > chi2        =    0.0000

                            (Std. Err. adjusted for clustering on newid_num)
------------------------------------------------------------------------------
                     |               Robust
              ln_inc |      Coef.   Std. Err.      z    P>|z|     [95% Conf.   Interval]
---------------------+--------------------------------------------------------
          yrssince06 |  -.0837022   .0480113    -1.74   0.081    -.1778026    .0103982
            2.gender |  -.6217098    .100058    -6.21   0.000    -.8178199   -.4255998
                     |
gender#c.yrssince06  |
                   2 |   .0742509   .0827539     0.90   0.370    -.0879438    .2364456
                     |
```

```
         yrs06sq |   .0050879   .0081957     0.62   0.535    -.0109753    .0211512
                 |
gender#c.yrs06sq |
               2 |  -.0076464    .013348    -0.57   0.567    -.0338081    .0185152
                 |
         stratum |
               2 |   .3158994   .5143065     0.61   0.539    -.6921227    1.323922
               3 |   .7467036   .5242659     1.42   0.154    -.2808386    1.774246
               4 |   -.444524   1.262352    -0.35   0.725    -2.918689    2.029641
               5 |   .7898456   1.618421     0.49   0.626    -2.382201    3.961892
               6 |   .8542305   .6791201     1.26   0.208    -.4768205    2.185282
               7 |   1.507295   .5454734     2.76   0.006     .4381867    2.576403
               8 |   .9551998   .5094052     1.88   0.061    -.0432159    1.953616
               9 |   .6773107   .5429146     1.25   0.212    -.3867823    1.741404
              10 |   1.467586    .492562     2.98   0.003     .5021822     2.43299
              11 |   .3092727   .7823276     0.40   0.693    -1.224061    1.842607
              12 |   1.187935   .5730078     2.07   0.038     .0648601    2.311009
              13 |   1.005892   .5650043     1.78   0.075    -.1014961     2.11328
              14 |     .97375   .4982294     1.95   0.051    -.0027617    1.950262
              15 |   .8222744   .5416592     1.52   0.129    -.2393582    1.883907
              16 |   .8456724   .4881629     1.73   0.083    -.1111093    1.802454
              17 |   1.276917   .5145329     2.48   0.013     .2684507    2.285383
              18 |   1.397814   .5267904     2.65   0.008     .3653236    2.430304
              19 |   1.055951   .6977541     1.51   0.130    -.3116214    2.423524
              20 |   1.124006   .7815293     1.44   0.150    -.4077634    2.655775
              21 |   1.154815    .484176     2.39   0.017     .2058479    2.103783
              22 |    1.13661   .5241062     2.17   0.030     .1093804    2.163839
              23 |   .5672964   .5454186     1.04   0.298    -.5017043    1.636297
              24 |   .9702661   .5499848     1.76   0.078    -.1076842    2.048216
              25 |   .9441888    .486515     1.94   0.052    -.0093632    1.897741
              26 |   1.251989    .559142     2.24   0.025      .156091    2.347887
              27 |   1.247647   .4964189     2.51   0.012     .2746839     2.22061
              28 |   1.012526   .4951085     2.05   0.041     .0421316    1.982921
              29 |   1.282404   .5053575     2.54   0.011     .2919212    2.272886
              30 |   .9693431   .5034691     1.93   0.054    -.0174383    1.956124
              31 |   1.161934   .4981921     2.33   0.020     .1854952    2.138372
              32 |   .9680243   .6344909     1.53   0.127    -.2755551    2.211604
              33 |   .9476448   .5865529     1.62   0.106    -.2019777    2.097267
              34 |    .423263   .4873485     0.87   0.385    -.5319225    1.378448
              35 |   1.167646   .5390566     2.17   0.030     .1111144    2.224177
              36 |   .7535923   .6609817     1.14   0.254     -.541908    2.049093
              37 |   .7030033   .4860949     1.45   0.148    -.2497252    1.655732
              38 |   1.383805   .5762668     2.40   0.016     .2543423    2.513267
              39 |   .9507585   .4916987     1.93   0.053    -.0129532     1.91447
              40 |   1.437053   .4904077     2.93   0.003     .4758718    2.398235
              41 |   1.322208     .53213     2.48   0.013     .2792521    2.365164
              42 |   .6620029    .503468     1.31   0.189    -.3247763    1.648782
              43 |   1.071578    .517171     2.07   0.038     .0579411    2.085214
              44 |   1.174418   .4946859     2.37   0.018     .2048518    2.143985
              45 |   1.382111   .4885863     2.83   0.005     .4244997    2.339723
              46 |   1.071244    .507519     2.11   0.035     .0765251    2.065963
              47 |    .693522   .4855557     1.43   0.153    -.2581497    1.645194
              48 |   .9355299   .4909906     1.91   0.057    -.0267939    1.897854
              49 |   .4250969   .6258578     0.68   0.497     -.801562    1.651756
              50 |   .8213465   .4904102     1.67   0.094    -.1398399    1.782533
              51 |   .9867888   .5205207     1.90   0.058     -.033413    2.006991
              52 |   .0818238   .5202392     0.16   0.875    -.9378264    1.101474
              53 |   1.028963     .50682     2.03   0.042     .0356143    2.022312
              54 |   .3523894   .5920819     0.60   0.552    -.8080697    1.512849
              55 |   .3059683    .550127     0.56   0.578    -.7722608    1.384197
              56 |   1.381775    .504372     2.74   0.006     .3932245    2.370326
                 |
           _cons |   9.808905   .4726353    20.75   0.000     8.882557    10.73525
------------------------------------------------------------------------------
```

```
Estimated within-newid_num correlation matrix R:

      |        c1         c2         c3         c4
------+--------------------------------------------
   r1 |         1
   r2 |  .4357663          1
   r3 |  .5732002   .5657337          1
   r4 |  .3024913   .3282373   .463225          1
```

We note marginal evidence of a negative linear trend over time, where income appears to be decreasing slightly, and this is consistent with our initial descriptive analyses. The overall gender gap in expected income remains strongly significant when employing the weighted GEE approach. Finally, we see evidence of varying correlations of the repeated measures over time when examining the weighted estimates of the correlations in the unstructured working correlation matrix.

As a reminder, one can also fit models using weighted GEE with time-varying weights (as opposed to the case-specific weights illustrated here). This can be done by applying the standard design-based regression modeling procedures illustrated in prior chapters to a data set with a "vertical" format (as shown in this example) and setting the individual respondent IDs as the sampling clusters for the purpose of variance estimation. This would invoke the equivalent of weighted GEE with an exchangeable working correlation structure for the repeated measurements on each individual. For examples, see Si et al. (2022) or West et al. (2024).

11.4 Concluding Remarks

We have presented a variety of state-of-the-art approaches to the analysis of longitudinal survey data in this chapter. In general, we recommend that readers carefully define the objective(s) of their longitudinal analysis, and determine whether simple descriptive approaches would suffice, or more complicated modeling approaches might be needed. If trajectory estimation is of interest, readers need to decide whether they are interested in estimating between-cluster or between-subject variance in trajectories (which would require a weighted multilevel modeling approach). If estimation of marginal trajectories is the only objective, weighted GEE approaches would likely suffice.

Other weighting approaches for longitudinal surveys have been proposed in the literature. One is the **generalized weight share method** described by Lavallee (1995, 2007). This method mainly has applications in longitudinal household surveys where the composition of households is frequently changing over time, and measures are collected from individuals within the households. For example, adult children may move out of households, households may be combined (via marriage, possibly), etc. This approach also has applications in longitudinal surveys where refreshment samples are added at later waves. Notably, this approach does not primarily address issues related to nonresponse adjustment and adjustment for attrition over time, but rather what happens when the sample composition is changing, and the individuals introduced into the sample may not necessarily have a direct link with the target population of interest. For more details on these approaches, we refer readers to Lavallee (2007).

11.5 Exercises

11.1) Briefly describe the main analytic objectives outlined in this chapter and provide a list of suggested analysis methods, given a particular objective. Make sure to clarify what each objective addresses analytically and discuss the key assumptions of each analysis method.

11.2) Open the 2006–2012 HRS longitudinal data set (available from the ASDA website) and obtain a list of variables contained in this data set. If needed, review the data set description provided in Section 11.3. Consider two options to perform descriptive analysis for a single time point: 1) complete case analysis with design variables/weights, and 2) weight adjustments similar to those of Section 11.3.1. Now, carry out these steps:

a. Use a software package of your choice, and restrict the data set to those individuals that were financial respondents in 2006 and also had a non-zero respondent level weight in 2006 (example Stata code: `keep if kfinr == 1 & kwgtr != 0)`.

b. Next, create a series of log-transformed variables from the H8ITOT-H11ITOT variables, representing total household income for each year: 2006, 2008, 2010, 2012. Use the same names as in the example from the chapter: LN_INC06, LN_INC08, LN_INC10, LN_INC12. Make sure to follow the coding logic presented just before Section 11.3.1 when creating these variables and check the distributions of the variables using a series of histograms. Do they appear to be relatively normal?

c. The analytic goal is to estimate mean household income in 2010. With that in mind, how many financial respondents provided income information in 2006, and how many financial respondents did the same in 2010? What percentage of the 2006 baseline respondents also gave information about household income in 2010?

d. Perform a complete-case analysis using a design-based approach, and obtain the number of observations, back-transformed (exponentiated) weighted estimates, and 95% confidence limits for the mean 2010 income. Place those results in the first row of a table like Table 11.2 and label this row "Complete Case Analysis."

e. Next, perform a weight adjustment to account for the missing data on the 2010 income variable, using the code presented in Example 11.3.1. Be sure to include the modal assignments of missing data on the model covariates. Fit a design-based logistic regression model using these same covariates, and then output the predicted values for use with the decile "smoothing" approach/weight adjustment. This process creates an adjusted weight called ADJ_KWGTR that accounts for possible bias due to wave nonresponse in 2010.

f. Using the adjusted weight from part e., reset the survey variables to use this adjusted weight (rather than the original weight) and perform design-based estimation of mean income in 2010. Place these results in the 2nd row of the table created in part d. and label this row "Response Propensity Adjustment". Title this table with a meaningful title like Table 11.2, and make sure the rows and columns have understandable labels.

g. Finally, include your finished table and write a short paragraph describing your results. Be sure to describe the two methods used in this analysis and whether the weight adjustment improved upon the complete case analysis.

h. **(Extra Credit)** Implement the multiple imputation method for estimated mean total household income in 2010 by following the code and logic from Example 11.3.1. Use statistical software of your choice that can perform multiple imputation (MI) and analyze imputed data sets with the correct MI combining rules (see Chapter 12 for details on imputation). Add these results to the table you created for parts a.-g. of this question. How do the MI results compare to the other methods used?

11.3) Open the HRS 2006–2012 longitudinal data set (available from the ASDA website) and obtain a list of variables contained in this data set. If needed, review the data set description provided in Section 11.3. This exercise examines change in total HH income between 2006 and 2012, using weight adjustments like those in Section 11.3.2. Use the logic of the Stata code from this section as a guide for answering this question. With that in mind, carry out these steps:

a) As in Question 1, use a software package of your choice and restrict the data set to those that were financial respondents in 2006 and also had a non-zero respondent level weight in 2006 (Stata code: `keep if kfinr == 1 & kwgtr != 0`).

b) Create a series of log-transformed variables from the H8ITOT-H11ITOT variables representing the total household income for each year: 2006, 2008, 2010, 2012. Use the same names as in the example and Question 2: LN_INC06, LN_INC08, LN_INC10, LN_INC12.

c) Generate a variable called INCDIFF_06_12, representing the difference in total household income from 2006 to 2012. Use the original variables for this step rather than the log transformed versions (sample Stata code: `gen incdiff_06_12 = H11ITOT - H8ITOT`).

d) Set or use the survey variables using the original 2006 respondent survey weight (KWGTR) and the stratum and cluster variables (STRATUM, SECU). Obtain the design-based estimate of the mean difference in total household income from 2006 to 2012. What is the estimated mean difference in total household income from 2006 to 2012, and what is the estimated standard error?

e) Compute a response indicator for income in 2012 using the variable LN_INC12 and the logic presented in Section 11.3.2 (the weight adjustment method). Also, use modal imputation of missing covariate values, again as presented in Section 11.3.2.

f) Fit a logistic regression model to the indicator of response on the log-transformed income variable for 2012 with the same predictors as in Section 11.3.2 and output the predicted probabilities from this model. Create deciles of response propensity, compute the mean response propensity for each decile, and finally, create an adjusted weight called ADJ_KWGTR1, as demonstrated in the example (and like the weight adjustment from Question 2).

g) Perform a design-based analysis of the mean difference in total household income (2006 to 2012) using the adjusted weight, ADJ_KWGTR1, and compare the results to those without a weight adjustment. What is the estimated mean and standard error from the analysis using the adjusted weight? How does the weight adjustment help "repair" the wave nonresponse?

11.4) This exercise asks you to replicate the analysis presented in Section 11.3.3, using the weighted multilevel modeling method. The analysis objective is to examine change in total household income over 4 waves of HRS data (2006, 2008, 2010, and 2012). The example works through data management steps including variable construction, weight adjustments/re-scaling, and reshaping the data from the wide to long format. (Make sure to save the "long" file prior to graphing since collapsing of data is done in the graphing!) Following data management, you will generate some graphics and finally fit a set of weighted multilevel models. Use the logic of the Stata code from this section as a guide for replicating the example outputs in a software package of your choice (note: not all software packages will have procedures enabling you to fit weighted multilevel models!). Your task is to perform all steps of the process and provide the following results and code based on your own work (from the software that you select for this task). Make sure that your outputs match those below or from Section 11.3.3:

a) Reproduce this output:

```
sum level1wgt_r

  Variable |        Obs        Mean      Std. Dev.       Min        Max
-------------+---------------------------------------------------------
level1wgt_r |     40,325           1      .2649942          0          4


list newid year level1wgt_r level2wgt ln_inc gender in 1/20

     +-----------------------------------------------------------+
     |     newid   year   level1~r   level2~t     ln_inc   gender |
     |-----------------------------------------------------------|
  1. | 000003010      1   .9055811       4093   10.59636        1 |
  2. | 000003010      2   .9485039       4093   10.69197        1 |
  3. | 000003010      3   1.063333       4093   11.01484        1 |
  4. | 000003010      4   1.082582       4093    10.4913        1 |
  5. | 010001010      1   1.181266       7434   9.229064        1 |
     |-----------------------------------------------------------|
  6. | 010001010      2   1.320145       7434   9.236106        1 |
  7. | 010001010      3   .7501688       7434   9.575053        1 |
  8. | 010001010      4   .7484209       7434   9.392745        1 |
  9. | 010004010      1   .9737449       5217   11.34865        1 |
 10. | 010004010      2   .9731849       5217   10.98191        1 |
     |-----------------------------------------------------------|
```

b) Reproduce Figure 11.1: Change the title for the Y Axis to "<Your Last Name>" plus the original title.

c) Reproduce Figure 11.2: Change the title for the Y Axis to "<Your Name>" plus the original title.

d) Reproduce the full output from the weighted multilevel model accounting for the design features, with your name inserted into the code using a comment. For example (Stata code):

```
mixed ln_inc c.yrssince06##i.gender c.yrs06sq##i.gender i.stratum ///
    [pweight = level1wgt_r] ///
    /* Output produced by <YOUR NAME> on <DATE> */ ///
    || newid_num: yrssince06 yrs06sq, variance cov(unstruct) ///
    pweight(level2wgt) pwscale(size) vce(cluster unique_secu)
```

e) Write a brief paragraph describing how the weighting methods used and inclusion of complex sample design features in the design-based model improve the quality of the analysis. What other analysis methods might be used given the analysis objective of this exercise?

Note

1 In the HRS, mortality is also a reason why responses may not be available from a given participant in later waves. The HRS provides indicators that allow data users to determine whether a participant was still alive (and thus able to respond) in a given wave. For the purposes of these illustrations, we assume that all of the participants indicated were available to provide a response in later waves.

12

Imputation of Missing Data: Practical Methods and Applications for Survey Analysts

12.1 Introduction

In the 14 years that have elapsed since the publication of the first edition of this book, there have been many published advances in the theory, methods, and tools for conducting statistical analyses of complex sample survey data with missing data. These have included:

- The method of Fractional Imputation (Kim & Fuller, 2004; Kim & Shao, 2014; SAS Institute Inc., 2023b);
- The Finite Population Bayesian Bootstrap (Zhou et al., 2016a, 2016b, 2016c; Raghunathan et al., 2022);
- Bayesian modeling in the presence of missing data (Goldstein et al., 2018);
- Maximum Likelihood methods (Chambers et al., 2012);
- Multiple Imputation (van Buuren, 2012; Carpenter & Kenward, 2013; Berglund & Heeringa, 2014; Raghunathan, 2016; Quartagno et al., 2020; StataCorp, 2023; Kalpourtzi et al., 2024; Tang et al., 2024);
- Handling attrition in longitudinal studies (e.g., Cai et al., 2023); and
- Analytic methods for non-ignorable missing data mechanisms (Andridge & Little, 2011; West & Little, 2013; Sullivan & Andridge, 2015; Andridge & Little, 2020; Little et al., 2020).

The sheer volume of recently published articles, new texts, and how-to guides on the subject testify to the pace of new developments and the importance that missing data plays in everyday statistical estimation and inference. The overall aim of this chapter is to bring the problem of item missing data forward and focus on it, examine its nature, assess its potential impact on estimation and inference, and offer practical solutions for dealing with "item missingness" in the course of analyzing complex sample survey data.

Missing data is a ubiquitous problem in the analysis of survey data. **Statistical imputation procedures** are techniques for assigning analyzable values to item missing data on survey variables. Although major Federal statistical programs such as the Current Population Survey (CPS) were employing imputation of item missing data as early as the late 1940s (Ono & Miller, 1969), it was rare to see applications of imputation to item missing data prior to the early 1970s. Further, when imputation methods were applied, the

DOI: 10.1201/9781003472094-13

techniques were often combined with post-survey editing, were very *ad hoc* in nature, and paid little attention to the theoretical implications of the methods for analysis and inference.

The U.S. National Academy of Sciences Panel on Incomplete Data (Madow & Olkin, 1983) brought together leading survey statisticians to consider item missing data as a true statistical problem and to lay the foundation for a more theoretically justified treatment of missing data in survey analysis. Since that time, the statistical science of imputation and other methods of analysis that address the missing data problem have flourished. A taste of the variety of methods that have been proposed can be found in the colorful names of the techniques: mean imputation, predictive mean matching, nearest neighbor, regression imputation, the hot deck, row and column method, multiple imputation, and fractional imputation. Kalton and Kasprzyk (1986), Durrant (2005), and Little and Rubin (2019) review the wide range of techniques, the methods and applications, and the relative strengths and weaknesses of these techniques.

Throughout the preceding discussion of statistical methods and examples in Chapters 5–10 of this book, the problem of **item missing data** for otherwise complete cases was allowed to remain in the background. As a case in point, recall the example analysis in Section 9.4 in which a negative binomial regression model was fitted to the HRS data for number of falls in the past two years. In addition to the dependent variable (count of falls), the model included six predictor variables. Table 12.1 below summarizes the item missing data rates for the count of falls and the six predictors in that model. The interviewed respondents in the 2012 HRS panel included n = 10,735 adults age 65+ who were eligible to be asked this question. After listwise deletion, Stata used only n = 10,026 complete data observations to estimate the regression model parameters and the standard errors of the parameter estimates—a 6.6% reduction in the nominal sample size. The example analysis incorporated the 2012 HRS weighting adjustments for baseline sample unit nonresponse and panel attrition, but did not attempt to compensate for the otherwise complete cases that were lost in the analysis due to the presence of item missing data on one or more of the analysis variables.

This chapter begins with an overview of important missing data concepts in Section 12.2. Section 12.3 then provides an overview of imputation-based approaches to analysis with missing data, outlining the general framework of method choice, model specification (where applicable), and practical methods for imputation, estimation, and inference for complex sample survey data. This is followed in Section 12.4 by an introduction to the **Multiple Imputation (MI)** method, including selection of the imputation model, multivariate MI methods and software, and estimation and inference with multiply imputed data. An introduction to **Fractional Imputation (FI)** theory, methods, and software then follows in Section 12.5. The chapter concludes in Section 12.6 with applications of the MI and FI methods to a missing data problem encountered in an analysis of the 2011–2012 NHANES data on diastolic blood pressure and hypertension.

TABLE 12.1

Item Missing Data Rates for Variables Included in the 2012 HRS Falls Model (n = 10,735 eligible respondents aged 65 and older, n_{com} = 10,026 with complete data for all six variables)

Variable	Falls	Age	Gender	Arthritis	Diabetes	BMI
% Missing	1.3%	<0.01%	0.0%	0.2%	0.1%	1.1%

12.2 Important Missing Data Concepts

12.2.1 Sources and Types of Missing Data

Respondents' participation in surveys is almost always a voluntary decision. Even in "mandatory" government survey programs such as the American Community Survey (ACS), some sampled units may not provide an interview. **Unit nonresponse** is the term commonly used by survey methodologists to refer to situations in which sampled units either refuse or are unable to participate in the survey interview (Groves et al., 2009). **Wave nonresponse** occurs when a panel member is not reinterviewed in one or more follow-up waves of a longitudinal survey program (Cai et al., 2023).

Increasingly, population-based studies such as NHANES or HRS have added supplemental components to collect biomarker data (saliva, blood, and body measurements), work histories, medical records information, and other information that requires special respondent consent and different data collection procedures. Failure to obtain respondent consent or cooperation in a phase of the study (**phase nonresponse**) results in missing data for variables from that entire phase of the study. For example, approximately 4.3% of adults who participated in the initial 2011–2012 NHANES interview did not agree to participate in a subsequent medical (MEC) examination. HRS requests special consent of respondents to link U.S. Social Security earnings records and Medicare records (for panel members age 65+) to their survey data. At any given wave, approximately 20% of respondents do not provide consent to this special record linkage.

Unit nonresponse or nonresponse to a longitudinal wave (or phase) of a survey data collection can be major sources of missing information in sample surveys, but they are not the only source. To ensure high levels of overall survey cooperation and to guarantee that basic human subject protections are met, respondents are often read or provided a statement to the following effect at the start of the interview: "Your participation is voluntary. If we come to a question that you do not wish to answer, you may skip it." Categories of "Don't Know" or "Refused" are often explicitly included in the response options for individual survey items. Missing data for individual variables can also occur due to mistakes in interview administration, although in today's world, computer assisted interviewing (CAI) and web-based survey administration minimizes many of the missing data problems associated with incorrect skips or missed item follow-ups. In the survey literature and throughout this text, the term **item missing data** is used to refer to missingness on one or more variables in a survey record that is otherwise complete.

12.2.2 Patterns of Item Missing Data in Surveys

If we consider a rectangular file of survey data with rows representing sample cases and columns representing the survey variables of interest, unit nonresponse produces a pattern of missing data that is illustrated in Figure 12.1.

Section 2.7.3 introduced the concept of weighting adjustments to compensate for potential bias due to unit nonresponse for significant fractions (e.g., 10%, 20%, 30% or more) of the probability sample that was selected to represent the survey population.

Complete nonresponse to an entire wave of a longitudinal survey or phase of a multiphase survey produces a **monotonic pattern** of item missing data (Figure 12.2)—core data are present for interviewed cases but data for subsequent waves/phases may be missing. Although imputation methods may be used to address monotonic patterns of item missing

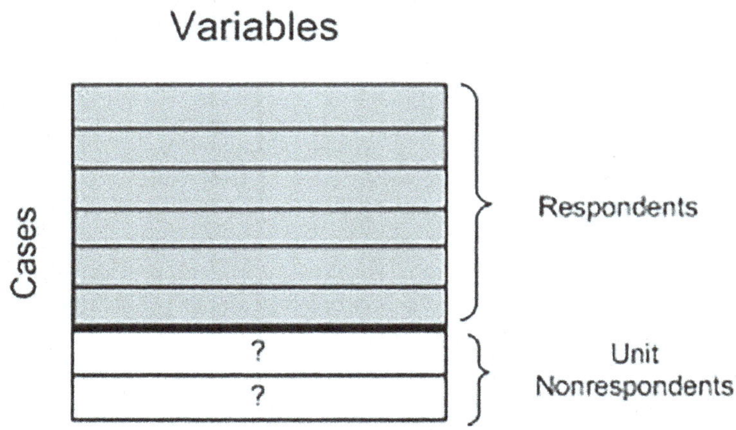

FIGURE 12.1
Unit Nonresponse Missing Data Pattern.

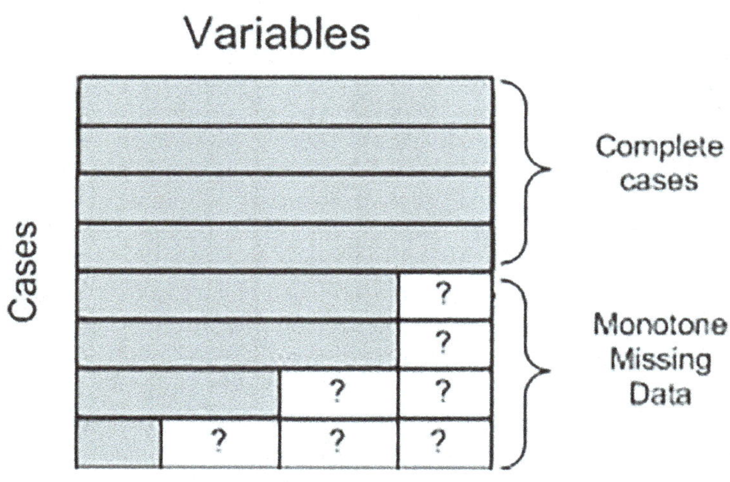

FIGURE 12.2
Monotonic Pattern of Missing Data.

data, including complete phase nonresponse (Schafer et al., 1996), it is common in survey practice to see nonresponse weighting adjustments applied to repair this problem.

As noted in the example of the HRS falls analysis (above), many variables measured in surveys are subject to low rates of item missing data, and more sensitive or difficult items (e.g., income measures) may have higher missing data rates (e.g., 5% – 15%). The arbitrary nature of these response outcomes typically produces a **generalized pattern** of item missing data, in which there is no particular hierarchical or monotonic trend in the missing data structure (see Figure 12.3).

Survey designers may also deliberately decide to use randomized procedures to permit item missing data on selected variables for subsets of respondents. The technique of **matrix sampling** or "missing by design" sampling can be employed with modularized

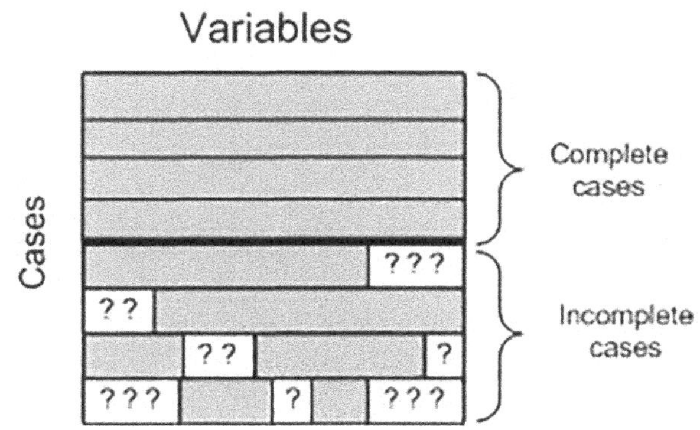

FIGURE 12.3
Generalized Pattern of Missing Data.

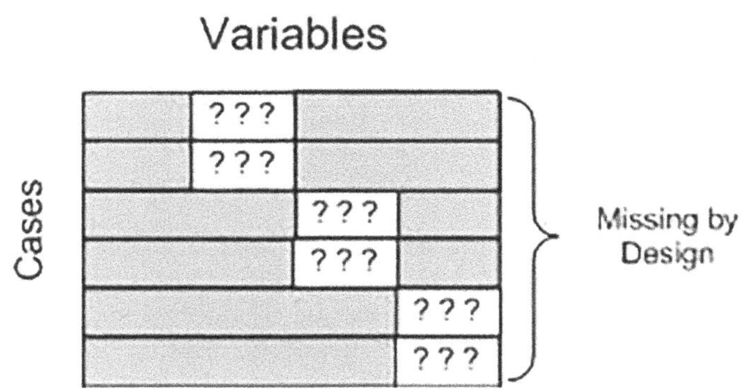

FIGURE 12.4
Matrix Sampling (Missing by Design).

sets of survey questions (Raghunathan & Grizzle, 1995; Thomas et al., 2006; Gonzalez & Eltinge, 2007; Chipperfield & Steel, 2011). In these designs, a battery of core questions is asked of all respondents and more in-depth modularized questionnaire components are randomly assigned to subsamples of survey participants. Matrix sampling designs tend to produce a non-monotonic missing data structure such as that illustrated in Figure 12.4, and the technique of **multiple imputation** (to be discussed later in this chapter) has typically been used to analyze these data.

12.2.3 Item Missing Data Mechanisms

In describing the multiple imputation method and other imputation techniques, we will employ the notation introduced by Little and Rubin (2019). Let Y denote a complete data matrix (i.e., the matrix of true values for all survey variables of interest, where there are no missing data). This is illustrated below for a simple data set of $i = 1, \ldots, 4$ cases and $j = 1, \ldots, 4$ variables. Note in the illustration below that the values ($\mathbf{y}_{12}$, $\mathbf{y}_{23}$, $\mathbf{y}_{31}$, $\mathbf{y}_{44}$) are boldfaced and italicized, to signify that these true underlying values will *not* be observed in the survey.

$$\mathbf{Y} = \begin{bmatrix} y_{11} & \boldsymbol{y_{12}} & y_{13} & y_{14} \\ y_{21} & y_{22} & \boldsymbol{y_{23}} & y_{24} \\ \boldsymbol{y_{31}} & y_{32} & y_{33} & y_{34} \\ y_{41} & y_{42} & y_{43} & \boldsymbol{y_{44}} \end{bmatrix}$$

This underlying set of true values of the variables of interest is decomposed into two subsets of values, $\mathbf{y} = \{\mathbf{y}_{obs}, \mathbf{y}_{mis}\}$, where $\mathbf{y}_{obs}$ are the values that are observed and $\mathbf{y}_{mis}$ are the values that are not observed and replaced by item missing values (e.g., a "." code in Stata). Readers should note that the "data" matrix Y can include design variables such as stratum and cluster codes and weights, which generally do not have any missing data.

Let M denote a missing data indicator matrix of the same dimensions as Y, where the value in row i and column j is equal to 1 if the corresponding value in Y is recorded as missing in an actual survey, and 0 if the value is observed. Returning to the simple example of four cases and four variables, the matrix of missing data indicators would look like this:

$$M = \begin{bmatrix} 0 & \boldsymbol{1} & 0 & 0 \\ 0 & 0 & \boldsymbol{1} & 0 \\ \boldsymbol{1} & 0 & 0 & 0 \\ 0 & 0 & 0 & \boldsymbol{1} \end{bmatrix}$$

A **missing data pattern** describes the distribution of the values in M. A **missing data mechanism** defines the conditional distribution of the values in M given the values in Y, $P(M \mid Y)$. The missing data mechanism can be thought of as a probability model for the missing data indicator. In other words, *why* are the missing data arising? Are reasons for missing data related to the study variables in any way? Survey analysts need to have a basic understanding of missing data mechanisms in order to evaluate the models and methods that are applied in compensating for missing data.

The missing data mechanism is defined to be **missing completely at random (MCAR)** if the probability that a respondent does not report an item value is completely independent of the true underlying values of all the observed and unobserved variables. Expressed more formally, $P(M \mid Y) = P(M)$ for all Y. Complete case analysis (listwise deletion) requires the assumption that the missing data are MCAR.

The mechanism is termed **missing at random (MAR)** if missingness depends only on the observed values of the variables in the survey; that is, $P(M \mid Y) = P(M \mid Y_{obs})$. In other words, the probability of a value being recorded as missing depends on *observed values of other variables*, but not on the missing values themselves. Using an example, the probability of having a missing value on height might depend on the observed value of the individual's fully-observed age, but *not* on the unobserved value of height. Finally, the mechanism is **missing not at random (MNAR)** if missingness depends on the unobserved values of the variables with missing data (in addition to the observed values) in the survey data set; that is, $P(M \mid Y) = P(M \mid Y_{obs}, Y_{mis})$. Again, using the example, the probability of having a missing value on height does depend on the actual height, and possibly also depends on the age of the respondent.

From the standpoint of being able to address item missing data effectively and practically in a survey data set, most imputation methods and software assume that the missing data mechanism is in fact MAR. Certainly, the MAR assumption may not strictly apply for all missing items. It is reasonable to expect that the probability of item missing data on survey variables such as personal income may depend, at least in part, on the underlying value even when other factors are accounted for. In applying imputation methods that assume MAR, the resulting imputations may not completely compensate for an MNAR missing data mechanism.

Unfortunately, missing data problems where the mechanism is strongly MNAR are difficult to address in a simple and straightforward manner. Techniques for analyzing data subject to non-ignorable missingness such as **selection models** (Heckman, 1979) or **pattern mixture models** (Andridge & Little, 2011; West & Little, 2013; Little & Rubin, 2019; Andridge & Little, 2020) have been proposed, but they typically require either independent assessment of the missing data model or interpretation based on a sensitivity analysis (Scharfstein et al., 1999; Little et al., 2020) conducted over a set of reasonable models postulated for the complete data and the MNAR mechanism.

12.2.4 Review of Strategies to Address Item Missing Data in Surveys

Before turning specifically to the subject of best practices for imputation of item missing data from complex sample surveys, it is useful to review the full range of options that are available to survey analysts.

12.2.4.1 Option 1: Complete case analysis

The simplest approach to handling item missing data is to do nothing and analyze only the available set of complete cases. Software that defaults to listwise deletion will automatically strike any case with item missing data on one or more variables from the analysis. Implicit in this approach is the assumption that the missing data are MCAR. Practically, if the individual and aggregate rates of item missing data for variables included in the analysis are very low (say, less than 2% of cases have missing values), analysis based on complete cases only should be acceptable. However, even when item missing data rates are low, the listwise deletion of cases with item missing data may result in small losses of sample precision, irregular counts of case inputs over different analyses, and potentially small biases if the missing data mechanism is MAR or MNAR.

12.2.4.2 Option 2: Use full information maximum likelihood (FIML) methods

This alternative approach generally involves the use of all available information in a survey data set (including values from cases with incomplete data) and applies maximum likelihood estimation techniques, given all the observed information, to compute parameter estimates of interest. The FIML technique assumes that the missing data are MAR. Applications of the FIML technique for complex sample survey data are not common, primarily due to the challenge of incorporating informative sample design features (stratification, cluster sampling, weighting) into the maximum likelihood analysis and the corresponding inferential techniques. Chambers et al. (2012) and Allison (2012a) are excellent references for analysts who wish to explore the application of maximum likelihood methods to complex sample survey data with missing values.

12.2.4.3 Option 3: Use the Expectation-Maximization (EM) algorithm

This approach to handling missing data, which is described in extensive detail in Little and Rubin (2019), is an iterative procedure defined by a two-step algorithm. The algorithm is initiated by a model of interest that the analyst wishes to fit to a data set that might include missing values on the variables of interest. First, in the expectation (E) step, the **EM algorithm** computes the expected value of the complete data log-likelihood function (given the model) based on the cases with *complete* data and the algorithm's estimates of the sufficient statistics for the missing data (given the model and the available data points for the incomplete cases). Then, in the maximization (M) step, the algorithm substitutes the expected values for the missing data obtained from the E step into the complete data likelihood and maximizes the likelihood function *as if no data were missing* to obtain new parameter estimates. The new parameter estimates are substituted back into the E step and the EM cycle repeats. The procedure iterates through these two steps until a pre-specified convergence criterion for the parameter estimates is satisfied (i.e., there is little change in the estimates at subsequent steps). This approach is currently employed by the Missing Values Analysis (MVA) module in the SPSS software but is limited in that standard errors and associated test statistics (e.g., t-tests) do not reflect the complex sample design or the uncertainty in estimating the missing values. Because of this limitation, imputation methods described below are generally considered more appropriate for survey data.

12.2.4.4 Option 4: Analyze complete cases but introduce additional weighting adjustments to compensate for item missing data on a key variable

The use of weighting to compensate for missing data is generally limited to monotonic patterns of missing data, in which large numbers of variables are associated with each "step" in the pattern. As described in Section 12.2.1 above, unit nonresponse, phase nonresponse, and longitudinal attrition in panel surveys produce missing data patterns where a weighting adjustment is the practical choice.

There is no theoretical barrier to employing a re-weighting approach to compensate for item missing data on key survey variables. Practically, though, attempting to adjust the available survey weight variables to address item missing data on single variables leads to difficulties. If the pattern of missing data is generalized (see Figure 12.3), different weighting adjustment factors would be needed for each target variable with missing data. Analytically, the weight-by-variable approach would work for univariate analyses, but which of the variable-specific weights would be chosen for a multivariable analysis? Simply put, imputation is the better strategy for addressing multivariate, generalized patterns of item missing data.

12.2.4.5 Option 5: Perform a single imputation of missing values, creating a "complete" data set

Public use survey data sets are often released with a single imputed value replacing missing data on key survey variables. Ideally, the data producers who performed the imputation also include a companion imputation flag variable that distinguishes actual from imputed values in the data set (see Section 4.4). When missing values are 'flagged" in a survey data set, data users may also choose to perform their own single imputations using an established stochastic imputation method such as the **hot deck** (Andridge & Little, 2010), **regression imputation**, or **predictive mean matching** (PMM; Little & Rubin, 2019). Stata, R, SAS, and other software packages include procedures that allow data users to perform imputations using these techniques. Use of deterministic procedures such as single mean, median, or modal value imputation is not encouraged unless the imputation is simply serving to fill in a small handful of missing values for an otherwise nearly complete survey variable (see Chapter 11 for examples of the modal approach).

The advantage of a singly imputed data set is that it is "complete," with missing values replaced by analyzable data entries. Provided that the imputation technique is multivariate and retains the stochastic properties in the observed data, a single imputation may address potential bias for a MAR missing data mechanism. The principal shortcoming in the standard design-based analysis of a singly-imputed data set is that it precludes estimation and inference that fully reflects the variance attributable to the item missing data imputations. Here again, if rates of item missing data for individual variables and in the aggregate are low, the resulting slight underestimation of variance may not be a true problem. For imputations performed using the hot deck method, Rao and Shao (1992) proposed a Jackknife technique for estimating variances from singly imputed data sets that include missing data flags, but the method has not been widely incorporated into survey data analysis procedures in the major software packages.

12.2.4.6 Option 6: Perform multiple imputations (MI) or fractional imputation (FI) for estimation and inference

A final alternative to address the problems of generalized patterns of missing data in survey analysis is to conduct the analysis in the model-based **Multiple Imputation (MI)** framework or alternatively using the design-based method of **Fractional Imputation (FI)**. The MI and FI methods will be explored in detail in Sections 12.4–12.6. Like single imputation approaches, MI and FI produce completed data sets for analysis. Although theoretically distinct in their approaches, both MI and FI use repetition of the imputation process or multiple possible outcomes for y_{mis} to enable estimation and inference that reflects the uncertainty due to imputation of missing values.

12.3 Factors to Consider in Choosing an Imputation Method

There is no single imputation method or statistical modeling technique that is optimal for all forms of item missing data problems. Survey analysts have a range of options to consider depending on the missing data pattern, the missing data mechanism, and rates of item missing values. Imputation techniques ranging from the simplest univariate mean substitution to a full application of multiple imputation or fractional imputation analysis share some basic attributes—for example, all imputation techniques produce a *completed* data set that can then be analyzed using standard software procedures for the analysis of complex sample survey data. However, in cases where the missing data problem is non-trivial, the goals of the imputation process should be more ambitious. The following is a list of those goals.

12.3.1 Goal 1: Stochastic

The imputation procedure should be *stochastic* — that is, based on **random draws** of donor values from observed cases (as in the **hot deck** or **predictive mean matching**), or of model parameters and error terms from the predictive distribution of y_{mis}. For example, in linear regression imputation of the missing values of a continuous variable y_k, the predictive function used to impute missing values based on the conditional predictive distribution for y_k may be written as $\hat{y}_{i,k,mis} = \hat{B}_0 + \hat{B}_{j\neq k} \cdot y_{i,j\neq k} + e_{i,k}$. In forming the imputed values of $y_{i,k,mis}$ for each sampled individual i, the individual predictions should incorporate random draws of the coefficients and independent draws of the errors ($e_{i,k}$) from their respective estimated distributions. In a hot deck imputation procedure or procedures such as predictive mean or propensity score matching imputation, the donor value for $y_{k,mis}$ is drawn at random from observed values in the same hot deck cell or in a matched "neighborhood" of the missing data case.

12.3.2 Goal 2: Multivariate

To preserve associations among the many variables that may be included in the analysis model, the imputation procedure should be multivariate in nature. The multiple imputation algorithms included in today's major statistical software packages preserve the multivariate properties of the data, through a sequence of conditional imputations (given monotonic missing data patterns) or using iterative methods that are designed to simulate draws of missing values from the joint posterior distribution of the multivariate set of survey items. Donor-based methods such as the fractional imputation hot deck can also be adapted to jointly impute missing data for multiple variables conditional on the observed data for the case.

12.3.3 Goal 3: "Model-Based"

The procedure for imputing missing values should be explicitly or implicitly based on a model for the data to ensure the statistical transparency and integrity of the imputation process. A process of arbitrary assignment of imputations, even if based on expert judgement, is not replicable in the scientific sense and will not preserve the statistical properties (means, variances, covariances) of the survey data. The model that underlies the imputation process is often an explicit distributional model, but good results may also be obtained using techniques where the imputation model is implicit (see Theory Box 12.1).

THEORY BOX 12.1: EXPLICIT VS. IMPLICIT MODELS FOR IMPUTATION OF ITEM MISSING DATA

Imputations of missing values for survey variables may be based on explicit distributional models or they may be based on techniques that "imply" an underlying model or set of assumptions concerning the missing data. In the example described in Section 12.4, the joint distribution for the three continuous variables (diastolic blood pressure, age, and BMI) is assumed to be multivariate normal: $f(y \mid \theta) = \text{MVN}(\mu, \Sigma)$. This is an example of an **explicit imputation model.** A modification of this explicit model might be to assume that diastolic BP, age, and BMI follow multivariate normal distributions that are conditioned on gender. This modified model is an example of the **general location model** (Schafer, 1999).

An example of an imputation procedure that incorporates an **implicit imputation model** is **hot deck imputation**. Following this approach, cases are grouped into cells based on common values for observed categorical variables (e.g., age category × gender), and within these cells missing values of another variable (e.g., diastolic blood pressure) are imputed by randomly drawing a replacement value from an observed "donor" in the same hot deck cell (Kalton & Kasprzyk, 1986; Andridge & Little, 2010). Although this simple hot deck method does not impute based on an explicit regression model for y, the model that is implied by the method might be the general location model or an ANOVA model.

12.3.4 Goal 4: "Congenial" to Intended Analyses

In the words of Meng (1994), we should avoid the mistake of choosing an imputation model such that ". . . a procedure for analyzing the imputed data sets cannot be derived from the model adopted for the imputation." This concept of **congeniality** between the model used for imputation and the analysis model/methods has been the main theme in critiques of multiple imputation as a general solution to item missingness in survey data (Fay, 1996; Kim et al., 2006). This concern is especially relevant in MI applications for public use data sets where the imputer and analyst are generally not the same individual. In imputing generalized patterns of missing data in a large survey data set, the imputer may have the advantage of being able to access data that may not be available to the analyst (e.g., detail on the structure of the sample design, or data from the sampling frame or an external administrative source). Including added information in the imputations that is not available to analysts may result in **"super-efficiency"** (Rubin, 1996) in the analysis of imputed data.

However, the general-purpose imputer is also disadvantaged in that they cannot know in advance all the analytic applications (Which variables? Which subpopulations? What are the true functional forms of the models?) for which the completed data sets will be used. To address the bias that can result if imputations are "uncongenial" to the ultimate analyses, this chapter will recommend two practical solutions. First, compared to the analysts' model, the imputation model should: 1) be broader in terms of variables; 2) include controls for major subpopulations or potential domains of analysis; and 3) to the extent possible, reflect the true non-linearities and interactions among the variables included in the imputation model (see Figure 12.6 later in this chapter). Second, and maybe most effective in preventing imputations that are not congenial with the analyst's model, if the survey data include appropriate imputation flags or standard missing value codes (e.g., −88) for missing values, recent software developments permit the analyst to perform multiple imputations directly as part of their analysis session, thereby reducing the prospect of serious uncongeniality between the imputation model and the analytic method/model.

A special issue of congeniality of the imputation model and analysis methods/models for survey data is how informative features of the complex sample design, including stratification, cluster sampling, and weighting, are factored into the imputation of missing values (Reiter et al., 2006). Guidance on this issue will be provided under the general overview of MI provided in Section 12.4. The Finite Population Bayesian Bootstrap method (Zhou et al., 2016a, 2016b, 2016c) and related weighted bootstrap MCMC approaches (Goldstein et al., 2018) are also specifically designed to solve the problem of multiply imputing missing values in data collected under an informative complex sample design.

12.3.5 Goal 5: Enable Analysis (Estimation and Inference) that Incorporates the Imputation Uncertainty

As noted above, the primary disadvantage of a single imputation approach to analysis with item missing data is that is does not provide an accessible means to incorporate the imputation uncertainty into the standard errors (and confidence intervals) that are critical to making correct inferences about population parameters. Multiple imputation (MI) and fractional imputation (FI) are two theoretically distinct approaches to the imputation of missing survey data that share the common feature that multiple imputed values are assigned to each missing data point. In replicating the imputation over multiple possible values of y_{mis}, both MI and FI enable estimation of the added variance that is due to the imputation process.

12.3.6 Goal 6: Robust

No imputation model or procedure will ever exactly match the true distributional assumptions for the underlying random variables or the assumed missing data mechanism. Imputation procedures applied to survey data sets by data producers or data users should be reasonably robust against modest departures from the underlying assumptions. Fortunately, empirical research has demonstrated that if the more demanding theoretical assumptions underlying MI are relaxed, applications to survey data can produce estimates and inferences that remain valid and robust (Herzog & Rubin, 1983). Likewise, Kim (2011) provides similar evidence of the robustness of the FI method, even under conditions in which the analysis model includes parameters that were omitted from the imputation model.

12.3.7 Goal 7: Usable

For applied survey data analysis, usability and software support is always a criterion that should be used in selecting a statistical procedure. In the most mathematically rigorous representations, the theory of multiple imputation or the fractional imputation method can be highly complex (Rubin, 1987; Schafer, 1997; Kim, 2011). For highly specific problems with complex missing data patterns or mechanisms, survey analysts who lack the mathematical sophistication are encouraged to consult with a survey statistician prior to conducting and publishing their analysis. However, for general missing data problems of the type illustrated in Section 12.7, today's software provides a usable platform for conducting a multiple imputation or fractional imputation analysis that when appropriately applied should generate robust inferences for the target population.

12.4 Multiple Imputation

12.4.1 Overview of Multiple Imputation and MI Phases

The ideas that underlie multiple imputation (MI) methods for missing data were formulated by Donald Rubin in the early 1970s (Rubin, 1976). Since that time, MI theory and methods have been advanced by Rubin and his students (Rubin, 1996; Yucel, 2011). MI is not simply a technique for imputing missing data. It is also a method for obtaining estimates and inferences for statistics ranging from simple descriptive statistics to the population parameters of complex multivariable models.

As illustrated in Figure 12.5, three distinct phases comprise a complete **multiple imputation analysis** of a survey data set. The first phase is the definition of the data and the distributional components of the **imputation model**. The second phase applies specific methods and algorithms to generate the multiple imputations of the missing values. The third and final phase is the calculation of MI estimates and standard errors, and the construction of confidence intervals and hypothesis test statistics for making inferences regarding population parameters and relationships.

12.4.2 Models for Multiple Imputation of Missing Data

The actual process of imputing item missing values is governed by the **imputation model**, which is defined by the set of values that are available to the imputation process,

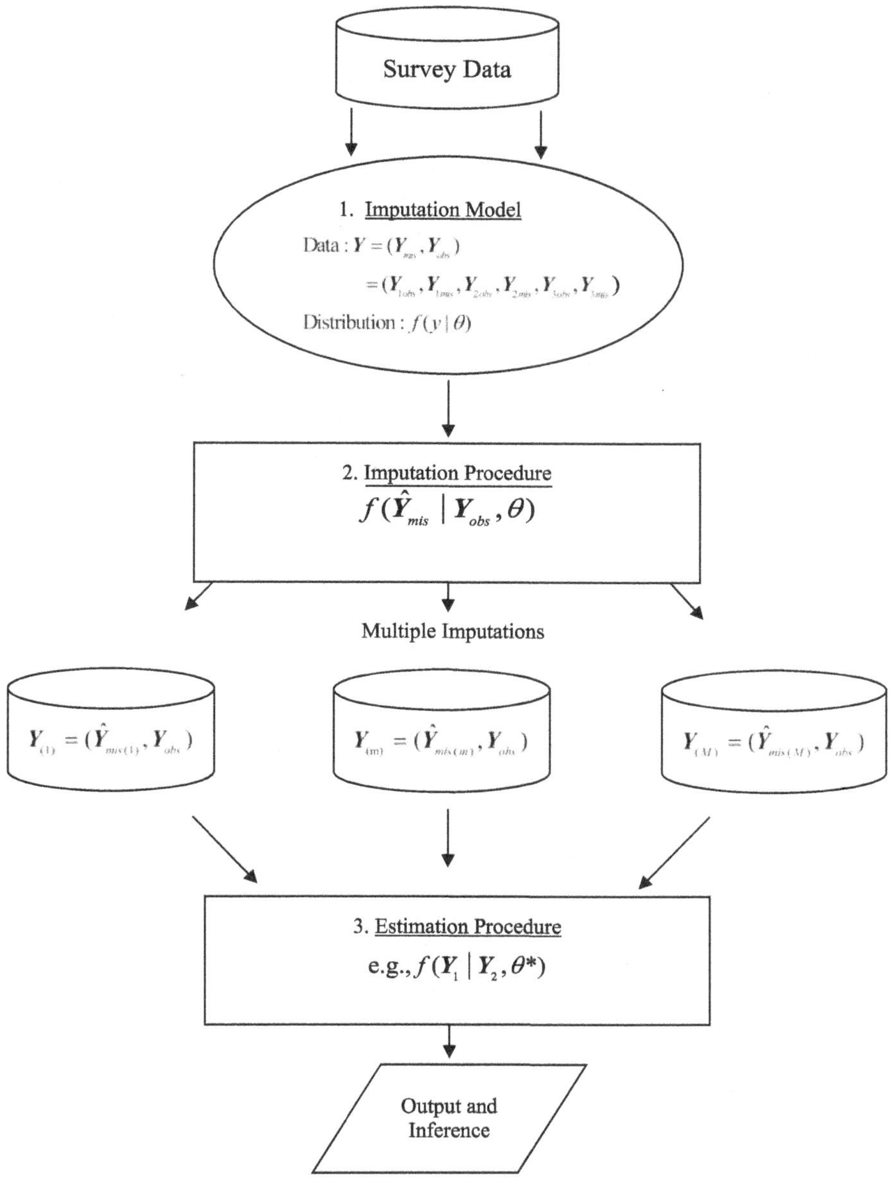

FIGURE 12.5
The Stages of Multiple Imputation.

$\mathbf{y} = \{\mathbf{y}_{obs}, \mathbf{y}_{mis}\}$, and the distributional assumptions, $f(y \mid \theta)$, for the multivariate relationships among the elements of **y**. As noted in the goals that we outlined for imputation methods (see Section 12.3), to ensure congeniality with a future analyst's modeling of the imputed data, the imputation model should:

- Include a broader set of variables than the eventual analytic model;
- Control for variables representing major subpopulations or domains of analysis;

- Reflect to the degree possible the true non-linearities and interactions among the variables included in the imputation model; and
- Utilize conditional predictive methods that conform to methods (e.g., logistic regression) that analysts will use to analyze the imputed data.

Consider a problem in which there are three continuous variables of interest in a 2011–2012 NHANES analysis of adult diastolic blood pressure (DBP): y_1 = diastolic BP (mm Hg); y_2 = age (years); and y_3 = Body Mass Index (kg/m^2). Further, let's assume that the item missing data rate is 9% for diastolic BP and 2% for Body Mass Index (BMI), and that following rigorous edit checks, cleaning, and follow-up by NHANES staff, age is observed for every case. One possible imputation model would be to include all three variables in the imputation process and assume that the joint distribution for these three continuous variables is multivariate normal: $f(\mathbf{y} \mid \boldsymbol{\theta}) = \text{MVN}(\boldsymbol{\mu}, \Sigma)$. Under this model, multiple imputations of the missing data for diastolic BP and BMI are easily performed using methods described in Schafer (1997). The imputations performed under this imputation model would serve for univariate MI estimation of the mean diastolic BP, μ_1, or mean BMI, μ_3, and they would also serve for the MI estimation of the regression of diastolic BP on age and BMI, for example, $E(y_1 \mid y_2, y_3) = B_0 + B_1 \cdot y_2 + B_2 \cdot y_3$. Van Buuren et al. (1999) illustrate the process of imputation model selection for an MI analysis of blood pressure in a cohort of persons age 85+ from Leiden in the Netherlands.

12.4.2.1 Choosing the Variables to Include in the Imputation Model

The choice of variables to include in the imputation model should not be limited to only variables that have item missing data or variables that are expected to be used in a subsequent analysis. As a rule of thumb, the set of variables included in the **imputation model** for an MI analysis should be much larger and broader in scope than the set of variables required for the **analytic model**. For example, if age and BMI are the chosen predictors in the analytic model for diastolic BP, the imputation of item missing data for diastolic BP and BMI might include many additional variables such as gender, race/ethnicity, marital status, height, weight, systolic BP, etc.

Figure 12.6 is a schematic representation of the desired relationship between the imputation model and the analytic model. The large outer oval represents the broad set of imputation model variables and joint distribution model that can contribute to the derivation of imputed values for missing items. Nested within this broad set of imputation model variables is the smaller set of variables and the conditional model that will be the focus of the analysis.

Obviously, it is not feasible to define an imputation model and perform multiple imputations using every possible variable in the survey data set. Based on recommendations from Schafer (1999) and Van Buuren (2012), some practical guidelines for choosing which variables to include in the imputation model are the following:

- All analysis variables (e.g., **Y1** and **Y2** in Figure 12.6)
- Other variables that are correlated or associated with the analysis variables (**Y3**)
- Variables that predict item missing data on the analytic variables (**Y4**).

Failure to include one or more analysis variables (**Y1** and **Y2**) in the imputation model can result in bias in the subsequent MI estimation and inference. Including additional

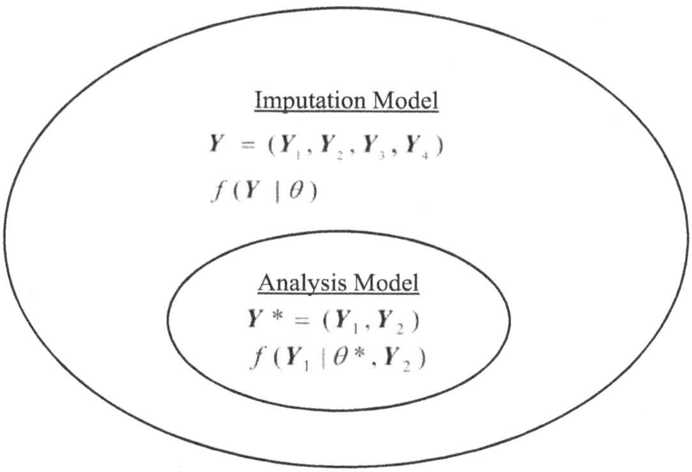

FIGURE 12.6
Relationship of the Imputation and Analysis Models.

variables that are good predictors of the analytic variables (e.g., **Y3**) improves the precision and accuracy of the imputation of item missing data. For example, when analyzing longitudinal data from a panel survey, **Y1**, **Y2** and **Y3** might represent measures of the same characteristic across three waves. An analytic model may be focused on estimating the relationship of **Y2** and **Y3**, but an analyst imputing missing values for **Y3** could certainly use **Y1** to improve the imputation model. Finally, under the assumption that item missing data are MAR, incorporating variables (**Y4**) that predict the propensity for response reduces bias associated with the item missing data mechanism.

Imputation of missing data collected under a complex sample design must also consider this question: How can informative design features such as stratification, cluster sampling, and weighting be incorporated into the imputation model? Early published work in this area (Rubin, 1996; Reiter et al., 2006) emphasized the importance of including variables identifying complex sample design features in the imputation models. Reiter et al. (2006) demonstrated that ignoring complex sample design variables in the imputation models can lead to bias in multiple imputation estimates when the design variables are *informative* or related to the survey variable of interest (which is often the case in practice). Including design variables in the imputation models when they are *not* related to the survey variable of interest can at worst lead to inefficient (and thus conservative) multiple imputation estimates.

Theoretically, the effects of sampling strata and randomly sampled clusters may be modeled using a multilevel model that includes fixed effects of the strata, γ_{str}, and random effects of the clusters randomly sampled within strata, $u_{cl(str)}$:

$$y_i = \beta x_i + \gamma_{str,i} + u_{cl(str),i} + \varepsilon_{(str,cl),i}$$

A hierarchical Bayes approach would be a flexible choice of method to multiply impute missing values of y under this mixed-effects regression model (van Buuren, 2012, Gelman et al., 2013). Recent software developments have made it possible to impute missing data using these types of multilevel models to explicitly account for complex sampling features;

examples include the contributed `pan` package and the contributed `micemd` package for the R software (Schafer & Yucel, 2002; Audigier et al., 2018).

As an alternative to this multilevel modeling approach, one could include the complex sample design strata, clusters, and weights as *fixed effects* in the imputation model (Reiter et al., 2006; Berglund & Heeringa, 2014). Empirically, the following reduced model (including fixed effects for the final survey weights and the stratum × cluster indicator) performs well and generally avoids the instability problems encountered in employing the full hierarchical mixed-effects model for design parameters:

$$y_i = \beta x_i + \delta w_i + \gamma_{\text{strat x cluster},i} + \varepsilon_i$$

For example, the following Stata command creates the composite stratum × cluster indicator variable, DESCODE, for the 2011–2012 NHANES data set:

```
gen descode = sdmvstra*10+sdmvpsu
```

Unfortunately, for complex sample designs with many strata and two or more clusters per stratum (e.g., the Health and Retirement Study, $H = 56$), this full "fixed effects" specification with indicator variables for the strata × cluster combinations may require estimation of too many parameters and result in an unstable model fit or convergence failure. In such cases, it may be necessary to include as predictors in the imputation model 1) fixed characteristics that differentiate strata and PSU population characteristics (e.g., geographic region, urban/rural, PSU size), 2) collapsed groups of similar strata, or 3) combined PSUs that form pseudo-PSUs.

The approaches outlined above explicitly include the complex sampling features as predictors in the imputation models. More recently, Zhou et al. (2016a, 2016b, 2016c) have demonstrated the benefits of using the **Finite Population Bayesian Bootstrap (FPBB)** to simulate synthetic populations based on informative complex sample design features, and then impute missing values in the simulated synthetic populations under simple random sampling assumptions, eliminating any further need to account for the complex sampling features in imputation and analysis. These authors provide R code for implementing their FPBB approach in Zhou et al. (2016b), and their method is also now available via the BBDESIGN command in the latest release of the IVEware software (Raghunathan et al., 2022). A fully worked example of using the FPBB approach to impute missing data in the 2011–2012 NHANES and conduct subsequent analyses of the imputed synthetic populations is available online as part of the IVEware documentation.[1]

Goldstein et al. (2018) described a related Bayesian MCMC method using weighted bootstrapping for imputation in the presence of informative survey weights and demonstrated via simulation that such an approach has good inferential properties. The algorithm outlined by Goldstein et al. (2018) is implemented in MATLAB, with code available from the authors. Kalpourtzi et al. (2024) demonstrated via simulation that a two-level MI approach works well for imputing missing data in multi-stage probability samples, especially when the auxiliary variables used for weighting adjustment or imputation are themselves subject to missingness. Tang et al. (2024) developed a "hybrid missingness model" for addressing both unit nonresponse and item nonresponse simultaneously, leveraging known marginal population distributions of auxiliary variables that are used in the models for both types of nonresponse.

Collectively, these important recent developments have provided analysts with principled tools and guidance for rigorously accounting for complex sampling features when imputing missing values in survey data sets.

12.4.2.2 Distributional Assumptions for the Imputation Model

Once the set of variables for the imputation model has been chosen, a decision must be made concerning the joint distributional model for the chosen variables. In theoretical discussions of multiple imputation methods, convenient choices of multivariate models for the joint distribution of the broad set of imputation variables might be multivariate normal (continuous), multinomial (categorical), or **general location** (mixed categorical and continuous) models. As described below in Section 12.4.2.3, many of the current software procedures explicitly assume that the variables in the imputation model follow one of these standard multivariate models.

Because the chosen imputation model may include many variables of different types (continuous, categorical, count, semi-continuous mixtures of zeroes and continuous values), it may be difficult if not impossible to specify an analytical form for the joint distribution of these variables. In such cases, several authors (Van Buuren et al., 1999; Raghunathan et al., 2001) have proposed using iterative **sequential regression** methods (see Section 12.4.3.3 for details) to approximate imputation draws from the unknown **joint posterior distribution** of the imputation model variables (Schafer, 1997). In these cases, the joint posterior distribution of the data is assumed to exist, but its exact form is never truly observed. A potential vulnerability of this technique is that the target distribution that is being simulated in the sequential regression process may not in fact exist for some missing data problems. Fortunately, empirical tests and simulation studies suggest that when applied to large survey data sets with moderate rates of item missing data, these generalized methods of multivariate imputation do in fact yield reasonable results even when the number and diversity of variables included in the imputation model is large.

12.4.3 Creating the Multiple Imputations

The second phase of a multiple imputation analysis is to generate $m = 1, \ldots, M$ **completed data sets** in which the missing values, y_{mis}, have been imputed. The $m = 1, \ldots, M$ independently imputed versions of the data set are termed **repetitions**. As described in Section 12.4.3.2 below, Bayesian MI imputes missing values by performing stochastic draws from the posterior predictive distribution, $f(y_{mis} \mid y_{obs})$, averaging over possible values of θ (or the parameters defining the imputation model). Using various algorithms, MI employs a sequence of imputation (I) and prediction (P) steps to approximate the posterior predictive distribution. After convergence of the algorithm, MI generates $m = 1, \ldots, M$ independent imputations of the y_{mis} through stochastic draws from the constructed posterior predictive distribution.

A variety of algorithms, often tailored to the specific missing data pattern and mechanism, are used to generate the imputations. Readers are referred to the comprehensive texts by Rubin (1987), Schafer (1997), and Gelman et al. (2013) and articles by Tanner and Wong (1997), Gelfand and Smith (1990), Gelfand et al. (1990), and King et al. (2001) for more in-depth coverage of specific algorithms for addressing missing data under explicit data models and theoretically exact Bayesian methods.

Here we consider the more typical survey data context where the imputation model is multivariate, the model includes variables of all types, and there is a generalized pattern of missing data across the variables in the model. In such cases of a "messy" pattern of missing data where exact methods do not strictly apply, the authors of multiple imputation software have generally followed one of three general approaches described in the following sections.

12.4.3.1 Transforming the Imputation Problem to Monotonic Missing Data

The first approach is to transform the pattern of item missing data to a monotonic pattern by first using simple imputation methods or a **Markov Chain Monte Carlo (MCMC)** posterior simulation approach to fill in the missing values for variables in the model that have very low rates of item missing data (Berglund & Heeringa, 2014). Under a specified model for the data, imputation of a true monotonic pattern of item missing data is greatly simplified: the imputation of missing data is reduced to a sequence of imputations for single variables. Each imputation requires only a single step—no iteration is required as in the MCMC approach described below. This approach works best when the generalized pattern of missing data is dominated by missing data for one or two variables. Through inspection of the pattern of missing data across individual cases, it would be possible to use simple methods to "fill in" selected missing values for the variables with small amounts of missing data, creating a monotonic pattern of missing data (Little & Rubin, 2019) and simplifying the imputation problem. Solas 5.0 (https://blog.statsols.com/overviews-of-multiple-imputation-in-solas-5.0) and PROC MI in SAS offer the analyst this option.

12.4.3.2 Specifying an Explicit Multivariate Model and Applying Exact Bayesian Posterior Simulation Methods

A second algorithmic approach to generating imputations for a generalized pattern of item missing data is to declare an explicit probability model for the data, for example, $f(y\,|\,\theta) = \text{MVN}(\mu, \Sigma)$ or $f(y\,|\,\theta) =$ general location model (Schafer, 1997). Individual imputations are then generated by random draws from the correct posterior distribution of the missing values $f(\hat{Y}_{mis}\,|\,Y_{obs}, \theta)$ under the explicit probability model.

Early versions of multivariate imputation programs, such as NORM (Schafer, 1997), assumed a multivariate normal distribution for all variables. Actual imputations for the multivariate model employed **data augmentation**, a form of the general class of MCMC posterior simulation algorithms. Later updates such as MIX (Schafer, 1997) incorporated the option of a general location model, more suitable to a mix of continuous and categorical variables. SAS PROC MI is another general-purpose MI program that offers the option to employ the MCMC methods of Schafer (1997) and can accommodate both continuous and categorical variables.

12.4.3.3 Sequential Regression or "Imputation by Chained Equations"

The third alternative for multiply imputing item missing data for large mixed sets of continuous, nominal, ordinal, count, and semi-continuous variables is a method that is variously labeled the **sequential regression (SR) algorithm**, the **imputation by chained equations (ICE)** technique, or the **fully conditional specification (FCS)** method. The sequential regression algorithm forms the basis for the %IMPUTE command in the IVEware software (Raghunathan et al., 2001). The ICE algorithm (Van Buuren et al., 1999) was later extended and enhanced (Royston, 2005; Carlin et al., 2008) and is now available in the `mi impute chained` command of Stata, in addition to various contributed R packages (e.g., `mice` and `mi`). SAS PROC MI now includes a sequential regression algorithm specified with an `fcs` statement. Appendix A provides an updated inventory of the features of these and other MI software.

Each of these sequential regression algorithms is based on an iterative **Gibbs Sampler** type of algorithm. Each iteration ($t = 1, \ldots, T$) of the algorithm moves one-by-one through the sequence of variables in the imputation model, $Y = \{Y_1, \ldots, Y_k, \ldots, Y_p\} = \{Y_1, Y_2, Y_3, Y_4\}$. In the very first iteration, the variable with the *smallest* amount of missing data is regressed on variables in the imputation model with complete data, using an appropriate model (e.g., logistic regression for a binary variable). Imputations for the missing values on the first variable are generated by stochastic draws from the predictive distribution defined by the regression model. Next, the variable with the second smallest amount of missing data is regressed on the first imputed variable and all other variables in the imputation model with complete data, and imputed values are generated. This algorithm proceeds until the variable with the most missing data is regressed on all other variables (some having imputed values) in the imputation model to develop the predictive distribution for the last variable.

Subsequent iterations use some or all of the variables in the imputation model (depending on the software) to re-impute the values that were originally missing for each of the variables. Based on the current (iteration t) values of the observed and imputed values for the imputation model variables, regression models are once again estimated for each target variable that originally had missing data, and as mentioned above, the regression models can be tailored to each specific variable. For example, the IVEware software uses linear regression (continuous variables), multinomial logistic regression (nominal/ordinal variables), Poisson regression (count variables), or two-stage logistic/linear regression (semi-continuous variables) to model the conditional predictive distributions of individual variables Y_k, $f(\widehat{Y}_k^{(t)} \mid \widehat{Y}_{j \neq k}^{(t)}, \theta^{(t)})$. Updated imputations, $\hat{Y}_k^{(t)}$, are then generated by stochastic draws from the predictive distribution defined by the updated regression model. Once the last variable in the sequence has been imputed, the algorithm again cycles through each variable. The iteration of the cycles stops when a user defined convergence criterion is met (e.g., $T = 10$ iterations in the Stata `mi` procedures).

Under an explicitly defined imputation model, $f(Y \mid \theta)$, and a suitable prior distribution, $g(\theta)$, for the distributional parameters, the sequential regression algorithm will, in theory, converge to the Bayesian joint posterior distribution for the data. Since the sequential regression method never explicitly defines $f(Y \mid \theta)$, the assumption must be made that the posterior distribution does exist and that the final imputations generated by the iterative algorithm do represent draws from an actual, albeit unknown, posterior distribution. Although the exact theoretical properties of the resulting imputations remain somewhat of an unknown, in most applications the sequential regression approach with its application of the Gibbs Sampler algorithm does converge to a stable joint distribution, and the multivariate imputations generated by the algorithm show reasonable distributional properties. Further, empirical studies have shown that this method produces results comparable to those for the EM algorithm and the exact methods of Bayesian posterior simulation (Heeringa et al., 2002).

Multiple imputation (MI) inference requires that the imputation process be independently repeated multiple times, $m = 1, \ldots, M$. In theory, MI inference is most efficient at recovering statistical information from incomplete cases when $M = \infty$. Obviously, analysts' tolerance for MI would have faded away quickly if each analysis required an infinite number of independently imputed data sets. Fortunately, research has shown that virtually all of the efficiency gains possible in an MI analysis can be achieved using as few as $M = 5$ to $M = 20$ independent repetitions of the imputation process (e.g., Rubin, 1986). More recent work has proposed a rule for how many imputations are needed (von Hippel, 2020), with a focus on being able to replicate estimated standard errors across replications of the multiple imputation procedure. This work includes two user-written commands that can be employed in SAS and Stata for these calculations.

Although there are no exact criteria to establish that an MCMC or sequential regression imputation algorithm has converged to a stable joint posterior distribution, graphical methods such as **trace plots** and **autocorrelation plots** can be used to monitor the progress of the algorithm (van Buuren, 2011). As the number of iterations progresses, trace plots should show the successive iterations' posterior estimates of means and variances converging to a stable distribution. Autocorrelation plots should show that the autocorrelation of sequential draws declines, ultimately varying randomly about zero. SAS PROC MI provides users with the option to output these diagnostic charts for MCMC and FCS imputations (Berglund and Heeringa, 2014).

The example exercise presented in Section 12.6 will demonstrate the application of the Stata `mi impute chained` sequential regression command to a general multivariate missing data problem.

12.4.4 Estimation and Inference for Multiply Imputed Data

Once the item missing data have been multiply imputed under a suitable imputation model and the $m = 1, \ldots, M$ completed replicate data sets have been stored in an appropriate data format, the next step in a complete MI analysis is to compute MI estimates of the population parameters and the variances of the MI estimates. The MI estimates and standard errors can then be used to construct confidence intervals for the population quantities of interest to the survey analyst.

Many of the statistical software packages now include a pair of procedures for conducting MI analysis: one procedure to perform the multiple imputations of item missing data, and a second to perform MI estimation and inference based on the multiple files that are produced by the imputation procedure. Examples of such pairs of procedures include Stata's `mi impute` and `mi estimate`, PROC MI and PROC MIANALYZE in SAS, `mice()` and `with()/pool()` in the `mice` package for R, and %IMPUTE and %DESCRIBE / %REGRESS / %SASMOD in IVEware.

Although the MI software mentioned above supports coordinated processing of both the imputation and estimation/inference phases of an MI analysis, provided that some care is taken in choosing the imputation model, the imputation and estimation phases can be performed separately by different persons (e.g., imputation by data producers, analysis by data users) or by using separate procedures (e.g., conducting imputations in IVEware, and then estimation and inference in PROC MIANALYZE).

12.4.4.1 Estimators for Population Parameters and Associated Variance Estimators

MI estimates of finite population parameters (denoted here by θ) are computed using a simple averaging of the parameter estimates obtained by performing design-based analyses of the $m = 1, \ldots, M$ completed data sets output from an MI algorithm:

$$\bar{\theta} = \frac{1}{M}\sum_{l=1}^{M}\hat{\theta}_l \tag{12.1}$$

where:

$\hat{\theta}_l$ = estimate of θ from the completed data set $l=1, \ldots, M$

The corresponding MI variance for the MI estimate $\bar{\theta}$ is estimated as a function of the average of the estimated sampling variances for each repetition estimate and a between-imputation variance component that captures the imputation variability over the M repetitions of the imputation process:

$$var(\bar{\theta}) = \bar{U} + \left(\frac{M+1}{M}\right) \cdot B \tag{12.2}$$

where:

$\bar{U}$ = Within-imputation variance = $\frac{1}{M}\sum_{l=1}^{M} U_l = \frac{1}{M}\sum_{l=1}^{M} var(\hat{\theta}_l)$; and

B = Between-imputation variance = $\frac{1}{M-1}\sum_{l=1}^{M}(\hat{\theta}_l - \bar{\theta})^2$

In applications involving complex sample survey data, the estimated sampling variance of each repetition estimate, $var(\hat{\theta}_l)$, is computed using an appropriate Taylor Series Linearization or replication variance estimator. When using the FPBB technique introduced earlier (Zhou et al., 2016a, 2016b, 2016c), the multiply imputed synthetic populations are analyzed like simple random samples, so specialized variance estimation procedures are not necessary.

12.4.4.2 Model Evaluation and Inference

Based on the estimates of the within and between components of the multiple imputation variance estimator, the **fraction of missing information** is computed as:

$$\hat{\gamma}_{mi} = \left[\frac{\left(\frac{M+1}{M}\right) \cdot B}{\left(\frac{M+1}{M}\right) \cdot B + \bar{U}}\right] = \left[\frac{\left(\frac{M+1}{M}\right) \cdot B}{T}\right] \tag{12.3}$$

If missing data for a single variable y were imputed based only on the distribution of the observed values of that same variable, the fraction of missing information would equal the missing data rate for y. However, more generally, when the model for imputing y conditions on observed values of other related variables (denoted here by x_1 and x_2), for example, $\hat{y}_{i,mis} = \hat{B}_0 + \hat{B}_1 \cdot x_{1i} + \hat{B}_2 \cdot x_{2i}$, the imputation borrows strength from the multivariate relationships and the fraction of information lost will often be reduced: $0 < \hat{\gamma}_{mi} <$ Missing rate for y (see Nishimura et al., 2016, for a discussion of exceptions). We note that the fraction of missing information is specific to an *estimate* of interest.

Confidence intervals for descriptive population parameters or single parameters in superpopulation regression models are constructed from the MI estimate, its standard error, and a critical value from the Student t distribution (Rubin & Schenker, 1986):

$$CI_{(1-\alpha)}(\theta) = \bar{\theta} \pm t_{\tilde{v}_{mi},1-\alpha/2} \cdot se(\bar{\theta}) \tag{12.4}$$

where:

$\tilde{v}_{mi}$ = the degrees of freedom for the MI variance estimate (see Theory Box 12.2).

The unique feature of the MI confidence interval for population parameters is the degrees of freedom determination for the Student t distribution. Current software uses the "small sample" method of Barnard and Rubin (1999) to determine the degrees of freedom, v, for constructing the confidence interval. Theory Box 12.2 provides a brief description of this degrees of freedom calculation. The MI confidence intervals are routinely reported for individual descriptive parameters or regression model parameters by the MI analysis programs in Stata (`mi estimate`) and SAS (PROC MIANALYZE). Simulation studies have demonstrated that for large sample sizes, this MI confidence interval provides true coverage of the population value that is very close to the nominal 100(1 – α)% coverage level.

THEORY BOX 12.2: DEGREES OF FREEDOM FOR MI INFERENCES FROM COMPLEX SAMPLE SURVEY DATA

In general, large-sample multiple imputation confidence intervals for population values are based on a Student *t* distribution with the following degrees of freedom:

$$v_{mi} = (M-1)\hat{\gamma}^{-2} \tag{12.5}$$

where:

$\hat{\gamma}$ = the estimated fraction of missing information (12.3).

Barnard and Rubin (1999) point out that this large-sample approximation may not be suitable for "small" samples. Even in small simple random samples, this formula can lead to an estimate of degrees of freedom that exceeds the number of complete observations ($df \sim n - 1$). For most complex sample designs, application of this large-sample formula results in values of v_{mi} that greatly exceed the nominal design-based degrees of freedom (#clusters–#strata). Barnard and Rubin propose the following adjustment to establish the degrees of freedom for small samples or designs with limited degrees of freedom:

$$\tilde{v}_{mi} = \left(\frac{1}{v_{mi}} + \frac{1}{\hat{v}_{obs}}\right)^{-1} \tag{12.6}$$

where:

v_{mi} = the large sample MI degrees of freedom;

$\hat{v}_{obs}$ = the degrees of freedom for the complete data,

$$= \left(\frac{v_{com}+1}{v_{com}+3}\right) \cdot v_{com} \cdot (1-\hat{\gamma}_{mi})$$

v_{com} = the degrees of freedom for the complete case analysis;

$\hat{\gamma}_{mi}$ = the estimated fraction of missing information.

In the analysis of a small simple random sample, the degrees of freedom for the Student *t* reference distribution would be $v_{com} = n-1$. The complete data degrees of freedom for an analysis of a complex sample survey data set would be the now familiar $v_{com} = v_{des} \approx \#clusters - \#strata$. Chapter 3 discussed the approximate nature of the current rules for determining design-based degrees of freedom for complex sample survey data. The same or greater level of approximation must hold in any MI inferences from complex sample survey data. Additional research in this area is certainly needed. The Barnard and Rubin approximation to the effective degrees of freedom is now incorporated in MI estimation software such as Stata `mi estimate` and SAS PROC MIANALYZE.

Li et al. (1991) have developed an MI *F*-test statistic that serves the function of a Wald statistic to test multi-parameter hypotheses. In Stata, this multi-parameter *F*-test is invoked using the `mi test` post-estimation command.

One can also compute predictions of dependent variables of interest for new cases based on models fit to multiply imputed data, but there are some important considerations when doing so (including how to quantify the uncertainty in the predictions). Miles (2016) discusses how to compute these predictions and measures of uncertainty for the predictions. This approach has been implemented via the `mi predict` command in Stata. See also https://stats.oarc.ucla.edu/stata/faq/how-can-i-get-margins-and-marginsplot-with-multiply-imputed-data/ for more details about how to use the `margins` command with multiply imputed data; this approach has been implemented in the user-written `mimrgns` command.

12.5 Fractional Imputation

12.5.1 Background

We have seen in Section 12.4 that the Bayesian approach to analysis with item missing data is embodied in the now highly developed literature on theory, methods, and algorithms for multiple imputation (MI). **Fractional imputation (FI)** has been developed as a "frequentist" or design-based alternative to Bayesian MI for handling item missing data in complex sample survey data sets. The basic concept of fractionally imputing missing values using multiple "donors" or draws and apportioning the full survey weight across the resulting multiple records for a case was first mentioned by Kalton and Kish (1984). Over the next four decades, theoretical development of FI was extended in further work by Fay (1996), Kim and Fuller (2004), Durrant and Skinner (2006), Kim (2011) and Kim and Yang (2014). Unfortunately, like many good ideas in statistics, applications of FI in actual survey analyses were delayed by a lack of software in major statistical software packages. This was rectified in 2015 when the SAS® system incorporated an FI hot deck imputation option into its SURVEYIMPUTE procedure.

The underlying theory and methods of the FI technique have now been developed for the case in which an explicit parametric imputation model (e.g. normal, multinomial) is assumed (Kim, 2011), as well as approaches in which the predictive distribution for the missing items is implicit, as in the hot deck imputation method (Kim & Fuller, 2004; Kim & Yang, 2014). The example application in Section 12.6 will focus on the nonparametric FI hot deck method. The following subsections will highlight key features of the FI approach to imputation model choice, the FI process, and how fractionally imputed data sets can be analyzed with standard design-based weighting and variance estimation methods.

12.5.2 Creating the Fractional Imputations

Fractional imputation is a "frequentist" and "design-based approach" to analyzing item missing data in complex sample surveys. FI creates fractional imputations using an Expectation-Maximization (EM) algorithm and an iterative I, W, P process (additional details on these three steps are provided below). Like the Bayesian MI methods, FI utilizes an imputation model for the data to obtain imputed values for y_{mis} and an iterative algorithm that results in a multiple records for the *i*-th case, $\mathbf{y}_{i(j)} = \{y_{mis,i(j)}, y_{obs,i}\}$, $j = 1, \ldots, M(i)$.

Note here that the number of fractionally imputed records for each case is not necessarily equal to a constant *M*, but can vary across cases and certainly does when an FI hot deck algorithm is used to create the imputations.

I Step: FI performs the I (Imputation) step only once, drawing *M(i)* imputed values for each missing value for the *i*-th case from a "proposal" or initializing distribution $h(y_{mis} \mid y_{obs})$ that assigns positive probability to each possible outcome for $y_{mis} \mid y_{obs}$. In Hot Deck FI (Kim & Fuller, 2004), imputations for $y_{mis,i}$ are obtained by drawing $j = 1, \ldots, M(i)$ donor values from imputation cells defined by the cross-classification of observed variables $y_{obs,i}$ for that case. **Fully efficient FI hot deck imputation** (FEFI; SAS®, 2015) creates an imputed record for each possible outcome for $y_{mis} \mid y_{obs}$ that is present in the full set of donor records.

W Step: Once the initial set of *M(i)* distinct values for $y_{mis} \mid y_{obs}$ has been chosen from the proposal distribution, $h(y_{mis} \mid y_{obs})$, or obtained from hot deck donors, an initial value of the **fractional weight** is calculated for each of the *M(i)* imputed values:

$$w_{ij}^{*(t)} = \frac{f(y_{obs,i}, y_{mis,i(j)}; \hat{\theta}^{(t)})}{h(y_{mis,i(j)} \mid y_{obs,i})}; \tag{12.7}$$

where:

$w_{ij}^{*(t)}$ = the fractional weight for imputed value j=1, . . ., $M(i)$ at iteration t;

$f(y_{obs,i}, y_{mis,i(j)}; \hat{\theta}^{(t)})$ = the complete data likelihood for observation i evaluated for $y_{obs,i}$, $y_{mis,i(j)}$ and current iteration parameter estimate, $\hat{\theta}^{(t)}$.

Together, the I step and the W step form the Expectation step in the EM algorithm.

P Step: The P step of the FI algorithm is the Maximization step. Given the updated values of w_{ij}^{*}, these fractional weights are used to update the weighted estimates of θ. For an explicit model $f(y \mid \theta)$, such as a normal or multinomial distribution, the updated estimates of θ for iteration t can be obtained by solving the weighted score equation corresponding to $f(y \mid \theta)$:

$$\sum_{i=1}^{n} w_i \sum_{j=1}^{M(i)} w_{i(j)}^{*(t)} S(\hat{\theta}^{(t)}, y_{obs,i}, y_{mis,i(j)}) = 0 \tag{12.8}$$

If all variables in **y** are categorical and the imputation model is implicit in hot deck draws from donor records, the joint probabilities for the vector of response values for the categorical variables that comprise the **y** can be estimated as:

$$\pi^{(t)}(y) = \frac{\sum_{i=1}^{n} w_i \sum_{j \in D_i} w_{ij}^{*(t)} \cdot I\{(y_{obs,i}, y_{mis,i(j)}) = y\}}{\sum_{i=1}^{n} w_i}; \tag{12.9}$$

where:

D_i is the set of all possible $y_{mis,i(j)}$ given $y_{obs,i}$.

Once the P step of the iterative algorithm is complete, the EM algorithm cycles back to the W step where the updated estimates of the distributional parameters (e.g., $\pi^{(t)}(y)$, $\hat{\theta}^{(t)}$) are used to produce updated fractional weights, $w_{ij}^{*(t+1)}$. These updated weights are then used in a new P step to update the parameters, that is, $\pi^{(t+1)}(y)$ and $\hat{\theta}^{(t+1)}$. The iterative cycle of W and P steps continues until specified convergence criteria are met for the fractional weight values, $w_{ij}^{*(\text{final})}$.

To summarize, FI uses an EM algorithm to solve for $j = 1, \ldots, M(i)$ fractional weight factors, w_{ij}^*, that can be applied to the final survey weight for a case i, w_i. These weighting factors are used to create compound **imputation weights**, $w_{ij}^* \cdot w_i$, to be used in standard design-based analysis of the resulting $n^* = \sum_{i=1}^{n} M(i)$ fractionally imputed records. Under the EM algorithm, the values of w_{ij}^* are refined subject to the following constraints:

$$(i)\ w_{ij}^* > 0;$$

$$(ii)\ \sum_{j=1}^{M(i)} w_{ij}^* = 1.0;\ \text{and}$$

$$(iii.a)\ \sum_{i=1}^{n} w_i \sum_{j=1}^{M(i)} w_{ij}^* S(\hat{\theta}, y_{obs,i}, y_{mis,i(j)}^*) = 0,$$

where $\sum_{i=1}^{n} w_i \sum_{j=1}^{M(i)} w_{ij}^* S(\hat{\theta}, y_{obs,i}, y_{mis,i(j)}^*) = 0$ *is the weighted score equation used to solve for the pseudo maximum likelihood estimate of the parameters of* $f(y \mid \theta)$*, or alternatively:*

$$(iii.b)\ \pi(y) = \frac{\sum_{i=1}^{n} w_i \sum_{j \in D_i} w_{ij}^* \cdot I\{(y_{obs,i}, y_{mis,i(j)}) = y\}}{\sum_{i=1}^{n} w_i};$$

when the fractional imputation is nonparametric, as in hot deck FI.

The consistency property of design-based estimates from fractional imputations of continuous variables assumes that the number of fractional imputations for each case, $M(i)$, is large, but Kim (2011) recommends $M = 10$ as a practical choice. The number of fractional imputations for categorical variables is limited by the discrete number of categories for $y_{mis} \mid y_{obs}$. When fractional imputations for categorical variables utilize a hot deck method, fully efficient fractional imputations (FEFI) are obtained by creating one fractionally imputed record for each possible value of y_{mis} (from the available donors) in the hot deck imputation cells.

12.5.3 Estimation and Inference with Fractionally Imputed Data

Survey data are collected from finite populations. As noted above, the FI method has been developed within the general design-based framework for finite population estimation and inference. Following Kim (2011), in this finite population context, a survey analyst may be interested in estimating a finite population parameter, say β, that is a function of a vector of variables $\mathbf{y} = \{y_1, \ldots, y_p\}$. As discussed in Chapter 7, we may assume that the finite population $\mathbf{y}$ is a realized set of values from a superpopulation model, $f(y \mid \theta)$. In the complete population of elements, the analyst's statistic of interest is the unique solution to a population estimating equation of the form:

$$\sum_{i=1}^{N} U(\beta; \mathbf{y}_i) = 0 \tag{12.10}$$

For example, if β is the slope parameter in the simple linear regression of y_1 on y_2, the estimating function U(β, **y**) can be written as is:

$$U(\beta, \mathbf{y}) = \sum_{i=1}^{N} [(y_{1i} - \bar{y}_1)\cdot(y_{2i} - \bar{y}_2) - \beta\cdot(y_{2i} - \bar{y}_2)^2] = 0 \tag{12.11}$$

Given a probability sample of n of the N population elements, a consistent estimator of β is obtained (assuming no missing data) by solving the weighted estimating equation:

$$\sum_{i=1}^{n} w_i \cdot U(\beta; y_{obs,i}) = 0; \tag{12.12}$$

where:

w_i = the full survey weight for case i=1, . . ., n.

When y is not fully observed, this weighted estimating equation can be extended using the fractionally weighted imputations for y_{mis}:

$$\sum_{i=1}^{n} w_i \sum_{j=1}^{M(i)} w_{ij}^* \cdot U(\beta; y_{obs,i}, y_{mis,i(j)}) = 0; \tag{12.13}$$

where:

$y_{mis,i(j)}$ = the jth imputed value of y_{mis}; *and*
w_{ij}^* = the fractional weight for j^{th} imputed value of y_{mis} for case i=1, . . ., n.

Under FI, estimates of the finite population parameter $\boldsymbol{\beta}$ are solved by a weighted averaging over the imputed values for y_{mis}.

In short, if the missing data have been fractionally imputed, there is no need for special combining rules such as those employed in MI estimation and inference. Using the composite imputation weights formed as the product of the final survey weight, w_i, and the fractional weight variable, w_{ij}^*, standard survey analysis software such as the Stata svy commands can be used for estimation. Using only the fractionally imputed records and the imputation weight for each case, standard TSL, JRR, BRR, or bootstrap methods for variance estimation can be applied. Confidence interval formation and Wald tests of hypotheses also follow the standard approaches described for complex sample survey data described in Chapters 5–11 of this text.

12.5.4 FI Software

The fully efficient FI (FEFI) hot deck method of imputation has now been implemented in SAS® PROC SURVEYIMPUTE. PROC SURVEYIMPUTE can be used to generate an output data set with imputation weights for each fractionally imputed record. The analyst may request that PROC SURVEYIMPUTE utilize the user-specified stratum and cluster codes to generate replicated imputation weights for subsequent use in JRR or BRR variance estimation. Section 12.6 (below) will illustrate an application of the SAS® PROC SURVEYIMPUTE FEFI option to the 2011–2012 NHANES data.

One can also perform fractional hot deck imputation using the contributed `FHDI` package for the R software. We have replicated the FI example in Section 12.6 using functions in this package on the ASDA website. We will report on any other software developments as they occur and provide appropriate links to any open source documentation and user guides that are available on the website.

12.6 Application of MI and FI Methods to the 2011–2012 NHANES Data

The imputation examples presented in this section focus on the 2011–2012 NHANES measures of diastolic blood pressure (DBP) for U.S. adults. The example will focus on MI and FEFI treatment of the item missing data for 2011–2012 NHANES adult respondents (AGE ≧ 18) who agreed to participate in the mobile examination center (MEC) phase of the study (WTMEC2YR > 0 and not missing). Recall that in this analysis, the nonresponse adjustment factor included in the NHANES MEC survey weight, WTMEC2YR, serves as compensation for the 4.3% of NHANES interview respondents who did not consent to participate in the MEC, but that this phase of nonresponse adjustment does not address problems with item missing data in the MEC phase.

12.6.1 Problem Definition

Table 12.2 provides an overview of item missing data rates for a selected set of items from the 2011–2012 NHANES data on adults who visited the MEC. Although actual missing data rates can vary substantially across surveys, the missing data rates displayed in this table show a typical pattern. Missing data on key demographic variables such as age, gender and race are generally extremely low (typically <1% of interviewed cases). Significantly higher rates of missing data occur for variables that measure personal or family income, assets, benefit amounts, or other such financial variables. The poverty index variable (INDFMPIR) is a constructed measure derived from survey measures of household income and family characteristics, and holds an 8.7% missing data rate. Missing data rates for the physical measures are 1.6% for BMI (BMXBMI) and 8.6% for diastolic blood pressure (BPXDI1).

The aim of this example is to impute the missing values for these seven MEC variables and perform two analyses of the imputed data, accounting for the complex sampling

TABLE 12.2

Item Missing Data Rates for Selected 2011–2012 NHANES Variables (n = 5,615)

		NHANES Adult MEC Respondents	
Variable	**Variable Name**	n	**% Missing**
Age	AGEC	5,615	0.0%
Gender	RIAGENDR	5,615	0.0%
Marital Status	MARCAT	5,315	5.3%
Race/Ethnicity	RIDRETH1	5,615	0.0%
Body Mass Index	BMXBMI	5,525	1.6%
Poverty Index	INDFMPIR	5,128	8.7%
Diastolic BP	BPXDI1	5,132	8.6%

features of the 2011–2012 NHANES throughout. The first analysis will be to estimate the population proportion of U.S. adults with diastolic hypertension, which we conservatively define as measured DBP $\geq$ 90 mmHg. The second analysis will involve estimating a simple logistic regression model of the probability of diastolic hypertension as a function of adults' age, gender, and race/ethnicity. Four approaches to the treatment of the item missing data will be compared: 1) analysis of complete cases; 2) multiple imputation of missing data using the Stata `mi impute chained` command (no design variables in the imputation model); 3) multiple imputation in which the imputation model includes fixed effects of the strata and sampled clusters (see Section 12.4.1); and 4) the fully efficient hot deck fractional imputation (FEFI) as implemented in SAS® PROC SURVEYIMPUTE. Conditional on each treatment of missing data, the analyses will utilize the available survey weights in estimation and incorporate sample stratification and cluster sampling in the estimation of standard errors for single analyses.

12.6.2 The Imputation Models for the NHANES Diastolic Blood Pressure Example

The first step in the analysis of diastolic blood pressure is to select the variables that will be included in the imputation model. For the MI analyses, the imputation model includes the following mixture of continuous and categorical variables:

Imputation model variables: Multiple Imputation method

Survey variables:

BPXDI1: Diastolic blood pressure
MARCAT: Marital Status
RIAGENDR: Gender
RIDRETH1: Race/Ethnicity Category
AGEC, AGECSQ: Linear and quadratic terms for centered age (years)
BMXBMI: Body mass index (kg/m^2)
INDFMPIR: Poverty Index (continuous scale with values in the range 1–5)

Optional sample design variables:

WTMEC2YR: NHANES MEC survey weight variable
DESCODE: Categorical variable formed by combining unique values of the NHANES primary stage stratum and cluster codes (see Section 12.4.2.1)

Note that the imputation model includes three variables (MARCAT, BMXBMI and INDFMPIR) that will not be incorporated in the final analysis model. Also, recall that the 2011–2012 NHANES MEC data have no missing values for the age, gender and race/ethnicity variables (Table 12.2). Therefore, only diastolic blood pressure, marital status, body mass index, and the poverty index variables require imputations. The linear and quadratic terms for age and the gender and race/ethnicity predictors are used by the `mi impute chained` command in Stata to build the regression models that are used to impute missing values for BPXDI1, MARCAT, BMXBMI, and INDFMPIR, but their observed distribution will not be altered across the $M = 5$ multiple imputations.

The imputation model for the fractional imputation approach utilizes the same variable set, but to facilitate the hot deck imputation, the four continuous variables were converted to the following grouped categorical variables prior to imputation (code not shown):

AGE4CAT (1=18–24; 2=25–44; 3=45–64; 4=65+)
POVCAT (1=0.00–0.99; 2=1.00–1.99; 3=2.00–2.99; 4=3.00–3.99; 5–4.00–4.99; 6=5.00+)
BMICAT (1=<18.5; 2=18.5–24.9; 3=25.0–29.9; 4=30+)
DBPCAT (1=<80.0; 2=80.0–89.9; 3=90+).

Before actually imputing the missing data, the Stata `misstable patterns` command is used to display the generalized pattern of missing data in the imputation model variable set:

```
misstable patterns BPXDI1 bmxbmi indfmpir ///
marcat riagendr RIDRETH1 agec agecsq WTMEC2YR descode
```

```
         |  Pattern
Percent  | 1  2  3  4
---------+-----------------
79%      | 1  1  1  1
         |
  7      | 1  1  1  0
  7      | 1  1  0  1
  4      | 1  0  1  1
  1      | 1  1  0  0
 <1      | 0  1  1  1
 <1      | 1  0  0  1
 <1      | 1  0  1  0
 <1      | 0  1  1  0
 <1      | 0  1  0  1
 <1      | 1  0  0  0
 <1      | 0  0  1  1
 <1      | 0  1  0  0
 <1      | 0  0  1  0
---------+-----------------
100%     |
Variables are (1)bmxbmi (2)marcat (3)indfmpir (4)bpxdi1
```

The `misstable patterns` output identifies 14 patterns of missingness across the four variables that have missing values (completely observed variables are not displayed).

The data set includes complete observation of all variables for 79% of cases, with an additional 7% of cases missing only the diastolic blood pressure measure, 7% missing only the poverty index, and a residual 7% displaying a generalized pattern of missingness on one or more variables.

12.6.3 Imputation of the Item Missing Data

12.6.3.1 Multiple Imputation

After examining the rates and patterns of item missing data for the variables included in the imputation model, the next step is to multiply impute the item missing data for BPXDI1, MARCAT, BMXBMI, and INDFMPIR. We first illustrate the MI imputation step using one general purpose program for imputation of item missing data: the Stata `mi impute chained` command (imputation by chained equations). This command is capable of multiply imputing generalized patterns of missing data for a multivariate vector of survey variables. Using the following commands, the missing data are independently imputed $M = 5$ times using the chained equations (i.e., sequential regression) technique, and five completed data sets with the observed and imputed values are created.

```
* set output data set to "full long" style
mi set flong

mi register imputed marcat bmxbmi indfmpir BPXDI1

mi register regular riagendr RIDRETH1 agec descode WTMEC2YR

mi impute chained ///
(mlogit) marcat ///
(regress) bmxbmi ///
(regress, include(i.descode WTMEC2YR)) BPXDI1 ///
(regress) indfmpir = i.riagendr i.RIDRETH1 agec agecsq, ///
add(5) rseed(2024) replace
```

Before the imputation is performed, the `mi set flong` command specifies that the output file should be in the `flong` format, with the original input record for each case plus M additional records for each case representing the $m = 1, \ldots, M$ imputation replicates for the case. The current data set will be expanded to contain $n \times (M + 1)$ records. On this file the generated system variable, _ mi _ m, will be set to 0 for the original record and values of _ mi _ m = 1, 2, . . ., M for the imputation replicate records. The `mi register regular` command that follows identifies the fully observed variables that should be included in the imputation model. This syntax includes the survey weight variable (WTMEC2YR) and the composite design stratum and cluster variable (DESCODE) in the predictive model for the DBP variable. As noted above, the multiple imputation was also run excluding these two "design variables" from the DBP prediction model.

Immediately following the `mi impute chained` command is the list of variables that will be included in the imputation process. Recall that only the variables BPXDI1, MARCAT, BMXBMI, and INDFMPIR have missing data. Preceding each variable, the form of the regression model to be used by the chained equations algorithm to predict the conditional distribution of the missing values is listed in parentheses, for example, (`mlogit`) for the nominal categorical variable MARCAT. Variables for age, gender, and race are included (after the equal sign) to be used as predictors in *each* of the regression models that `mi impute chained` will use to predict (impute) missing values of the other four variables. We note that the addition of the `include()` option when specifying the form of the regression model for DBP (`regress`) will only include fixed effects of the design variables in the imputation model for DBP, and not in the imputation models for the other three variables. Using the "`i.`" syntax, gender, race, and descode are represented as categorical variables with a reference category parameterization.

The initial output from the `mi impute chained` command is displayed below. The first portion of this output is a summary identifying the exact form of each regression model and the independent variables used in the imputation models for each variable. The second part of the output table summarizes the distribution of cases by the counts of missing values. As noted above, Stata `mi impute chained` provides the user with the option to control the exact set of predictors for the regression model that will be used to impute missing data for individual variables.

```
Conditional models:
            bmxbmi: regress bmxbmi i.marcat BPXDI1 indfmpir i.riagendr
                    i.RIDRETH1 agec agecsq
            marcat: mlogit marcat bmxbmi BPXDI1 indfmpir i.riagendr
                    i.RIDRETH1 agec agecsq
            BPXDI1: regress BPXDI1 bmxbmi i.marcat indfmpir i.descode
                    WTMEC2YR i.riagendr i.RIDRETH1 agec agecsq
          indfmpir: regress indfmpir bmxbmi i.marcat BPXDI1 i.riagendr
                    i.RIDRETH1 agec agecsq

Performing chained iterations . . .

Multivariate imputation                 Imputations =          5
Chained equations                             added =          5
Imputed: m=1 through m=5                    updated =          0

Initialization: monotone                 Iterations =         50
                                            burn-in =         10

         marcat: multinomial logistic regression
         bmxbmi: linear regression
         BPXDI1: linear regression
       indfmpir: linear regression
------------------------------------------------------------------
               |  Observations per m
               |--------------------------------------------------
      Variable |    Complete   Incomplete   Imputed |     Total
---------------+-----------------------------------+--------------
        marcat |        5315          300       300 |      5615
        bmxbmi |        5525           90        90 |      5615
        BPXDI1 |        5132          483       483 |      5615
      indfmpir |        5128          487       487 |      5615
------------------------------------------------------------------
(Complete + Incomplete = Total; Imputed is the minimum across m
 of the number of filled-in observations.)
```

We strongly recommend that users visually examine and compare the key statistics for the original and imputed versions of variables included in the MI run. To display weighted estimates of selected means from the data set that is created by the `mi impute chained` command, the following Stata descriptive command can be used:

```
mean bmxbmi indfmpir BPXDI1 [pweight=WTMEC2YR], ///
over( _mi _m)
```

Table 12.3 displays the weighted estimates of means for BMI, poverty index, and diastolic blood pressure from each of the $M = 5$ imputation replicates as well as the estimated mean based only on analysis of observed values for each variable.

12.6.3.2 Fully Efficient Fractional Imputation (FEFI): Hot Deck Method

As noted in Section 12.5, fully efficient fractional imputation (FEFI) by the hot deck method is now accessible to analysts through the user-friendly SAS® PROC SURVEYIMPUTE command. The following command syntax initiates a fractional imputation of the missing data in the 2011–2012 NHANES data extract used for this example (available in SAS format on the ASDA website):

```
proc surveyimpute data=c12 method=FEFI varmethod=Jackknife;
   id seqn;
   class age4cat povcat riagendr marcat ridreth1 bmicat dbpcat;
   var age4cat povcat riagendr marcat ridreth1 bmicat dbpcat;
   impjoint povcat bmicat;
   strata sdmvstra;
   cluster sdmvpsu;
   weight wtmec2yr;
   output out=nhanesFEFI outjkcoefs=nhanesJKCOEFS;
run;
```

The FEFI method is selected as part of the command statement using the `method=FEFI` option. To obtain the vector of replicated imputation weights suitable for design-based variance estimation using the delete-one JRR methods (see Chapter 3), the `varmethod=Jackknife` option is also specified on the command line. The `class` statement identifies the categorical variables that will be used to form hot deck cells to identify the set of potential donor values for each fractional imputation. The `var` statement lists the variables for which any missing values will be imputed by the FEFI hot deck method. The `impjoint` statement in the command sequence specifies combinations of variables (here Poverty Index and BMI category, just for illustration) that must be jointly imputed, thereby preserving the joint distribution of these two variables based on the observed data. Univariate hot deck imputation will be performed for all single variables that are not specified as an `impjoint` group.

The standard design variables are specified in the `strata`, `cluster`, and `weight` statements. PROC SURVEYIMPUTE requires the `weight` variable for construction of the

TABLE 12.3

Weighted Estimates of Means for Selected Variables Before and After Multiple Imputation in Stata (among adults who completed the 2011–2012 NHANES Interview and MEC surveys)

		Estimates Before Imputation		Multiple Imputation Repetition Estimates* (n = 5,615)				
Variable	**Statistic**	n_{obs}	**Value**	$m = 1$	$m = 2$	$m = 3$	$m = 4$	$m = 5$
BMI	Mean	5,525	28.62	28.63	28.63	28.64	28.63	28.63
Poverty Index	Mean	5,128	2.89	2.83	2.83	2.83	2.83	2.81
Diastolic BP	Mean	5,132	71.51	71.43	71.35	71.26	71.41	71.44

Note: *Estimates from imputation model for Diastolic BP with design variables and weight included.

imputation weights for fractionally imputed cases, and the `stratum` and `cluster` variables are required for the construction of the JRR replicate imputation weights. The output statement specifies the output file that will contain the expanded data file with the fractionally imputed cases, as well as the output file name for the JRR replicate weights that are produced when `varmethod=Jackknife` is specified.

The following sample of output illustrates some of the header information included in the PROC SURVEYIMPUTE results file. Additional output (not shown) summarizes: 1) the pattern of missing data across the imputation model variables (like the Stata `misstable` output); 2) the distribution (proportions) of the observed variable categories for each of the 14 missing data patterns; and 3) an iteration/convergence history for the EM algorithm used to derive the final fractional weights for each case.

```
                         The SURVEYIMPUTE Procedure

                            Imputation Information

Data Set                    WORK.C12
Weight Variable             wtmec2yr        Full sample 2 year MEC exam weight
Stratum Variable            sdmvstra        Masked variance pseudo-stratum
Cluster Variable            sdmvpsu         Masked variance pseudo-PSU
Imputation Method           FEFI


Number of Observations Read           5615
Number of Observations Used           5615
Sum of Weights Read                 2.32E8
Sum of Weights Used                 2.32E8


    Class Level Information
Class        Levels     Values
age4cat           4     1 2 3 4
povcat            6     1 2 3 4 5 6
riagendr          2     1 2
marcat            3     1 2 3
ridreth1          5     1 2 3 4 5
bmicat            4     1 2 3 4
dbpcat            3     1 2 3


       Design Summary

Number of Strata         14
Number of Clusters       31
```

To explore the contents of the FEFI output file for the 2011–2012 NHANES data set, PROC SURVEYFREQ is used to display the weighted and unweighted distributions of the imputed variables (shown here for DBPCAT, the original trinomial categorization of the continuous diastolic blood pressure measure):

```
proc surveyfreq data=nhanesFEFI varmethod=jackknife;
weight impwt;
```

```
repweights imprepwt _ : / jkcoefs=nhanesJKCOEFS;
tables dbpcat;
run;
```

Here are the results:

Dbpcat	Frequency	Weighted Frequency	Std Err of Wgt Freq	Percent	Std Err of Percent
1	7382	177105202	9960915	76.3376	1.2397
2	2204	40830620	3673002	17.5992	0.9164
3	1437	14066718	2227071	6.0632	0.7940
Total	11023	232002539	14006079	100.000C	

Following the FEFI of the missing items, the original data set of $n = 5{,}615$ cases is expanded to a total of 11,023 records, each with a new imputation weight. Based on the fractionally imputed data, the estimated percentage of the U.S. adult population with measured diastolic blood pressure >= 90 mmHg is 6.063%, with a design-based JRR estimate of the standard error equal to 0.794.

12.6.4 Estimation and Inference

12.6.4.1 Multiple Imputation

Once the multiple imputations are complete, the data are ready for analysis. In this example, two MI analyses of the imputed data are performed. The first analysis computes the MI estimate and 95% CI for the proportion of U.S. adults who have diastolic blood pressure ≥ 90mmHg. The second analysis uses MI methods to fit a logistic regression model of the probability of high diastolic blood pressure (HIGH_DIASTOLIC) as a simple additive function of RIAGENDR (Gender), RIDRETH1 (Race/Ethnicity), AGEC, and AGECSQ (linear and quadratic terms for age in years).

Each analysis uses the $M = 5$ multiply imputed data sets and applies the general procedures for MI estimation and inference outlined in Section 12.4. In Stata, including the `mi` prefix with the `estimate` modifier invokes an MI analysis for the multiple imputed data sets produced by the `mi impute chained` command, or independently developed MI data sets that include the correct variable coding (i.e., _mi_m) for the M repetitions.

12.6.4.1.1 Multiple Imputation Analysis 1: Estimation of the proportion of U.S. adults with diastolic blood pressure ≥ 90 mmHg

The Stata commands required to generate the MI estimate of the proportion of U.S. adults with high diastolic blood pressure are as follows:

```
mi svyset sdmvpsu [pweight=WTMEC2YR], strata(sdmvstra)
gen high _ diastolic = 0
replace high _ diastolic = 1 if BPXDI1 >= 90
mi estimate: svy: prop high _ diastolic
```

TABLE 12.4

Estimates of the Proportion of U.S. Adults with DBP ≥ 90. Comparison of Complete Case and Imputation-Based Analyses

Analysis Type	n	p	$se(p)$	95% CI (p)	df
Complete Case	5,132	0.061	0.008	(0.046–0.080)	17
MI-No Design	5,615	0.060	0.007	(0.044–0.075)	15.16
MI-Model Design	5,615	0.062	0.008	(0.045–0.079)	15.07
FEFI-Hot Deck*	5,615	0.061	0.008	(0.044–0.077)	31

(*Source:* 2011–2012 NHANES).

Note: *Imputed data set includes n* = 11,023 records with fractional weights.

The active Stata data set for this analysis is the output file generated by the Stata `mi impute chained` MI procedure. The initial command is the now familiar `svyset` command in which the stratum, cluster, and survey weight variables for the 2011–2012 NHANES are specified, only now specifically for the MI analysis (note that the command is now `mi svyset`). The second and third command lines use the imputed data for the continuous diastolic blood pressure measures to derive the binary indicator of high diastolic blood pressure for all cases in the new data set, including cases with imputed values of DBP.

The fourth command in the sequence includes the `mi estimate` modifier that informs Stata that the active data set includes multiple imputations and that MI computations for estimates, standard errors, degrees of freedom, and 95% CIs are desired. Following the `mi estimate` modifier is the standard Stata `svy: prop` command for design-based estimation of the proportion and its standard error. Table 12.4 provides the MI estimate of the proportion of U.S adults with DBP >= 90 mmHg, its standard error, and a 95% CI for the population proportion. For comparison purposes, the table also reports the same statistics for a standard complete case analysis.

Following expression (12.1) above, the final MI estimate of the proportion with high diastolic blood pressure is 0.062 (0.008) for imputations of diastolic blood pressure that incorporated the design variables, and 0.060 (0.007) when the design variables are excluded from the imputation model for diastolic blood pressure.

Following Barnard and Rubin (1999), the approximate degrees of freedom (12.6) for the Student t distribution used to construct the confidence interval for the proportion with high diastolic blood pressure are 15.07, which is slightly less than the 17 degrees of freedom assumed for a design-based analysis of the 2011–2012 NHANES complex sample survey data.

12.6.4.1.2 *Multiple Imputation Analysis 2: Estimation of the Logistic Regression Model for High Diastolic Blood Pressure*

The next step in the example exercise is to extend the analysis to a logistic regression model in which the probability of high diastolic blood pressure is modeled as a function of the selected covariates. The Stata command required to fit the logistic regression model to the multiply imputed data sets is:

```
mi estimate: svy: logistic high_diastolic i.RIDRETH1 ///
i.riagendr agec agecsq
```

The standard Stata `svy: logistic` command is preceded by the `mi estimate:` modifier to invoke the MI analysis of the $M = 5$ multiply imputed data sets. Table 12.5 (further below)

provides the MI estimates of the regression parameters and standard errors along with the p-values for the t-tests of the null hypothesis $H_0 : B_j = 0$ for each parameter indexed by j.

Finally, to illustrate the MI multi-parameter test of Li et al. (1991), consider a joint test of the null hypothesis that the effects of gender and race/ethnicity on high diastolic blood pressure levels are not significantly different from zero. To perform this test following the MI logistic regression analysis, the Stata `mi test` command is used:

```
mi test 2.riagendr 2.RIDRETH1 3.RIDRETH1 ///
 4.RIDRETH1 5.RIDRETH1
```

The Stata output generated by this test command (for the analysis including the design variables in the imputation model for diastolic blood pressure) is shown below.

```
note: assuming equal fractions of missing information.

 ( 1) [high_diastolic]2.riagendr = 0
 ( 2) [high_diastolic]2.RIDRETH1 = 0
 ( 3) [high_diastolic]3.RIDRETH1 = 0
 ( 4) [high_diastolic]4.RIDRETH1 = 0
 ( 5) [high_diastolic]5.RIDRETH1 = 0

       F( 5, 14.2) =    8.81
          Prob > F =    0.0006
```

Based on the test result, the null hypothesis that both gender and race/ethnicity are unrelated to diastolic blood pressure is rejected.

12.6.4.2 Fractional Imputation

12.6.4.2.1 *Fractional Imputation Analysis 1: Estimation of the proportion of U.S. adults with diastolic blood pressure ≥90 mmHg*

After FEFI of the missing data, the following SAS syntax is used to generate estimates of the proportion of U.S adults with high diastolic blood pressure:

```
data imputed;
 set nhanesfefi;
 if dbpcat=3 then high_dbp=1; else high_dbp=0;
run;

proc surveyfreq data=imputed varmethod=jackknife;
 weight impwt;
 repweights imprepwt_: / jkcoefs=nhanesJKCOEFS;
 tables high_dbp / cl;
run;
```

A SAS data step is used to collapse the three category DBPCAT variable into a new binary indicator (HIGH_DBP) for diastolic blood pressure ≥ 90 mmHg. The PROC SURVEYFREQ command is then run to generate weighted estimates and design-based standard errors for the proportion of interest. Note the use of the `varmethod=jackknife` option and `repweights` statement, which collectively direct PROC SURVEYFREQ to use the delete-one JRR method for variance estimation. The `jkcoefs=nhanesJKCOEFS` option on the `repweights` statement identifies the SAS data file that contains the imputation replicate weights previously output by PROC SURVEYIMPUTE. The estimated proportion of U.S. adults with high DBP and its JRR standard error are displayed in Table 12.4.

12.6.4.2.2 *Fractional Imputation Analysis 2: Estimation of the Logistic Regression Model for High Diastolic Blood Pressure*

Following the FEFI imputation of the missing values, the logistic regression model for the probability of high diastolic blood pressure is estimated using SAS PROC SURVEYLOGISTIC:

```
proc surveylogistic data=imputed varmethod=Jackknife;
class riagendr (ref='1') ridreth1 (ref='1') / param=ref;
model high _ dbp (event='1') = ridreth1 riagendr agec
agecsq;
weight ImpWt;
repweights ImpRepWt _ : / jkcoefs=nhanesJKCOEFS;
testeth _ gender: test ridreth12, ridreth13, ridreth14,
ridreth15, riagendr2 ;
run;
```

The parameter estimates and standard errors for the fitted model are presented in the lower right panel of Table 12.5. Using standard SAS test syntax for specifying a Wald test of a multi-parameter null hypothesis produces the following result:

```
                    Linear Hypotheses Testing Results
Label              F Value       Num DF        Den DF       Pr > F
testeth_gender       11.40            5            31       <.0001
```

Comparing these test results to those obtained from the MI test of the same hypothesis (see above), the values of the *F* statistic and the denominator degrees of freedom for the reference distribution differ; however, both tests conclusively reject the null hypothesis that race/ethnicity and gender are not significant predictors of high diastolic blood pressure in U.S. adults.

We note that with appropriate conversion of the PROC SURVEYIMPUTE output data sets to Stata format, the fractionally imputed data could also be analyzed using the Stata commands `svy: prop` and `svy: logistic`.

12.6.5 Comparison of Example Results from Complete Case Analysis, MI, and FEFI

Tables 12.4 and 12.5 provide a comparative summary of the results for the two example analyses under complete case analysis (listwise deletion), MI, and FEFI approaches to the missing data problem. Looking first at Table 12.4, to three decimal places, a comparison of estimates and confidence intervals for the proportion of adults with high DBP shows only very minor differences. The 95% CIs based on the MI and FEFI data sets are slightly

TABLE 12.5

Estimated Logistic Regression Models for Diastolic Hypertension (DBP ≥ 90) of U.S. Adults (Comparisons of Complete Case and Imputation-Based Analyses);

		Complete Case Analysis (n = 5,132)			Multiple Imputation: Ignore Design (n = 5,615)		
Predictor*	**Category**	$\hat{B}_j$	$se(\hat{B}_j)$	p-value	$\hat{B}_j$	$se(\hat{B}_j)$	p-value
Intercept	Constant	−2.249	0.198	<.001	−2.271	0.188	<.001
Race/Ethnicity	Other Hisp.	−0.724	0.245	0.009	−0.662	0.237	0.014
	White	0.131	0.224	0.567	0.144	0.193	0.468
	Black	0.655	0.247	0.017	0.652	0.227	0.012
	Other	0.051	0.245	0.838	0.073	0.234	0.762
Gender	Female	−0.550	0.207	0.017	−0.551	0.206	0.019
Age (Centered)	Continuous	0.008	0.007	0.249	0.009	0.007	0.217
Age (Centered) Squared	Continuous	−0.002	<0.001	<.001	−0.002	<0.001	<.001

		Multiple Imputation: Include Design (n = 5,615)			FE Fractional Imputation (n = 11,023**)		
Predictor*	**Category**	$\hat{B}_j$	$se(\hat{B}_j)$	p-value	$\hat{B}_j$	$se(\hat{B}_j)$	p-value
Intercept	Constant	−2.284	0.204	<.001	−2.282	0.183	<.001
Race/Ethnicity	Other Hisp.	−0.520	0.293	0.110	−0.619	0.211	0.006
	White	0.197	0.232	0.412	0.118	0.209	0.576
	Black	0.690	0.256	0.019	0.599	0.232	0.015
	Other	0.077	0.257	0.769	0.044	0.223	0.847
Gender	Female	−0.520	0.202	0.024	−0.503	0.189	0.012
Age (Centered)	Continuous	0.007	0.007	0.312	0.007	0.006	0.260
Age (Centered) Squared	Continuous	−0.002	<0.001	<.001	−0.001	<0.001	<.001

Source: 2011–2012 NHANES

Note: MI degrees of freedom are determined separately for each estimate. *Reference categories are Mexican (Race/Ethnicity) and Male (Gender). **Imputed data set includes n^* = 11,023 records with fractional weights.

narrower than those for the complete case analysis, but practically speaking, the implications for inference concerning the population proportion are negligible.

Turning to Table 12.5, the upper left panel provides weighted estimates of the logistic regression model parameters (on the log odds scale) and their design-adjusted standard errors for a complete case analysis in which there is listwise deletion of cases with missing values on the dependent variable or one or more predictors. Comparable estimates based on the two MI approaches and the FEFI are provided in the remaining three panels. On first observation of the comparative results in Table 12.5, the estimated coefficients in the fitted regression models are not identical—an expected outcome given that the imputation procedures have permitted several additional cases with at least one original missing value to enter the analysis. Furthermore, the MI and FEFI methods differ in their approach to weighting the imputed data and combining the imputed data to produce final estimates. A comparison of the standard errors for the estimated model coefficients for the complete case analysis and the two MI analyses shows no consistent pattern. However, we do observe that standard errors based on the FEFI method are generally smaller than those for the complete case or MI approaches, indicating increased efficiency engendered by this approach.

In this example, the differences in results for the MI and FEFI treatments of the missing data are generally not large enough to lead an analyst to reverse their decisions concerning the statistical significance of the individual predictors in the model. We do observe that inference concerning the difference between Other Hispanic individuals and Mexican individuals would vary somewhat between the MI (including the design variables in the imputation model for DBP) and FEFI approaches when adjusting for the other covariates; this is where the increased efficiency in the estimates introduced by the FEFI approach may be helpful before reporting final inferences based on this model. Readers should bear in mind that the results of these comparative analyses do not necessarily transfer to other missing data problems with different variables, rates of missing data, or strengths of associations among variables in the imputation models.

12.7 Exercises

12.1) These exercises consider data from the 2011–2012 NHANES. The objective of this set of exercises is to perform a typical combination of imputation and subsequent analysis. The steps include examination of missing data, imputation of missing data values using a multiple imputation software procedure of choice, and analysis of the imputed data sets using a companion software procedure of choice that is capable of handling multiply imputed data sets.

Begin by downloading the subset of data from the 2011–2012 NHANES: nhanes1112_c12.dta or nhanes1112_c12.sas7bdat (available from the ASDA website). Note that this data set is limited to adults 18+ years of age who completed the NHANES medical examination (n = 5,615).

The variables used in the imputation and subsequent analysis include gender (RIAGENDR), Body Mass Index (BMXBMI), race/ethnicity (RIDRETH1), age (AGE), total cholesterol (LBXTC), and systolic blood pressure (BPXSY1). The data set also contains the NHANES complex sample design variables and the final survey weight for examined individuals (SDMVSTRA, SDMVPSU, WTMEC2YR).

Examine simple descriptive statistics for these variables (means, proportions, ranges, and counts of missing values), keeping in mind that the full n is 5,615. Use a software procedure of your choice for this step. Pay close attention to the types of the variables with missing data (continuous, ordinal, binary or nominal) and the amount of missing data, that is, what percent are missing on each variable. Prepare a table like Table 12.2 to outline the missing data problem.

12.2) Using your software of choice and the data from 12.1), examine the missing data patterns that exist in the file.

a) How many unique missing data patterns exist?

b) Which variables have some missing data?

c) Which variables have complete data?

12.3) Prepare a combined stratum and PSU variable as demonstrated previously in this chapter and call this variable STRATA_PSU. This variable should have 31 unique values and should be treated as a categorical variable in your imputation. Prepare and execute syntax to impute the missing values using the software of your choosing. Use the Sequential Regression/FCS/Chained Equation approach for multiple imputation. Keep in mind what type of model should be used for the imputation, depending on the type of variable to be imputed. For example, linear regression for continuous variables, ordinal or cumulative logit regression for ordered variables, binary logistic regression for binary variables, and so forth. Use $M = 10$ to prepare 10 imputed data sets during this step and be sure to use a seed value so that your results can be replicated later.

a) Provide output from the imputation process and make sure it verifies that 10 imputations were performed and lists the types of imputation models used for the imputed variables.

b) What is the name of the variable that identifies the imputation number 1–10? Note that this will depend on the software procedure that you are using.

c) How many observations does this "concatenated" file contain? If you changed the number of imputations, how would that impact the number of observations in this file?

d) Why might we include the STRATA_PSU and WTMEC2YR variables in the imputation models?

12.4) Perform an appropriate design-based descriptive analysis of total cholesterol (LBXTC) within each imputed data set (1–10) contained in the concatenated imputed data set. For example, if the imputed variable is continuous, estimate means within each imputed data set using the correct survey weight (WTMEC2YR) and design variables (SDMVSTRA, SDMVPSU), and estimate appropriate standard errors for the estimated means using Taylor Series Linearization or a replication technique. Create a table showing the estimated means and proportions (and estimated standard errors) for LBXTC from each imputed data set.

a) Why do the means change over the 10 data sets? Is this expected?

b) Why is this analysis not yet complete and what else should be accounted for in the analysis of imputed data sets?

12.5) Next, perform a multiple imputation analysis to estimate the mean LBXTC and compute a variance estimate for the estimated mean that incorporates the design-based standard errors from the analysis of each imputed data set in addition to the between-imputation variance in the estimates. (Use a command designed to analyze multiply imputed data sets and incorporate the complex sample design features in addition to the variability introduced by imputing the missing values multiple times.)

a) What is the overall combined estimate of the mean LBXTC?

b) What is the standard error of the estimated mean, and what has been accounted for in the standard error when using the MI estimation procedure?

c) Obtain the within- and between-imputation variance components by either calculating these statistics manually or requesting them from the software package of choice (see Section 12.6.1).

12.6) Finally, fit a linear regression model with total cholesterol as the dependent variable and age, gender, race/ethnicity, and BMI as predictors. Fit the model first using a standard design-based analysis of complete cases only, and then perform a multiple imputation analysis using the software of your choice.

a) Prepare a table comparing estimated regression parameters and standard errors under the two analysis approaches (complete cases only and multiple imputation).

b) Have the multiple imputations of the item missing values and subsequent multiple imputation analysis changed any of your conclusions about the significance of the predictor variables?

Note

1 https://smponline.isr.umich.edu/software/iveware-user-guide-chapter-9-iveware-and-sas/#93bbdesign

13

Advanced Topics in the Analysis of Survey Data

13.1 Introduction

This final chapter provides coverage of advanced and emerging topics in the analysis of survey data. Over the past two decades, applications of **generalized linear mixed models (GLMMs)** to multilevel modeling of hierarchical data (e.g., schools, classrooms, students) have grown to be increasingly important. As was discussed in Chapter 11, these models, along with models fitted using **generalized estimating equations (GEE)**, are important in the analysis of longitudinal survey data and other survey data sets containing repeated measurements over time from the sampled units. In Section 13.2, we consider multilevel models for *cross-sectional* complex sample survey data with multiple levels of substantive interest.

Section 13.3 provides a summary of recent work on the applications of latent variable models, such as **structural equation models**, to complex sample survey data, and presents examples of fitting SEMs to complex sample survey data using state-of-the-art software procedures. Section 13.4 presents an overview of modern methods for the analysis of data from **non-probability samples**, which lack formal probability sampling mechanisms and therefore cannot be analyzed using classical design-based approaches. Section 13.5 discusses recent research on the integration of design-based approaches into **machine learning techniques**, such as recursive partitioning and random forests, where finite population inference is the objective. A review of recent developments in the applications of survey data to **small area estimation** problems is provided in Section 13.6. Finally, **nonparametric methods** have always been important statistical tools for data analysts. Recently, there has been increased interest in theoretical developments and applied research on the application of nonparametric analyses to complex sample survey data. Section 13.7 reviews this recent work on nonparametric methods.

13.2 Generalized Linear Mixed Models (GLMMs) in Survey Data Analysis

13.2.1 Overview of GLMMs

Generalized linear mixed models (GLMMs) comprise a broad class of regression-type models that include several specific analytic forms, such as **hierarchical linear models (HLMs)** or **multilevel models**, models for **repeated measures** on individuals, or **growth curve models** of longitudinal change in individual attributes (see Chapter 11). We use the adjective "generalized" to acknowledge that linear mixed modeling techniques are

DOI: 10.1201/9781003472094-14

now available for continuous, nominal, ordinal, and count-type data. GLMMs are **subject-specific models** that analyze the influence of both **fixed effects** and **random effects** on individual attributes. The label "subject-specific" refers to the fact that GLMMs are modeling the relationships of specific covariates with a dependent variable of interest *for a given higher-level unit* (see Figure 13.1 below), by explicitly controlling for the random effects of randomly sampled individuals (higher-level units in a longitudinal study) or clusters of observations (higher-level units in a cross-sectional study).

GLMMs contrast with the **marginal modeling** or **population averaged** approach that is employed by the **generalized estimating equations (GEE)** technique. This modeling technique is an alternative for the analysis of repeated measures or other dependent sets of observations on population units that does *not* explicitly incorporate random subject effects, and adjusts standard errors for the clustering present in the observations in a manner like Taylor Series Linearization (using **robust sandwich estimates** of standard errors). See Chapter 11 for more details on the GEE methodology.

Following Fitzmaurice et al. (2004), a general expression for the GLMM in the context of longitudinal data (repeated observations nested within subjects) is:

$$g\{E(Y_{it} \mid \boldsymbol{b}_i)\} = \eta_{it} = \boldsymbol{X}_{it}\boldsymbol{\beta} + \boldsymbol{Z}_{it}\boldsymbol{b}_i \tag{13.1}$$

where:

$g(\cdot)$ = a known link function, often identity (Normal) but also log (Poisson) or logit (see Chapters 8 and 9);

$\boldsymbol{X}_{it}$ = row vector of j=1, . . ., p covariates

$\boldsymbol{\beta}$ = vector of j=1, . . . p fixed effects regression parameters;

$\boldsymbol{Z}_{it}$ = row vector of k=1, . . ., q covariates for subject i, at observation t=1, . . ., T;

$\boldsymbol{b}_i$ = vector of k=1, . . ., q random effects.

From this general expression, more specific expressions for individual models may be derived. For example, a simple **random intercept model** for change in a continuous random variable, y, over times t = 1, . . ., T, can be written as follows:

$$\begin{aligned} y_{it} &= \boldsymbol{X}_{it}\boldsymbol{\beta} + b_i + e_{it} \\ &= (\beta_0 + b_i) + \beta_1 x_{it,1} + \cdots + \beta_p x_{it,p} + e_{it} \end{aligned} \tag{13.2}$$

where:

$b_i \sim N(0, \tau^2)$ = random effect for individual i;

$e_{it} \sim N(0, \sigma^2)$ = random error term; and

$cov(b_i, e_{it}) = 0$.

The notation in the general expression (13.2) suggests a GLMM for modeling multiple dependent observations on a single individual, such as longitudinal measures in a panel survey (Fitzmaurice et al., 2009; see also Chapter 11). However, by using slightly different subscripts, we can show that the GLMM also represents a model where the dependency is not among repeated observations on a single individual but among units that are clustered in a hierarchical data set (e.g., survey respondents nested within sampling clusters).

Consider for example a survey such as the National Assessment of Educational Progress (NAEP), where measures on individual students (denoted by i) are nested within randomly sampled schools (denoted by s). The random intercept model can be rewritten to model fixed effects on individual student test performance, $y_{i(s)}$, while accounting for the random effects of the school clusters:

$$\begin{aligned} y_{i(s)} &= \mathbf{X}_{i(s)}^T \boldsymbol{\beta} + b_s + e_{i(s)} \\ &= (\beta_0 + b_s) + \beta_1 x_{i(s),1} + \cdots + \beta_p x_{i(s),p} + e_{i(s)} \end{aligned} \tag{13.3}$$

where:

$b_s \sim N(0,\tau^2)$ = random effect for school s;

$e_{i(s)} \sim N(0,\sigma^2)$ = random error term; and

$cov(b_i, e_{i(s)}) = 0$

The literature generally refers to models such as (13.3) that incorporate dependence among observations due to hierarchical clustering of the ultimate observational units as **hierarchical linear models (HLMs)** or **multilevel models** (Raudenbush & Bryk, 2002; Goldstein, 2003). Because the dependence among observations results from hierarchical levels of clustering, HLMs are often described by decomposing the fixed and random components for the overall GLMM into models for the individual levels. For example, consider the simple model (13.3) for test scores where students are nested within randomly sampled schools. The Level 1 model is the model of the student test outcome, controlling for the effects (which may be school-specific) of covariates for that student. Note that the Level 1 model below allows the intercept and the effects of the student-level covariates to be unique to the sampled schools:

$$\begin{aligned} &\text{Level 1:} \\ &y_{i(s)} = \beta_{0(s)} + \beta_{1(s)} x_{i(s),1} + \cdots + \beta_{p(s)} x_{i(s),p} + e_{i(s)} \end{aligned} \tag{13.4}$$

Level 2 of the multilevel specification defines equations for each of the random school-specific coefficients in the Level 1 model. In the case of (13.3), only the intercept randomly varies across sampled schools. For example, suppose that the school-specific intercept is defined by the fixed overall intercept, β_0, and the random school effect, b_s, while the effects of the covariates are fixed across schools (i.e., defined by fixed effect parameters only):

$$\begin{aligned} &\text{Level 2:} \\ &\beta_{0(s)} = \beta_0 + b_s \\ &\beta_{1(s)} = \beta_1 \\ &\ldots \\ &\beta_{p(s)} = \beta_p \end{aligned} \tag{13.5}$$

A complete treatment of GLMMs is beyond the scope of this intermediate-level book on applied survey data analysis. There are many texts that describe the theory and applications of GLMMs in detail, including (in no order) Diggle et al. (2002), Gelman and

Hill (2006), Fitzmaurice et al. (2004), Verbeke and Molenberghs (2000), Molenberghs and Verbeke (2005), West et al. (2022), Raudenbush and Bryk (2002) and Goldstein (2003). Our objective here is to acquaint the reader with GLMMs, introduce several key issues in the applications of these models to complex sample survey data, and describe current methodological research that addresses these issues.

Throughout the following discussion, we will focus on cross-sectional data arising from a stratified multi-stage sample design. We will refer to individual survey respondents as Level 1 units, and the sampled clusters within which the survey respondents are nested as Level 2 units. Readers interested in multilevel models for longitudinal survey data should refer to Chapter 11.

13.2.2 GLMMs and Complex Sample Survey Data

As was mentioned above, GLMMs are specifically designed to address non-independence or "dependency" of observations. In HLMs, the dependency arises because observational units are hierarchically clustered, such as students within classrooms and classrooms with schools. In repeated measures or longitudinal models, the dependency arises because observations are clustered within individuals—for example, daily diaries of food intake for NHANES respondents, or longitudinal measures of household assets for HRS panel respondents (see Chapter 11).

This problem of lack of independence for observations should sound familiar. The survey data analysis techniques that have been the core subject of this book are designed to address the intraclass correlation among observations in sampled clusters. It is natural to draw the analogy between these two estimation problems. Take for example two parallel sets of data. The first data set records scores on a test administered to students who are clustered within classrooms, schools, and school districts. The second data set includes test scores for the same standardized test administered to a probability sample of students in their homes, where households were selected within area segments and primary sampling units (PSUs) of a multi-stage national sample design. In both cases, there is a hierarchical ordering of units—districts, schools, classes, and students in the first case, and PSUs, area segments, households, and students in the second case.

If the survey analyst was only interested in inferences concerning national student performance on the standardized test, robust inferences could be obtained using standard design-based estimates of population parameters (such as the mean test performance) and their standard errors (see, for example, Chapter 5). School districts would define the ultimate clusters in the first data set, and PSUs would form the ultimate clusters of the second. However, the survey analyst using the first data set may have broader analytic goals. Specifically, he/she may be interested in estimating the proportion of variance in test scores that is attributable to the student, the class, the school, and the district. The analyst working with the second data set may not be especially interested in the **components of variance** associated with PSUs, secondary sampling units (SSUs) within PSUs, households, and individuals. The first analyst will likely pursue a **model-based analysis**, specifying an HLM form of the GLMM.

Following this approach to the analysis of complex sample survey data, the first analyst would specify the best possible probability model for the data set in hand (possibly using a Bayesian approach), which includes the specification of an appropriate random effects structure to reflect the multi-stage cluster sampling and estimate the variance components of interest (Kim et al., 2022). The variance of the estimated parameters in this case would be estimated with respect to the specified model, and if the weights and stratum codes were associated with the dependent variable of interest, these would also need to be accounted for in the model specification. The second analyst will be satisfied with a standard **design-based analysis**, in which weighted estimates of the parameters of interest are computed, and Taylor Series Linearization or replication estimates of the overall variance of the sampling distribution (computed with respect to the sample design) are used to develop confidence intervals and test statistics. Interested readers can refer to Hansen et al. (1983) for more discussion of the differences between model-based and design-based approaches.

Consider another problem in which a cluster sample of individuals is asked whether they have experienced any hay fever symptoms in the past week. The dependent variable for each individual is an indicator of whether they experienced hay fever symptoms ($y_{ij} = 1$) or did not ($y_{ij} = 0$), where i indexes individuals and j indexes ultimate clusters. The independent variables in the analysis might include age, gender, and an indicator of a previous allergy diagnosis. To estimate the relationship of hay fever symptoms with age, the survey analyst could choose among three approaches that all use variants of logistic regression modeling:

1) A cluster-specific model-based analysis using a GLMM, in which the logit $[P(y_{ij} = 1)]$ is modeled as a function of **fixed effects** that include the constant effects of age, gender, and previous allergy diagnosis, and **random effects** of the randomly sampled ultimate clusters (enabling estimation of between-cluster variance);
2) A marginal model-based analysis using a GEE model, in which the logit model relating symptoms to the covariates is estimated using GEE methods and the covariance matrix for the model coefficients is separately estimated using the robust Huber-White **sandwich estimator** (Diggle et al., 2002); or
3) A marginal design-based analysis using a program/command/procedure such as Stata's `svy: logit` command, with a single record (row) in the data set for each individual, multiple individuals (rows in the data set) per ultimate cluster, and the ultimate clusters identified as the "clusters" for variance estimation purposes (see, for example, Chapter 8).

The GLMM analysis is a cluster-specific analysis, explicitly controlling for the random cluster effects, and would yield estimates of the fixed effects (or regression parameters) associated with the covariates *in addition to* an estimate of the variance of the random cluster effects. The GEE and `svy: logit` approaches are population-averaged (or marginal) modeling techniques and would provide comparable estimates of robust standard errors for the estimated logistic regression coefficients for the covariates. The GEE and design-based population-averaged approaches would *not* separately estimate the variance of the random cluster effects or their contributions to the total sampling variability. This is the key distinction between these alternative approaches to analyzing clustered or longitudinal data: GLMMs enable analysts to make inferences about between-subject (or between-cluster) variance, based on the variances of the subject-specific (or cluster-specific) random effects explicitly included in the models, while GEE and design-based modeling approaches are only concerned with overall estimates of parameters and their total sampling variance. See Hsu et al. (2018) for an in-depth simulation study that clarifies this distinction.

It is increasingly common to see survey samples designed to be optimal for analysis under a GLMM-type model. For example, a multi-stage national probability sample of school districts, schools, classrooms, and students could be consistent with a GLMM model that would enable the education researcher to study the influence of each level in this hierarchy on student outcomes (see Kim et al., 2022). To achieve data collection efficiency, the sampling statistician designing the sample of individuals for the hay fever study could select a primary stage sample of U.S. counties and then a sample of individuals within each primary stage county. The result would be a two-level data set with individuals nested within the sampled counties.

However, even when care is taken to design a probability sample to support multilevel or longitudinal analysis using GLMM-type models, there are several theoretical and practical issues to be addressed when fitting these models to complex sample survey data:

1. **Stratification.** If the stratification used in the sample selection (regions, urban/rural classification, population size, etc.) is **informative**, that is, associated with the survey variables of interest (which is typically the case in practice), fixed effects of the stratum identifiers, or at a minimum the major variables used to form the strata, will need to be included in the model.
2. **Cluster Sampling.** In the general sample design context, clustering of population elements serves to reduce the costs of data collection. The increase in sampling variance attributable to the intraclass correlation of characteristics within the cluster groupings is considered a "nuisance," inflating standard errors of estimates to no analytical benefit. In the context of a specific analysis involving a hierarchical linear (or multilevel) model, the clustering of observations must be reflected (or incorporated) to obtain stable estimates of the effects and variance components at each level of the hierarchy. Ideally, the sample design clusters may be integrated into the natural hierarchy of the GLMM and the cluster effects on outcomes can be directly modeled using additional levels of random effects—say nesting students within classrooms, classrooms within schools, and schools within county PSUs.
3. **Weighting.** Conceptually, one of the more difficult problems in the application of GLMMs to complex sample survey data is how to handle the survey weights. Theoretically, if the weights were **noninformative**, they could be safely omitted from the model estimation. If the sample design and associated weights are **informative** for the analysis, fixed effects of the variables used to build the weights or appropriate functional forms of the weight values themselves could be included in the model (DuMouchel & Duncan, 1983; Little, 1991; Korn & Graubard, 1999, Section 4.5; Fuller, 2009, Ch. 6). However, there are other complications with weighting in GLMMs, as we discussed in Chapter 11. Attrition adjustments that are often included in longitudinal survey weights can yield different weight values for each measurement time point. Clusters in a multi-stage sample design may not enter the sample with equal probability. Even in a national equal-probability multi-stage sample of students, the most efficient samples require that counties and school units are selected with probability proportionate to size (PPS). Conditional on a given stage of sampling, the observed units would enter with varying probability. What role should this information play in the estimation of GLMMs?
4. **Subpopulations.** As was discussed in Chapter 4, *unconditional* subpopulation analyses are important when conducting design-based analyses of complex sample survey data and focusing inference on subclasses of interest. Do the same issues still apply when fitting GLMMs? For example, if one wanted to make inference about variance components due to classrooms and schools in lower-income school districts exclusively, and district income was not an explicit stratification variable in the original sample design, how should one approach the analysis?

> Koziol (2019) addressed this question directly via simulation and found that the answer depends on the sizes of the subpopulation samples in the clusters (assuming that the sample design is informative, and weights should be used). When the sample sizes are small (e.g., less than 15 per cluster), standard design-based approaches to the subpopulation analysis ("single-level" approaches) are recommended (e.g., the regression models described in Chapters 7–9). When the sample sizes are larger, multilevel approaches or single-level approaches can be used.

In this section, we focus on approaches for fitting GLMMs to cross-sectional complex sample survey data that use the weights associated with different randomly sampled units to define the (pseudo) likelihood function for the observed data. These approaches are identical to the approaches outlined in Chapter 11 for weighted multilevel modeling, with the difference being that Level 1 units are generally survey respondents (rather than different time points), and the Level 2 units are generally ultimate clusters (rather than individuals in a longitudinal survey):

- Survey weights for the Level 1 respondents need to reflect probabilities of selection *conditional* on the higher-level ultimate cluster to which the respondent belongs being selected into the sample (including possible adjustments for nonresponse);
- The (adjusted) conditional weights for Level 1 respondents need to be appropriately *scaled* within Level 2 ultimate clusters, with the "normalizing" approach (which ensures that the sum of the scaled weights within a given ultimate cluster is equal to the sample size for that cluster) generally serving as a good multi-purpose approach for estimation (Pfeffermann et al., 1998; Rabe-Hesketh & Skrondal, 2006; Carle, 2009);
- Survey weights for the sampled Level 2 ultimate clusters, reflecting the probabilities of selection for the clusters, need to be accounted for in the (pseudo) likelihood function as well (much like the "baseline" weights for sampled individuals at Level 2 in Chapter 11); see Pfeffermann et al. (1998), Rabe-Hesketh and Skrondal (2006), Rao et al. (2013), or Kim et al. (2022) for the technical details underlying this requirement.

We use the weights associated with different levels of the data hierarchy to compute unbiased estimates of both the fixed effects and the variance components in a GLMM (see West et al., 2015, for a case study). Ideally, such estimates will be unbiased with respect to both the sample design and the model being specified (Pfeffermann, 1993). A failure to use the weights associated with Level 2 ultimate clusters, for example, could lead to biased estimates of the variance components describing between-cluster variability in the larger population. This makes it essential for these weights to be included in a survey data set if analysts of those data will be interested in fitting GLMMs. The use of the weights in estimation also provides some protection against model misspecification, as discussed in Chapter 7; estimates of the fixed effects and variance components in a GLMM will still be unbiased with respect to the sample design if a model has been misspecified.

Importantly, the weights associated with Level 1 respondents need to be *conditional* and *scaled*, as indicated above. A failure to use these conditional, scaled weights (e.g., if an analyst were to use the *overall* survey weights reflecting all stages of sample selection that are typically provided in public-use survey data sets) could lead to bias in the estimates of multilevel model parameters (Pfeffermann et al., 1998). If a data producer is willing to provide both Level 1 and Level 2 weights, *in addition to* joint probabilities of inclusion for Level 1 and Level 2 units, then inferences can be improved even further (Rao et al., 2013).

When using the weights associated with all randomly sampled units to define the pseudo-likelihood function used to estimate the parameters in a GLMM, the variances of the maximum likelihood estimates can be computed using a variation of Taylor Series Linearization developed by Binder (1981, 1983) for generalized linear models fitted to complex sample survey data (see Chapter 8). This variance estimation approach can explicitly account for multi-stage stratified cluster sampling, and is sometimes referred to as a robust, sandwich-type approach to computing standard errors in the software that implements these approaches. Pfeffermann et al. (1998) and Rabe-Hesketh and Skrondal (2006) provide technical details regarding these variance estimation approaches.

Recent years have seen additional developments related to optimal estimation approaches for analysts interested in fitting these types of GLMMs to complex sample survey data. These have included Bayesian approaches for implementing the methods described above. Silva and Moura (2022) and Savitsky and Williams (2022) describe Bayesian approaches to the survey-weighted estimation of GLMMs and demonstrate via simulation and an empirical evaluation that this approach has improved inferential properties relative to the other approaches introduced above. The authors provide data and code demonstrating how to implement this approach using the `rstan` package in the R software, and we have included this code on the ASDA website.

13.2.3 Alternative Approaches to Fitting GLMMs to Survey Data: The PISA Example

We now consider an example of fitting a GLMM to cross-sectional survey data arising from a complex sample design. These data come from the 2000 Programme for International Student Assessment (PISA) and were also analyzed by Rabe-Hesketh and Skrondal (2006) using the `gllamm` command that they developed for Stata (see www.gllamm.org). The dependent variable of interest in this case (ISEI) is continuous and measures the socio-economic status (SES) of a given student. The predictor variable of interest in this case is COLLEGE (an indicator of whether the highest level of education for either parent is college). Also included in the data set (freely available online for Stata users) is an ID code for the sampled school (the PSU in the case of this sample design), the overall final student weight (not conditional!), or W_FSTUWT, and a final school weight (WNRSCHBW). The data can be loaded into Stata's working memory using the following command:

```
use http://www.stata-press.com/data/r13/pisa2000
```

We desire to fit the following GLMM to these data:

$$ISEI_{ij} = \beta_0 + \beta_1 COLLEGE_{ij} + u_{0j} + u_{1j} COLLEGE_{ij} + e_{ij}$$

$$\begin{pmatrix} u_{0j} \\ u_{1j} \end{pmatrix} \sim N\left(\begin{pmatrix} 0 \\ 0 \end{pmatrix}, \begin{pmatrix} \sigma_0^2 & 0 \\ 0 & \sigma_1^2 \end{pmatrix} \equiv D \right)$$

$$e_{ij} \sim N(0, \sigma^2)$$

We note that this model includes two random effects (u_{0j} and u_{1j}), allowing each school (denoted by j) to have a unique intercept and a unique relationship of COLLEGE with ISEI. The random effects are assumed to follow a bivariate normal distribution, with mean vector 0, unique variance components for each random effect, and zero covariance. The random errors are assumed to follow a normal distribution with constant variance. We desire to compute weighted estimates of the two fixed effects (β_0 and β_1) and the two variance components (σ_0^2 and σ_1^2) for the PISA target population. The intercept represents the expected ISEI for students who do not have any college-educated parents, the fixed effect for COLLEGE represents the change in the expected ISEI for students with college-educated parents, and the variance components describe variation among the schools in the intercepts and the college effects, respectively.

We first use the `mixed` command in Stata (appropriate for the approximately normal dependent variable) to fit a standard GLMM for comparison purposes, completely ignoring the weights in the estimation:

```
mixed isei college || id_school: college, ///
 covariance(independent) variance
```

We note in this syntax that the dependent variable (ISEI) is listed first, followed by the predictor variable. The two vertical bars indicate the first level of clustering, which is defined by the values of ID_SCHOOL. The inclusion of COLLEGE after a colon following the cluster identifier means that the coefficient for this predictor is allowed to randomly vary across clusters (schools), and a random intercept will be included in the model by default. The covariance structure for the two random effects is independent (meaning that they have zero covariance), and estimates of variance components are desired in the output (instead of the default standard deviations). The estimates of the parameters in this model are included in Table 13.1.

Next, we fit a model including a fixed effect of the final overall student weight, but not using the weights to estimate the parameters of interest:

```
mixed isei college w_fstuwt || id_school: college, ///
 covariance(independent) variance
```

Estimates of the parameters in this model are included in Table 13.1 as well.

TABLE 13.1

Parameter estimates (and standard errors in parentheses) from the different multilevel modelling approaches for the 2000 PISA data

Parameter	No Weights	Weights as a Covariate	Scaling Method 1: Effective	Scaling Method 2: Size
FIXED EFFECTS				
Intercept	38.79 (0.62)	36.82 (1.14)	35.89 (0.91)	35.89 (0.91)
COLLEGE	12.65 (0.90)	12.60 (0.90)	14.28 (1.42)	14.28 (1.42)
WEIGHT		<0.01 (<0.01)		
VARIANCE COMPONENTS				
Var(Intercepts)	16.14 (5.45)	13.94 (5.22)	17.74 (6.43)	17.79 (6.43)
Var(College)	43.12 (10.85)	42.26 (10.58)	41.03 (13.74)	41.06 (13.73)
Var(Residuals)	219.33 (7.20)	219.92 (7.22)	214.96 (12.82)	214.92 (12.84)
Pseudo Log(L)	−8611.88	−8609.91	−1439307.8	−1443258.0

We now fit the model of interest using the weights to define the pseudo-likelihood function (and thus computing weighted estimates of the parameters of interest). We first compute the conditional student weight by dividing the final student weight (accounting for all probabilities of selection) by the final school weight. (In fact, one can show that this step is not necessary if the Level-1 weights are normalized within clusters; one could obtain the same results in this case by simply inputting the overall respondent weight into the command, *provided that the appropriate option for normalizing these weights is included as well.*) We then fit the model of interest, specifying 1) the conditional student weights at Level 1, via the [`pweight` =. . .] option; 2) the final school weights at Level 2, via the `pweight()` option after the cluster identifier; and 3) the weight scaling method, which in this case is "`size`" for the "normalizing" method. We also fitted the model using the "`effective`" weight scaling method ("Method 1" in Pfeffermann et al., 1998), and the results were virtually identical (as suggested by Carle, 2009).

```
gen conwt = w _ fstuwt / wnrschbw

mixed isei college [pw = conwt] || id _ school: college, ///
covariance(independent) variance pweight(wnrschbw) ///
pwscale(size)
```

The results from fitting this model are also included in Table 13.1. Examining the results from the different approaches, we can make the following conclusions:

- Parental college education has a strong positive relationship with mean SES, regardless of the method used;
- The weighted estimates of the fixed effects are different from the unweighted estimates, especially so for the intercept (i.e., the mean for students with non-college educated parents);
- Including the final student-level weight as a covariate changes interpretation of the parameters: the intercept now corresponds to the mean for someone with weight equal to zero, which is meaningless (using weight deciles, for example, may be a more effective approach);
- The weighted estimates of the variance components also differ:
    - There is more evidence of variability across schools in the means for students with non-college educated parents (i.e., the random intercepts) when computing weighted estimates; and
    - There is less variability in college vs. non-college gaps (i.e., the random coefficients) across the sampled schools.
- Different weight scaling methods do not result in different estimates or conclusions; in this case, one would use the "size" method (Method 2), per Carle (2009); and
- The robust, sandwich-type standard errors for the weighted estimates are generally larger, as expected, but this does not change the overall inferences.

We provide code for fitting the same model using other software on the ASDA website. At the time that this third edition is being written, LMMs for continuous dependent variables like the model fitted above (dropping the 'G') can also be fitted to complex sample survey data using procedures available in R. The contributed `svylme` package, developed by Thomas Lumley, and the contributed `pwlmm` package, developed by Veiga et al. (2014), currently only fit LMMs. Mplus, SAS (PROC GLIMMIX), and the HLM and MLwiN software packages are each capable of fitting GLMMs to data from complex samples.

This worked example has illustrated the importance of examining the sensitivity of GLMM results to alternative estimation approaches. In this example, our inferences were generally quite similar whether the weights were used or not, suggesting that these weights were not especially informative about the parameters in this model. This will not always be the case for different surveys (e.g., West et al., 2015) or different analyses of the same data set. For example, a series of simulation studies by Zheng and Yang (2016) in an **item response theory (IRT)** context demonstrates the importance of accounting for survey weights when using multilevel models for IRT investigations. Existing software makes it very straightforward to perform these kinds of analyses examining the sensitivity of estimated parameters in multilevel models to the use of survey weights (at each level) for estimation.

In addition to assessing changes in parameter estimates due to the use of weights, assessment of changes in estimated standard errors when fully accounting for the complex sampling features is also important. In this regard, Stapleton and Kang (2018) evaluate design effects for weighted parameter estimates in multilevel models fitted to educational survey data. These authors found that ignoring complex sampling features (i.e., stratification and cluster sampling) in variance estimation *above and beyond* the levels being modeled (e.g., fitting a weighted multilevel model with students nested within schools, but ignoring higher-level primary sampling units, such as school districts or counties; see Rabe-Hesketh & Skrondal, 2006) generally has a minor impact on inferences, and that accounting for the fixed effects of stratum identifiers at Level 2 of a given multilevel model (for higher-level clustering units, such as schools) is a reasonable strategy for inferential purposes.

13.3 Fitting Structural Equation Models to Complex Sample Survey Data

Analysts of survey data, and particularly social scientists, are often interested in fitting **structural equation models (SEMs)** to the data, exploring complicated direct and indirect relationships between variables and testing causal hypotheses in complex theoretical models. There are several complete texts devoted to structural equation modeling, latent variable modeling, and related methods (including **confirmatory factor analysis**, or CFA, and **exploratory factor analysis**, or EFA). For a good practical introduction to these techniques, we recommend Kline (2023) or Schumacker and Lomax (2004). For more theoretical background on SEMs, readers can turn to the classic text from Bollen (1989).

Relatively recent statistical and technical developments have enabled analysts to fit structural equation models to complex sample survey data sets using readily available statistical software, and to make inferences based on those models that correctly adjust for complex sampling features. Much of the pioneering work in this area has been done by the authors of the **Mplus statistical software** (http://www.statmodel.com). Some of the most fundamental work introducing methods for fitting SEMs to complex sample survey data can be found in an article by Muthén and Satorra (1995), who established a theoretical framework for fitting design-adjusted models and making inferences based on those models. These authors describe how the final survey weights should be used in estimation of SEMs, and outline methods for computing design-based standard errors for weighted estimates of SEM parameters that reflect complex sampling features. Stapleton (2008) provides a practical overview of these approaches, including the relevant theory and examples using the LISREL software. In this section, we present illustrations of fitting SEMs to complex sample survey data using existing procedures in Mplus and Stata, and we also discuss procedures in other packages that can be used to fit these models.

The approach to computing weighted estimates of SEM parameters involves three broad steps. When fitting SEMs, one tries to find estimates of the SEM parameters that minimize the differences between the variances and covariances of observed variables *implied* by the parameters in the specified SEM and the *actual* sample variances and covariances of the observed variables. When fitting these models to complex sample survey data, the first step involves computing *weighted* estimates of the variances and covariances of the observed variables for use in estimation. Then, given a hypothesized SEM, one defines a weighted estimate of the population log-likelihood function for the observed data, where each individual respondent's contribution to the log-likelihood is weighted by their final survey weight (similar to the pseudo-likelihood approach outlined in Chapters 8 and 9). This population log-likelihood function will include the weighted estimates of the variances and covariances of the observed variables (essentially the "observed data" that is input to the estimation process). Minimizing the log-likelihood essentially amounts to finding the parameter estimates that minimize the aforementioned difference between the model-implied and observed covariance matrices. The key difference is that the weights are accounted for when defining this log-likelihood function, given the specified model and the observed variables. See Stapleton (2008) for additional details on this estimation process.

Given weighted estimates of the parameters defining an SEM (e.g., path coefficients, variances of error terms, covariances of latent variables, etc.), one then needs to compute estimates of standard errors for the weighted estimates that reflect the complex sampling features. Muthén and Satorra (1995) outlined procedures for variance estimation based on Taylor Series Linearization (TSL), and demonstrated the superiority of their proposed approach over competing variance estimation approaches. A key message arising from that work was that variance estimates for weighted estimates of parameters in a given SEM can be *extremely biased* if the complex sampling features are not correctly accounted for (e.g., assuming that one is working with a simple random sample). Stapleton (2008) used simulation studies to reiterate the advantages of this pseudo-likelihood approach and subsequent nonparametric variance estimation for minimizing the bias in the parameters estimates and ensuring that variance estimates reflect complex sampling features.

Fitting SEMs also involves the assessment of several different types of model information criteria and/or model fit statistics. The standard chi-square test statistic used to assess the overall differences between the model-implied variances and covariances and the direct weighted estimates of the variances and covariances needs to be adjusted in a manner similar to that described in Chapter 6 for categorical data analysis. This adjustment involves division of the test statistic by the "average design effect" for the estimates involved in the analysis, following the Rao-Scott methodology introduced in Chapter 6. Other commonly used measures of goodness of fit for SEMs (e.g., RMSEA, CFI) can be easily adapted to fully account for complex sampling features, and we will demonstrate computation of these measures in the forthcoming examples.

In terms of available software implementing these approaches for fitting SEMs to complex sample survey data, we are aware of only six packages that make this possible at the time that this third edition is being written: Mplus, Latent GOLD, Stata, LISREL, the contributed `lavaan.survey` package in R[1] (Oberski, 2014), and PROC LCA in SAS (for latent class analysis). As mentioned earlier, we will now consider some examples using both Mplus and Stata; to the extent possible, these examples have been replicated using other software procedures on the ASDA website.

13.3.1 An SEM Example: Analysis of ESS Data from Belgium

We now consider an example of fitting an SEM using a design-based approach. We analyze data collected from a probability sample of 1,703 adults as part of the European Social Survey in Belgium, and we focus on a specific subset of all of the variables measured. Included in this data set (immig.dat, available on the ASDA website) is a survey weight variable representing the inverse of an area-specific response rate (WEIGHT), along with a sampling cluster identifier for variance estimation purposes (PSU). Our overall research question concerns whether a latent construct representing a person's overall trust in politicians and political parties is predictive of a second latent construct representing a person's general attitude toward immigration.

The latent trust in politicians/political parties construct is thought to be indicated by two observed variables in the ESS: trust in politicians (TRSTPLT, ranging from 0 to 10 with 88 representing missing data), and trust in political parties (TRSTPRT, ranging from 0 to 10 as well). The latent construct capturing a person's general attitude toward immigration is thought to be indicated by three observed variables in the ESS: IMSMETN (the country should allow few/many immigrants of the same race/ethnicity as the majority, where 1 = few, 4 = many, and 8 = missing), IMDFETN (. . . of different race/ethnicity . . .), and IMPCNTR (. . . from poorer countries outside Europe . . .). The overall SEM of interest is displayed in Figure 13.1.

Before fitting this SEM, we first apply the method of Raykov et al. (2015) to compute a weighted estimate of **Cronbach's alpha coefficient** for each set of observed variables. This weighted estimate of the coefficient alpha captures the approximate composite reliability of the items indicating each construct that one would see if the whole target population had been measured (see the additional discussion about alpha following the illustration below). This method also enables computation of a 95% confidence interval for the coefficient alpha for each set of observed variables. This approach is possible using the Mplus software. We include the Mplus code below from Raykov et al. (2015) that can be used to compute these two weighted estimates of alpha, along with supplemental R code to form confidence intervals for the two alpha coefficients given the weighted estimates of alpha and their (linearized) standard errors. First, we focus on estimation of Cronbach's alpha for the three observed immigration attitude variables:

```
TITLE: WEIGHTED ESTIMATION OF IMMIGRATION ALPHA;
DATA:         FILE = M:\immig.dat;
VARIABLE:     NAMES = psu trstplt trstprt imsmetn imdfetn impcntr weight;
       USEV = psu imsmetn imdfetn impcntr weight;
       MISSING = ALL(8);
```

```
        WEIGHT = weight;
        CLUSTER = psu;
ANALYSIS:     TYPE = COMPLEX;
        ESTIMATOR = MLR;
MODEL:
        F1 BY imsmetn@1;
        F2 BY imdfetn@1;
        F3 BY impcntr@1;
        imsmetn@0;
        imdfetn@0;
        impcntr@0;
        F1-F3 (P1-P3);
        F1 WITH F2-F3 (P12-P13);
        F2 WITH F3 (P23);
MODEL CONSTRAINT:
        NEW(ALPHA);
        ALPHA = 1.5*2*(P12+P13+P23)/(P1+P2+P3+2*(P12+P13+P23));
OUTPUT:       CINTERVAL;
```

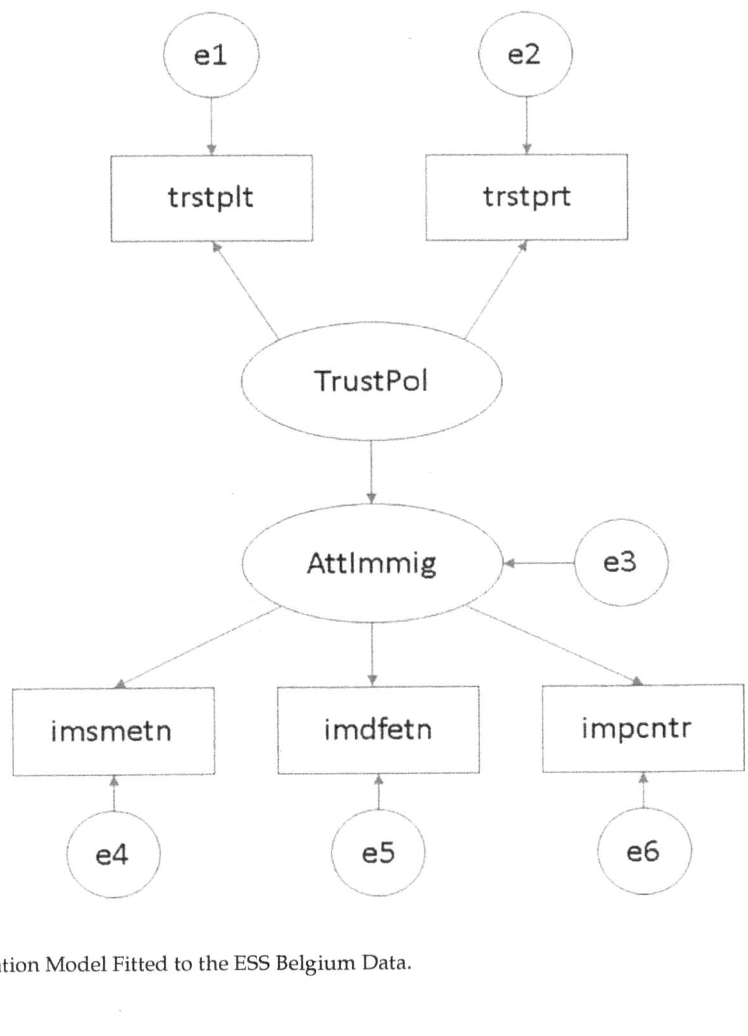

FIGURE 13.1
Structural Equation Model Fitted to the ESS Belgium Data.

Running this code in Mplus produces the output shown below:

```
New/Additional Parameters
   ALPHA                 0.880        0.010      91.631        0.000
```

The weighted estimate of alpha for the immigration variables is 0.880, with a linearized standard error of 0.010. We then use the R code below to form a 95% confidence interval for this coefficient alpha:

```
ci.alpha_in_cds = function(alpha, se)
{
   l = log(alpha/(1-alpha))   # use logit
                              # transformation first
   sel = se/(alpha*(1-alpha)) # get SE of logit of alpha
   ci_l_lo = l-1.96*sel       # lower and upper endpoint
                              # of 95%-CI
   ci_l_up = l+1.96*sel       # for logit of alpha
   ci_lo = 1/(1+exp(-ci_l_lo)) # transform back both ends
   ci_up = 1/(1+exp(-ci_l_up)) # of a CI for
                               # original alpha
   ci = c(ci_lo, ci_up) # print the resulting CI
   ci                   # of coefficient alpha to screen
}
```

Inputting the estimates from Mplus into this function produces the following output:

```
> ci.alpha_in_cds(0.880,0.010)
[1] 0.8589776 0.8982597
```

We see clear evidence of high reliability of these three immigration attitude variables in the target ESS population: these are correlated, reliable indicators of attitudes toward immigration.

We repeat this process for the two trust in politics variables:

```
TITLE: WEIGHTED ESTIMATION OF TRUST ALPHA;
DATA:         FILE = M:\immig.dat;
VARIABLE:     NAMES = psu trstplt trstprt imsmetn imdfetn impcntr weight;
        USEV = psu trstplt trstprt weight;
        MISSING = ALL(88);
        WEIGHT = weight;
        CLUSTER = psu;
ANALYSIS:     TYPE = COMPLEX;
        ESTIMATOR = MLR;
MODEL:
        F1 BY trstplt@1;
        F2 BY trstprt@1;
        trstplt@0;
        trstprt@0;
        F1-F2 (P1-P2);
```

```
        F1 WITH F2 (P12);
MODEL CONSTRAINT:
        NEW(ALPHA);
        ALPHA = 1.5*2*(P12)/(P1+P2+2*(P12));
OUTPUT:     CINTERVAL;
```

The weighted estimate of coefficient alpha for the two trust variables is 0.689, with a linearized standard error of 0.006. We then use the same R function to form a 95% confidence interval for this alpha coefficient:

```
> ci.alpha_in_cds(0.689,0.006)
[1] 0.6771198 0.7006364
```

While this analysis provides weaker evidence of high reliability for these two measures, we still see that the 95% confidence interval covers 0.7, which would generally be considered an acceptable value of Cronbach's alpha when justifying that observed indicators are reliably indicating the same construct. We therefore have faith in the quality of the **measurement model** that defines the latent constructs being analyzed in the larger SEM (Figure 13.1).

Importantly, despite the popularity of alpha as a tool for scale construction and reliability assessment in the social sciences, the limitations of alpha as an approximate measure of the composite reliability of the items forming a hypothetical scale (or indicating a hypothetical construct) have been well-established for decades. Referring to alpha as a composite measure of reliability is really only justified if: 1) the observed items are unidimensional (i.e., the items are not measuring different constructs); 2) the factor loadings for the latent construct of interest are uniformly high; and 3) the errors associated with the observed variables in the factor model are uncorrelated (Raykov, 1997). If these properties are not apparent when examining the coefficient alpha for a set of observed variables, more general composite measures of reliability should be presented. Raykov and Traynor (2016) outline a method similar to that described above using a combination of the Mplus and R software for estimating the composite reliability of a set of observed items following a design-based approach. We will consider whether the observed items have these essential properties when evaluating our larger overall SEM.

Next, we fit the full SEM using both Mplus and Stata, where our focus is on the path coefficient in the **structural model** that relates the latent trust construct to the latent immigration construct (see Figure 13.1). First, we consider the following Mplus code:

```
TITLE:        ESS SEM Example, Complex Sampling Features
DATA:         FILE = M:\immig.dat;
VARIABLE:     NAMES = psu trstplt trstprt imsmetn imdfetn impcntr weight;
        USEV = psu trstplt trstprt imsmetn imdfetn impcntr weight;
        MISSING = trstplt (88) trstprt (88)
           imsmetn (8) imdfetn (8) impcntr (8);
        WEIGHT = weight;
        CLUSTER = psu;
ANALYSIS:   TYPE = COMPLEX;
MODEL:      F1 BY imsmetn imdfetn impcntr;
            F2 BY trstplt trstprt;
            F1 ON F2;
OUTPUT:     STANDARDIZED SAMPSTAT;
```

We note in the Mplus code how the complex sampling features (weight and cluster variables) are explicitly indicated, the analysis type is set to COMPLEX for the design-based approach, and the SEM defines two latent variables (F1, indicated by the three immigration attitude variables, and F2, indicated by the two trust variables), with F1 being regressed on F2 in the structural portion of the model. By default, Mplus uses **full information maximum likelihood (FIML) estimation**, drawing on all available information from each of the 1,703 sampled cases when estimating the model parameters (including those cases with some missing data). For more detail on FIML as an approach for analyzing data sets with missing values, we refer readers to Allison (2012a). Executing this code in Mplus produces the selected output indicated below.

```
Chi-Square Test of Model Fit

          Value                                     1.103*
          Degrees of Freedom                            4
          P-Value                                  0.8939
          Scaling Correction Factor                1.4195
            for MLR

RMSEA (Root Mean Square Error Of Approximation)
          Estimate                                 0.000
          90 Percent C.I.                          0.000    0.016
          Probability RMSEA <= .05                 1.000

CFI/TLI
          CFI                                      1.000
          TLI                                      1.000
```

The various design-adjusted goodness of fit measures (model chi-square, RMSEA, and CFI/TLI indices) all suggest that the fit of this model is quite good, with the model chi-square being small and non-significant, the RMSEA value being less than 0.05, and the CFI and TLI values equal to 1.0. Next, we consider the weighted estimates of the path coefficients in the model in Figure 13.1 and their linearized standard errors:

```
MODEL RESULTS

                                                          Two-Tailed
                    Estimate       S.E.  Est./S.E.       P-Value

 F1       BY
    IMSMETN            1.000      0.000    999.000       999.000
    IMDFETN            1.316      0.054     24.559         0.000
    IMPCNTR            1.122      0.047     23.964         0.000
```

```
  F2         BY
     TRSTPLT              1.000        0.000      999.000      999.000
     TRSTPRT              1.007        0.063       15.916        0.000

  F1         ON
     F2                  -0.071        0.011       -6.407        0.000

  Intercepts
     TRSTPLT              3.868        0.068       56.653        0.000
     TRSTPRT              3.838        0.064       59.937        0.000
     IMSMETN              2.203        0.030       73.711        0.000
     IMDFETN              2.511        0.031       80.704        0.000
     IMPCNTR              2.504        0.028       88.621        0.000

  Variances
     F2                   3.958        0.298       13.271        0.000

  Residual Variances
     TRSTPLT              0.765        0.252        3.034        0.002
     TRSTPRT              0.649        0.277        2.345        0.019
     IMSMETN              0.266        0.016       16.619        0.000
     IMDFETN              0.096        0.020        4.890        0.000
     IMPCNTR              0.223        0.022       10.337        0.000
     F1                   0.363        0.038        9.506        0.000
```

We note how Mplus scales the latent variables by default by setting the path coefficients from the latent variables to one of the observed variables equal to 1.0. We also note that the estimated path coefficient from regressing the latent immigration attitude variable (F1) on the latent trust variable (F2) is negative and strongly significant, suggesting that increased trust in politics and politicians results in less favorable attitudes toward immigration. Mplus also produces standardized versions of these weighted estimates:

```
STANDARDIZED MODEL RESULTS
STDYX Standardization

                                                           Two-Tailed
                    Estimate       S.E.  Est./S.E.    P-Value

  F1         BY
     IMSMETN              0.768        0.022       35.544        0.000
     IMDFETN              0.934        0.014       67.751        0.000
     IMPCNTR              0.827        0.019       43.047        0.000

  F2         BY
     TRSTPLT              0.915        0.029       31.096        0.000
     TRSTPRT              0.928        0.032       29.398        0.000
```

```
F1        ON
   F2                  -0.228      0.032      -7.091      0.000
```

We see that the items indicating each latent construct tend to be highly correlated with that construct, with the loadings being uniformly high and no evidence of multi-dimensionality in each case. This supports our use of coefficient alpha in describing the initial reliability. Examination of modification indices for this model (not shown) also suggests that allowing the errors associated with the observed variables to be correlated would not result in substantial reductions in the overall model chi-square, further supporting the use of coefficient alpha. We therefore have confidence in making inference about the negative relationship between trust in politics and politicians and attitudes toward immigration in the larger Belgian population based on this SEM.

Next, we consider the code for fitting this same SEM in Stata. We start by setting the complex sampling features using the `svyset` command, and then we fit the model using the `svy: sem` command (which is designed for multivariate normal observed variables):

```
use "X:\. . .\immig.dta", clear

svyset psu [pweight = weight]

svy: sem ///
(Trust -> trstplt trstprt) ///
(Immig -> imsmetn imdfetn impcntr) ///
(Immig <- Trust) ///
if trstplt != 88 & trstprt != 88 & ///
imsmetn != 8 & imdfetn != 8 & impcntr != 8
```

We note how the `svy: sem` command allows users to indicate the names of the latent variables in a given SEM (which are required to start with a capital letter), and then use arrows to indicate those observed variables that are indicating a given latent variable, or which latent variable is predicting another latent variable in the structural portion of the model.

The resulting weighted estimates of the path coefficients and their linearized standard errors (not shown) are quite similar to those produced by Mplus, with slight differences emerging because Stata is using listwise deletion to handle missing values (rather than FIML). This results in a slight reduction in the number of cases analyzed (1,676). An important difference between the two packages at the time of this writing is that Stata does not yet compute design-adjusted goodness of fit measures (e.g., RMSEA, CFI), unlike Mplus.

We include examples of code from other packages and procedures (e.g., the `lavaan.survey` package in R) that can be used to fit this model using a design-based approach on the ASDA website. Importantly, both Mplus and Stata (and selected other packages as well) can also fit generalized SEMs to observed variables that are not necessarily normal using design-based approaches (e.g., using the `svy: gsem` command in Stata). The brief illustration above is designed to demonstrate the ease with which analysts can fit SEMs to data collected from complex samples; we do not consider model diagnostics or model selection in great detail here, and instead refer readers to many other excellent texts on these aspects of fitting SEMs (e.g., Kline, 2023). The important point demonstrated here is that one should use software that correctly accounts for the complex sampling features when fitting these models if population inferences are the objective of a given SEM analysis.

Several additional technical papers on the topic of fitting SEMs to complex sample survey data are available at the Mplus web site: http://www.statmodel.com/papers.shtml. For readers interested in published examples of these types of applications using the Mplus software, we suggest Zajacova et al. (2009), Stapleton (2008), or Striegel-Moore et al. (2008). We also note that social scientists are often interested in fitting **item response theory (IRT)** models to refine survey questionnaires and reduce respondent burden. Recent work in this area has discussed methods for fitting IRT models to complex sample survey data using the methods described in this section (Zheng & Yang, 2016). Stata users can now fit IRT models to complex sample survey data using the `svy: irt` command. Social scientists are also often interested in testing hypotheses about **mediation** via the fitting of SEMs. Qin et al. (2019) describe how to carry out tests of mediation when using SEMs to analyze complex sample survey data.

13.4 Analytic Methods for Non-Probability Samples

Recent years have seen a proliferation of methodological research into appropriate methods for making finite population inferences based on **non-probability samples** with undefined sampling and/or selection mechanisms (Elliott & Valliant, 2017; Buelens et al., 2018; Zhang, 2019; Valliant, 2020; Rueda et al., 2020; Chen et al., 2020; Rafei et al., 2022; Castro-Martin et al., 2022; Chrostowski & Beręsewicz, 2024; Beaumont et al., 2024; Li, 2024). Examples of such samples include opt-in online polls, convenience samples, volunteer samples, quota samples, snowball samples, and the like. The common feature of each of these samples is a lack of a well-defined probability sampling mechanism that enables the design-based estimation and inference methods discussed throughout this book; this absence of a probability sampling mechanism prevents the data user from making finite population inferences with measurable sampling error.

These methods have become crucial for users of survey data, as many carefully defined probability samples with extremely low response rates may (unfortunately) take on the characteristics of non-probability samples. Methods with the ability to correctly adjust for any types of selection bias that may occur in these types of samples while also providing reasonable estimates of sampling error are important to ensure that users of data from non-probability samples are not reporting biased and/or inefficient estimates.

We begin this section with overviews of state-of-the-art, theoretically grounded approaches for making finite population inferences based on non-probability samples. As a general comment, given population control totals on key auxiliary variables of interest that are related to the survey variables of interest and available in a non-probability sample, one could simply apply calibration weights to the non-probability sample (as discussed in Chapter 2). However, this procedure would only remove selection bias if the selection was occurring at random as a function of the auxiliary variables (Bethlehem, 2010), which is impossible to demonstrate in most cases. We therefore do not revisit that approach here. We instead focus on illustrations of modern approaches, using available software, in settings where the true population parameters are known, enabling comparisons of the performance of the alternative approaches.

13.4.1 Quasi-Randomization (QR)

The **quasi-randomization (QR) method** of inference for non-probability samples, also referred to in this context as **inverse probability weighting** (Chen et al., 2020) or **propensity score adjustment** (Rueda et al., 2020), involves combining the non-probability sample of respondents with a probability sample of respondents, where both samples collected measures on a common set of covariates from the respondents (and the covariates were measured in an identical fashion). We then think of the non-probability sample as an additional random sample drawn from the same finite population as the probability sample, with the only problem being that we don't know the probabilities of selection for the non-probability sample. We need to estimate these probabilities of selection based on the combined sample, and assuming that we can compute unbiased estimates of these probabilities, we can then use the estimates to compute "quasi" selection weights and subsequently apply design-based methods to the non-probability sample.

This QR approach proceeds as follows. After combining the two samples, the respondents in the probability sample will have their associated survey weights (generally incorporating additional adjustments, per Chapter 2), and the respondents in the non-probability sample are each assigned a value of 1 on the variable containing the survey weights for the probability sample. One then fits a *weighted* regression model (or applies a machine learning technique incorporating the weights) to the combined sample, modeling an indicator of belonging to the non-probability sample (1 = yes, 0 = no) as a general function of the common covariates (possibly including interactions, non-linear terms, etc.), and using the weights to estimate the model. See Figure 13.2 for the necessary data structure; Beaumont et al. (2024) provide additional guidance on modeling and variable selection.

This model could be a logistic regression model, a probit model, etc. However, standard weighted maximum likelihood estimation of this model in this context (e.g., Chapter 8) will generally produce biased estimates of the coefficients used to estimate the probabilities of being selected for the non-probability sample (Valliant & Dever, 2011). An additional adjustment to the likelihood function is needed to ensure unbiased estimation, and this adjustment is detailed in Chen et al. (2020). Machine learning techniques incorporating the combined weight variable could also be applied to compute the predicted probabilities or create strata based on them (Buelens et al., 2018; Rueda et al., 2020; Beaumont et al., 2024).

After applying this adjustment to the estimation procedure, one then computes predicted probabilities of being selected for the non-probability sample based on the fitted model and inverts them to form the estimated base selection weights. These estimated selection weights can then be calibrated to known population control totals (Valliant & Dever, 2011), or the calibration constraints can be incorporated into the maximum likelihood estimation procedure (Chen et al., 2020). The resulting (adjusted) weights for the non-probability sample are then input to standard design-based estimation procedures, and replication techniques (e.g., JRR, bootstrapping) can be used for variance estimation (so that uncertainty in the estimated weights can be incorporated into the variance estimates). The QR procedure can also be applied when microdata are not available for the probability sample; see Chrostowski and Beręsewicz (2024) for details. Importantly, the QR procedure assumes **conditional exchangeability**, which means that the probability of participating in the non-probability sample for a member of the finite population of interest is not related to the outcome variables of interest, after conditioning on all of the covariates used to compute the predicted probabilities. Li

(2024) evaluated the performance of this approach when this assumption is violated and proposed a methodology that can correct for the potential bias introduced when this assumption is violated.

At the time of this writing, the QR procedure is implemented in the contributed R packages `NonProbEst` and `nonprobsvy`, the latter of which implements the additional adjustment described by Chen et al. (2020). We illustrate both packages in the later subsections presenting examples of this approach.

13.4.2 Superpopulation Modeling

The **superpopulation modeling (SM)** approach to finite population inference based on non-probability samples generally relies on two steps:

1) fitting a model to the outcome variable of interest that is only measured in the non-probability sample; and
2) using the fitted model and the same covariates, either available a) in aggregate for the entire finite population, b) for all individual elements in the finite population, or c) for all individual elements in a large probability sample from the finite population (also known as **sample matching**; see Rivers, 2007) to make predictions for the outcome variable of interest, and then subsequently make inference related to population parameters that are a function of that outcome variable.

This model-driven prediction approach is also known as **mass imputation**, as values on the survey variable of interest that was only measured in the non-probability sample are effectively "imputed" for all cases in the finite population, enabling inference related to that variable in the larger population (Yang et al., 2021). The effectiveness of this approach therefore relies heavily on sound model specification and informative covariates that are available in both the non-probability sample and (in some form) for the finite population of interest. In the absence of knowledge about the structure of the best possible model, machine learning techniques can also be very effective for making predictions in this context (Buelens et al., 2018). The approach also relies heavily on an assumption of ignorable selection into the non-probability sample, given the common covariates; that is, the same prediction model derived from the non-probability sample also holds for the larger finite population.

Common Cov. 1	Common Cov. 2	Common Cov. 3	Non-Prob. Sample	Survey Weight
1	23	1	1	1
0	55	3	1	1
...	...	...	...	...
1	68	2	1	1
0	44	3	0	w_1
1	23	3	0	w_2
...	...	...	...	...
1	56	2	0	w_n

FIGURE 13.2
Necessary structure of the combined data set for the quasi-randomization (QR) approach, including the non-probability and probability samples (where there are n respondents in the probability sample) and three hypothetical common covariates. An ellipsis indicates additional data records not shown.

13.4.3 Doubly Robust Methods

Doubly robust (DR) methods for making finite population inference from non-probability samples (Valliant, 2020; Yang et al., 2020; Chen et al., 2020) combine ideas from the QR and SM approaches in the development of a single consistent estimator of a finite population mean. Following this approach, the estimated mean of a variable of interest in the finite population of interest is the sum of:

1) a weighted mean *residual* based on the model fitted to the non-probability sample for the outcome variable of interest, where the finite population weights for computing the population estimate of the mean residual are derived from the QR approach and the residuals for each case are based on the fitted model; and
2) the weighted mean of the predicted values for the probability sample (in the case of sample matching) or the mean of the predicted values (if full microdata are available for the finite population).

This approach is "doubly robust" in that it produces a consistent estimator of the finite population mean if only one of the two models underlying the estimator (the model for the probability of being selected for the non-probability sample, or the model for the outcome variable of interest) is correctly specified. Chen et al. (2020) provide proof of this property, which is attractive for secondary analysts who may not have detailed insight into the best possible models for the selection indicator and the outcome variable of interest, along with a variance estimator for the DR estimate. As of this writing, the DR estimator and its variance estimator have only been implemented in the contributed `nonprobsvy` package for R.

13.4.4 Adjustments Based on Measures of Selection Bias

Recent years have also seen a significant amount of research into measures of the selection bias in a variety of estimators that may be introduced by deviations from selection at random (which often characterize non-probability samples or low response rate probability samples). This family of measures generally relies on bivariate normal **pattern-mixture models** for both the variable of interest and an auxiliary "proxy" variable that represents the best possible prediction of that variable based on a model fitted to the selected sample. At this point in Section 13.4, readers should note that making good predictions of the variable of interest that is only measured in the non-probability sample based on informative auxiliary variables that are also available for the target population (or a large probability sample from it) is *crucial* to all these methods.

These pattern-mixture models are also defined by a sensitivity parameter, generally denoted by φ, that ranges from 0 to 1. This parameter governs the dependence of selection into the non-probability sample on the outcome variable of interest and the auxiliary proxy. When this parameter is equal to 1, selection is completely non-ignorable and depends entirely on the variable of interest. When this parameter is equal to 0, selection occurs completely at random and is a function of the auxiliary proxy only. Little et al. (2020) suggest that a value of 0.5 for this parameter, indicating the selection is equally dependent on both the auxiliary proxy and the variable of interest, provides sound indications of bias, but there is no information in the available data about this sensitivity parameter.

Measures of selection bias based on this approach have now been derived for proportions (Andridge et al., 2019), means (Little et al., 2020), and the coefficients in both linear and probit regression models (West et al., 2021). Simulations in Boonstra et al. (2021),

Hammon (2023), and Gómez-Echeverry (2023) provide significant evidence of the effectiveness of these measures of bias. Recent work has also suggested that the measures can be very effective in terms of adjusting pre-election polling estimates for selection bias (West & Andridge, 2023; Jackson, 2023), adjusting estimates of COVID-19 vaccination rates (Andridge, 2024), and adjusting for selection bias in off-wave mail surveys in ongoing panel studies (Schroeder & West, 2024).

Importantly, one can also use Bayesian methods when computing these measures and making inference about the selection bias, and the measures of bias can also be used to compute adjusted estimates of finite population parameters in a way that accounts for the selection bias (including Bayesian credible intervals for the parameter(s) of interest). These methods generally assume non-informative Jeffreys priors for the parameters defining the underlying pattern-mixture model, but relevant prior distributions can be accommodated if available. The measures also only require aggregate information (means, variances, and covariances) on the covariates used to form the auxiliary proxy for the finite population of interest, and these can easily be derived from large probability samples from the target population. We illustrate the use of these methods for proportions below, using freely available R software provided by West and Andridge (2023).

13.4.5 Case Study: Comparisons of Methods

For this case study, we analyze data that were previously examined by West and Andridge (2023). Our non-probability sample is defined by selected respondents to a pre-election poll for the 2020 U.S. presidential election that was conducted in the state of Wisconsin by ABC and the *Washington Post* (with data collection performed in September and October of 2020 by Abt Associates). We specifically focus on likely voters from this pre-election poll, which was conducted using dual-frame probability sampling of landline and cellular telephone numbers. Despite the probability-based design, we consider this as a non-probability sample for the following reasons: 1) the response rate (AAPOR RR4) was under 7%, and 2) we only focus on a selected subsample of likely voters who either: a) already voted, b) were registered to vote, or c) indicated that they would register to vote by election day, and self-reported at least a 50/50 chance of actually voting on election day. The resulting sample was of size n = 803, with each individual providing an indicator of whether they would vote for candidate Donald Trump (1 = yes, 0 = no), in addition to values on other covariates described below. We consider the Trump indicator as the outcome variable of interest in this case study, and our focus is on estimating the proportion of likely voters in Wisconsin who would vote for candidate Trump in the 2020 election.

Final survey weights were computed for the poll respondents according to their probabilities of selection based on self-reported landline and cell-phone ownership, and these weights were then calibrated to known population control totals on selected demographic variables, derived from the National Health Interview Survey (NHIS) and the U.S. Census Bureau. Additional details on the weighting methodology are available online.[2] For purposes of this case study, we evaluate unweighted estimates based on the final sample of likely voters, along with weighted estimates according to the final survey weights provided with the data. No other information about the probability sample design is available in the public-use data files from this poll, which can also be accessed online.[3]

We aim to examine whether the methods described earlier for non-probability samples produce adjusted estimates that approach the actual 2020 election outcome in the state of Wisconsin. In 2020, 48.8% of age-eligible Wisconsin voters voted for candidate Trump, per the Massachusetts Institute of Technology Election Data and Science Lab.[4] In our analysis, we compare the unweighted and weighted estimates based on the selected sample (each of which assume ignorable selection into the final selected sample, conditional on the available covariates) to adjusted estimates computed using the methods described in Sections 13.4.1 to 13.4.4.

As a source of information about the population of likely voters in Wisconsin, we focus on a probability sample of likely voters from the state that was measured prior to the election by the American National Election Study (ANES) in 2020. This probability sample collected data on the exact same covariates measured in the ABC/*Washington Post* poll, including sex (1 = male, 0 = female), age (18–24, 25–29, 30–39, 40–49, 50–64, 65+), education (high school or less, some college, college, and post-college), race/ethnicity (white, black, Hispanic, other), ideology (liberal, moderate, conservative), and party identification (Democrat, Republican, Independent/other), each of which in theory are highly relevant predictors of voting patterns in U.S. presidential elections (Kennedy et al., 2018). We created binary indicators for each of these categories for purposes of the analyses illustrated here. The ANES data set also provides final survey weights based on the probabilities of selection and calibration adjustments for the sake of computing finite population estimates. Both the ABC/*Washington Post* and ANES data sets from Wisconsin are available on the ASDA website.

We begin our case study by examining the unweighted estimate based on the selected sample, using Stata to compute the point estimate and a 95% confidence interval (CI) for the proportion of age-eligible likely voters in Wisconsin that will vote for candidate Trump:

```
use "X:\. . .\abc _ wisc.dta"
prop trumpind
```

This approach assumes that we are working with a simple random sample. The resulting point estimate of the proportion based on the selected sample is 0.384, with a 95% CI of (0.351, 0.418). Clearly, the unweighted estimate has significant negative bias relative to the known population truth (0.488). Do the weighting adjustments applied to the selected sample, assuming ignorable selection, help? We first declare the weights in Stata, and then implement the design-based approach using Taylor Series Linearization for variance estimation (we remind readers that no other information about the sample design is available in the data set):

```
svyset [pweight = weight]
svy: prop trumpind
```

The resulting weighted point estimate is 0.394, with a 95% CI of (0.356, 0.434). While some of the negative bias in the unweighted estimate has been removed by the weighting adjustments, a pollster attempting to estimate this proportion using the design-based approach, with calibrated weights and assuming ignorable selection, would still be "missing" the actual proportion of the likely voter population in Wisconsin voting for candidate Trump (0.488).

We now illustrate the QR approach discussed in Section 13.4.1. We first consider functions implementing this approach in the contributed `NonProbEst` package within R. We start by loading this library:

```
library(NonProbEst)
```

Next, we explicitly identify the vector of common covariate names to the `propensities()` function within this contributed package, along with the non-probability and probability sample data frames (with any cases having missing data excluded from the ANES data set, which is required), and the algorithm that will be used to predict selection probabilities for the non-probability sample (the poll in this case):

```
# read data for probability sample (ANES) from population
popdata <- read.csv("X:/ . . ./anes _ wisc.csv", h=T)
# remove cases with missing values
popdata2 <- popdata[complete.cases(popdata),]

# read data from WI poll, with common covariates
poll _ nomiss <- read.csv("X:/ . . ./abc _ wisc.csv", h=T)

covariates <- c("male", "age1", "age2", "age3", "age4", "age5",
"somecoll", "coll", "postcoll", "black", "hisp", "other",
"demparty", "otherparty", "lib", "con")

pi <- propensities(poll _ nomiss, popdata2, covariates,
algorithm = "knn", smooth = T, tuneGrid = data.frame(k = seq(3, 11, by = 2)))
```

We note that this implementation of the QR approach in `NonProbEst` does *not* recognize the final survey weights in ANES in estimation of the response propensities (or the additional adjustments for bias correction described by Chen et al., 2020). While `NonProbEst` makes a variety of novel machine learning algorithms available for predicting the probabilities of being selected for the non-probability sample (k-Nearest Neighbors was used in this case, and all algorithms in the `caret` package are available), the final survey weights are needed for unbiased estimation of the probabilities.

We proceed with computing the weights for the non-probability sample based on the predicted probabilities, illustrating the basic approach of first inverting the response propensities (Valliant, 2020; implemented in the `valliant _ weights` function) and then using the post-stratification method outlined in Valliant and Dever (2011) (implemented in the `vd _ weights` function, which is another one of several additional estimation options that `NonProbEst` provides). We then compute our adjusted estimate of the proportion voting for Trump using these weights (we use the weights based on the first estimation approach to start), along with a jackknife variance estimate for this proportion (which requires the size of the target population as input, which in this case is the number of individuals who voted in Wisconsin in 2020):

```
# calculate weights for NP sample per Valliant (2020)
wiv <- valliant _ weights(propensities = pi$convenience)

# calculate weights for NP sample per Valliant and Dever (2011)
wivd <- vd _ weights(convenience _ propensities = pi$convenience,
reference _ propensities = pi$reference)

summary(wiv)

mean _ estimation(sample = poll _ nomiss, weights = wiv,
estimated _ vars = "trumpind")

# size of population for WI
n1 <- 3298041
```

```
# Estimation of the 95% confidence interval
estim <- mean_estimation(sample = poll_nomiss, weights = wiv,
estimated_vars = "trumpind")

std_dev <- fast_jackknife_variance(sample = poll_nomiss,
weights = wiv, estimated_vars = "trumpind", N = n1)

confidence_interval(estimation = estim, std_dev = std_dev,
confidence = 0.95)
```

The resulting adjusted estimate is 0.3942, with a jackknife 95% CI of (0.3936, 0.3948). This approach therefore did not result in a significant adjustment of the unweighted proportion, and the confidence interval seems far too narrow. This is likely because the jackknife variance estimator used by NonProbEst in this case assumes that the adjusted weights are on the population scale (i.e., they would sum to an estimate of the population size), and that is not the case here, since we are simply inverting estimated response propensities.

Valliant and Dever (2011) emphasize the importance of applying additional calibration adjustments to these initial weights based on the estimated probabilities of inclusion in the non-probability sample, using population control totals on auxiliary variables that are relevant predictors of the outcome variable of interest, to further reduce bias in the adjusted estimate. This would also put the estimated weights on a population scale if that were not already the case. The NonProbEst package also provides tools for further calibrating the estimated weights to known population control totals prior to variance estimation, which we now illustrate.

We suppose that population control totals on the male indicator (50% of the likely voter population is male), the Democratic party indicator (40% of the likely voter population identifies as Democratic), and the other party indicator (20% of the likely voter population supports some other party) are available from some external source, and we calibrate the initial weights to these control totals prior to variance estimation using the calib_weights() function (specifically using the raking technique; see Chapter 2):

```
totals <- data.frame(male = 0.5*3298041, demparty = 0.4*3298041,
otherparty = 0.2*3298041)
totals <- colSums(totals)

aux.vars <- c("male", "demparty", "otherparty")

initial_weights <- wiv # weights from initial code above

wiv2 <- calib_weights(poll_nomiss[, aux.vars], totals,
initial_weights, N = 3298041, method = "raking")
```

Now that the calibrated weights are available, we re-compute the estimated proportion (now specifying the known population size based on the external information), and re-apply the jackknife variance estimation:

```
mean_estimation(sample = poll_nomiss, weights = wiv2,
estimated_vars = "trumpind", N = 3298041)

v <- fast_jackknife_variance(poll_nomiss, wiv2,
estimated_vars = "trumpind", N = 3298041)
```

```
ci <- confidence_interval(mean_estimation(sample = poll_nomiss,
weights = wiv2, estimated_vars = "trumpind", N = 3298041), sqrt(v))
```

The updated point estimate after calibration is 0.4097, with a 95% confidence interval of (0.3706, 0.4488), which seems far more reasonable (and the upper limit of the CI is now approaching the truth). In this case, the additional calibration adjustment was critical for multiple reasons: additional bias reduction, given that sex and political party preference are strong correlates of vote choice, and ensuring that the estimated weights are on the population scale.

What if the model for selection probability was misspecified in the QR approach, but we feel reasonably confident in a substantive model for the probability of voting for candidate Trump? In such settings, the doubly robust (DR) approach combines the benefits of the QR and SM approaches and generates a consistent estimator of the finite population parameter of interest if one of the two models is misspecified (Chen et al., 2020). Next, we illustrate use of the DR approach in the `nonprobsvy` package. First, assuming that the package has already been installed from CRAN, we load the package and its dependencies (which include the contributed `survey` package):

```
library(nonprobsvy)
```

Next, we implement the DR approach using the fairly straightforward `nonprob()` function in this package. In the setting where microdata are available for both the selected sample and the reference probability sample (which is the case here), we need to specify: 1) the model formula for selection into the non-probability sample, which will be the same as used in the QR approach above; 2) the substantive model formula for the outcome of interest (the Trump indicator); 3) the data frame object containing the data for the selected sample; 4) a survey design object for the reference probability sample, from the `survey` package; 5) the method to be used for modeling the outcome of interest (a generalized linear model, or GLM, in this case); and 6) the family of GLM to use (binomial).

```
# survey design object for reference sample
pop.des <- svydesign(ids = ~1, data = popdata2, weights = ~ weight)

nonprob(
  selection = ~ male + age1 + age2 + age3 + age4 + age5 +
somecoll + coll + postcoll + black + hisp + other +
demparty + otherparty + lib + con,
  outcome = trumpind ~ male + age1 + age2 + age3 + age4 + age5
+ somecoll + coll + postcoll + black + hisp + other + demparty
+ otherparty + lib + con,
  data = poll_nomiss,
  svydesign = pop.des,
  method_outcome = "glm",
  family_outcome = "binomial"
)
```

The computed estimate based on the DR approach is 0.5205, with a 95% CI of (0.4427, 0.5983). Here, we see that the doubly robust estimator has effectively eliminated the negative bias in the unweighted and design-based estimates, and produced a 95% CI that covers the population truth. We note that the nonprobsvy package provides many other related estimation approaches based on the current literature; we chose to illustrate the approach above as we believe that this will be a common setting for researchers conducting secondary analyses of data from non-probability samples. See Chrostowski and Beręsewicz (2024) for more details.

Next, we consider adjustments based on the **measure of unadjusted bias for proportions, or MUBP** (Andridge et al., 2019; West & Andridge, 2023). This code is only available for the R software. We first load the necessary contributed packages and functions within R:

```
# Load necessary libraries
library(tidyverse)
library(broom)
library(survey)

# Load necessary functions for Bayesian MUBP approach
source("https://github.com/bradytwest/IndicesOfNISB/raw/master/
proxyDrawsMUBP_sumstats.R")
source("https://github.com/bradytwest/IndicesOfNISB/raw/master/
mubp_functions.R")
```

Next, we enter a control total representing the size of the population of interest (in this case, the total number of voters in the state of Wisconsin, based on the MIT lab data from 2020), and read in the population microdata from the ANES sample:

```
# size of non-selected sample for WI
n0 <- 3298041
# read data from probability sample (ANES) from population
popdata <- read_csv("X:/ . . ./anes_wisc.csv")
```

Next, we obtain population estimates of means, variances, and covariances for each of the common covariates in the ANES probability sample (note that we work with either continuous covariates or binary indicators of the non-reference categories in the model of interest):

```
# we only need to account for the weights to get pop. estimates
pop.des <- svydesign(ids = ~1, data = popdata, weights = ~ weight)
means.pop <- c(svymean(~male, pop.des, na.rm=T)[1],      #ref=female
               svymean(~age1, pop.des, na.rm=T)[1],      #ref=age6
               svymean(~age2, pop.des, na.rm=T)[1],
               svymean(~age3, pop.des, na.rm=T)[1],
               svymean(~age4, pop.des, na.rm=T)[1],
               svymean(~age5, pop.des, na.rm=T)[1],
               svymean(~somecoll, pop.des, na.rm=T)[1], #ref=hsless
               svymean(~coll, pop.des, na.rm=T)[1],
               svymean(~postcoll, pop.des, na.rm=T)[1],
               svymean(~black, pop.des, na.rm=T)[1],     #ref=white
               svymean(~hisp, pop.des, na.rm=T)[1],
               svymean(~other, pop.des, na.rm=T)[1],
               svymean(~demparty, pop.des, na.rm=T)[1], #ref=repparty
               svymean(~otherparty, pop.des, na.rm=T)[1],
               svymean(~lib, pop.des, na.rm=T)[1],       #ref=mod/noideo
               svymean(~con, pop.des, na.rm=T)[1])
```

```
# variance-covariance matrix
varcov.pop <- svyvar(~ male + age1 + age2 + age3 + age4 + age5 + somecoll +
coll + postcoll + black + hisp + other + demparty + otherparty + lib + con,
na.rm=TRUE, design=pop.des)
```

Now, we read in the data for the selected sample, identify the dependent variable of interest and the sample size, and fit the probit model to the indicator of voting for candidate Trump that is required to create the auxiliary proxy in the selected sample (using all of the common covariates), saving the design matrix from this model fit:

```
poll_nomiss <- read_csv("X:/ . . ./abc_wisc.csv")

###################
# MUBP calculations
###################

# Selected sample Y
y1 <- poll_nomiss$trumpind
# Selected sample size
n1 <- length(y1)

# Probit model using selected sample [Y|Z,S=1]
fit <- glm(trumpind ~ male + age1 + age2 + age3 + age4 + age5 +
somecoll + coll + postcoll + black + hisp + other + demparty +
otherparty + lib + con, family=binomial(link="probit"),
data=poll_nomiss)

# Predictors Z
z1_withInt <- model.matrix(fit)
z1 <- z1_withInt[,-1] # remove intercept (added later)
```

Finally, given the necessary ingredients for the Bayesian MUBP calculations (including the population estimates computed earlier from the ANES), we first set a seed to ensure that these results can be replicated, and then generate 1,000 posterior draws of the MUBP for the proportion of interest, starting with random UNIFORM(0,1) draws of the sensitivity parameter (φ) and assuming non-informative Jeffreys priors for the parameters defining the pattern-mixture model:

```
set.seed(41279)

draws <- proxyDrawsMSB(y1, z1, n1, means.pop, varcov.pop,
n0-n1, phi=0, drawphi=TRUE, scaleX=TRUE, nreps=1000)
```

Given the posterior draws of the MUBP (which is a measure of selection bias in the unadjusted estimate of the proportion based on the selected sample), we then compute the adjusted estimate of the proportion based on the posterior median of the MUBP draws, along with a 95% credible interval for the proportion (see Andridge et al., 2019, for details):

```
ymean_mubpadj_p50 = mean(y1) - quantile(draws$msb,0.5)
ymean_mubpadj_p50
```

TABLE 13.2

Comparison of performance of methods for making finite population inference based on non-probability samples, focusing on the estimated proportion of likely voters in the state of Wisconsin who would vote for President Trump in 2020 (Data Sources: ABC/*Washington Post* pre-election poll, ANES; True population parameter = 0.4882)

Method	Point Estimate	95% CI	Cover Truth?
Unweighted	0.3836	(0.3505, 0.4178)	No
Design-Based	0.3943	(0.3563, 0.4336)	No
QR (NonProbEst)*	0.4097	(0.3706, 0.4488)	No
DR (nonprobsvy)	0.5205	(0.4427, 0.5983)	Yes
MUBP	0.5091	(0.4702, 0.5541)**	Yes

Notes: QR = Quasi-Randomization method. DR = Doubly Robust method. MUBP = Measure of Unadjusted Bias for Proportions (Andridge et al., 2019). * Results based on additional calibration adjustments. ** Denotes a 95% credible interval based on the fully Bayesian approach. Results may vary if the same seed is not used.

```
ymean_mubpadj_lb = mean(y1) - quantile(draws$msb,0.975)
ymean_mubpadj_lb
ymean_mubpadj_ub = mean(y1) - quantile(draws$msb,0.025)
ymean_mubpadj_ub
```

The resulting adjusted point estimate is 0.5091, with a 95% credible interval of (0.4702, 0.5541), which covers the population truth (0.4882). Adjustments based on the MUBP, which relies on highly relevant covariates and explicitly allows for non-ignorable selection into the selected sample, seemed to be effective in this case.

Table 13.2 summarizes the performance of these alternative methods for computing estimates based on the selected sample in this case study. These results suggest that the doubly robust and MUBP approaches were able to correct the rather substantial negative bias in the estimated proportion from the selected sample in this case study (which was evident even when using the weights provided for the poll and following a standard design-based approach). The MUBP approach also appears to do so in a more efficient manner.

We end this case study by noting that the methods illustrated here are designed entirely to address the potential non-ignorable *selection* bias in survey estimates that may be engendered by non-probability sampling. Very recent work in this area has suggested that non-probability samples may also be significantly affected by *measurement error* due to bogus responding (see Kennedy et al., 2024, and the related discussions). We anticipate that future work in this area will focus on the development of statistical methods for addressing both types of error (selection and measurement) simultaneously, and we will keep readers updated regarding these developments on the ASDA website.

13.5 Machine Learning with Complex Sample Survey Data

A solid understanding of the computational techniques that can be used to carry out **machine learning** has become essential for the modern survey data analyst (Buskirk et al.,

2018). Machine learning refers to the use of computational algorithms to uncover patterns in data and complex relationships among variables that can be used to make predictions of key outcome variables for new observations. While a variety of statistical and data science tools fall under the general umbrella of machine learning, research into the adaptation of these techniques to complex sample survey data has primarily focused on the subset of techniques known as **classification and regression trees (CART)**. These tree-based methods are used to build decision trees that enable predictions of outcomes (classification → categorical outcomes, regression → continuous outcomes) for new cases and describe complex relationships among variables, including interactions that likely would not be detected when using classical regression modeling. For general primers on tree-based methods and **ensemble methods** (including **random forests**) that employ modal predictions based on collections of trees constructed from many bootstrap samples selected from the observed data, see Breiman (2001), Strobl et al. (2009), or Buskirk (2018).

As has been the case with nearly all the methods discussed in this book, the goal in adapting these machine learning techniques for use with complex sample survey data is design-consistent estimation and inference, where in expectation the methods would yield results that would be seen if we had information available on the entire finite population from which the designed probability sample was selected. This is often accomplished via the available survey weights, which as we discussed in Chapter 3 may or may not be informative about the regression model, classification tree, or other representation of the data that is of interest to the analyst. With respect to classification and regression trees, fundamental work in this area has been done by Toth and Eltinge (2011), who developed design-consistent modifications of the recursive partitioning algorithm that is often used to build classification and regression trees. These authors showed that ignoring survey weights in tree construction can have severe negative implications for tree construction (in terms of significant bias for the arbitrary regression function being estimated). This methodology has been implemented in the contributed R package `rpms` (Toth, 2021), which we will demonstrate later in this section.

Additional design-consistent tree-based estimators have since been proposed by McConville and Toth (2019), who focused on the case of identifying post-strata for weighting adjustment; Dagdoug et al. (2023) and Toth and McConville (2024), who focused on the case of random forests; Opsomer and Riddles (2023), who proposed a design-based version of the **Chi-squared automatic interaction detection (CHAID) algorithm** originally developed by Kass (1980)[5] for constructing decision trees; and Beaumont et al. (2024), who developed design-adjusted methods for pruning trees in the context of estimating selection probabilities for non-probability samples. Bilton et al. (2017) proposed a tree-based methodology that also accommodates the effects of randomly sampled clusters in a small area estimation context, and Toth (2020) developed a unique permutation test for identifying the importance of predictors when developing splits for trees that accounts for unequal probabilities of selection and sample clustering. This approach has been implemented in the `rpms` package for R, which can produce both individual trees (via the `rpms()` function) and random forests (via the `rpms_forest()` function) fully accounting for complex sampling features.

Classification trees are often used to predict response propensities when computing nonresponse adjustments to sampling weights (e.g., Earp et al., 2018, 2022), and Jones et al. (2021) evaluated alternative approaches to accounting for complex sampling features (including survey weights and cluster sampling) when modeling response propensity with classification trees. These authors did not find significant differences in performance between the algorithms considered in the response propensity modeling context. In their specific simulation, there did not appear to be any harm introduced by using the rpms approach in R. Given the past literature demonstrating how ignoring survey weights in the building of classification trees can be quite problematic depending on the context (Toth & Eltinge, 2011), especially when modeling response propensity (Flores Cervantes & Brick, 2016), we would encourage researchers modeling response propensity to account for these complex sample design features when building classification trees.

We now consider an example of building a regression tree while accounting for complex sample design features, using the rpms package in R. In this example, we will consider the construction of a regression tree based on recursive partitioning with and without accounting for the complex sampling features of the 2011–2012 NHANES data. The overall objective is to develop a regression tree for predicting a person's mean systolic blood pressure as a function of selected socio-demographics. We will contrast the tree based on the design-consistent methods implemented in the rpms package of R with the tree that would be constructed when ignoring the complex sampling features and assuming simple random sampling.

First, we load the rpms package in R, along with the 2011–2012 NHANES data:

```
library(rpms)
load("X:/. . ./nhanes1112.rdata")
```

Next, we build a regression tree using rpms(), specifying all the complex sample design variables in a reduced version of the NHANES data set with complete cases only. The predictors retained in this data set include body mass index (BMI), age, indicators of being white, black, or "other" (where the reference category of race includes those of Mexican or other Hispanic descent), and a continuous measure of family poverty to income ratio.

```
nhanes <- nhanes1112[, c(7, 8, 9, 10, 11, 19, 24, 34, 35, 36)]
nhanes <- nhanes[complete.cases(nhanes),]

dtree <- rpms(bpxsy1 ~ bmxbmi + age + white + black + other +
indfmpir, data = nhanes, weights = ~wtmec2yr, strata = ~sdmvstra,
clusters = ~sdmvpsu)
```

Note the explicit specification of the complex sample design features as arguments to the function. We examine the output from the resulting tree as follows:

```
dtree$partition
```

```
   node    n
1    16  175
2    17  341
3     9  364
4    80  156
5    81  146
6   164  306
7   165  195
8    83  204
9    21  696
10   11 1562
```

```
11    6 1434
12   14  294
13   15  286

splits
1       age <= 50.5 & bmxbmi <= 19.95 & age <= 12.5 & bmxbmi <= 16.15
2       age <= 50.5 & bmxbmi <= 19.95 & age <= 12.5 & bmxbmi > 16.15
3       age <= 50.5 & bmxbmi <= 19.95 & age > 12.5
4       age <= 50.5 & bmxbmi > 19.95 & bmxbmi <= 26.85 & age <= 27.5 &
         age <= 13.5 & bmxbmi <= 22.55
5       age <= 50.5 & bmxbmi > 19.95 & bmxbmi <= 26.85 & age <= 27.5 &
         age <= 13.5 & bmxbmi > 22.55
6       age <= 50.5 & bmxbmi > 19.95 & bmxbmi <= 26.85 & age <= 27.5 &
         age > 13.5 & black <= 0.5 & white <= 0.5
7       age <= 50.5 & bmxbmi > 19.95 & bmxbmi <= 26.85 & age <= 27.5 &
         age > 13.5 & black <= 0.5 & white > 0.5
8       age <= 50.5 & bmxbmi > 19.95 & bmxbmi <= 26.85 & age <= 27.5 &
         age > 13.5 & black > 0.5
9       age <= 50.5 & bmxbmi > 19.95 & bmxbmi <= 26.85 & age > 27.5
10      age <= 50.5 & bmxbmi > 19.95 & bmxbmi > 26.85
11      age > 50.5 & age <= 70.5
12      age > 50.5 & age > 70.5 & age <= 78.5
13      age > 50.5 & age > 70.5 & age > 78.5

      value         cvar      mean    N_hat      p_se
1  97.85448 1.560343e-05  97.85448  3650921  7.918817
2  101.6138 4.124817e-05 101.61376  7665499  9.204525
3  106.8738 1.177065e-05 106.87378 11310735 10.633436
4  102.7499 2.048956e-05 102.74987  3663154  8.522023
5  107.9064          Inf 107.90640  3032237       Inf
6  109.231           Inf 109.23098  6874978       Inf
7  112.0833          Inf 112.08331 15819741       Inf
8  114.1625          Inf 114.16249  3473886       Inf
9  114.7567 3.565868e-06 114.75665 32421548 11.904868
10 118.2921 2.818634e-06 118.29208 65652655 13.051515
11 127.4496 4.869742e-06 127.44959 62580901 17.587106
12 133.4925 3.467798e-05 133.49247 10116992 17.420158
13 144.0165 8.442347e-05 144.01648  9378667 22.028991
```

We see evidence of 13 terminal nodes, along with the unweighted sample size assigned to each node. We then see detailed descriptions of each terminal node based on the input predictors; it would appear that age and BMI were the dominant predictors of systolic blood pressure in this case. Finally, we see weighted estimates of means in each terminal node, along with the estimated size of the population in each node (N_hat). A simple examination of the `dtree` object indicates that about 30.95% of the variance in systolic blood pressure in the target population was explained by this tree solution (other output not shown):

```
dtree
```

```
[1] "R-squared of model: 0.309518339039549"
```

Next, as a simple contrast, we completely ignore all of the NHANES design features (equivalent to assuming simple random sampling):

```
stree <- rpms(bpxsy1 ~ bmxbmi + age + white + black + other + indfmpir,
data = nhanes)
```

Examining the output in the same way, we see a vastly different tree: 30 terminal nodes, with an R-squared value of 39.31% and white, black, and family poverty to income ratio all emerging as important predictors. The design features are clearly informative about the regression function being estimated here, as several additional predictors emerge as significant when ignoring the design features. It would appear that ignoring the design features significantly overstates our ability to predict systolic blood pressure in the NHANES target population.

We have focused exclusively on recent developments in tree-based methods for complex sample survey data in this section. As we mentioned initially, this is only one subset of methods under the more general machine learning umbrella. Other very recent work has found that it is important to account for survey weights in **gradient boosting** to improve predictions for target populations (especially with larger sample sizes; MacNell et al., 2023). We anticipate additional developments related to other machine learning techniques in the very near future, such as the **LASSO** (e.g., Signorino & Kirchner, 2018; Iparragirre et al., 2023), **cross-validation** (Wieczorek et al., 2022), and **neural networks** (e.g., Eck, 2018), and we will continue to track these developments on the ASDA website.

13.6 Small Area Estimation and Complex Sample Survey Data

Chapters 4 and 5 introduced estimation and inference for subclasses of the survey population. Included under the general umbrella of subclass analysis is estimation for geographic domains of the survey population. Surveys such as the NCS-R, NHANES, and HRS generally support separate estimation and analysis for large geographic areas such as Census Regions or Divisions; however, the potential to directly estimate statistics for smaller geographic units such as individual states, counties, or municipalities is limited by poor precision due to the size and distribution of the multi-stage sample design. In cases such as the estimation of unemployment rates where direct state or metropolitan level estimates are important, major survey programs such as the U.S. Current Population Survey (CPS) expand the stratification, primary stage sample allocation, and overall sample size of the survey to enable stand-alone estimation for these critical geographic reporting units.

Over the past 50 years, a class of statistical techniques labeled **small area estimation methods** (Rao & Molina, 2015) has been developed to address the problem of estimating population characteristics for small areas. Most of these techniques blend survey data, ancillary population data, and a population model to estimate small area statistics. For larger domains, methods such as generalized regression (GREG) estimation or calibration methods (Deville & Sarndal, 1992) simply use population data to adjust direct regression estimates computed from the survey data. Other techniques such as synthetic estimation or structure preserving estimation (SPREE) use survey data to model the structure of relationships in the survey population and then combine that structure with administrative or population data for the target small area to generate a small area estimate. Additional techniques such as composite estimation or the James-Stein method create an estimate that is a weighted function of a direct survey estimate for the small area and a model-based estimate or a direct estimate for a larger domain.

Computation of the direct estimates or regression coefficients used to form small area estimates should generally incorporate survey weights, which will ensure design consistency of the estimates, and the variances of small area estimates should reflect the sampling distributions of the (weighted) direct survey estimates, given any complex sampling features (You & Rao, 2002, 2003; Ghosh & Maiti, 2004; Pfeffermann & Sverchkov, 2007). The probabilities of selection, if available, could also be used as covariates in the models used to compute small area estimates (Verret et al., 2015). Related work by Stefan and Hidiroglou (2020) extended this idea by demonstrating the benefits of models using smooth functions of the inclusion probabilities when the sample design is "highly informative".

The use of weighted estimators is especially crucial for informative sampling designs, where the probability of selection is associated with the outcome variable of interest that is the focus of the small area estimation. Guadarrama et al. (2018) describe an extension of the **empirical best (EB) method** for parameter estimation in small areas that incorporates informative survey weights, and demonstrate via simulation that 1) these weighted estimators significantly reduce the bias associated with unweighted EB estimators, and 2) losses in efficiency of the estimates when the design is non-informative are small. R code implementing this approach is available with the supplemental materials for that article. Zimmermann and Münnich (2018) further extended the EB method to account for the skewed distribution of data common in small area estimation problems focused on business data. These authors handled the case of informative sampling by including covariates containing sample design information in the models, similar to Verret et al. (2015).

Methodological advances for small area estimation with complex samples have generally centered on the application of generalized linear mixed models (GLMMs) of the form in (13.1) (Jiang & Lahiri, 2006) and Bayesian methods (e.g., Datta et al., 2024; see also Section 13.2). In particular, Chen et al. (2014) and Vandendijck et al. (2016) discussed how to account for survey weights when using hierarchical Bayesian models for small area estimation, and provide empirical evidence of bias and variance reduction when using their approaches. Both of these studies include supplemental code demonstrating the use of R software for this type of small area estimation. Chandra et al. (2019) describe methods for accommodating survey weights and cluster sampling information in area-level GLMMs used for small area estimation, demonstrating significant gains in efficiency when using these area-level methods to compute small area estimates of proportions.

Other recent advances have focused on:

1. Improving **spatial smoothing methods** for small area estimation by accounting for informative sampling and survey nonresponse (Mercer et al., 2014);
2. Simultaneously accounting for informative sampling and (potentially non-ignorable) unit nonresponse in small area estimation, including use of the bootstrap for variance estimation (Sverchkov & Pfeffermann, 2018);

3. Estimation of quantiles for small areas, including the use of resampling methods to estimating sampling variability for these estimated quantiles (Chen & Liu, 2019);
4. Model-based small area estimation based on *grouped data*, where the available information corresponds to frequencies of pre-specified groups in each area (as opposed to unit-level data; Kawakubo & Kobayashi, 2023);
5. Improving small area estimates by combining multiple surveys with overlapping content using area-level models (Franco & Maitra, 2023); and
6. Joint Bayesian modeling of the survey weights and the outcome of interest, combined with post-stratification, to improve small area estimates (Gelman et al., 2024).

Parker et al. (2023) recently provided a comprehensive review of alternative approaches to performing unit-level modeling for small area estimation in the context of informative sampling. We encourage readers to consult this review for additional information on state-of-the-art approaches to fitting models to unit-level microdata when small area estimation is the goal. Collectively, the newest literature in this area at the time that this third edition is being written provides compelling evidence of the need to account for informative probability sampling in all types of small area estimation. Jiang and Rao (2020, Section 6) provide a review of software tools that are available for small area estimation, and readers are encouraged to consult the supplemental materials for many of the newer articles cited in this section for additional code implementing these newer methods.

13.7 Nonparametric Methods for Complex Sample Survey Data

Another area that is (at present) wide open for additional research is the application of **nonparametric statistical methods** to the analysis of complex sample survey data (Montanari & Ranalli, 2005; Breidt & Opsomer, 2009). Indeed, of all the topics discussed in this book, this literature is probably the most nascent, with important developments only dating back to the turn of the century. Whittaker and Scott (1999) discussed the Averaged Shifted Histogram (ASH) approach for estimating multivariate nonparametric regression models with complex sample survey data. Bellhouse and Stafford (2001) then discussed how to handle complex sampling features when performing local polynomial regression.

In Chapter 11 of *Analysis of Survey Data*, Chambers et al. (2003) provide a detailed but largely theoretical discussion of possible methods for incorporating complex design features into **nonparametric regression methods** designed for exploratory data analysis (EDA), such as localized **smoothing models**. Their focus was on "scatterplot smoothing," or modeling the functional relationship between some dependent variable y and some independent variable x. Chambers et al. performed a series of simulations evaluating different methods for incorporating complex sample design features into smoothing methods, but their results were not clear in terms of optimal methods or whether survey weights should be incorporated in the analyses. Further, only the case of stratification was considered, and the authors indicated that extensions of their theory for cluster sampling designs were needed. The authors concluded that their proposed design adjustments to standard smoothing methods can be beneficial when sampling schemes are **informative** (i.e., probability of selection is related to the survey variable of interest), and that there would be little loss of efficiency when using their suggested design adjustments if a sampling scheme was not informative.

Opsomer and Miller (2005) provided additional guidance on the amount of smoothing that should be done in nonparametric regression applications with survey data, and Wang and Bellhouse (2009) extended prior developments for nonparametric regression modeling to describe a *semiparametric* model for complex sample survey data. Harms and Duchesne (2010) presented an evaluation of design adjustments for the local linear regression estimator, and Zhang et al. (2015) then described smoothing spline estimators for the general case of nonparametric regression modeling of complex sample survey data, demonstrating the improved performance of weighted estimators of smoothing splines. Mostafa and Shan (2019) later showed that combining parametric models with nonparametric regression smoothers can help to improve the model-assisted estimation of finite population totals; see also McConville and Breidt (2013), who studied the asymptotics of penalized spline regression estimators in the context of this type of model-assisted estimation. Design-adjusted smoothing models can now be fitted to complex sample survey data using the `svysmooth()` function in the R `survey` package (Lumley, 2010); see Section 5.4.1 for an example.

In terms of **nonparametric density estimation**, which is used for comparing population subgroups in terms of distributions on continuous survey variables, estimating quantiles (see Chapter 5), and generating smooth estimates of density functions without relying on parametric models, Buskirk and Lohr (2005) examined the finite-sample and asymptotic properties of a modified kernel density estimator incorporating both sampling weights and kernel weights. These authors presented regularity conditions under which the sample estimator of the density function is consistent and normal under various modes of inference used with sample survey data. They also presented methods for incorporating clustering when computing design-based confidence bands for a density function. The authors present applications of their design-adjusted density estimation methods to the National Crime Victimization Survey (NCVS) and NHANES, but unfortunately, their methodology has yet to be programmed in any of the more general-purpose statistical software packages.

These methods for nonparametric density estimation based on complex sample survey data were then extended by Bonnéry et al. (2017), who developed kernel density estimation of a probability density function in the case of informative sampling from a finite population and provided R code enabling replications of their approach in a GitHub repository.[6] Mostafa and Ahmad (2019) also examined model-assisted kernel density estimation as a function of available auxiliary information. Via simulation studies, Mostafa and Ahmad showed that the proposed estimators making use of auxiliary information had improved performance over more standard kernel density estimators that did not leverage this information. Unfortunately, to our knowledge at the time of this writing, these estimators have yet to be programmed in any general-purpose software packages.

Our hope is that these promising nonparametric methods, which to date have largely been evaluated using theoretical simulations and only a handful of applications, will continue to be studied and will be more widely available for analysts of survey data as statistical software for survey data analysis continues to develop. As with all other methods discussed in this book, we aim to update readers with any software or methodological developments on the ASDA website.

Notes

1 This package was removed from CRAN in 2023 for reasons described here: https://cran.r-project.org/web/packages/lavaan.survey/index.html. Archived versions of the package are still available.
2 https://abcnews.go.com/US/PollVault/abc-news-polling-methodology-standards/story?id=145373
3 https://2020electionpolls.parc.us.com/client/index.html#/
4 https://doi.org/10.7910/DVN/42MVDX
5 Jim Morgan [from the Institute for Social Research (ISR) at the University of Michigan] and the ISR OSIRIS team originally developed the AID software in the OSIRIS system. This software was used by Jim and others at ISR throughout the 1970s.
6 https://github.com/DanielBonnery/BonneryBreidtCoquetMetronKDE

Appendix A: Software Overview

A.1 Introduction

Today, most analysts can access procedures for survey data analysis either from within the major statistical software packages such as Stata, R, SAS/STAT (referred to as SAS from this point forward), SPSS, and SUDAAN (SAS-callable version), or via stand-alone and/or more specialized software such as SUDAAN (stand-alone version), Mplus, WesVAR, and IVEware. In the overview and evaluation of software for survey data analysis presented in this appendix, we make a distinction between software packages with integrated survey procedures (e.g., Stata) and software that is more specifically tailored to complex sample survey data analysis (SUDAAN). The reason for this distinction is clear: programs that cannot perform a full range of data management, analysis, graphing, and other tasks require more work and setup costs for the research analyst to perform survey data analyses correctly. Although this penalty is often minor, research analysts generally prefer to work efficiently within a single, unified software system and minimize the amount of time and effort needed for data management and data setup prior to analysis.

For this reason, as well as general availability, we focus on five major commercial software packages (R, Stata, SAS, SPSS, and SUDAAN) heavily used by a wide range of analysts working in varied fields. Use of the freely-available Base R or RStudio software, in combination with the R `survey` package and related extensions, has increased tremendously since the release of the second edition of ASDA in 2017. As a result, we offer more examples of R tools for analysis of complex sample survey data in the third edition of the book and this appendix. Although SUDAAN is primarily a data analysis package for complex sample survey data, the fact that it functions as both a stand-alone and SAS-callable tool essentially makes it a SAS-like product and therefore, by extension, it is an add-on to the SAS suite of procedures and functionality. In addition to the five packages mentioned, we also present brief summaries of the survey data analysis capabilities in other software packages including IVEware, Mplus, and WesVar. For those readers interested in replicating the example analyses in Chapters 5–13 using Stata, SAS, SPSS, SUDAAN, IVEware, WesVar, R, and Mplus, please refer to the ASDA website: http://websites.umich.edu/~surveymethod/asda.

This appendix presents an overview of modern software packages that correctly perform survey data analysis and evaluates the relative strengths and weaknesses of each. We consider key features, such as the overall ability to analyze survey data (including sample design structures accommodated), available variance estimation methods, range of analytic techniques offered, hypothesis testing ability, effectiveness in dealing with common issues such as subpopulation analysis, and ease of graphical display of results (including model diagnostics). Although cost is an obvious consideration for most users, the wide variability in pricing and purchasing options makes this type of financial comparison impractical. Given anticipated version changes and rapid software development, we plan to regularly update our website examples as software developments emerge. Rather than recommend a particular tool as the "best" solution, our goal is to objectively evaluate and consider the pertinent features of the major software currently in use.

A.1.1 Historical Perspective

Prior to the late 1990s, most survey data analysts used specialized software tools for correct estimation of population parameters and variance estimation tasks. Some of the more popular tools in the 1990s were SUDAAN, IVEware, WesVar, and early Stata and PC SAS routines. Going back further in time, packages such as OSIRIS (PSALMS and REPERR; University of Michigan, 1982), Super CARP/PC CARP (Fuller et al., 1989), and other standalone programs were in use in the 1970s and 1980s (see Chapter 1).

Only during the past two decades have most major statistical software producers incorporated options to analyze complex sample survey data as part of their overall package. The early survey data analysis tools were generally developed for the individual needs of a group of researchers analyzing survey data and for their time were often very well-designed. Indeed, each of the major software packages in use today owes a debt to the early software and work of the researchers who developed these tools and in turn promoted the use of correct analytic techniques for survey data analysis. However, most of these early packages required extra data management steps and data transfer from package to package, resulting in additional time demands on the analyst. With the advent of survey analysis procedures included within major software packages, the survey analyst was able to easily do the necessary work without the extra burden of switching between software packages—data management and survey analysis capabilities were integrated into a single software system.

A.1.2 Software for Analysis of Complex Sample Survey Data

The software reviewed in detail in this appendix includes five commonly used data analysis tools: Base R and the `survey` package (version 4.42), along with key extensions to these tools such as `svydiags`, `csSampling`, and their dependent packages; SAS/STAT (SAS version 9.4 and SAS/STAT 15.3); Stata (version 18.5); SPSS Complex Samples (version 29.0.1); and SUDAAN (version 11.0.1). We limit our review to the PC platform, but many of these tools are also available on UNIX/LINUX/MAC platforms as well. As mentioned earlier, the additional software packages reviewed in less detail are WesVar (version 5.1), Mplus (version 8.11) and IVEware (version 0.3). Please see the ASDA website for code examples for each software package as well as links to external websites for software, documentation, and statistical assistance.

For a systematic approach to the selection of a survey data analysis tool, consider the software and respective features outlined in Tables A.1 through A.4. For each software package, these tables present the types of complex sample design structures that can be specified, available variance estimation methods, ability to perform common analytic techniques such as those detailed in the third edition of this book, and key options such as hypothesis testing, subpopulation analysis, and the ability to impute and/or analyze multiply imputed data sets. Although these tables are not intended to provide readers with a complete list of every possible analytic technique, they include common procedures routinely used by survey data analysts. Subsequent sections provide detailed reviews of the features summarized in Tables A.1 through A.4 for each software package, and available features will continue to be updated on the ASDA website as new developments occur.

TABLE A.1

Ability of Software to Accommodate Various Complex Sample Designs

	Stata	SAS	SUDAAN	SPSS	IVEware	WesVar	MPlus	R
Sample Design								
With and Without Replacement	Yes	Yes	Yes	Yes	Yes	Yes	Yes	Yes
Equal and Unequal Weighting	Yes	Yes	Yes	Yes	Yes	Yes	Yes	Yes
Non-Stratified and Stratified	Yes	Yes	Yes	Yes	Yes	Yes	Yes	Yes
Single and Multistage	Yes	No	Yes	Yes	No	Yes	Yes	Yes

TABLE A.2

Variance Estimation Capabilities and Additional Analysis Features of the Software Packages

	Stata	SAS	SUDAAN	SPSS	IVEware	WesVar	MPlus	R
Variance Estimation Method*								
TSL (Taylor Series)	Yes	Yes	Yes	Yes	Yes	No	Yes	Yes
JRR (Delete One)	Yes	Yes	Yes	No	Yes	Yes	Yes	Yes
JRR (Replicate Weights)	Yes	Yes	Yes	No	No	Yes	Yes	Yes
BRR	Yes	Yes	Yes	No	No	Yes	Yes	Yes
BRR (w/Fay's Adjustment)	Yes	Yes	Yes	No	No	Yes	Yes	Yes
Bootstrap	Yes	Yes	No	No	Yes	No	Yes	Yes
Single Cluster per Stratum	Yes	Yes	Yes	Yes	No	Yes	Yes	Yes
Features of Survey Analysis								
Subpopulation Analysis	Yes	Yes	Yes	Yes	No	Yes	Yes	Yes
Finite Population Correction for SWOR Designs	Yes	Yes	Yes	Yes	No	Yes	Yes	Yes

Note: *See Chapter 3 for a discussion of each of these variance estimation methods.

A.2 Overview of Stata® Version 18.5

Stata (www.stata.com) is an outstanding software tool for the survey data analyst. This package offers the ability to handle numerous sample design structures, a full range of variance estimation methods, an extensive library of survey analysis commands, the ability to perform correct subpopulation analyses for every command, and many theoretically advanced options for significance testing (e.g., second-order design corrections for chi-square statistics; see Chapter 6, and goodness of fit tests; see Chapters 8 and 9). Another valuable feature is the ability to create and analyze multiply imputed data sets; see Chapter 12. Other modeling techniques such as structural equation modeling (SEM) are available in Stata; see Chapter 13.

TABLE A.3

Available Analytic Techniques in the Software Packages

	Stata	SAS	SUDAAN	SPSS	IVEware	WesVar	MPlus	R
Analytic Technique								
Descriptive								
Means	Yes	Yes	Yes	Yes	Yes	Yes	NA#	Yes++
Totals	Yes	Yes	Yes	Yes	Yes	Yes	NA#	Yes++
Ratios	Yes	Yes	Yes	Yes	Yes	Yes	NA#	Yes++
Percentiles	No	Yes	Yes	No	No	Yes	NA#	Yes++
Contingency Tables	Yes	Yes	Yes	Yes	Yes	Yes	NA#	Yes++
Regression								
Linear	Yes	Yes	Yes	Yes	Yes	Yes	Yes$$	Yes++!!
Binary Logistic	Yes	Yes	Yes	Yes	Yes	Yes	Yes$$	Yes++
Ordinal Logistic	Yes	Yes	Yes	Yes	Yes*	No	Yes$$	Yes++
Multinomial Logistic	Yes	Yes	Yes	Yes	Yes	Yes	Yes$$	No++
Poisson Regression	Yes	No	Yes	No	Yes	No	Yes$$	Yes++
Probit	Yes	Yes	No	Yes	Yes*	No	Yes$$	Yes++
C-log-log	Yes	Yes	No	Yes	Yes*	No	No	Yes++
Survival Analysis								
Cox Proportional Hazards Model	Yes	Yes	Yes	Yes	Yes	No	Yes$$	Yes
Kaplan-Meier Estimation	No	No	Yes	Yes	No	No	Yes	Yes
Missing Data								
Multiple Imputation of Missing Data	Yes	Yes	Yes	Yes^	Yes	Yes**	Yes	Yes
Analysis of Multiply Imputed Data Sets	Yes	Yes	Yes	Yes^	Yes	Yes**	Yes	Yes
Advanced Techniques								
SEM	Yes	No	No	No	No	No	Yes	Yes
Doubly Robust Estimation for Non-Probability Samples	No	No	No	No	No	No	No	Yes
Quasi-Randomization	No	No	No	No	No	No	No	Yes
Machine Learning	No	No	No	No	No	No	No	Yes

Notes: # Not applicable; this product is designed as a modeling tool.

* Can be done using the SASMOD feature of IVEware.

** Limited to tables analysis only, not available for regression.

^ Limited to a subset of survey commands to analyze completed data sets.

++ Bayesian Approach Available.

$$ Bayesian Approach Available only when using a model-based approach such as a two-level model with estimator = Bayes in Mplus.

!! Regression Model Diagnostics Available.

TABLE A.4

Hypothesis Testing Capabilities in the Software Packages

	Stata	SAS	SUDAAN	SPSS	IVEware	Wesvar	Mplus	R
Means, Totals, and Ratios								
Confidence Limits and *t* Tests*	Yes	Yes	Yes	Yes	Yes	Yes	NA	Yes
Contingency Tables								
Rao-Scott Adjusted *F***	Yes	Yes	Yes	Yes	No	Yes	NA	Yes
Wald and Pearson Chi-square**	Yes	Yes	Yes	Yes	No	Yes	NA	Yes
Wald and Rao-Scott Likelihood Ratio**	Yes	Yes	Yes	Yes	No	No	NA	Yes
Regression Modeling								
Linear								
Linear Contrasts***	Yes	Yes	Yes	Yes	No	Yes	Yes	Yes
Wald *F* and Adjusted *F*****	Yes	Yes	Yes	Yes	No	Yes	No	Yes
Wald Chi-square and Adjusted Chi-square****	Yes	Yes	Yes	Yes	No	No	Yes	No
Generalized Linear Models (Binary, Ordinal, Multinomial, Poisson, C-log-log, Probit)								
Linear Contrasts***	Yes	Yes	Yes	Yes	No	Yes	Yes	Yes
Wald *F* and Adjusted *F*****	Yes	Yes	Yes	Yes	No	Yes	No	No
Wald Chi-square and Adjusted Chi-square****	Yes	Yes	Yes	Yes	No	No	Yes	Yes
Archer and Lemeshow GOF Test*****	Yes	No	No	No	No	No	No	No
Fagerland and Hosmer GOF Test (Multinomial Logistic Regression)*****	Yes	No	No	No	No	No	No	No
Test of Parallel Lines Assumption (Ordinal Logistic Regression)******	Yes	Yes^	No	Yes	No	No	No	No
Survival Analysis								
Cox PH Model								
Linear Contrasts***	Yes	Yes	Yes	Yes	No	No	Yes	Yes
Wald *F* and Wald Chi-square****	Yes	Yes	Yes	Yes	No	No	Yes	Yes
Survival Curves								
Corrected Standard Errors for Estimates	No	No	Yes	Yes	No	No	Yes	Yes

Notes: * Test of means significantly different from 0 and test of difference between means (see Sections 5.6 and 5.6.1).

** Tests for independence of rows and columns in contingency table analysis or differences in population proportions (see Sections 6.4.3 and 6.4.4).

*** Test of individual parameter = 0, test of linear contrasts between parameters in regression models, and,

**** Test of multiple parameters in the model (overall or partial tests, see Section 7.3.4.1 and similar sections of Chapters 8–10).

***** Test goodness of fit (adjusted for complex sample design) for overall model (see Section 8.5.2).

****** Tests assumption of proportional odds or "equal slopes" in ordinal logistic regression (adjusted for complex sample design; see Section 9.3.6).

^ Test can be performed using data step coding, see SAS PROC SURVEYPHREG documentation/examples.

A key feature in Stata is the convenient method of declaring the "survey" or complex sample design variables one time prior to analysis. This approach allows the analyst to specify the complex design variables and probability (or sampling) weights through use of the `svyset` command, and once they are declared, these variables will be in effect for the entire survey data analysis session or until changed. The `svyset` command allows users to specify different forms of sampling weights, such as probability or replicate weights. In addition, there are lesser-used options to declare variables for finite population corrections and explicit post-stratification adjustments within the `svyset` command syntax.

In practice, analysts may encounter public-use data sets that employ a sampling error calculation model (see Chapter 4) where each stratum includes only a single cluster for sampling error computations, as opposed to more traditional sampling error calculation models specifying two or more sampling error clusters per stratum. Most survey data analysis software is designed to recognize multiple sampling error clusters in each stratum for variance estimation, and the "singleton" or "lonely" sampling error cluster can interfere with linearized variance estimation procedures. Fortunately, this problem can be addressed in Stata through use of the `singleunit` option in the `svyset` command.

This option allows the Stata user to declare that the sampling error calculation model for the survey data set includes only one sampling error cluster in at least one of the sampling error strata, and offers four options for how variance estimation should proceed in this situation: `singleunit(missing)`, `singleunit(certainty)`, `singleunit(scaled)`, and `singleunit(centered)`. The `singleunit(missing)` option is the default option and simply causes the program to not report standard errors; `singleunit(certainty)` informs Stata to include the "singleton" PSU with certainty (i.e., the PSU does not contribute to variance estimation); `singleunit(scaled)` uses the average sampling variance among all strata with multiple units for the single-unit stratum; and `singleunit(centered)` uses deviations from the grand mean across PSUs to calculate the variance contribution from a stratum with a single PSU. See Chapter 4 and the Stata help function for `svyset` for more details on this topic.

Variance estimation methods available for each of Stata's survey data analysis procedures include the Taylor Series Linearization method, Jackknife Repeated Replication (JRR), Balanced Repeated Replication (BRR) with an option for Fay's adjustment, Successive Differences using Replicate weights (SDR), and Bootstrap using bootstrap weights. The ability to use these varied approaches is a distinct advantage in situations that may call for a replicated variance estimation approach, such as JRR, rather than a Linearization method. There are numerous reasons for a replication approach. For example, one situation that often arises with public use data files is the availability of "replicate weights" instead of the more typical primary stage stratum and cluster code variables in addition to a final survey weight variable. This is generally due to confidentiality concerns and the desire to publish "masked" variables for public use that represent the design structure without jeopardizing confidentiality. Stata users can specify replicate weights for JRR or BRR variance estimation within the `svyset` command. See Chapter 3 for more details on variance estimation methods and the use of replicate weights.

The range of Stata survey commands is extensive. Every analytic command includes design-based significance testing options if statistically appropriate. An additional advantage of the Stata software is the consistency of syntax between the specialized survey commands and standard commands, thus allowing a Stata user new to survey data analysis a nearly immediate knowledge of the syntax required to execute the survey analysis commands. At the simplest level, a Stata user merely needs to add the `svy:` modifier in front of a standard analysis command to invoke design-based estimation, provided that the relevant design variables have been identified using the `svyset` command.

All of Stata's survey procedures also provide users with the correct unconditional approach to the analysis of subpopulations via the `subpop()` option. An unconditional subpopulation analysis invoked via use of the `subpop()` option ensures that 1) the full complex sample design structure will be recognized during the process of linearized variance estimation, and 2) the subpopulation sample size is treated as a random variable. See Section 4.5 for more details.

Given the array of Stata survey analysis procedures and the way the commands are organized (one for each separate type of analysis, instead of a single procedure/command that optionally does many things), it would be impractical to list all the survey commands here. In terms of descriptive analyses (see Chapter 5 and Chapter 6), users can estimate means, proportions, ratios, totals, and one-way frequency tables or two-way cross-tabulations. For linear regression modeling (see Chapter 7), linear regression for continuous outcomes and several related techniques, such as constrained linear regression and Tobit regression, are available. For binary outcome variables, logistic and probit regression modeling procedures are available (see Chapter 8), along with several convenient options such as skewed logistic regression and complementary log-log regression. For categorical outcomes, ordinal logistic regression, probit regression, multinomial logistic regression, conditional logistic regression, and stereotype logistic regression can be performed (see Chapter 9). In terms of count outcomes, Stata offers Poisson regression, negative binomial regression, and generalized negative binomial regression, along with zero-inflated or zero-truncated options as appropriate for these models. Analysts working with time-to-event (or survival) survey data can also fit design-based Cox proportional hazards models or other parametric survival models (see Chapter 10). Stata offers additional options for advanced analyses that perform analysis of survey data. Examples include structural equation models, IRT, and mixture models (see Chapter 13). For a complete list of options, see the Stata documentation.

Each of these survey commands has additional analytic options available (many of which are detailed in the chapters of this book). In addition, a variety of post-estimation commands are also available, such as those that compute design effects (DEFF/DEFT), misspecification effects (MEFF), subgroup contrasts, design-adjusted Wald tests for multiple regression parameters, goodness of fit tests for logistic regression models, subpopulation sizes, and other useful statistics. For example, after fitting a linear regression model, residuals and model diagnostics are available post-estimation, and examples of using these commands are presented in Chapter 7. Given that Stata code is presented throughout the text, we refrain from providing detailed examples of Stata commands here. Please see the ASDA website for sample programs and the Stata website (www.stata.com) for full documentation.

In summary, Stata offers an extensive range of design-based survey analysis capabilities, all common variance estimation methods, simple specifications for subpopulation analysis, and numerous tools for design-adjusted hypothesis testing. In addition, Stata includes multiple imputation capabilities. The `mi` suite of commands enables both multiple imputation with a variety of methods and analysis of multiply imputed data sets while incorporating Stata `svy` commands. This combination is especially convenient for survey analysts faced with dealing with missing data. We demonstrate use of the `mi` commands in detail in Chapter 12.

A.3 Overview of SAS® Version 9.4

The current version of SAS (SAS version 9.4 (TS1M8) and SAS/STAT version 15.3, www.sas.com) offers seven survey procedures, including six survey analysis procedures and one sample selection procedure. There are numerous sampling/analytic techniques available within each procedure, rather than the one command per analytic technique approach implemented by Stata. This is essentially a difference in style and structure and does not necessarily represent a loss of capability overall. Although SAS does not offer as many survey procedures or analytic techniques when compared to Stata, it can perform most common complex sample survey analyses using either repeated replication, bootstrap, or Taylor Series Linearization methods for variance estimation.

The seven procedures currently offered for sample selection and survey data analysis are PROC SURVEYSELECT, PROC SURVEYMEANS, PROC SURVEYFREQ, PROC SURVEYREG, PROC SURVEYLOGISTIC, PROC SURVEYPHREG, and PROC SURVEYIMPUTE. PROC SURVEYSELECT is primarily a sampling tool offering the ability to select probability samples from **sample frame** databases. Because this is not an analysis procedure, we do not focus on PROC SURVEYSELECT in this appendix. PROC SURVEYMEANS and PROC SURVEYFREQ offer descriptive analytic capabilities for means, ratios, totals, quantiles, contingency tables, and other types of descriptive analyses. For regression analysis, the two main procedures are PROC SURVEYREG (for linear regression) and PROC SURVEYLOGISTIC (for binary, ordinal, and multinomial logistic regression). Each of these analysis procedures offers the option of JRR, BRR, or Taylor Series Linearization for variance estimation as well as a DOMAIN or implied DOMAIN statement for correct unconditional subpopulation analysis (see Chapter 4).

Starting with SAS v9.4 and SAS/STAT 14.1, PROC SURVEYIMPUTE could impute item missing data using fully efficient fractional imputation (FEFI), fully hot-deck imputation (FHDI), and other traditional hot-deck options. PROC SURVEYIMPUTE also includes options for generation of imputation-adjusted replicate weights and can output a completed data set with replicate imputation weights designed for subsequent analysis of a fractionally imputed data set. Chapter 12 provides a worked example using the PROC SURVEYIMPUTE FEFI method. SAS software is also capable of multiply imputing and analyzing data sets through use of PROC MI (to impute missing data), survey procedures (estimation for imputed data sets), and PROC MIANALYZE (to combine MI analyses and produce final MI estimates and inferential statistics).

By default, the SAS survey analysis procedures provide a method to deal with a single PSU per stratum when performing linearized variance estimation: they will collapse strata with a single PSU for the variance estimation step. Alternatively, the analyst can opt not to collapse all single PSUs into one stratum for variance estimation via use of the "NOCOLLAPSE" option in PROC SURVEYREG only. In this situation, the stratum with the single PSU contributes a zero to the estimate of variance.

The SAS survey analysis procedures can analyze data arising from various types of complex samples using the sampling with replacement, "ultimate cluster" sampling error calculation model and formulas described in Chapter 3. The design is defined in each procedure using the STRATUM and CLUSTER statements. The final survey weight variable is declared through use of the WEIGHT statement. For example, typical PROC SURVEYMEANS syntax is as follows:

```
proc surveymeans data=one;
strata str;
cluster cluster;
weight weight;
var income;
run;
```

Several optional features for survey data analysts are in place. For example, replicate weights can be accommodated by declaring the variables containing them in the REPWEIGHTS statement. The analyst can also specify the variance estimation method in the procedure statement through use of the `varmethod=jackknife` or `varmethod=brr` options (the default variance estimation method for each analysis procedure is Taylor Series Linearization). The example syntax below illustrates typical use of SAS PROC SURVEYMEANS, with one final survey weight (FINWGT), the JRR variance estimation method, and four replicate weights:

```
proc surveymeans varmethod=jackknife data=one;
weight finwgt;
repweights rwgt1 rwgt2 rwgt3 rwgt4;
var income;
run;
```

Subpopulation analyses can be correctly performed in all SAS survey analysis procedures with the use of the domain statement or the "implied domain" statement (for PROC SURVEYFREQ). In the case of PROC SURVEYFREQ, use of the domain variable as the first variable in the TABLES statement functions as a DOMAIN statement. Multiply imputed data sets can be used with all SAS survey procedures with required output read into and analyzed by PROC MIANALYZE. Although PROC MIANALYZE is not considered a "survey" procedure, it can analyze output from each of the survey procedures outlined in this section. PROC MIANALYZE requires the design-based estimates of the standard errors and/or the variance-covariance matrix for proper analysis, and with this approach, the analyst can account for both the complex sample and the variability in estimates introduced by multiple imputation (Berglund & Heeringa, 2014). As an alternative to multiple imputation, PROC SURVEYIMPUTE offers several imputation methods such as fully efficient fractional imputation and more traditional hot-deck methods along with built-in generation of imputation-adjusted replicate weights and output data sets for analysis of imputed data sets. See Chapter 12 for more on imputation techniques for survey data and the SAS documentation for PROC SURVEYIMPUTE/PROC MI/PROC MIANALYZE for more SAS-related details.

A.3.1 The SAS SURVEY Procedures

The core analytic SAS procedures can be separated into descriptive procedures (PROC SURVEYMEANS and PROC SURVEYFREQ), regression procedures (PROC SURVEYREG, PROC SURVEYPHREG, and PROC SURVEYLOGISTIC), and an imputation procedure (PROC SURVEYIMPUTE). PROC SURVEYMEANS offers the ability to perform survey data analysis for means, means within subpopulations, percentiles, ratios, and totals, all taking into account the complex sample design variables and the final survey weights. One common analytic technique is a significance test for a linear combination of means (e.g., a difference between means) including a correct standard error for the difference. Starting with SAS v9.4/STAT 15.3, this feature was included directly in PROC SURVEYMEANS for comparisons of domain means, via this syntax: domain Grade / diffmeans;. Note that this test can also be done with use of contrasts in PROC SURVEYREG (see the SAS documentation for PROC SURVEYREG for details). We recommend either approach for domain contrasts.

The PROC SURVEYMEANS default output includes an estimated mean, design-based standard error, sample size (*n*), and design-based 95% confidence limits for the mean. When using the ratio option in the procedure, the defaults change to comply with the usual output for a ratio rather than a mean (see the ASDA website for examples). Here is basic PROC SURVEYMEANS syntax with statements specifying the strata, cluster, weight, and domain variables:

```
proc surveymeans data=one;
strata str;
cluster cluster;
weight weight;
var income;
domain sex;
run;
```

PROC SURVEYFREQ performs contingency table analyses including estimation of proportions and population totals with design-based standard errors. The CHISQ option for the TABLES statement requests bivariate tests of association through use of the Rao-Scott chi-square test and the likelihood ratio test (Rao & Scott, 1984), and the Wald chi-square test (see Chapter 6). First-order corrections in Rao-Scott chi-square tests are the default, but second-order tests are available as well using the CHISQ(secondorder) option in the TABLES statement. The default output from PROC SURVEYFREQ includes weighted and unweighted frequencies, a standard error for the weighted frequency, and weighted estimates of percentages and their standard errors. Other output such as "DEFF" for design effects can be requested within the procedure.

The SURVEYFREQ procedure includes an implied domain feature rather than an explicit statement for subpopulation analyses, as illustrated in the example syntax below (where SEX defines the domain variable, EDUCAT defines the row variable of the contingency table, and MARCAT is the column variable):

```
proc surveyfreq data=one;
strata strata;
cluster cluster;
weight weight;
tables sex*educat*marcat;
run;
```

PROC SURVEYREG is the SAS tool for linear regression with survey data (see Chapter 7). This procedure computes weighted estimates of the parameters in specified linear regression models, along with design-adjusted standard errors and variance-covariance matrices computed using the user-specified variance estimation option (TSL is the default). Subpopulation analyses can be defined via the DOMAIN statement. For hypothesis tests, the analyst can request linear contrasts and tests of model effects. Significance testing can be performed with the use of either the ESTIMATE or CONTRAST statement and custom hypothesis testing of linear combinations of regression parameters is available (see the ASDA website for examples). The procedure also includes design-based Wald *F*-tests for regression parameters, design effects and confidence limits for parameter estimates, along with many other options. See Chapter 7 for more details on variance estimation and significance testing for linear regression analysis of survey data. We include a simple example of PROC SURVEYREG code below:

```
proc surveyreg data=one;
strata strata;
cluster cluster;
weight weight;
model income=sex age educationyrs;
run;
```

The SAS survey analysis procedure for logistic regression is PROC SURVEYLOGISTIC, which performs logistic regression for discrete categorical outcomes, including binary, ordinal, and nominal dependent variables. This procedure utilizes the pseudo maximum likelihood method of parameter estimation (Binder, 1983) along with the usual variance estimation options, an optional DOMAIN statement, and significance testing options for survey data analysis. Multi-parameter hypothesis tests are available with the TEST and/or CONTRAST statements. Various extensions of logistic regression modeling are available through use of the "link" option on the model statement. The four types of links are cloglog, glogit, logit, and probit. Here is generic syntax for PROC SURVEYLOGISTIC with the default logit link (more examples are available on the ASDA website):

```
proc surveylogistic data=one;
strata strata;
cluster cluster;
weight weight;
model diabetes (event='1')=sex age educat;
run;
```

The SURVEYPHREG procedure performs survival analysis based on the Cox proportional hazards model for "time to event" data derived from a complex sample design. This procedure produces standard survival model output including hazard ratios, output data sets, and ODS GRAPHICS plots. Subpopulation analysis is specified using an optional DOMAIN statement. Multi-parameter hypothesis tests are available with the TEST and/or CONTRAST statements. See Chapter 10 for statistical details on fitting these models and relevant adjustments for complex sample survey data. Here is generic syntax for PROC SURVEYPHREG:

```
proc surveyphreg data = depress
strata strata;
cluster psu;
weight analysiswt;
model agedepressed*dep_ind(0) = bodyweight smoke;
run;
```

The fully efficient fractional hot-deck imputation method along with FHDI and other traditional hot-deck imputation methods are available in PROC SURVEYIMPUTE. The following generic syntax uses the fully efficient fractional imputation method (FEFI) with the default Jackknife Repeated Replication (JK) variance estimation method. An output data set called "survey_imputed" with output JK weights is then ready for subsequent analysis:

```
proc surveyimpute data=one method=fefi varmethod=jk;
class Department Response;
var Department Response;
strata stratum;
cluster cluster;
weight analysiswgt;
output out=survey _ imputed outjkcoefs=DrugAbuseJKCOEFS;
run;
```

Given the range of the analysis capabilities of the SURVEY procedures, SAS is an excellent choice as a survey data analysis tool, especially for analysts who are already familiar with the data management and standard analysis procedures of the SAS system.

A.4 Overview of SUDAAN® 11.0.1

SUDAAN® (https://www.rti.org/impact/sudaanr-statistical-software-analyzing-correlated-data) is another excellent tool for the survey data analyst. The history of this package is long and impressive. It began as a single-procedure tool during the early 1970s and has since evolved into a key software tool capable of analyzing many types of correlated data sets. The software can be executed as either a SAS-callable or stand-alone product and offers a wide range of features and commands. One important benefit of the SUDAAN package is that the syntax very closely mirrors that of SAS and the experienced SAS user will find learning and invoking the program relatively easy. On the other hand, knowledge of SAS command syntax is not required to use the program to full effect as it is well-organized with excellent documentation containing thorough language and example manuals.

SUDAAN can handle single primary sampling units per stratum using the MISSUNIT option on the NEST statement. The SUDAAN MISSUNIT option informs the software to use deviations of values for the "single" sampling error clusters from overall means to compute the contributions of those single-PSU strata to variance estimates (see the SUDAAN documentation for more details).

The three commonly used methods for variance estimation are included: Taylor Series Linearization, JRR, and BRR. The list of SUDAAN analysis commands is lengthy, with descriptive (PROC DESCRIPT, PROC CROSSTAB, PROC RATIO, PROC VARGEN), regression (PROC REGRESS, PROC LOGIST/RLOGIST, PROC MULTILOG, AND PROC LOGLINK), survival analysis (PROC KAPMEIER, PROC SURVIVAL), and weight adjustment and imputation (PROC WTADJUST, PROC WTADJX, PROC IMPUTE) commands included. Each of the analysis commands listed includes hypothesis testing tools appropriate for the corresponding analytic technique (see the SUDAAN language and example guides for complete details). SUDAAN can also impute missing data using PROC IMPUTE and optionally analyze multiply imputed data sets within any of its procedures.

One potential drawback of SUDAAN is the lack of data management tools, but then again, this tool is not promoted as a general-purpose statistical software package. In practice, it is often used for the analysis of survey data rather than data set construction, and given its ability to read in various types of external data sets and integrate seamlessly with SAS, this is not a major concern. Another common task is production of graphics based on survey analyses, and although SUDAAN does not contain a graphics capability, its SAS interface enables easy movement between SUDAAN and SAS statistical graphing procedures and the SAS/ODS (Output Delivery System). The analyst need only save the SUDAAN output as a SAS type file and can then use the output data set with the key SAS tools for graphing, reporting, or further data analysis. Alternatively, the analyst could save output in another format such as an ASCII file and use the file with other software as needed.

A.4.1 The SUDAAN Procedures

There are four descriptive procedures in SUDAAN. The DESCRIPT procedure is designed for estimation of means, totals, proportions, percentages, and quantiles (see Chapter 5) with design-based standard errors estimated using the selected variance estimation method. The DESCRIPT procedure also includes the ability to perform linear contrasts of estimates for individual levels of categorical variables (e.g., testing the difference of means for two subpopulations). Strata and cluster codes that enable estimation of complex sample standard errors are declared in the NEST statement, and the optional DESIGN= specification on the procedure statement allows the analyst to specify the appropriate sample design. The example code below illustrates SUDAAN syntax using the default of DESIGN=WR (With-Replacement selection of first-stage primary sampling units, or "ultimate clusters"; see Chapter 3), with the use of the FILETYPE=SAS for reading a SAS data set and the SUBPOPN statement for unconditional identification of a specific subpopulation for analysis. Subpopulation analysis is available on every SUDAAN procedure and provides correct subpopulation analyses similar to the Stata `subpop()` or SAS DOMAIN statements.

```
proc descript data=one filetype=sas;
nest strata cluster;
weight weight;
var income;
subpopn sex=1;
run;
```

The RATIO procedure is designed to estimate weighted ratios of selected observed variables and includes options like those of the DESCRIPT procedure. The PROC RATIO syntax is very similar to the preceding DESCRIPT commands but requires the addition of the numerator and denominator variables defining the ratio of interest:

```
proc ratio data=one;
nest strata cluster;
weight weight;
numer depressedwomen;
denom depressedtotal;
run;
```

The third descriptive procedure, PROC CROSSTAB, is used for estimation of frequency and contingency tables, percentages, odds ratios, and risk ratios, all with design-based estimates of variances and standard errors. The analyst can also obtain complex design adjusted chi-square and Cochran-Mantel-Haenszel (CMH) tests (see Chapter 6). SUDAAN also offers design-corrected CMH tests for trends, a convenient tool for analysts working with categorical variables.

Sample PROC CROSSTAB syntax presented below includes the usual design variable specification in the NEST statement, a SUBPOPN statement to restrict analysis to the subpopulation of women (SEX = 1) and a TABLES statement that defines the cross-tabulation of interest. This example also includes the TEST statement requesting a design-based chi-square test of association. The design-adjusted test statistics available for testing hypotheses about bivariate associations between categorical variables include: Wald *F* and adjusted Wald *F*, Wald chi-square and adjusted chi-square, and Satterthwaite (1946) tests.

```
proc crosstab data=one filetype=sas;
nest strata cluster;
weight weight;
class marcat;
subpopn sex=1;
tables depressed*marcat;
test chisq;
run;
```

The fourth descriptive procedure, PROC VARGEN, was new in SUDAAN Release 11. The VARGEN procedure adds the ability to calculate point estimates and standard errors for functions of estimated means, totals, ratios, percentages, and other statistics. For examples of PROC VARGEN features, see the SUDAAN Language and Examples guides (SUDAAN 11.0.1, Language Guide, RTI (2012)).

Sample VARGEN code below demonstrates basic usage of this command. After the XPER statistics are compiled (weighted estimates of the population % who have ever drunk alcohol or ever smoked), a PARAMETER statement is used to compute the difference between the weighted estimates of percents, EVERDRINK_PER and EVERSMOKE_PER. The CLASS statement is used to separate the output by males and females using the variable SEX, coded 1 if the individual is a male and 2 if female. The CLASS statement is required if the DIFFVAR statement is called. In this procedure, a DIFFVAR statement is used to generate contrasts between males and females for all statistics that are created from the XPER and PARAMETER statistics:

```
proc vargen data = one ;
weight analysiswt ;
nest stratum cluster ;
class sex ;
diffvar sex=(1 2) / name = "Male vs. Female";
xper everdrink _ per : everdrink / value = 1 name = "Percent Ever Drink";
xper eversmoke _ per : eversmoke /value = 1 name = "Percent That Have
Ever Smoked";
parameter drink _ smoke: everdrink _ per - eversmoke _ per /
name="Percent Ever Drink minus Percent Ever Smoke";
run;
```

SUDAAN regression procedures include PROC REGRESS (used for linear regression), PROC LOGIST (RLOGIST for SAS-callable SUDAAN, to avoid confusion with SAS PROC LOGISTIC) for logistic regression with binary dependent variables, PROC MULTILOG for logistic regression with ordinal or nominal outcomes, and PROC LOGLINK for count outcomes. Each of the SUDAAN regression procedures uses GEE (Generalized Estimating Equations) for efficient estimation of robust and complex design corrected variances and standard errors. (Refer to the SUDAAN documentation for details on the implementation of the GEE estimation technique.) These procedures all include appropriate hypothesis testing capabilities.

Linear regression is implemented through PROC REGRESS. This procedure outputs the usual weighted estimates of regression parameters along with robust and complex design corrected standard errors as well as a design-corrected variance-covariance matrix, model tests, and parameter tests via the TEST statement. Each SUDAAN modeling procedure includes the ability to estimate linear contrasts of regression parameters by using either the CONTRAST or EFFECTS statements as well as optional tests of groups of parameters via the TEST statement. The REGRESS procedure also includes various printed output options such as design effects as well as output data sets for further analysis. Although the SUDAAN SAS-callable version runs within the SAS program, it does not interface with the SAS Output Delivery System directly, but instead produces SAS-type output files. These files can then be used directly within SAS for subsequent analysis.

One typical use of SUDAAN output within SAS might be examination of regression diagnostics and associated graphics using SAS ODS GRAPHICS. Here is syntax for a typical SUDAAN PROC REGRESS example, with a TEST statement (design-adjusted Wald *F*-test) to test the null hypothesis that the regression parameters for both AGE and SEX are equal to 0, and an OUTPUT statement for producing a SAS data file including estimated parameters (beta) and their standard errors:

```
proc regress data=one;
nest strata cluster;
weight weight;
model income=age sex;
test waldf;
effects sex age / name="test effects of age and sex";
output beta sebeta / filename=outestimate
filetype=SAS;
run;
```

Logistic regression analysis is performed using either PROC LOGISTIC or PROC RLOGIST (for SAS-callable SUDAAN). Sample design and weight variables are declared in the usual manner for SUDAAN. Tests of contrasts or "chunk" parameter testing is included through use of the EFFECTS statement or the CONTRAST statement. The user should declare the categorical variables in their model using a CLASS statement prior to the MODEL statement. Beginning with SUDAAN Version 9, PROC LOGISTIC/RLOGIST offered the Hosmer-Lemeshow Goodness of Fit test (Hosmer et al., 2013). Here is example code for SAS-callable SUDAAN using the JRR variance estimation method, an EFFECTS statement for testing the parameters associated with the indicator variables AGE1, AGE2, and AGE3 against 0, and the HLTEST statement for the H-L Goodness of Fit test:

```
proc rlogist data=one method=jrr;
nest str secu;
weight weight;
class maritalcat;
```

```
model depressed=maritalcat female age1 age2 age3;
effects age1 age2 age3 / name="age dummy test";
print / hltest=default;
run;
```

Ordinal and multinomial logistic regression is available in PROC MULTILOG. SUDAAN uses a cumulative logit modeling (proportional odds) approach for ordinal dependent variables, and a generalized logit modeling approach for nominal categorical outcomes. See Chapter 9 for details and statistical background on these methods. The user must declare the method in the MODEL statement and the outcome as a categorical variable in the CLASS or SUBGROUP statement of the procedure. For example, the following SUDAAN code uses PROC MULTILOG with an ordinal outcome (self-rated health status, ranging from 1–5) and the same predictors as the previous logistic example, along with a test specification for Wald chi-square tests rather than the default Wald *F*-test:

```
proc multilog data=one;
nest str secu;
weight weight;
class maritalcat healthstatus;
model healthstatus=maritalcat female age1 age2 age3 /
cumlogit;
test waldchi;
effects age1 age2 age3 / name="age dummy test";
run;
```

For a nominal categorical outcome, the syntax would differ in that the categorical variable must be declared in the CLASS statement (for example, marital status) and the use of "/ GENLOGIT" is required on the model statement, indicating a generalized logit estimation method rather than the cumulative logit model for ordinal regression. Like all SUDAAN procedures, the MULTILOG procedure can read in multiply imputed data sets and account for the added variability introduced by imputation as well as the complex sample design.

For Poisson or count-type outcomes (see Chapter 9), SUDAAN offers PROC LOGLINK, which like PROC LOGIST and PROC MULTILOG uses GEE for estimation and includes relevant design-adjusted hypothesis tests. The dependent variable must be continuous and greater than or equal to 0 for count-type regression models and an offset or logarithm of the offset can also be specified for fitting rate models (see Chapter 10). The predictor variables can be either continuous or categorical. Categorical variables are declared in the CLASS statement prior to modeling. The LOGLINK procedure offers a full range of tests and output options. It can also handle multiply imputed data sets and subpopulation analyses. The example syntax below illustrates the use of PROC LOGLINK with a count outcome (the number of incidents of asthma per year), a TEST statement for a Wald *F*-test of the null hypothesis that the regression parameters are equal to 0, and use of the SUBPOPN statement for a subpopulation analysis of black females:

```
proc loglink data=one;
nest str secu;
weight weight;
subpopn female=1 & black=1;
class race;
model numincidents=age race;
test waldf;
run;
```

The SURVIVAL procedure in SUDAAN is used for survival (or time-to-event) analysis with discrete and continuous proportional hazards models that include censored data. This procedure can treat time as either continuous or discrete. For continuous time, SUDAAN uses the Cox proportional hazards model and for discrete time, the program uses a log-likelihood function able to handle ties for time to event (see Chapter 10). Estimated hazard ratios and design-adjusted estimates of standard errors are produced along with optional hypothesis tests for regression parameters and linear contrasts specified by the analyst. The SURVIVAL procedure can also handle time-dependent covariates correctly. Any of the major variance estimation methods can be selected along with the full array of sample design types. One important note is that PROC SURVIVAL does not currently handle multiply imputed data sets. For goodness-of-fit evaluation, Martingale residuals (Therneau et al., 1990) and normalized Schoenfeld (1982) and score residuals are available output options. Here is an example of typical syntax where the event of interest is onset of depression (MDE), the outcome is years until event or censoring (AGEMDE), and the predictors are all time-invariant and categorical (SEX, RACECAT, and EDUCATION).

```
proc survival data=one;
nest str secu;
weight weight;
event mde;
class sex racecat education;
model agemde=sex racecat education;
run;
```

For Kaplan-Meier estimation of survival curves, SUDAAN offers the KAPMEIER procedure (see Chapter 10). The procedure will produce weighted estimates and design-based estimates of standard errors for survival functions (see Chapter 10). SUDAAN's easy output of data facilitates graphical display of survival curves using software such as SAS, Stata, or R. Use of an OUTPUT statement is required with PROC KAPMEIER and the TIME and EVENT statements are also mandatory. Optional features include unconditional restriction of estimation to a subpopulation via the SUBPOPN statement, alternative methods for variance estimation (JRR, BRR, or Taylor Series), and the STRHAZ statement for estimation of survival rates for subpopulations. Here is an example of SUDAAN syntax for PROC KAPMEIER. The code illustrates the use of the EVENT, TIME, CLASS, STRHAZ, and SUBPOPN statements along with the required OUTPUT statement:

```
proc kapmeier data=one design=wr;
nest str secu;
weight weight;
subpopn women=1;
event depression;
time onsetage;
class agegroup;
strhaz agegroup;
output / kapmeier=all filename="c:\km _ all _ out"
filetype=sas replace;
run;
```

Imputation of missing data and analysis of imputed data sets can be performed in SUDAAN. PROC IMPUTE was introduced in Release 11 and serves as a replacement for the former HOTDECK procedure (from Release 10). This procedure includes four methods of item missing data imputation: 1. weighted sequential hot deck imputation, discussed in Cox (1980) and Iannacchione (1982); 2. cell mean imputation (see Korn & Graubard, 1999); 3. linear regression imputation; and 4. logistic regression imputation.

Generic syntax for IMPUTE below details use of a required PROC statement, a WEIGHT statement to declare the final survey weight, an IMPBY statement to define the variables whose cross-classification defines the imputation classes, and an IMPVAR statement to list the variables that require imputation. As usual, many other optional statements and features are available:

```
PROC IMPUTE DATA=one;
WEIGHT wgtvar;
IMPBY variable(s);
IMPVAR variable(s) / [MULTIMP=integer];
RUN;
```

Most SUDAAN analysis procedures support analysis using multiply imputed data, except for VARGEN, SURVIVAL, and KAPMEIER.

In summary, SUDAAN is an excellent tool for survey data analysis. It accommodates a wide range of sample designs and offers a variety of variance estimation methods, numerous analytic tools, and frequent updates as new statistical theory and methods emerge. It is relatively easy to learn given the similarity to SAS syntax and produces output files that are ready to use in other more general packages for graphing and reporting or further analysis. See the ASDA website for the SUDAAN code that replicates the example analyses in Chapters 5–13.

A.5 Overview of the R Survey Package (Version 4.42) and Related Extensions

The R `survey` package (visit http://www.r-project.org/, where links to CRAN web sites can be used to download the specific package) is a free software tool that offers a full range of survey data analysis techniques. The package includes the Taylor Series Linearization approach, repeated replication techniques (JRR or BRR), and bootstrap variance estimation options. Base R or RStudio can be used for data management in a command driven or interactive manner, while the `survey` package and its numerous extensions (presented throughout the book chapters) combine to provide users with an excellent package of tools for accurate and efficient analysis of survey data.

A convenient feature of the `survey` package is the frequent updates by the developer (Lumley, 2024). This package handles multi-stage cluster sample designs and unequally weighted sample designs, and offers estimation of descriptive statistics, linear and generalized linear models, and the Cox proportional hazards model for survival analysis. All descriptive analysis and modeling commands include appropriate hypothesis testing options. Also included are design-adjusted graphics and options such as correct analysis of subpopulations and special survey weighting adjustments that employ raking, calibration, and post-stratification to external controls. Multiple imputation and analysis of multiply imputed data is available in the R `mi tools` suite of commands. For a complete guide to use of the package with complex sample data, see Lumley (2024).

Several important extensions to the survey package have emerged since the publication of the second edition of ASDA. In this appendix, we provide a summary of selected extensions along with general code examples for common descriptive and regression analyses. The extensions we discuss here are the csSampling package (https://github.com/RyanHornby/csSampling) and related packages rstan (https://cran.r-project.org/package=rstan) for a link to STAN, and brms (https://cran.r-project.org/package=brms) for Bayesian Regression Models with STAN, all used together for an alternative Bayesian approach to descriptive and regression analyses (see Chapters 5 through 9).

There are many other extensions presented as worked examples in Chapters 7 and 13 but not reviewed in detail here. For linear regression diagnostics, the svydiags package (https://cran.r-project.org/package=svydiags) includes a number of tools and is discussed and demonstrated in detail in Chapter 7. Selected extensions highlighted in Chapter 13 include NonProbEst (https://cran.r-project.org/package=NonProbEst) for the analysis of data from non-probability samples, nonprobsvy (https://cran.r-project.org/package=nonprobsvy) to implement doubly robust methods for the analysis of non-probability samples, mubp _ functions (https://github.com/bradytwest/IndicesOfNISB/raw/master/proxyDrawsMUBP_sumstats.R, https://github.com/bradytwest/IndicesOfNISB/raw/master/mubp_functions.R) for a measure of unadjusted bias for proportions (MUBP), and rpms (https://cran.r-project.org/package=rpms) for Recursive Partitioning for Modeling Survey Data.

A.5.1 The R Survey Functions

Given the extensive list of R functions available for analysis of complex sample survey data, we limit the examples here to just a few of the commonly used functions, along with simple examples of descriptive and regression analyses. For more information and detailed worked examples, see Chapters 5–13 and the ASDA website for analysis example replication using R.

A.5.2 Descriptive Analysis Examples

The following example illustrates the use of the svydesign command to declare the Strata, ID (or cluster), and Weight variables from the National Comorbidity Survey-Replication data set. Once this step is complete, we present use of the svytotal and svyby functions for a design-based analysis of population and subpopulation totals. There are many other functions appropriate for descriptive analysis and associated graphics, such as svymean, svyboxplot, svyratio, and svyhist.

```
# Set 1: Total MDE - NCS-R Data
# Set survey design variables and weights, using ncsr data
ncsrsvypop <- svydesign(strata=~SESTRAT, id=~SECLUSTR,
weights=~popweight, data=ncsr, nest=T)

# Example of svytotal command with confidence intervals
svytotal (~mde, ncsrsvypop, deff=T)
confint(svytotal(~mde, ncsrsvypop))

# Analysis of Major Depressive Episode by Marital Status
ex53 <- svyby (~mde, ~mar3catc, ncsrsvypop, svytotal, deff=T)
confint(ex53) # for design-adjusted confidence intervals
```

As an alternative analysis approach, we present sample syntax using the csSampling package along with needed dependent packages for a Bayesian approach to analyze mean family income in 2011, based on the 2012 Health and Retirement Study data. The analysis setup requires loading packages, formatting the data and weights correctly, and defining a model using the rstan, brms, and survey packages.

```
# Set 2: Bayesian Approach for Total Income
# load needed packages
library(csSampling)
library(rstan)
library(brms)
library(survey)
rstan_options(auto_write = TRUE)

# read in the full HRS data set for the third edition
load("P:\\ASDA3\\Data Sets for Analysis Examples and Stata R
Code\\hrs12.rdata")

# create complete data set with variables of interest
hrs.red <- hrs12[hrs12$nfinr == 1, c("h11itot", "secu", "stratum", "nwgthh")]
hrs.red <- hrs.red[complete.cases(hrs.red),]

# need to normalize HRS weights to match what is done for Stan modeling
hrs.red$wtsc <- hrs.red$nwgthh / mean(hrs.red$nwgthh)

# survey design object
hrs.des <- svydesign(id = ~secu, strata = ~stratum, weights = ~wtsc,
nest = T, data = hrs.red)

# Bayesian approach, flat prior (non-informative), which is default
set.seed(41279)
model_formula <- formula("h11itot|weights(wtsc) ~ 1")
mod.brms <- cs_sampling_brms(svydes = hrs.des,
brmsmod = brmsformula(model_formula), data = hrs.red, family = gaussian())
```

A.5.3 Regression Analyses Using Linear and Logistic Regression

The first example below demonstrates the use of the svydesign function to set up the 2011–2012 NHANES data for analysis of complex sample survey data using the key design variables: SDMVSTRA, SDMVPSU, and WTMEC2YR. We also employ the subset option for an unconditional subpopulation analysis of adults aged 18+ who participated in the NHANES medical examination. With the data ready for analysis, the following example shows use of a bivariate linear regression model predicting diastolic blood pressure with race used as a "factor" type predictor. The initial syntax fits the linear regression model using the svyglm function and uses the regTermTest function for a multi-parameter test of the null hypothesis that all of the race coefficients are equal to zero. The second example illustrates use of the alternative Bayesian method for a linear regression model with a continuous dependent variable.

```
# Set 1: Linear Regression Analysis using the Design-Based Method
# Set Strata, ID, and Weight Variables
nhanessvy2 <- svydesign(strata=~SDMVSTRA, id=~SDMVPSU,
weights=~WTMEC2YR, data=nhanesdata, nest=T)

# Prepare subset, adults 18+ that Participated in Medical Examination
subnhanes <- subset(nhanessvy2 , age >= 18)

# Bivariate Linear Regression with Test of Race
summary(ex75_race <- svyglm(bpxdi1_1 ~racec, design=subnhanes))
regTermTest(ex75_race, ~racec)

# Set 2: Linear Regression, Bayesian Approach
# need to normalize NHANES weights to match what is done for Stan modeling
nhanes.red$wtsc <- nhanes.red$wtmec2yr / mean(nhanes.red$wtmec2yr)

# survey design object
nhanes.des <- svydesign(id = ~sdmvpsu, strata = ~sdmvstra, weights = ~wtsc,
nest = T, data = nhanes.red)

# Bayesian approach, flat prior
set.seed(41279)
model_formula <- formula("bpxdi1.1|weights(wtsc) ~ as.factor(ridreth1) +
as.factor(riagendr) + as.factor(marcat) + agec + agec2 +
as.factor(ridreth1):agec + as.factor(ridreth1):agec2 +
as.factor(riagendr):agec + as.factor(riagendr):agec2")
mod.brms <- cs_sampling_brms(svydes = nhanes.des, brmsmod =
brmsformula(model_formula), data = nhanes.red, family = gaussian())
```

For other types of regression analyses such as logistic, probit or Poisson models, use of the `svyglm` command with the correct `link=` and `family=` options is possible. In this section, we include examples for logistic regression using a design-based approach and the alternative Bayesian approach as well.

The following syntax demonstrates use of the `svyglm` command with the correct link and family options for binary logistic regression. This example uses the NCS-R Part 2 data. The code sets the data for survey analysis using the strata=SESTRAT, id=SECLUSTR, and weights=NCSRWTLG variables provided by the data producer. Next, we fit a multivariable logistic regression model predicting the binary outcome of major depressive episode by selected factor/indicator variables: age, gender, alcohol dependence, education, and marital status.

```
# Set 1: Logistic Regression Example Using a Design-Based Method
# Set design variables and weight
srsvyp2 <- svydesign(strata=~SESTRAT, id=~SECLUSTR, weights=~NCSRWTLG,
data=ncsrp2, nest=T)

# Fit logistic regression model
# Obtain coefficients and odds ratios from model
mod81 <- svyglm(mde ~ factor(ag4catc) + sexm + ald + factor(ed4catc) +
factor(mar3catc), family=quasibinomial, design=ncsrsvyp2)
summary(mod81)
```

```
# Exponentiated versions of coefficients (odds ratios)
exp(mod81$coef)
```

Next, we present syntax for a similar logistic regression analysis using the alternate Bayesian approach discussed in Chapter 8.

```
# Set 2: Alternative Bayesian Approach
# Create complete data set with variables of interest
ncsr.red <- ncsr[, c("mde", "marcatc", "edcatc", "sexc", "agcatc",
"ald", "seclustr", "sestrat", "ncsrwtlg")]
ncsr.red <- ncsr.red[complete.cases(ncsr.red),]

# Normalize NCS-R weights to match what is done for Stan modeling
ncsr.red$wtsc <- ncsr.red$ncsrwtlg / mean(ncsr.red$ncsrwtlg)

# survey design object
ncsr.des <- svydesign(id = ~seclustr, strata = ~sestrat, weights = ~wtsc,
nest = T, data = ncsr.red)

# Bayesian approach, flat prior
set.seed(41279)
model_formula <- formula("mde|weights(wtsc) ~ agcatc + sexc + ald +
edcatc + marcatc")
mod.brms <- cs_sampling_brms(svydes = ncsr.des, brmsmod =
brmsformula(model_formula), data = ncsr.red, family = bernoulli())
mod.brms$stan_fit
```

In summary, R offers a wide range of packages and rapidly developing extensions to the essential survey package, thus making it an excellent choice for analysis of survey data. An additional advantage is the open source policy without licensing fees, which is a definite advantage for many researchers.

A.6 Overview of IBM/SPSS® V29

Next, we review IBM/SPSS Version 29 (https://www.ibm.com/spss) with a focus on the Complex Samples (CS) module. The CS module is included with the basic SPSS installation and offers many survey data analysis tools as well as the full range of usual SPSS features to perform data management, analysis, and graphing tasks (via commands that are a part of the main SPSS package). An added feature of this software is that it offers a dual option of running the procedures with either a point-and-click approach (and optionally saving the generated syntax) or running the procedures directly via generation of command syntax by the analyst. The tutorials for each of the CS commands are excellent and provide a step-by-step path through the command options as well as interpretation and examples of model fit and other considerations that arise during the analysis process.

All survey data utility and analysis commands are contained in the Complex Samples section of the installation. SPSS V29 offers a number of complex samples routines, including sampling plan (CSSample) and analysis plan (CSPlan) "wizards" for sample selection and data preparation prior to analysis, along with a number of "Complex Samples" analysis commands: frequencies (CSTABULATE), descriptives (CSDESCRIPTIVES), crosstabs (CSTABULATE), ratios (CSRATIOS), general linear models (CSGLM), logistic regression with binary or multinomial outcomes (CSLOGISTIC), ordinal regression (CSORDINAL), and Cox regression (CSCOX) for survival analysis.

The analysis plan wizard allows the analyst to specify several complex sample designs, such as With-Replacement selection of ultimate clusters (WR), equal-size clusters Without Replacement (WOR), and unequal-size clusters Without Replacement (WOR), along with the finite population correction ability (FPC). All analysis commands use the Taylor Series Linearization method for variance estimation and allow the use of a subpopulation indicator, the ability to perform appropriate design-based hypothesis tests, and additional options such as estimation of design effects and other optional statistics. Version 29 of SPSS does offer the option to use multiple imputation for missing data problems. However, it does not offer the ability to analyze multiply imputed data sets with the Complex Samples procedures (standard SPSS procedures can be used with multiply imputed data sets). This limitation effectively means that SPSS can be used to impute missing data in survey data sets, but another software tool would be required to correctly analyze these multiply imputed data sets.

A.6.1 The SPSS Complex Samples Commands

Two commands comprise sampling and analysis plan setup utilities: CSSAMPLE and CSPLAN. The CSSAMPLE procedure provides the ability to generate probability samples from existing sample frames (similar to PROC SURVEYSELECT in SAS), while the CSPLAN routine is designed to establish the complex sample "analysis plan" in preparation for analysis. The CSPLAN command asks the analyst to specify the sampling weight, stratum, and PSU (or cluster) variables required for sampling error calculations to the program. This command generates a "plan" file that is then saved for use with any of the SPSS CS analysis commands. Here is an example of a typical CSPLAN command, including the setup of the complex sample design features and identification of the sampling weight and the stratum and cluster codes for sampling error calculations. The file generated by running this command (csplan1.csaplan) can then be used throughout the analysis session and/or edited during the session if desired.

```
* Analysis Preparation Wizard.
CSPLAN ANALYSIS
  /PLAN FILE='C:\csplan1.csaplan'
  /PLANVARS ANALYSISWEIGHT=NCSRWTLG
  /SRSESTIMATOR TYPE=WR
  /PRINT PLAN
  /DESIGN STRATA=SESTRAT CLUSTER=SECLUSTR
  /ESTIMATOR TYPE=WR.
```

There are four SPSS "CS" commands for descriptive analysis and four for various forms of regression analysis. In the descriptive group, CSFREQUENCIES and CSCROSSTABS are used for analysis of categorical data, CSDESCRIPTIVES for continuous variables, and CSRATIOS for ratios. CSTABULATE drives both CSFREQUENCIES and CSCROSSTABS for one- and two-way table analysis and generates unweighted and weighted frequency counts, weighted estimates of proportions, linearized standard errors, confidence limits, design effects, and a Wald *F*-test of equal cell proportions. There are also options for subpopulation analysis and for handling missing data with either listwise or table-by-table deletion. The

SPSS program menus list two options that use the CSTABULATE command: frequencies and crosstabs. Here is typical SPSS syntax for this command including options for estimated population cell size, linearized standard errors, confidence limits, design effects, and a test for cell homogeneity for a single categorical measure of obesity (OBESE6CA):

```
* COMPLEX SAMPLES FREQUENCIES.
CSTABULATE
  /PLAN FILE='C:\CSPLAN1.CSAPLAN'
  /TABLES VARIABLES=OBESE6CA
  /CELLS POPSIZE
  /STATISTICS SE CIN(95) DEFF
  /TEST HOMOGENEITY
  /MISSING SCOPE=TABLE CLASSMISSING=EXCLUDE.
```

Note that the analysis command refers to the "plan" file generated earlier, describing the complex sample design features.

The CSDESCRIPTIVES command performs descriptive analyses of continuous variables and includes similar options to CSTABULATE except for the test of equal cell proportions. The following syntax example illustrates the use of a subpopulation indicator variable for women (SEXF) and requests the square root of the design effect (or DEFT) along with the standard output of standard errors, confidence limits, and a design effect. A *t*-test of the null hypothesis that the mean of AGE is equal to 0 is also requested:

```
* COMPLEX SAMPLES DESCRIPTIVES.
CSDESCRIPTIVES
  /PLAN FILE='C:\CSPLAN1.CSAPLAN'
  /SUMMARY VARIABLES=AGE
  /SUBPOP TABLE=SEXF DISPLAY=LAYERED
  /MEAN TTEST=.05
  /STATISTICS SE DEFF DEFFSQRT CIN(95)
  /MISSING SCOPE=ANALYSIS CLASSMISSING=EXCLUDE.
```

The CSCROSSTABS command is designed for the design-based analysis of two-way or *n*-way contingency tables. This code uses a variant of the CSTABULATE command for *n*-way contingency table analysis and includes options for weighted estimates of row and column percentages, linearized standard errors, DEFF, DEFT, confidence limits, and the coefficient of variation (for design corrections; see Chapter 6), along with a design-based test for independence of rows and columns. For 2 × 2 tables, the analyst can request odds ratio and risk ratio statistics. This example illustrates a 2 × 4 cross-tabulation of SEX by REGION, along with many of the available table options:

```
* Complex Samples Crosstabs.
CSTABULATE
  /PLAN FILE='C:\csplan1.csaplan'
  /TABLES VARIABLES=SEX BY REGION
  /CELLS POPSIZE ROWPCT COLPCT
  /STATISTICS SE CV CIN(95) DEFF
  /TEST INDEPENDENCE
  /MISSING SCOPE=TABLE CLASSMISSING=EXCLUDE.
```

The CSRATIOS command utilizes the CSDESCRIPTIVES command and performs a ratio analysis (see Chapter 5) by declaring the numerator and denominator variables for the ratio statistic. This example illustrates estimation of the ratio of taxable income (TAXINC) to all income (ALLINC) and requests that SPSS output linearized standard errors and 95% confidence limits using the default missing data option of table-by-table exclusion:

```
* Complex Samples Ratios.
CSDESCRIPTIVES
  /PLAN FILE='c:\csplan1.csaplan'
  /RATIO NUMERATOR=TAXINC DENOMINATOR=ALLINC
  /STATISTICS SE CIN(95)
  /MISSING SCOPE=ANALYSIS CLASSMISSING=EXCLUDE.
```

The SPSS regression commands for complex sample survey data include CSGLM for continuous outcomes, CSLOGISTIC for binary or nominal outcomes, CSORDINAL for ordinal logistic regression, and CSCOXREG for survival analysis with the Cox Proportional Hazards model (Cox, 1972; see Chapter 10). All these commands include subpopulation analysis options, complex design corrected standard errors, and hypothesis testing appropriate for the procedure used.

The CSGLM command performs linear regression analyses for continuous dependent variables and offers the usual features such as design-based standard errors for parameter estimates, design-based confidence limits for parameters, design effects and square roots of design effects, subpopulation analysis, and hypothesis testing. Specifically, the hypothesis tests include unadjusted and adjusted Wald *F* and chi-square tests for model parameters, and degrees of freedom specifications based on either the sample design or a fixed number of degrees of freedom. There is an optional estimated means technique for estimation of marginal effects for factors included in the model. This example shows syntax for a basic linear regression model for household income (HHINC) with a number of options specified (confidence limits, design effects, and *F*-tests):

```
* COMPLEX SAMPLES GENERAL LINEAR MODEL.
CSGLM HHINC WITH AGE REGION
  /PLAN FILE='C:\CSPLAN1.CSAPLAN'
  /MODEL AGE REGION
  /INTERCEPT INCLUDE=YES SHOW=YES
  /STATISTICS PARAMETER SE CINTERVAL DEFF
  /PRINT SUMMARY VARIABLEINFO SAMPLEINFO
  /TEST TYPE=F PADJUST=LSD
  /MISSING CLASSMISSING=EXCLUDE
  /CRITERIA CILEVEL=95
```

See the ASDA website for replication of example analyses presented in Chapter 7 using the CSGLM procedure.

The CSLOGISTIC command is used to fit design-based logistic regression models to binary or multinomial outcomes. This command offers options and hypothesis testing specific to logistic regression modeling. The command also includes two options for assessing model fit: a classification approach or calculation of Pseudo R-Square values (Cox & Snell, 1989). Analysts can request that estimates of odds ratios and confidence limits for the ORs be displayed, in addition to hypothesis tests including the Wald *F* and chi-square tests, both design-adjusted and unadjusted. The Pseudo R-Square statistics are based on calculations suggested by Cox and Snell (1989), Nagelkerke (1991), and McFadden (1974), but are not design-adjusted in this version of SPSS. Other available options include display of correlations and covariances among parameter estimates, subpopulation analysis, and design effects. The CSLOGISTIC command can also perform pairwise comparisons of predicted probabilities between levels of categorical predictor variables for fixed values of model covariates (see Chapter 8 and the SPSS documentation for details). Here is example syntax to fit a logistic regression model to a dichotomous outcome (MDE), with common options of odds ratios, confidence limits, design effects, and specification of the reference category for the outcome variable (see the ASDA website for additional examples):

```
* COMPLEX SAMPLES LOGISTIC REGRESSION.
CSLOGISTIC MDE(LOW) WITH SEXF BMI AGECENTERED
  /PLAN FILE='C:\CSPLAN1.CSAPLAN'
  /MODEL SEXF BMI AGECENTERED
  /INTERCEPT INCLUDE=YES SHOW=YES
  /STATISTICS PARAMETER EXP SE CINTERVAL DEFF
  /TEST TYPE=F PADJUST=LSD
  /MISSING CLASSMISSING=EXCLUDE
  /CRITERIA MXITER=100 MXSTEP=5 PCONVERGE=[1E-006 RELATIVE]
LCONVERGE=[0] CHKSEP=20 CILEVEL=95
  /PRINT SUMMARY VARIABLEINFO SAMPLEINFO.
```

The CSORDINAL command is designed for fitting ordinal logistic regression models. There are several link options within the CSORDINAL command, including logit, complementary log-log, and probit. The command includes options specific to this type of logistic regression analysis for ordinal outcomes, such as calculation of Pseudo R-Square statistics (not design-adjusted) (Cox & Snell, 1989) for model fit, a design-adjusted test for the proportional odds assumption (see Chapter 9) as well as the usual options for any type of logistic regression (hypothesis tests and output options noted above). Here is an example of the syntax for the CSORDINAL command:

```
* COMPLEX SAMPLES ORDINAL REGRESSION.
CSORDINAL ED4CAT (ASCENDING) WITH MDE BMI
  /PLAN FILE='C:\CSPLAN1.CSAPLAN'
  /LINK FUNCTION=LOGIT
  /MODEL MDE BMI
  /STATISTICS PARAMETER EXP SE CINTERVAL
  /NONPARALLEL TEST
  /TEST TYPE=F PADJUST=LSD
  /MISSING CLASSMISSING=EXCLUDE
  /CRITERIA MXITER=100 MXSTEP=5 PCONVERGE=[1E-006 RELATIVE]
LCONVERGE=[0] METHOD=NEWTON CHKSEP=20 CILEVEL=95
  /PRINT SUMMARY VARIABLEINFO SAMPLEINFO.
```

For design-based survival analyses, SPSS includes the CSCOXREG command. This command models hazard rates using the Cox Proportional Hazards model for outcomes representing times to events and can handle both time-varying and time-invariant covariates. One key difference between CSCOXREG and the other SPSS CS commands is that the structure of the data set is often organized with multiple and varying numbers of records per individual. (See the discussion of discrete time survival models, "person-time" data structures, and time-varying covariates in Chapter 10.) The "person-time" data structure is used to reflect discrete observational times when events of interest, such as year one of life to onset of an illness or censoring, can be observed. The analyst can provide a multiple-record file or a one-record-per-individual file, or create the needed data set within the command directly.

The CSCOXREG command includes subpopulation analysis, hypothesis testing, control over the degrees of freedom used for significance testing (either based on the sample design or a fixed value), and the ability to perform multiple comparisons of hazard rates for groups defined by categorical variables. Like all SPSS regression commands, predictors that are categorical are defined as "factors" and "covariates" are treated as continuous predictors. Other options include declaration of the "event" of interest, a variable representing time at which the event occurred (for non-censored cases), and a test of the proportional hazards assumption (see Chapter 10 for details). Options for displaying estimated survival plots for models with and without covariates are also built into the command, which is a nice feature, and simple Kaplan-Meier curves can be displayed as well. Here is an example of the use of the CSCOXREG command for modeling time until onset of depression or time to censoring if no event occurred (AGEEVENT):

```
* COMPLEX SAMPLES COX REGRESSION.
CSCOXREG AGEEVENT BY OBESE6CA WITH SEXF
  /PLAN FILE='C:\CSPLAN1.CSAPLAN'
  /VARIABLES STATUS=MDE(1)
  /MODEL OBESE6CA SEXF
  /PRINT SAMPLEINFO EVENTINFO
  /STATISTICS PARAMETER EXP SE CINTERVAL
  /TESTASSUMPTIONS PROPHAZARD=KM
  /PLOT SURVIVAL CI=NO
  /TEST TYPE=F PADJUST=LSD
  /CRITERIA MXITER=100 MXSTEP=5 PCONVERGE=[1E-006 RELATIVE]
LCONVERGE=[0] TIES=EFRON CILEVEL=95
  /SURVIVALMETHOD BASELINE=EFRON CI=LOG
  /MISSING CLASSMISSING=EXCLUDE.
```

As previously mentioned, the CSCOXREG command can also produce Kaplan-Meier survival curves when used as a survival model without predictors. This approach will output a survival curve and baseline survival and cumulative hazards tables along with design-adjusted standard errors, and offers numerous types of plots, such as survival, failure, hazard, and log-log of the survival function. This code illustrates how to request a K-M survival curve from the CSCOXREG command.

```
* Complex Samples K-M Survival Curve.
CSCOXREG AGEEVENT
/PLAN FILE='c:\ncsr _ p1wt.csaplan'
/VARIABLES STATUS=mde(1)
/PRINT SAMPLEINFO EVENTINFO BASELINE
/STATISTICS PARAMETER SE
/TESTASSUMPTIONS PROPHAZARD=KM
/PLOT SURVIVAL CI=YES
/TEST TYPE=F PADJUST=LSD
/CRITERIA MXITER=100 MXSTEP=5 PCONVERGE=[1E-006 RELATIVE]
LCONVERGE=[0] TIES=EFRON CILEVEL=95
/SURVIVALMETHOD BASELINE=EFRON CI=LOG
/MISSING CLASSMISSING=EXCLUDE.
```

A.7 Overview of Additional Software

In this section, we present an abbreviated overview of three additional software tools for complex sample survey data analysis: WesVar (version 5.1), IVEware (version 0.3), and Mplus (version 8.11). For each of these software packages, we provide general guidance on overall features and recommended usage but omit syntax examples. Please refer to the ASDA website for analysis examples and code samples.

A.7.1 WesVar

WesVar Version 5.1 is an excellent software tool for survey data analysis and is produced by the Westat organization (www.westat.com/wesvar). WesVar is primarily a repeated replication tool for the analysis of survey data (in terms of variance estimation) and is available free of charge. It features a point-and-click interface, organizes projects in "workbooks" for shared use, and is quite flexible in terms of data types accepted. For example, the software can read in data sets from SPSS, SAS, Stata, SPlus/R, Excel/Access, ASCII, and relational database products. A key strength of WesVar is its ability to create replicate weights within the program as well as handle existing replicate and probability-based weights. It also offers a full range of repeated replication methods for variance estimation, such as JRR for two-per-stratum or *n*-per-stratum cluster samples, BRR, and BRR with Fay's adjustment. Subpopulation analyses can be performed with the use of a subgroup variable statement in each procedure. The analytic procedures of WesVar include the usual descriptive analyses plus the optional estimation of percentiles adjusting for complex design features. In terms of procedures for regression analysis, WesVar can fit linear regression models to continuous outcomes and logistic regression models to dichotomous or unordered outcomes. All analytic techniques include design-adjusted hypothesis tests as well as the ability to use multiply imputed data sets. Westat provides excellent online documentation rich with examples for readers interested in using WesVar.

A.7.2 IVEware (Imputation and Variance Estimation Software)

IVEware V0.3 is a free software tool produced and maintained by the University of Michigan Survey Methodology Program (https://smponline.isr.umich.edu/software). IVEware runs either as a stand-alone tool or as an external software-callable tool. Though the software was originally based on SAS macros, Version 0.3 can now be executed from an XML based editor by calling SAS, R, SPSS, Stata, or the Stand-Alone version of IVEware. Alternatively, the software can be run from within the SAS, R, SPSS, or Stata packages. Given the wide range of execution options, please see the IVEware V0.3 User's Guide and website and for additional examples (Raghunathan et al., 2022).

This software offers imputation and variance estimation for complex sample survey data analyses through the following macros: %IMPUTE, %DESCRIBE, and %REGRESS. In addition, the package offers %SASMOD, which allows the SAS user to perform complex sample survey data analyses for additional SAS procedures not included in the %DESCRIBE or %REGRESS modules (see the user documentation for a list of %SASMOD procedures available). Additional macros include %SYNTHESIZE, which produces synthesized and optionally imputed data sets which address data security concerns, %BBDESIGN, a macro that uses a weighted finite population Bayesian Bootstrap to expand data sets to create population data sets that reflect the complex sample design, and %COMBINE, used to combine multiple data sources for imputation and/or subsequent analysis.

The %DESCRIBE, %REGRESS, and %SASMOD modules are used for survey data analysis. All three analysis modules can utilize multiple imputed (%IMPUTE), synthesized (%SYNTHESIZE), or weighted finite population Bayesian Bootstrap data sets (%BBDESIGN), or standard survey data sets, and they analyze the survey data using appropriate design-based techniques for variance estimation. The %DESCRIBE module will perform various descriptive analyses such as estimation of means, contingency tables, and ratios with design-based variances estimated using the Taylor Series Linearization method. The %REGRESS module provides the user with a variety of regression techniques, including linear, logistic (binary, multinomial, ordinal), Tobit, Poisson, and proportional hazards modeling for survival analysis. The %SASMOD command enables use of SAS procedures with the JRR variance estimation approach and permits design-based estimation for SAS procedures not covered by the SAS SURVEY procedures. An advantage of IVEware is the ability to impute missing data using the flexible multivariate sequential regression technique of Raghunathan et al. (2001) and then analyze the multiply imputed data sets using the correct design-based variance estimation methods.

A.7.3 Mplus

Mplus Version 8.11 (https://www.statmodel.com/) is a relatively new (initially developed during the late 1990s) and advanced analysis tool designed primarily for complex statistical modeling. Mplus enables complex sample design adjustments for every analytic technique in the package, including advanced structural equation modeling (see Chapter 13). These techniques are analytically advanced, including single- and multilevel models, observed and latent variables, and approaches for cross-sectional and longitudinal data. Mplus offers the ability to handle multiply imputed data sets or files with missing data and can analyze a wide range of outcomes: continuous, categorical (binary, nominal, ordinal), count, censored, or various combinations of these variables.

A very useful feature of Mplus is the flexibility to approach survey data analysis in two ways. The first approach, termed the "design-based approach" in this book and elsewhere, uses variables that represent the stratification and clustering inherent to the sample design and subsequently adjusts variances taking these design features (and sample weighting) into account. The second approach incorporates design features directly within the multiple levels of the model framework (multilevel modeling) and accounts for stratification and clustering within the model specification. Note that use of the "model-based" approach with a two or three level model and `estimator=bayes` permits a Bayesian analysis of survey data. This approach is not permitted using the "design-based" method since use of `type=complex` and `estimator=bayes` is not available. See the MPlus documentation for more on how to use the software.

Multiply imputed data sets can be used with all Mplus routines and subpopulation options are also available. Though Mplus is a very useful and advanced addition to the set of software for survey data analysis, most analysts would perform data management and descriptive analyses in other software because Mplus does not offer a direct or simple way to accomplish these tasks (by design).

A.7.4 Summary

This appendix has presented a brief overview of current statistical software packages enabling the analysis of complex sample survey data, with the goal of providing practical guidance on basic capabilities and options for the survey data analyst. Each of the reviewed software packages has basic procedures for analyzing complex sample survey data and most include important features such as subpopulation analysis, hypothesis testing, ability to handle strata with single PSUs, multiple imputation analysis, and other key survey data analysis features. Although the software alternatives may differ in terms of program offerings for more specialized analyses or advanced modeling (e.g., Structural Equation Modeling in MPlus, Kaplan-Meier survival function estimation in SUDAAN and SPSS, Fractional Imputation in SAS, Bayesian modeling in R), each of the packages (Stata, SAS, SPSS, SUDAAN, WesVar, IVEWare, R, and MPlus) is capable of performing the near totality of analyses that survey practitioners will need to use on a day-to-day basis. The ultimate choice of software is a complex decision that will also involve data management considerations, existing familiarity with the software system, and other factors specific to the individual analyst. In the end though, given the common underlying theory and methods, basic features, and usability inherent in all these software packages, it would be difficult to make a bad choice.

References

Agresti, A. (2012). *Categorical data analysis* (3rd ed.). John Wiley & Sons.

Ai, C., & Norton, E. C. (2003). Interaction terms in logit and probit models. *Economic Letters, 80*(1), 123–129.

Alfons, A., & Templ, M. (2013). Estimation of social exclusion indicators from complex surveys: The R package laeken. *Journal of Statistical Software, 54*(15), 1–25.

Allison, P. D. (1982). Discrete-time methods for the analysis of event histories. *Sociological Methodology, 13*, 61–98.

Allison, P. D. (2010). *Survival analysis using SAS: A practical guide* (2nd ed.). SAS Institute, Inc.

Allison, P. D. (2012a). *Handling missing data by maximum likelihood* (pp. 312–2012). SAS Global Forum.

Allison, P. D. (2012b). *Logistic regression using SAS: Theory and application* (2nd ed.). SAS Institute, Inc.

Andersson, P. G. (2019). "Optimal" calibration weights under unit nonresponse in survey sampling. *Survey Methodology, 45*(3), 533–543.

Andridge, R. R. (2024). Using proxy pattern-mixture models to explain bias in estimates of COVID-19 vaccine uptake from two large surveys. *Journal of the Royal Statistical Society, Series A, 187*(3), 831–843.

Andridge, R. R., & Little, R. J. (2010). A review of hot deck imputation for survey non-response. *International Statistical Review, 78*(1), 40–64.

Andridge, R. R., & Little, R. J. (2011). Proxy pattern-mixture analysis for survey nonresponse. *Journal of Official Statistics, 27*(2), 153–180.

Andridge, R. R., & Little, R. J. (2020). Proxy pattern-mixture analysis for a binary variable subject to nonresponse. *Journal of Official Statistics, 36*(3), 703–728.

Andridge, R. R., West, B. T., Little, R. J., Boonstra, P. S., & Alvarado-Leiton, F. (2019). Indices of non-ignorable selection bias for proportions estimated from non-probability samples. *Journal of the Royal Statistical Society, Series C: Applied Statistics, 68*(5), 1465–1483.

Archer, K. J., & Lemeshow, S. (2006). Goodness-of-fit test for a logistic regression model estimated using survey sample data. *Stata Journal, 6*(1), 97–105.

Archer, K. J., Lemeshow, S., & Hosmer, D. W. (2007). Goodness-of-fit tests for logistic regression models when data are collected using a complex sample design. *Computational Statistics and Data Analysis, 51*, 4450–4464.

Audigier, V., White, I. R., Jolani, S., Debray, T. P., Quartagno, M., Carpenter, J., Van Buuren, S., & Resche-Rigon, M. (2018). Multiple imputation for multilevel data with continuous and binary variables. *Statistical Science, 33*(2), 160–183.

Baker, R., Brick, J. M., Bates, N. A., Battaglia, M., Couper, M. P., Dever, J. A., Gile, K. J., & Tourangeau, R. (2013). Summary report of the AAPOR task force on non-probability sampling. *Journal of Survey Statistics and Methodology, 1*(2), 90–143.

Barnard, J., & Rubin, D. B. (1999). Small-sample degrees of freedom with multiple imputation. *Biometrika, 86*(4), 948–955.

Bauer, G. (2015). Graphical display of regression results. In H. Best & C. Wolf (Eds.), *Regression analysis and causal inference* (pp. 205–224). Sage Publications.

Beatty, P. C. (2023). Reclaiming the narrative and charting our course through the new normal of public opinion research. *Public Opinion Quarterly, 87*(3), 847–861.

Beaumont, J. F. (2020). Are probability surveys bound to disappear for the production of official statistics? *Survey Methodology, 46*(1), 1–28.

Beaumont, J. F., Bosa, K., Brennan, A., Charlebois, J., & Chu, K. (2024). Handling non-probability samples through inverse probability weighting with an application to statistics Canada's crowdsourcing data. *Survey Methodology, 50*(1), 77–106.

Bellhouse, D. R., & Stafford, J. E. (2001). Local polynomial regression in complex surveys. *Survey Methodology, 27*(2), 197–203.

Belli, R. F. (2000). Computerized event history calendar methods: Facilitating autobiographical recall. In *Proceedings of the section on survey research methods* (pp. 471–475). The American Statistical Association.

Berglund, P. A., & Heeringa, S. G. (2014). *Multiple imputation of missing data using SAS®*. SAS Institute Inc.

Bethlehem, J. (2010). Selection bias in web surveys. *International Statistical Review, 78*(2), 161–188.

Bhattacharya, D. (2007). Inference in inequality from household survey data, *Journal of Econometrics, 137*, 674–707.

Biemer, P. P., Groves, R. M., Lyberg, L. E., Mathiowetz, N. A., & Sudman, S. (Eds.). (1991). *Measurement errors in surveys*. John Wiley & Sons.

Bilton, P., Jones, G., Ganesh, S., & Haslett, S. (2017). Classification trees for poverty mapping. *Computational Statistics & Data Analysis*, 115, 53–66.

Binder, D. A. (1981). On the variances of asymptotically normal estimators from complex surveys. *Survey Methodology, 7*, 157–170.

Binder, D. A. (1983). On the variances of asymptotically normal estimators from complex surveys, *International Statistical Review*, 51, 279–292.

Binder, D. A. (1991). Use of estimating functions for interval estimation from complex surveys. *Paper presented at the international statistical institute meetings in Cairo.*

Binder, D. A. (1992). Fitting Cox's proportional hazards model from survey data. *Biometrika, 79*, 139–147.

Binder, D. A. (1998). Longitudinal surveys: Why are these surveys different from all other surveys? *Survey Methodology, 24*(2), 101–108.

Binder, D. A. (2011). Estimating model parameters from a complex survey under a model-design randomization framework. *Pakistan Journal of Statistics, 27*(4), 371–390.

Binder, D. A., & Kovacevic, M. S. (1995). Estimating some measures of income inequality from survey data: An application of the estimating equations approach. *Survey Methodology, 21*(2), 137–145.

Bishop, Y. M. M., Feinberg, S. E., & Holland, P. W. (1975). *Discrete multivariate analysis*. MIT Press.

Bollen, K. A. (1989). *Structural equations with latent variables*. Wiley-Interscience.

Bollen, K. A., Biemer, P. P., Karr, A. F., Tueller, S., & Berzofsky, M. E. (2016). Are survey weights needed? A review of diagnostic tests in regression analysis. *Annual Review of Statistics and Its Application, 3*, 375–392.

Bonnéry, D., Breidt, F. J., & Coquet, F. (2017). Kernel estimation for a superpopulation probability density function under informative selection. *Metron, 75*, 301–318.

Boonstra, P. S., Little, R. J., West, B. T., Andridge, R. R., & Alvarado-Leiton, F. (2021). A simulation study of diagnostics for selection bias. *Journal of Official Statistics, 37*(3), 751–769.

Bowley, A. L. (1906). Address to the economic science and statistics section of the British association for the advancement of science. *Journal of the Royal Statistical Society, 69*, 548–557.

Breidt, F. J., & Opsomer, J. D. (2009). Nonparametric and semiparametric estimation in complex surveys. In C. R. Rao & D. Pfeffermann (Eds.), *Sample surveys: Theory, methods and inference, handbook of statistics* (Vol. 29). Elsevier.

Breiman, L. (2001). Random forests. *Machine Learning, 45*, 5–32.

Breiman, L., & Friedman, J. H. (1985). Estimating optimal transformations for multiple regression and correlation. *Journal of the American Statistical Association, 80*(391), 580–598.

Brewer, K. R. W., & Mellor, R. W. (1973). The effects of sample structure on analytic surveys. *The Australian Journal of Statistics, 15*, 145–152.

Brick, J. M., & Tourangeau, R. (2017). Responsive survey designs for reducing nonresponse bias. *Journal of Official Statistics, 33*(3), 735–752.

Briesacher, B. A., Tjia, J., Doubeni, C. A., Chen, Y., & Rao, S. R. (2012). Methodological issues in using multiple years of the medicare current beneficiary survey. *Medicare & Medicaid Research Review, 2*(1), E1–E20.

Buchanan, A. L., Hudgens, M. G., Cole, S. R., Mollan, K. R., Sax, P. E., Daar, E. S., Adimora, A. A. Eron, J. J., & Mugavero, M. J. (2018). Generalizing evidence from randomized trials using inverse probability of sampling weights. *Journal of the Royal Statistical Society, Series A: Statistics in Society, 181*(4), 1193–1209.

Buelens, B., Burger, J., & van den Brakel, J. A. (2018). Comparing inference methods for non-probability samples. *International Statistical Review, 86*(2), 322–343.

Bulmer, M. (2001). History of social survey. In N. J. Smeltser & P. B. Baltes (Eds.), *International encyclopedia of the social and behavioral sciences* (Vol. 21, pp. 14469–14473). Elsevier.

Burgard, J. P., & Dörr, P. (2022). Generalized linear mixed models with crossed effects and unit-specific survey weights. *Journal of Computational and Graphical Statistics, 31*(4), 1037–1050.

Burns, C. J., Laing, T. J., Gillespie, B. W., Heeringa, S. G., Alcser, K. H., Mayes, M. D., Wasko, M. C., Cooper, B. C., Garabrant, D. H., & Schottenfeld, D. (1996). The epidemiology of scleroderma among women: Assessment of risk from exposure to silicone and silica. *The Journal of Rheumatology, 23*(11), 1904–1912.

Buskirk, T., & Lohr, S. (2005). Asymptotic properties of kernel density estimation with complex survey data. *Journal of Statistical Planning and Inference, 128*, 165–190.

Buskirk, T. D. (2018). Surveying the forests and sampling the trees: An overview of classification and regression trees and random forests with applications in survey research. *Survey Practice, 11*(1).

Buskirk, T. D., Kirchner, A., Eck, A., & Signorino, C. S. (2018). An introduction to machine learning methods for survey researchers. *Survey Practice, 11*(1).

Cai, J., Zeng, D., Li, H., Butera, N. M., Baldoni, P. L., Maitra, P., & Dong, L. (2023). Comparisons of statistical methods for handling attrition in a follow-up visit with complex survey sampling. *Statistics in Medicine, 42*(11), 1641–1668.

Carle, A. C. (2009). Fitting multilevel models in complex survey data with weights: Recommendations. *BMC Medical Research Methodology, 9*. https://doi.org/10.1186/1471-2288-9-49

Carlin, J. B., Galati, J. C., & Royston, P. (2008). A new framework for managing and analyzing multiply imputed data in Stata. *The Stata Journal, 8*(1), 49–67.

Carpenter, J. R., & Kenward. M. G. (2013). *Multiple imputation and its application*. John Wiley and Sons.

Carrillo, I. A., Chen, J., & Wu, C. (2010). The pseudo-GEE approach to the analysis of longitudinal surveys. *Canadian Journal of Statistics, 38*, 540–554.

Carrillo, I. A., & Karr, A. F. (2013). Combining cohorts in longitudinal surveys. *Survey Methodology, 39*, 149–182.

Castro-Martin, L., del Mar Rueda, M., & Ferri-Garcia, R. (2022). Combining statistical matching and propensity score adjustment for inference from non-probability surveys. *Journal of Computational and Applied Mathematics, 404*, 113414.

Chambers, R. L., Dorfman, A. H., & Sverchkov, M. Y. (2003). Nonparametric regression with complex sample survey data. In R. L. Chambers & C. J. Skinner, (Eds.), *Analysis of survey data*. John Wiley & Sons.

Chambers, R. L., & Skinner, C. J. (Eds.). (2003). *Analysis of survey data*. John Wiley & Sons.

Chambers, R. L., Steel, D. G., Wang, S., & Welsh, A. H. (2012). *Maximum likelihood estimation for sample surveys*. Chapman & Hall.

Chandra, H., Chambers, R., & Salvati, N. (2019). Small area estimation of survey weighted counts under aggregated level spatial model. *Survey Methodology, 45*(1), 31–59.

Chen, C., Wakefield, J., & Lumley, T. (2014). The use of sampling weights in Bayesian hierarchical models for small area estimation. *Spatial and Spatio-Temporal Epidemiology, 11*, 33–43.

Chen, J., & Liu, Y. (2019). Small area quantile estimation. *International Statistical Review, 87*, S219–S238.

Chen, Q., Elliott, M. R., Haziza, D., Yang, Y., Ghosh, M., Little, R. J., Sedransk, J., & Thompson, M. (2017). Approaches to improving survey-weighted estimates. *Statistical Science, 32*(2), 227–248.

Chen, S., & Rust, K. (2017). An extension of Kish's formula for design effects to two- and three-stage designs with stratification. *Journal of Survey Statistics and Methodology, 5*(2), 111–130.

Chen, S., & Zhao, Y. D. (2019). Quantile regression analysis of survey data under informative sampling. *Journal of Survey Statistics and Methodology, 7*(2), 157–174.

Chen, Y., Li, P., & Wu, C. (2020). Doubly robust inference with nonprobability survey samples. *Journal of the American Statistical Association, 115*(532), 2011–2021.

Chipperfield, J. O., & Steel, D. G. (2011). Efficiency of split questionnaire surveys. *Journal of Statistical Planning and Inference, 141*, 1925–1932.

Chowdhury, S. (2013). *A comparison of Taylor linearization and balanced repeated replication methods for variance estimation in medical expenditure panel survey*. Agency for Healthcare Research and Quality.

Chrostowski, L., & Beręsewicz, M. (2024). *Nonprobsvy: An R package for modern statistical inference methods based on non-probability samples*. https://github.com/ncn-foreigners/nonprobsvy.

Clark, R. G., & Steel, D. G. (2022). Sample design for analysis using high-influence probability sampling. *Journal of the Royal Statistical Society, Series A: Statistics in Society, 185*(4), 1733–1756.

Cleveland, W. S. (1993). *Visualizing data*. Hobart Press.

Cleves, M. A., Gould, W. W., & Marchenko, Y. V. (2016). *An introduction to survival analysis using stata* (3rd ed.). Stata Press.

Cochran, W. G. (1954). Some methods for strengthening the common $\chi 2$ tests. *Biometrics, 10*(4), 417–451.

Cochran, W. G. (1977). *Sampling techniques* (3rd ed.). John Wiley & Sons.

Conrad, F. G., Keusch, F., & Schober, M. F. (2021). New data in social and behavioral research. *Public Opinion Quarterly, 85*(S1), 253–263.

Converse, J. M. (1987). *Survey research in the United States: Roots and emergence*. University of California Press.

Cooney, K. A., Strawderman, M. S., Wojno, K. J., Doerr, K. M., Taylor, A., Alcser, K. H., Heeringa, S. G., Taylor, J. M., Wei, J. T., Montie, J. E., & Schottenfeld, D. (2001). Age-specific distribution of serum prostate-specific antigen in a community-based study of African-American men. *Urology, 57*, 91–96.

Couper, M. P. (2013). Is the Sky falling? New technology, changing media and the future of surveys. *Survey Research Methods, 7*(3), 145–156.

Cox, B. G. (1980). The weighted sequential hot deck imputation procedure. In *Proceedings of the survey research methods section* (pp. 721–726). American Statistical Association.

Cox, D. R. (1972). Regression models and life tables. *Journal of the Royal Statistical Society, Series B, 34*, 187–220.

Cox, D. R. (2007). Applied statistics: A review. *The Annals of Applied Statistics, 1*(1), 1–16.

Cox, D. R., & Snell, E. J. (1989). *The analysis of binary data* (2nd ed.). Chapman & Hall.

Curtin, R., Presser, S., & Singer, E. (2000). The effects of response rate changes on the index of consumer sentiment. *Public Opinion Quarterly, 64*(4), 413–428.

Dagdoug, M., Goga, C., & Haziza, D. (2023). Model-assisted estimation through random forests in finite population sampling. *Journal of the American Statistical Association, 118*(542), 1234–1251.

Da Silva, D. N., & Skinner, C. (2011). A semiparametric propensity score weighting method to adjust for nonresponse using multivariate auxiliary information. In *International Statistical Institute: Proceedings of the 58th world statistical congress* (Session CPS002).

Datta, G., Lee, J., and Li, J. (2024). Pseudo-Bayesian small-area estimation. *Journal of Survey Statistics and Methodology, 12*(2), 343–368.

Dean, N., & Pagano, M. (2015). Evaluating confidence interval methods for binomial proportions in clustered surveys. *Journal of Survey Statistics and Methodology, 3*(4), 484–503.

DeMaris, A. (2004). *Regression with social data*. John Wiley & Sons.

Deming, W. E. (1950). *Some theory of sampling*. John Wiley & Sons.

Deville, J.-C., & Sarndal, C.-E. (1992). Calibration estimators in survey sampling. *Journal of the American Statistical Association, 87*, 376–382.

Diggle, P. J., Heagerty, P., Liang, K.-Y, & Zeger, S. L. (2002). *Analysis of longitudinal data* (2nd ed.). Calrendon Press.

Dong, Q., Elliott, M. R., & Raghunathan, T. E. (2014). A nonparametric method to generate synthetic populations to adjust for complex sampling design features. *Survey Methodology, 40*(1), 29.

Draper, N. R., & Smith, H. (1981). *Applied regression analysis* (2nd ed.). John Wiley & Sons.

DuMouchel, W. H., & Duncan, G. S. (1983). Using sample survey weights in multiple regression analyses of stratified samples. *Journal of the American Statistical Association, 78*, 535–543.

Durrant, G. B. (2005). Imputation methods for handling item-nonresponse in the social sciences: A methodological review. In *NCRM methods review papers, 002*. University of Southampton: ESRC National Centre for Research Methods.

Durrant, G. B., & Skinner, C. (2006). Using data augmentation to correct for non-ignorable non-response when surrogate data are available: an application to the distribution of hourly pay. *Journal of the Royal Statistical Society Series A: Statistics in Society*, *169*(3), 605–623.

Earp, M., Kaplan, R., & Toth, D. (2022). Modeling the relationship between proxy measures of respondent burden and survey response rates in a household panel survey. *Journal of Official Statistics*, *38*(4), 1145–1175.

Earp, M., Toth, D., Phipps, P., & Oslund, C. (2018). Assessing nonresponse in a longitudinal establishment survey using regression trees. *Journal of Official Statistics*, *34*(2), 463–481.

Eck, A. (2018). Neural networks for survey researchers. *Survey Practice*, *11*(1).

Elliott, M. R. (2007). Bayesian weight trimming for generalized linear regression models. *Survey Methodology*, *33*(1), 23–34.

Elliott, M. R., & Little, R. J. A. (2000). Model-based approaches to weight trimming. *Journal of Official Statistics*, *16*, 191–210.

Elliott, M. R., & Valliant, R. (2017). Inference for nonprobability samples. *Statistical Science*, *32*(2), 249–264.

Elliott, M. R., & Xia, X. (2021). Weighted dirichlet process mixture models to accommodate complex sample designs for linear and quantile regression. *Journal of Official Statistics*, *37*(1), 71–95.

Fagerland, M. W., & Hosmer, D. W. (2012). A generalized hosmer-lemeshow goodness-of-fit test for multinomial logistic regression models. *The Stata Journal*, *12*(3), 447–453.

Faraway, J. J. (2014). *Linear models with R* (2nd ed.). Chapman & Hall/CRC.

Faraway, J. J. (2016). *Extending the linear model with R: Generalized linear, mixed effects and nonparametric regression models* (2nd ed.). Chapman & Hall/CRC.

Fay, R. E. (1979). On adjusting the pearson chi-square statistic for cluster sampling. In *Proceedings of the social statistics section* (pp. 402–405). American Statistical Association.

Fay, R. E. (1985). A jack-knifed chi-squared test for complex samples. *Journal of the American Statistical Association*, *80*, 148–157.

Fay, R. E. (1996). Alternative paradigms for the analysis of imputed survey data. *Journal of the American Statistical Association*, *91*(434), 490–498.

Feder, M., Nathan, G., & Pfeffermann, D. (2000). Multilevel modelling of complex survey longitudinal data with time varying random effects. *Survey Methodology*, *26*, 53–65.

Fellegi, I. P. (1980). Approximate tests of independence and goodness of fit based on stratified multistage samples. *Journal of the American Statistical Association*, *75*, 261–268.

Fisher, R. A. (1925). *Statistical methods for research workers*. Oliver and Boyd.

Fitzmaurice, G. M., Davidian, M., Verbeke, G., & Molenberghs, G. (Eds.). (2009). *Longitudinal data analysis*. John Wiley & Sons.

Fitzmaurice, G. M., Laird, N. M., & Ware, J. H. (2004). *Applied longitudinal analysis*. John Wiley & Sons.

Fitzmaurice, G. M., Molenberghs, G., & Lipsitz, S. R. (1995). Regression models for longitudinal binary responses with informative drop-outs. *Journal of the Royal Statistical Society, Series B (Methodological)*, *57*(4), 691–704.

Flores Cervantes, I., & Brick, J. M. (2016). Nonresponse adjustments with misspecified models in stratified designs. *Survey Methodology*, *42*(1), 161–177.

Foster, I., Ghani, R., Jarmin, R. S., Kreuter, F., & Lane, J. (2016). *Big data and social science*. CRC Press.

Fox, J. (2008). *Applied regression analysis and generalized linear model* (2nd ed.). Sage Publications.

Franco, C., Little, R. J. A., Louis, T. A., & Slud, E. V. (2019). Comparative study of confidence intervals for proportions in complex sample surveys. *Journal of Survey Statistics and Methodology*, *7*(3), 334–364.

Franco, C., & Maitra, P. (2023). Combining surveys in small area estimation using area-level models. *Wiley Interdisciplinary Reviews: Computational Statistics*, *15*(6), e1613.

Freedman, D. A. (2006). On the so-called "huber sandwich estimator" and "robust standard errors". *The American Statistician*, *60*(4), 299–302.

Freedman, D. A. (2008). Survival analysis: A primer. *The American Statistician*, *62*, 110–119.

Fuller, W. A. (1975). Regression analysis for sample survey. *Sankyha, Series C*, *37*, 117–132.

Fuller, W. A. (1984). Least squares and related analyses for complex survey designs. *Survey Methodology*, *10*, 97–118.

Fuller, W. A. (1987). *Measurement error models*. John Wiley & Sons.

Fuller, W. A. (2002). Regression estimation for survey samples (with discussion). *Survey Methodology, 28*(1), 5–23.

Fuller, W. A. (2009). *Sampling statistics*. Wiley.

Fuller, W. A. (2022). Post-strata based on sample quantiles. *Journal of the Royal Statistical Society, Series A: Statistics in Society, 185*(4), 1504–1521.

Fuller, W. A., Kennedy, W., Schnell, D., Sullivan, G., & Park, H. J. (1989). *PC CARP*. Iowa State University, Statistical Laboratory.

Gabler, S., Hader, S., & Lynn, P. (2006). Design effects for multiple design samples. *Survey Methodology, 32*(1), 115–120.

Gardiner, J. C. (2015). Survival analysis with survey data. In *SAS Global Forum Proceedings*, Paper 2040-2015. http://support.sas.com/resources/papers/proceedings15/2040-2015.pdf.

Gelfand, A. E., Hills, S. E., Racine-Poon, A., & Smith, A. F. M. (1990). Illustration of Bayesian inference in normal data models using Gibbs sampling. *Journal of the American Statistical Association, 85*, 972–985.

Gelfand, A. E., & Smith, A. F. M. (1990). Sampling-based approaches to calculating marginal densities. *Journal of the American Statistical Association, 85*, 398–409.

Gelman, A. (2007). Struggles with survey weighting and regression modeling. *Statistical Science, 22*(2), 153–164.

Gelman, A., Carlin, J. B., Stern, H. S., & Rubin, D. B. (2013). *Bayesian data analysis* (3rd ed.). Chapman & Hall/CRC Press.

Gelman, A., & Hill, J. (2006). *Data analysis using regression and multilevel/hierarchical models*. Cambridge University Press.

Gelman, A., Si, Y., & West, B. T. (2024). Regression, poststratification, and small-area estimation with sampling weights. In *Unpublished working paper*. www.stat.columbia.edu/~gelman/research/unpublished/weight_regression.pdf

Ghosh, M., & Maiti, T. (2004). Small-area estimation based on natural exponential family quadratic function models and survey weights. *Biometrika, 91*(1), 95–112.

Goldstein, H. (2003). *Multilevel statistical models* (3rd ed.). Arnold.

Goldstein, H., Carpenter, J., & Kenward, M. G. (2018). Bayesian models for weighted data with missing values: A bootstrap approach. *Journal of the Royal Statistical Society, Series C: Applied Statistics, 67*(4), 1071–1081.

Gómez-Echeverry, S. (2023, July 20). The sensitivity of selection bias estimators: A diagnostic based on a case study and a simulation. In *Paper presented at the 2023 conference of the European Survey Research Association*.

Gonzalez, J. M., & Eltinge, J. L. (2007). Multiple matrix sampling: A review. In *Proceedings of the section on survey research methods*(pp. 3069–3075). American Statistical Association.

Grau, E., Potter, F., Williams, S., & Diaz-Tena, N. (2006). Nonresponse adjustment using logistic regression: To weight or not to weight. In *Proceedings of the survey research methods section of the American Statistical Association* (pp. 3073–3080). American Statistical Association.

Graubard, B. I., Flegal, K. M., Williamson, D. F., & Gail, M. H. (2007). Estimation of attributable number of deaths and standard errors from simple and complex sampled cohorts. *Statistics in Medicine, 26*, 2639–2649.

Greenland, S. (2004). Interval estimation by simulation as an alternative to and extension of confidence intervals. *International Journal of Epidemiology, 33*, 1389–1397.

Greenwood, M. (1926). The 'error of sampling' of the survivorship tables. *Reports on public health and medical subjects* (No. 33, Appendix 1). H.M. Stationery Office.

Grizzle, J., Starmer, F., & Koch, G. (1969). Analysis of categorical data by linear models. *Biometrics, 25*, 489–504.

Groves, R. M. (2004). *Survey errors and survey costs* (2nd ed.). John Wiley & Sons.

Groves, R. M. (2011). Three eras of survey research. *Public Opinion Quarterly, 75*(5), 861–871.

Groves, R. M., Fowler, F. J., Couper, M. P., Lepkowski, J. M., Singer, E., & Tourangeau, R. (2009). *Survey methodology* (2nd ed.). John Wiley & Sons.

Groves, R. M., Habermann, H., Kennedy, C., & Lahiri, P. (2017). A conversation with Robert Groves. *Statistical Science, 32*(1), 128–137.

Groves, R. M., & Heeringa, S. G. (2006). Responsive design for household surveys: Tools for actively controlling survey errors and costs. *Journal of the Royal Statistical Society Series A: Statistics in Society, 169*(3), 439–457.

Guadarrama, M., Molina, I., & Rao, J. N. K. (2018). Small area estimation of general parameters under complex sampling designs. *Computational Statistics & Data Analysis, 121*, 20–40.

Hahs-Vaughn, D. L., & Onwuegbuzie, A. J. (2006). Estimating and using propensity score analysis with complex samples. *The Journal of Experimental Education, 75*(1), 31–65.

Hammon, A. (2023, July 20). Validating an index of selection bias for proportions in non-probability samples. In *Paper presented at the 2023 conference of the European Survey Research Association.*

Hansen, M. H., Hurwitz, W. N., & Madow, W. G. (1953). *Sample survey methods and theory, volumes I and II*. John Wiley & Sons.

Hansen, M. H., Madow, W. G., & Tepping, B. J. (1983). An evaluation of model-dependent and probability-sampling inferences in sample surveys. *Journal of the American Statistical Association, 78*, 776–793.

Harms, T., & Duchesne, P. (2010). On kernel nonparametric regression designed for complex survey data. *Metrika, 72*, 111–138.

Harrell, F. E. Jr. (2015). *Regression modeling strategies, with applications to linear models, logistic regression, and survival analysis* (2nd ed.). Springer-Verlag.

Heckman, J. (1979). Sample selection bias as a specification error. *Econometrica, 47*(1), 153–161.

Heeringa, S. G., Alcser, K. H., Doerr, K., Strawderman, M., Cooney, K., Medberry, B., & Schottenfeld, D. (2001). Potential selection bias in a community-based study of PSA Levels in African-American men. *Journal of Clinical Epidemiology, 54*(2), 142–148.

Heeringa, S. G., Berglund, P. A., West, B. T., Mellipilan, E., & Portier, K. (2015a). Attributable fraction estimation from complex sample survey data. *Annals of Epidemiology, 25*(3), 174–178.

Heeringa, S. G., & Connor, J. (1995). Technical description of the Health and Retirement Survey sample design. In *Technical report, Survey Research Center*. University of Michigan. Retrieved June 2009 from http://hrsonline.isr.umich.edu/sitedocs/userg/HRSSAMP.pdf

Heeringa, S. G., Little, R. J. A., & Raghunathan, T. (2002). Multivariate imputation of coarsened survey data on household wealth. In R. M. Groves, D. A. Dillman, J. L. Eltinge, & R. J. A. Little (Eds.), *Survey nonresponse*. John Wiley & Sons.

Heeringa, S. G., & O'Muircheartaigh, C. (2010). Sample design for cross-national, cross-cultural survey programs. In J. Harkness, M. Braun, B. Edwards, T. Johnson, L. Lyberg, P. Mohler, B.-E. Pennell, & T. W. Smith (Eds.), *Survey methods in multinational, multiregional, and multicultural contexts* (pp. 251–267). John Wiley & Sons.

Heeringa, S. G, Wagner, J., Torres, M., Duan, N., Adams, T., and Berglund, P. (2004). Sample designs and sampling methods for the Collaborative Psychiatric Epidemiology Studies (CPES). *International Journal of Methods in Psychiatric Research, 13*(4), 221–239.

Heeringa, S. G., West, B. T., & Berglund, P. (2015b). Regression with complex samples. In H. Best & C. Wolf (Eds.), *Regression analysis and causal inference* (pp. 225–248). Sage Publications.

Henry, K., & Valliant, R. (2015). A design effect measure for calibration weighting in single-stage samples. *Survey Methodology, 41*(2), 315–331.

Herzog, T., & Rubin, D. B. (1983). Using multiple imputations to handle nonresponse in sample surveys. In W. G. Madow, I. Olkin, & D. B. Rubin (Eds.), *Incomplete data in sample surveys, volume 2: Theory and bibliography*. Academic Press.

Hilbe, J. M. (2007). *Negative binomial regression*. Cambridge University Press.

Hill, M. S. (1992). *The panel study of income dynamics: A user's guide*. Sage Publications.

Hoerl, A. E., & Kennard, R. W. (1970). Ridge regression: Biased estimation for nonorthogonal problems. *Technometrics, 12*, 55–67.

Holt, D., & Smith, T. M. F. (1979). Post stratification. *Journal of the Royal Statistical Society, Series A (General), 142*(1), 33–46.

Hornby, R., Williams, M. R., Savitsky, T. D., & Elkasabi, M. (2023). *csSampling: An R package for Bayesian models for complex survey data*. arXiv:2308.06845.

Horvitz, D. G., & Thompson, D. J. (1952). A generalization of sampling without replacement from a finite universe. *Journal of the American Statistical Association*, *47*, 663–685.

Hosmer, D. W., Lemeshow, S., & May, S. (2008). *Applied survival analysis: Regression modeling of time to event data* (2nd ed.). John Wiley & Sons.

Hosmer, D. W., Lemeshow, S., & Sturdivant, R. X. (2013). *Applied logistic regression* (3rd ed.). John Wiley & Sons.

Hotz, V. J., Bollinger, C. R., Komarova, T., Manski, C. F., Moffitt, R. A., Nekipelov, D., Sojourner, A., & Spencer, B. D. (2022). Balancing data privacy and usability in the federal statistical system. *Proceedings of the National Academy of Sciences*, *119*(31), e2104906119.

House, J. S., Juster, F. T., Kahn, R. L., Schuman, H., & Singer, E. (2004). *A telescope on society: Survey research and social science at the University of Michigan and beyond*. The University of Michigan Press.

Hsu, H. Y., Lin, J. J., & Skidmore, S. T. (2018). Analyzing individual growth with clustered longitudinal data: A comparison between model-based and design-based multilevel approaches. *Behavior research methods*, *50*, 786–803.

Hyman, H. H. (1955). *Survey design and analysis*. Free Press.

Iannacchione, V. G. (1982). Weighted sequential hot deck imputation macros. In *Proceedings of the SAS users group international conference (SUGI 1982)* (pp. 759–763).

Iparragirre, A., Lumley, T., Barrio, I., & Arostegui, I. (2023). Variable selection with LASSO regression for complex survey data. *Stat*, *12*(1), e578.

Jackson, M. (2023, May 12). Can new metrics help us get a handle on partisan nonresponse bias? Evidence from state-level 2022 polling. In *Paper presented at the 2023 conference of the American Association for Public Opinion Research*.

Jamieson, K. H., Lupia, A., Amaya, A., Brady, H. E., Bautista, R., Clinton, J. D., Dever, J. A., Dutwin, D., Goroff, D. L., Hillygus, D. S., Kennedy, C., Langer, G., Lapinski, J. S., Link, M., Philpot, T., Prewitt, K., Rivers, D., Vavreck, L., Wilson, D. C., & McNutt, M. K. (2023). Protecting the integrity of survey research. *PNAS Nexus*, *2*(3), pgad049.

Jann, B. (2008). Multinomial goodness of fit: Large-sample tests with survey design correction and exact tests for small samples. *The Stata Journal*, *8*(2), 147–169.

Jenkins, S. P. (2006). Estimation and interpretation of measures of inequality, poverty, and social welfare using Stata. In *Presentation at the North American Stata users' group meetings*.

Jiang, J., & Lahiri, P. (2006). Mixed model prediction and small area estimation. *Test*, *15*(1), 1–96.

Jiang, J., & Rao, J. S. (2020). Robust small area estimation: An overview. *Annual Review of Statistics and its Application*, *7*, 337–360.

Jones, M., Cecere, W., Lin, T.-H., & Kali, J. (2021). Modeling survey nonresponse under a cluster sample design: Classification and regression tree methodologies compared. In *Proceedings of the survey research methods section, joint statistical meetings*.

Judge, G. G., Griffiths, W. E., Hill, R. C., & Lee, T.-C. (1985). *The theory and practice of econometrics* (2nd ed.). John Wiley & Sons.

Judkins, D. R. (1990). Fay's method for variance estimation. *Journal of Official Statistics*, *6*, 223–239.

Juster, F. T., & Suzman, R. (1995). The health and retirement study: An overview. *Journal of Human Resources*, *30*(Suppl), S7–S56.

Kaier, A. N. (1895). Observations et experiences concernant des denombrements representatives. Discussion appears in Liv. 1, XCIII-XCVII. *Bulletin of the International Statistical Institute*, *9*(Liv. 2), 176–183.

Kalaycioglu, O., Copas, A., King, M., & Omar, R. Z. (2016). A comparison of multiple-imputation methods for handling missing data in repeated measurements observational studies. *Journal of the Royal Statistical Society, Series A: Statistics in Society*, *179*(3), 683–706.

Kalbfleisch, J. D., & Prentice, R. L. (2002). *The statistical analysis of failure time data* (2nd ed.). John Wiley & Sons.

Kalpourtzi, N., Carpenter, J. R., & Touloumi, G. (2024). Handling missing values in surveys with complex study design: A simulation study. *Journal of Survey Statistics and Methodology*, *12*(1), 105–129.

Kalton, G. (1983). *Introduction to survey sampling*. Sage Publications.
Kalton, G. (1986). Handling wave nonresponse in panel surveys. *Journal of Official Statistics, 2*(3), 303–314.
Kalton, G. (2019). Developments in survey research over the past 60 years: A personal perspective. *International Statistical Review, 87*, S10–S30.
Kalton, G., & Citro, C. (1993). Panel surveys: Adding the fourth dimension. *Survey Methodology, 19*, 205–215.
Kalton, G., & Kasprzyk, D. (1986). The treatment of missing survey data. *Survey Methodology, 12*(1), 1–16.
Kalton, G., & Kish, L. (1984). Some efficient random imputation methods. *Communications in Statistics-Theory and Methods, 13*(16), 1919–1939.
Kass, G. V. (1980). An exploratory technique for investigating large quantities of categorical data. *Applied Statistics, 29*(2), 119–127.
Kavoussi, S. K., West, B. T., Taylor, G. W., & Lebovic, D. I. (2009). Periodontal disease and endometriosis: Analysis of the national health and nutrition examination survey. *Fertility & Sterility, 91*(2), 335–342.
Kawakubo, Y., & Kobayashi, G. (2023). Small area estimation of general finite-population parameters based on grouped data. *Computational Statistics & Data Analysis, 184*, 107741.
Keeter, S., Miller, C., Kohut, A., Groves, R. M., & Presser, S. (2000). Consequences of reducing nonresponse in a national telephone survey. *Public Opinion Quarterly, 64*, 125–148.
Kendall, P. L., & Lazarsfeld, P. F. (1950). Problems of survey analysis. In R. K. Merton & P. F. Lazaarsfeld (Eds.), *Continuities in social research: Studies in the scope and method of 'the American soldier'*. Free Press.
Kennedy, C., Blumenthal, M., Clement, S., Clinton, J. D., Durand, C., Franklin, C., McGeeney, K., Miringoff, L., Olson, K., Rivers, D., Saad, L., Witt, G. E., & Wlezien, C. (2018). An evaluation of the 2016 election polls in the United States. *Public Opinion Quarterly, 82*, 1–33.
Kennedy, C., Mercer, A., & Lau, A. (2024). Exploring the assumption that commercial online nonprobability survey respondents are answering in good faith. *Survey Methodology, 50*(1), 3–21.
Kennickell, A. B. (1998). *Multiple imputation in the survey of consumer finances* (Paper 78). Federal Reserve Board.
Kessler, R. C., Berglund, P., Chiu, W. T., Demler, O., Heeringa, S. G., Hiripi, E., Jin, R., Pennell, B.-E., Walters, E. E., Zaslavsky, A., & Zheng, H. (2004). The US National Comorbidity Survey Replication (NCS-R): Design and field procedures. *International Journal of Methods in Psychiatric Research, 13*(2), 69–92.
Khera, R., Angraal, S., Couch, T., Welsh, J. W., Nallamothu, B. K., Girotra, S., Chan, P. S., & Krumholz, H. M. (2017). Adherence to methodological standards in research using the national inpatient sample. *JAMA, 318*(20), 2011–2018.
Kim, J. K. (2011). Parametric fractional imputation for missing data analysis. *Biometrika, 98*(1), 119–132.
Kim, J. K., Brick, M. J., Fuller, W. A., & Kalton, G. (2006). On the bias of the multiple-imputation variance estimator in survey sampling. *Journal of the Royal Statistical Society, Series B, 68*(3), 509–521.
Kim, J. K., & Fuller, W. A. (2004). Fractional hotdeck imputation. *Biometrika, 89*, 470–477.
Kim, J. K., Rao, J. N. K., & Kwon, Y. (2022). Analysis of clustered survey data based on two-stage informative sampling and associated two-level models. *Journal of the Royal Statistical Society, Series A: Statistics in Society, 185*(4), 1522–1540.
Kim, J. K., Rao, J. N. K., & Wang, Z. (2023). Hypotheses testing from complex survey data using bootstrap weights: A unified approach. *Journal of the American Statistical Association, 119*(546), 1–23. https://doi.org/10.1080/01621459.2023.2183130
Kim, J. K., & Shao, J. (2014). *Statistical methods for handling incomplete data*. Chapman & Hall.
Kim, J. K., & Yang, S. (2014). Fractional hot-deck imputation for robust inference under item nonresponse in survey sampling. *Survey Methodology, 49*(2), 211–230.
King, G., Honacker, J., Joseph, A., & Scheve, K. (2001). Analyzing incomplete political science data: An alternative algorithm for multiple imputation. *American Political Science Review, 95*(1), 49–69.

Kish, L. (1949). A procedure for objective respondent selection within the household. *Journal of the American Statistical Association, 44*, 380387.

Kish, L. (1995). Methods for Design Effects. *Journal of Official Statistics, 11*(1), 55–77.

Kish, L. (1987). *Statistical design for research*. John Wiley & Sons.

Kish, L. (1965). *Survey sampling*. John Wiley & Sons.

Kish, L., & Frankel, M. R. (1974). Inference from complex samples. *Journal of the Royal Statistical Society, Series B, 36*, 1–37.

Kish, L., Groves, R. M., & Krotki, K. (1976). *Sampling errors for fertility surveys* (Occasional papers no. 17). World Fertility Survey.

Kish, L., & Hess, I. (1959). On variances of ratios and their differences in multistage samples. *Journal of the American Statistical Association, 54*, 416446.

Klein, L. R., & Morgan, J. N. (1951). Results of alternative statistical treatment of sample survey data. *Journal of the American Statistical Association, 46*, 442–460.

Kleinbaum, D., Kupper, L., & Muller, K. (1988). *Applied regression analysis and other multivariable methods* (2nd ed.). Duxbury Press.

Kline, R. B. (2023). *Principles and practice of structural equation modeling* (5th ed.). The Guilford Press.

Koch, G. G., & Lemeshow, S. (1972). An application of multivariate analysis to complex sample survey data. *Journal of the American Statistical Association, 54*, 59–78.

Kolenikov, S. (2010). Resampling variance estimation for complex survey data. *The Stata Journal, 10*(2), 165–199.

Kolenikov, S., West, B. T., & Lugtig, P. J. (2020). A checklist for assessing the analysis documentation for public-use complex sample survey data sets. *The Survey Statistician, 81*, 50–62. http://isi-iass.org/home/wp-content/uploads/Survey_Statistician_2020_January_N81.pdf

Korn, E. L., & Graubard, B. I. (1990). Simultaneous testing of regression coefficients with complex survey data: Use of bonferroni t statistics. *American Statistician, 44*, 270–276.

Korn, E. L., & Graubard, B. I. (1999). *Analysis of health surveys*. John Wiley & Sons.

Kott, P., and Liao, D. (2012). Providing double protection for unit nonresponse with a nonlinear calibration-weighting routine. *Survey Research Methods, 6*(2), 105–111.

Kott, P. S. (1991). A model-based look at linear regression with survey data. *The American Statistician, 45*, 107–112.

Kott, P. S. (2012). Why one should incorporate the design weights when adjusting for unit nonresponse using response homogeneity groups. *Survey Methodology, 38*(1), 95–99.

Kott, P. S. (2018). A design-sensitive approach to fitting regression models with complex survey data. *Statistics Surveys, 12*, 1–17.

Kott, P. S., & Carr, D. A. (1997). Developing an estimation strategy for a pesticide data program. *Journal of Official Statistics, 13*(4), 367–383.

Kovar, J. G., Rao, J. N. K., and Wu, C. F. J. (1988). Bootstrap and other methods to measure errors in survey estimates. *Canadian Journal of Statistics, 16*(Suppl), 25–45.

Koziol, N. A. (2019). Weighted multilevel versus robust single-level methods for analyzing subpopulation data. *Methodology, 15*, 67–76.

Kreuter, F., Olson, K., Wagner, J., Yan, T., Ezzati-Rice, T. M., Casas-Cordero, C., Lemay, M., Peytchev, A., Groves, R. M., & Raghunathan, T. E. (2010). Using proxy measures and other correlates of survey outcomes to adjust for non-response: Examples from multiple surveys. *Journal of the Royal Statistical Society, Series A, 173*(2), 389–407.

Krueger, B. S., & West, B. T. (Authors Alphabetical). (2014). Assessing the potential of paradata and other auxiliary information for nonresponse adjustments. *Public Opinion Quarterly, 78*(4), 795–831.

Landis, R. J., Stanish, W. M., Freeman, J. L., & Koch, G. G. (1976). A computer program for the generalized chi-square analysis of categorical data using weighted least squares (GENCAT). *Computer Programs in Biomedicine, 6*, 196–231.

Landsman, V., & Graubard, B. I. (2013). Efficient analysis of case-control studies with sample weights. *Statistics in Medicine, 32*(2), 347–360.

Langel, M., & Tille, Y. (2013). Variance estimation of the gini index: Revisting a result several times published. *Journal of the Royal Statistical Society, Series A, 176*, 521–540.

Lavallee, P. (1995). Cross-sectional weighting of longitudinal surveys of individuals and households using the weight share method. *Survey Methodology, 21*(1), 25–32.
Lavallee, P. (2007). *Indirect sampling*. Springer.
Lavrakas, P. J., Traugott, M. W., Kennedy, C., Holbrook, A. L., de Leeuw, E. D., & West, B. T. (Eds.). (2019). *Experimental methods in survey research: Techniques that combine random sampling with random assignment*. John Wiley & Sons.
Lee, E. T. (1992). *Statistical methods for survival analysis*. John Wiley & Sons.
León-Novelo, L. G., & Savitsky, T. D. (2019). Fully Bayesian estimation under informative sampling. *Electronic Journal of Statistics, 13*(1), 1608–1645.
Lepkowski, J. M., & Couper, M. P. (2002). Nonresponse in the second wave of longitudinal household surveys. In R. M. Groves, D. A. Dillman, J. L. Eltinge, & R. J. A. Little (Eds.), *Survey nonresponse* (pp. 259–271). John Wiley & Sons.
Lessler, J. T., & Kalsbeek, W. D. (1992). *Nonsampling errors in surveys*. John Wiley & Sons.
Levy, P. S., & Lemeshow, S. (2007). *Sampling of populations: Methods and applications* (4th ed.). John Wiley & Sons.
Li, J., & Valliant, R. (2009). Survey weighted hat matrix and leverages. *Survey Methodology, 35*(1), 15–24.
Li, J., & Valliant, R. (2011a). Detecting groups of influential observations in linear regression using survey data—adapting the forward search method. *Pakistan Journal of Statistics, 27*(4), 507–528.
Li, J., & Valliant, R. (2011b). Linear regression influence diagnostics for unclustered survey data. *Journal of Official Statistics, 27*(1), 99–119.
Li, J., & Valliant, R. (2015). Linear regression diagnostics in cluster samples. *Journal of Official Statistics, 31*(1), 61–75.
Li, K. H., Raghunathan, T. E., & Rubin, D. B. (1991). Large sample significance levels from multiply-imputed data using moment-based statistics and an F reference distribution. *Journal of the American Statistical Association, 86*, 1065–1073.
Li, Y. (2024). Exchangeability assumption in propensity-score based adjustment methods for population mean estimation using non-probability samples. *Survey Methodology, 50*(1), 37–55.
Liang, K.-Y., & Zeger, S. L. (1986). Longitudinal data analysis using generalized linear models. *Biometrika, 73*(1), 13–22.
Liao, D., & Valliant, R. (2012a). Variance inflation factors in the analysis of complex survey data. *Survey Methodology, 38*(1), 53–62.
Liao, D., & Valliant, R. (2012b). Condition indexes and variance decompositions for diagnosing collinearity in linear model analysis of survey data. *Survey Methodology, 38*(2), 189–202.
Lin, D., Wang, L., & Li, Y. (2016). Haplotype-based statistical inference for population-based case–control and cross-sectional studies with complex sample designs. *Journal of Survey Statistics and Methodology, 4*(2), 188–214.
Little, R. J. (2003). Chapter 4: The Bayesian approach to sample survey inference. In R. Chambers & C. J. Skinner (Eds.), *Analysis of survey data*. John Wiley & Sons.
Little, R. J. (2015). Calibrated Bayes: An inferential paradigm for official statistics in the era of big data. *Statistical Journal of the IAOS, 31*, 555–563.
Little, R. J., & Vartivarian, S. (2003). On weighting the rates in non-response weights. *Statistics in Medicine, 22*(9), 1589–1599.
Little, R. J., & Vartivarian, S. (2005). Does weighting for nonresponse increase the variance of survey means? *Survey Methodology, 31*(2), 161–168.
Little, R. J., West, B. T., Boonstra, P. S., & Hu, J. (2020). Measures of the degree of departure from ignorable sample selection. *Journal of Survey Statistics and Methodology, 8*(5), 932–964.
Little, R. J. A. (1991). Inference with survey weights. *Journal of Official Statistics, 7*, 405–424.
Little, R. J. A. (2004). To model or not to model? Competing modes of inference for finite population sampling. *Journal of the American Statistical Association, 99*(466), 546–556.
Little, R. J. A., & Rubin, D. B. (2019). *Statistical analysis with missing data* (3rd ed.). John Wiley & Sons.
Liu, X. (2016). Fitting proportional odds models for complex sample survey data with SAS, IBM SPSS, Stata, and R. *General Linear Model Journal, 42*(2), 1–39.

Liu, Y., & Chen, Q. (2020). Bayesian inference of finite population quantiles for skewed survey data using skew-normal penalized spline regression. *Journal of Survey Statistics and Methodology, 8*(4), 792–816.

Lohr, S. L. (2014). Design effects for a regression slope in a cluster sample. *Journal of Survey Statistics and Methodology, 2*(2), 97–125.

Lohr, S. L. (2021). *Sampling: Design and analysis* (3rd ed.). CRC Press.

Long, J. S., & Freese, J. (2006). *Regression models for categorical dependent variables using stata*. STATA Press Publication.

Loomis, D., Richardson, D. B., & Elliott, L. (2005). Poisson regression analysis of ungrouped data. *Occupational and Environmental Medicine, 62*, 325–329.

Lorenz, M. O. (1905). Measuring the concentration of wealth. *Publications of the American Statistical Association, 9*(70), 209–219.

Luchman, J. N. (2015). Determining subgroup difference importance with complex survey designs: An application of weighted dominance analysis. *Survey Practice, 8*(5).

Lumley, T., & Scott, A. J. (2014). Tests for regression models fitted to survey data. *Australian & New Zealand Journal of Statistics, 56*, 1–14.

Lumley, T., & Scott, A. (2015). AIC and BIC for modeling with complex survey data. *Journal of Survey Statistics and Methodology, 3*(1), 1–18.

Lumley, T. S. (2010). *Complex surveys: A guide to analysis using R*. John Wiley & Sons.

Lumley, T. S. (2024). R software from the R project. In *V4.42 analysis of complex survey samples*. University of Washington. www.r-project.org/

Lyberg, L., & Heeringa, S. G. (2021). A changing survey landscape. In U. Engel, A. Quan-Haase, S. Liu, & L. E. Lyberg (Eds.), *Handbook of computational social science* (Vol. 1, pp. 83–99). Routledge.

Lynn, P. (Ed.). (2009). *Methodology of longitudinal surveys*. John Wiley & Sons.

Lynn, P. (Ed.). (2021). *Advances in longitudinal survey methodology*. John Wiley & Sons.

Lynn, P., & Watson, N. (2021). Issues in weighting for longitudinal surveys. In P. Lynn (Ed.), *Advances in longitudinal survey methodology* (pp. 447–468). John Wiley & Sons.

MacNell, N., Feinstein, L., Wilkerson, J., Salo, P. M., Molsberry, S. A., Fessler, M. B, Thorne, P. S., Motsinger-Reif, A. A., & Zeldin, D. C. (2023). Implementing machine learning methods with complex survey data: Lessons learned on the impacts of accounting sampling weights in gradient boosting. *PLoS One, 18*(1), e0280387. https://doi.org/10.1371/journal.pone.0280387

Madow, W. G., & Olkin, I. (Eds.). (1983). Incomplete data in sample surveys. In *Proceedings of the symposium* (Vol. 3). Academic Press.

Mahalanobis, P. C. (1946). Recent experiments in statistical sampling in the Indian statistical institute. *Journal of the Royal Statistical Society, 109*, 325–370.

Maindonald, J. H., & Braun, W. J. (2007). *Data analysis and graphics using R: An example-based approach* (2nd ed.). Cambridge University Press.

Mantel, N., & Haenszel, W. (1959). Statistical aspects of the analysis of data from retrospective studies of disease. *Journal of the National Cancer Institute, 22*(4), 719–748.

McCarthy, P. J. (1969). Pseudoreplication: Half samples. *Review of the International Statistical Institute, 37*, 239–264.

McConville, K. S., & Breidt, F. J. (2013). Survey design asymptotics for the model-assisted penalised spline regression estimator. *Journal of Nonparametric Statistics, 25*(3), 745–763.

McConville, K. S., & Toth, D. (2019). Automated selection of post-strata using a model-assisted regression tree estimator. *Scandinavian Journal of Statistics, 46*(2), 389–413.

McCullagh, P., & Nelder, J. A. (1989). *Generalized linear models* (2nd ed.). Chapman & Hall.

McCulloch, C. E., & Neuhaus, J. M. (2011). Misspecifying the shape of a random effects distribution: Why getting it wrong may not matter. *Statistical Science, 26*(3), 388–402.

McCulloch, C. E., Searle, S. R., & Neuhaus, J. M. (2008). *Generalized, linear and mixed models* (2nd ed.). John Wiley & Sons.

McFadden, D. (1974). Conditional logit analysis of qualitative choice behavior. In P. Zarembka (Ed.), *Frontiers in economics*. Academic Press.

Menard, S. W. (Ed.). (2008). *Handbook of longitudinal research*. Academic Press.

Mendelson, J., & Huang, P. (2016). Weighting for nonresponse in the 2014 overseas citizen population survey. *Presented at the 2016 Joint Statistical Meetings.*

Meng, X. L. (1994). Multiple-imputation inferences with uncongenial sources of input. *Statistical Science*, 538–558.

Mercer, L., Wakefield, J., Chen, C., & Lumley, T. (2014). A comparison of spatial smoothing methods for small area estimation with sampling weights. *Spatial Statistics, 8*, 69–85.

Miech, R. A., Johnston, L. D., O'Malley, P. M., Bachman, J. G., & Schulenberg, J. E. (2016). *Monitoring the future national survey results on drug use, 1975–2015: Volume I, secondary school students.* Institute for Social Research, The University of Michigan. http://monitoringthefuture.org/pubs.html#monographs

Miles, A. (2016). Obtaining predictions from models fit to multiply imputed data. *Sociological Methods & Research, 45*(1), 175–185.

Miller, R. (1981). *Survival analysis.* John Wiley & Sons.

Miller, R. G. (1974). The jackknife—a review. *Biometrika, 61*, 1–15.

Mitchell, M. N. (2022). *A visual guide to Stata graphics* (4th ed.). Stata Press.

Mohadjer, L., & Curtin, L. R. (2008). NHANES: Balancing sample design goals for the national health and nutrition examination survey. *Survey Methodology, 34*(1), 119–126.

Molenberghs, G., & Verbeke, G. (2005). *Models for discrete longitudinal data.* Springer.

Montanari, G. E., & Ranalli, M. G. (2005). Nonparametric methods in survey sampling. In *New developments in classification and data analysis: Proceedings of the meeting of the classification and data analysis group (CLADAG) of the Italian statistical society* (pp. 203–210). University of Bologna, Springer, Berlin, Heidelberg. September 22–24, 2003.

Morel, G. (1989). Logistic regression under complex survey designs. *Survey Methodology, 15*, 202–223.

Mostafa, S. A., & Ahmad, I. A. (2019). Kernel density estimation from complex surveys in the presence of complete auxiliary information. *Metrika, 82*, 295–338.

Mostafa, S. A., & Shan, Q. (2019). Finite population model-assisted estimation using combined parametric and nonparametric regression smoothers. *Journal of Statistical Theory and Practice, 13*, 1–24.

Mukerjee, R., Dasgupta, T., & Rubin, D. B. (2018). Using standard tools from finite population sampling to improve causal inference for complex experiments. *Journal of the American Statistical Association, 113*(522), 868–881.

Muthén, B. O., & Satorra, A. (1995). Complex sample data in structural equation modeling. *Sociological Methodology, 25*, 267–316.

Muthén, L. K., & Muthén, B. O. (1998–2017). *Mplus user's guide* (8th ed.). Muthén & Muthén.

Nagelkerke, N. J. D. (1991). A note on the general definition of the coefficient of determination. *Biometrika, 78*(3), 691–692.

Nascimento, M. L., & Gonçalves, K. C. M. (2024). Bayesian quantile regression models for complex survey data under informative sampling. *Journal of Survey Statistics and Methodology, 12*(4), smae015. https://doi.org/10.1093/jssam/smae015

National Academies of Sciences, Engineering, and Medicine. (2024). *Toward a 21st century national data infrastructure: Managing privacy and confidentiality risks with blended data.* The National Academies Press. https://doi.org/10.17226/27335

Nattino, G., & Lu, B. (2018, July 31). *Estimating causal effects with propensity score in cluster sample surveys.* Paper presented at the 2018 joint statistical meetings.

Neter, J., Kutner, M. H., Wasserman, W., & Nachtsheim, C. J. (1996). *Applied linear statistical models* (4th ed.). McGraw-Hill/Irwin.

Neyman, J. (1934). On the two different aspects of the representative method: The method of stratified sampling and the method of purposive selection. *Journal of the Royal Statistical Society, 97*, 558–606.

Nishimura, R., Wagner, J., & Elliott, M. (2016). Alternative indicators for the risk of nonresponse bias: A simulation study. *International Statistical Review, 84*(1), 43–62.

Nuzzo, R. (2014). Statistical errors. *Nature, 506*, 150–152.

Nygard, F., & Sandstrom, A. (1981). *Measuring income inequality.* Almqvist and Wiksell International.

Oberski, D. L. (2014). lavaan.survey: An R package for complex survey analysis of structural equation models. *Journal of Statistical Software, 57*(1), 1–27.

Oberski, D. L., & Kreuter, F. (2020). Differential privacy and social science: An urgent puzzle. *Harvard Data Science Review*, *2*(1), 1–21.

O'Muircheartaigh, C., & Wong, S. T. (1981). The impact of sampling theory on survey practices: A review. In *Bulletin of the international statistical institute* (pp. 465–493). International Statistical Institute.

Ono, M., & Miller, H. P. (1969). Income nonresponses in the current population survey. In *ASA proceedings of the social statistics section* (pp. 277–288). American Statistical Association.

Opsomer, J., & Riddles, M. (2023, May 31). *Fitting classification trees to complex survey data*. Webinar presented to the International Association of Survey Statisticians.

Opsomer, J. D., & Miller, C. P. (2005). Selecting the amount of smoothing in nonparametric regression estimation for complex surveys. *Journal of Nonparametric Statistics*, *17*(5), 593–611.

Paczkowski, W. R. (2022). *Modern survey analysis: Using python for deeper insights*. Springer Nature.

Park, I., & Lee, H. (2004). Design effects for the weighted mean and total estimators under complex survey sampling. *Survey Methodology*, *30*(2), 183–193.

Parker, P. A., Janicki, R., & Holan, S. H. (2023). A comprehensive overview of unit-level modeling of survey data for small area estimation under informative sampling. *Journal of Survey Statistics and Methodology*, *11*(4), 829–857.

Paterson, M., & McDonagh, M. (2018). Data protection in an era of big data: The challenges posed by big personal data. *Monash University Law Review*, *44*(1), 1–31.

Peterson, B., & Harrell, F. (1990). Partial proportional odds models for ordinal response variables. *Applied Statistics*, *39*, 205–217.

Pfeffermann, D. (1993). The role of sampling weights when modelling survey data. *International Statistical Review*, *61*, 317–337.

Pfeffermann, D. (2011). Modelling of complex survey data: Why model? Why is it a problem? How can we approach it? *Survey Methodology*, *37*(2), 115–136.

Pfeffermann, D., & Holmes, D. J. (1985). Robustness considerations in the choice of method of inference for regression analysis of survey data. *Journal of the Royal Statistical Society, Series A*, *148*, 268–278.

Pfeffermann, D., Skinner, C. J., Holmes, D. J., Goldstein, H., & Rasbash, J. (1998). Weighting for unequal selection probabilities in multilevel models. *Journal of the Royal Statistical Society, Series B*, *60*(1), 23–40.

Pfeffermann, D., & Sverchkov, M. (2007). Small-area estimation under informative probability sampling of areas and within the selected areas. *Journal of the American Statistical Association*, *102*(480), 1427–1439.

Potter, F. (1990). A study of procedures to identify and trim extreme sample weights. In *Proceedings of the survey research methods section* (pp. 225–230). American Statistical Association.

Qin, X., Hong, G., Deutsch, J., & Bein, E. (2019). Multisite causal mediation analysis in the presence of complex sample and survey designs and non-random non-response. *Journal of the Royal Statistical Society, Series A: Statistics in Society*, *182*(4), 1343–1370.

Quartagno, M., Carpenter, J. R., & Goldstein, H. (2020). Multiple imputation with survey weights: A multilevel approach. *Journal of Survey Statistics and Methodology*, *8*(5), 965–989.

Rabe-Hesketh, S., & Skrondal, A. (2006). Multilevel modelling of complex survey data. *Journal of the Royal Statistical Society, Series A*, *169*, 805–827.

Rafei, A., Flannagan, C. A., West, B. T., & Elliott, M. R. (2022). Robust Bayesian inference for big data: Combining sensor-based records with traditional survey data. *The Annals of Applied Statistics*, *16*(2), 1038–1070.

Raghunathan, T. E. (2016). *Missing data analysis in practice*. Chapman & Hall/CRC Interdisciplinary Statistics.

Raghunathan, T. E. (2021). Synthetic data. *Annual Review of Statistics and Its Application*, *8*, 129–140.

Raghunathan, T. E., & Grizzle, J. E. (1995). A split questionnaire survey design. *Journal of the American Statistical Association*, *90*(429), 54–63.

Raghunathan, T. E., Lepkowski, J. M., Van Hoewyk, J., & Solenberger, P. (2001). A multivariate technique for multiply imputing missing values using a sequence of regression models. *Survey Methodology*, *27*(1), 85–95.

Raghunathan, T. E., Solenberger, P., Berglund, P., & van Hoewyk, J. (2022). *IVEWare: Imputation and variance estimation software*. University of Michigan. www.iveware.org

Rao, J. N. K. (2003). *Small area estimation*. Wiley Series in Survey Methodology. John Wiley & Sons.

Rao, J. N. K. (2016, March 22–24). Hypothesis testing for complex survey data using bootstrap weights: A unified approach. In *Presented at statistics Canada's international symposium on methodological issues*.

Rao, J. N. K., & Molina, I. (2015). *Small area estimation* (2nd ed.). Wiley.

Rao, J. N. K., & Scott, A. J. (1981). The analysis of categorical data from complex sample surveys: Chi-squared tests for goodness of fit and independence in two-way tables. *Journal of the American Statistical Association, 76*, 221–230.

Rao, J. N. K., & Scott, A. J. (1984). On chi-squared test for multiway contingency tables with cell proportions estimated from survey data. *The Annals of Statistics, 12*, 46–60.

Rao, J. N. K., & Shao, J. (1992). Jackknife variance estimation with survey data under hot deck imputation. *Biometrika, 79*, 811–822.

Rao, J. N. K., & Thomas, D. R. (1988). The analysis of cross-classified categorical data from complex sample surveys. *Sociological Methodology, 18*, 213–269.

Rao, J. N. K., Verret, F., & Hidiroglou, M. A. (2013). A weighted composite likelihood approach to inference for two-level models from survey data. *Survey Methodology, 39*(2), 263–282.

Rao, J. N. K., & Wu, C. F. J. (1985). Inference from stratified samples: Second order analysis of three methods for nonlinear statistics. *Journal of the American Statistical Association, 80*, 620–630.

Rao, J. N. K., Wu, C. F. J., & Yue, K. (1992). Some recent work on re-sampling methods for complex surveys. *Survey Methodology, 18*, 209–217.

Raudenbush, S. W. (2000). Synthesizing results for NAEP trial state assessment. In D. W. Grissmer & M. Ross (Eds.), *Analytic issues in the assessment of student achievement*. National Center for Educational Statistics.

Raudenbush, S. W., & Bryk, A. S. (2002). *Hierarchical linear models: Applications and data analysis methods* (2nd ed.). Sage Publications.

Raykov, T. (1997). Scale reliability, Cronbach's coefficient alpha, and violations of essential Tau equivalence with fixed congeneric components. *Multivariate Behavioral Research, 32*, 329–353.

Raykov, T., & Traynor, A. (2016). Evaluation of multicomponent measuring instrument reliability in complex design studies. *Structural Equation Modeling: A Multidisciplinary Journal, 23*(2), 270–277.

Raykov, T., West, B. T., & Traynor, A. (2015). Evaluation of coefficient alpha for multiple component measuring instruments in complex sample designs. *Structural Equation Modeling: A Multidisciplinary Journal, 22*(3), 429–438.

Reiter, J. P., Raghunathan, T. E., & Kinney, S. K. (2006). The importance of modeling the sampling design in multiple imputation for missing data. *Survey Methodology, 32*(2), 143–149.

Research Triangle Institute (RTI). (2012). *SUDAAN 11.0.1 user's manual: Software for statistical analysis of correlated data*. Research Triangle Institute.

Riddles, M. K., Kim, J. K., & Im, J. (2016). A propensity-score-adjustment method for nonignorable nonresponse. *Journal of Survey Statistics and Methodology, 4*(2), 215–245.

Rivers, D. (2007). *Sampling for web surveys*. Presented at the joint statistical meetings.

Roberts, G., Rao, J. N. K., & Kumar, S. (1987). Logistic regression analysis of sample survey data. *Biometrika, 74*, 1–12.

Roberts, G., Ren, Q., & Rao, J. N. K. (2009). Using marginal mean models for data from longitudinal surveys with a complex design: Some advances in methods. In P. Lynn (Ed.), *Methodology of longitudinal surveys* (pp. 351–366). Wiley.

Robins, J. M., Rotnitzky, A., & Zhao, L. P. (1995). Analysis of semiparametric regression models for repeated outcomes in the presence of missing data. *Journal of the American Statistical Association, 90*, 106–121.

Rosenbaum, P. R., & Rubin, D. B. (1983). The central role of the propensity score in observational studies for causal effects. *Biometrika, 70*(1), 41–55.

Rothman, K. J. (1988). *Causal inference*. Epidemiology Resources.

Royston, P. (2005). Multiple imputation of missing values. *Stata Technical Journal, 5*(4), 527–536.

Rubin, D. B. (1976). Inference and missing data. *Biometrika, 63*(3), 581–592.

Rubin, D. B. (1986). Basic ideas of multiple imputation for nonresponse. *Survey Methodology, 12*(1), 37–47.

Rubin, D. B. (1987). *Multiple imputation for nonresponse in surveys*. John Wiley & Sons.

Rubin, D. B. (1996). Multiple imputation after 18+ years. *Journal of the American Statistical Association, 91*(434), 473–489.

Rubin, D. B., & Schenker, N. (1986). Multiple imputation for interval estimation from simple random samples with ignorable nonresponse. *Journal of the American Statistical Association, 81*, 366–374.

Rueda, M., Ferri-Garcia, R., & Castro, L. (2020). The R package NonProbEst for estimation in non-probability surveys. *The R Journal, 12*(1).

Rust, K. (1985). Variance estimation for complex estimators in sample surveys. *Journal of Official Statistics, 1*, 381–397.

Rust, K., & Hsu, V. (2007). Confidence intervals for statistics for categorical variables from complex samples. In *Proceedings of the survey research methods section* (pp. 3134–3141). American Statistical Association.

Rust, K. F., & Rao, J. N. K. (1996). Variance estimation for complex surveys using replication techniques. *Statistical Methods in Medical Research, 5*, 283–310.

Ryan, B. L., Koval, J., Corbett, B., Thind, A., Campbell, M. K., & Stewart, M. (2015, March 24). Assessing the impact of potentially influential observations in weighted logistic regression. *Information and Technical Bulletin of the Statistics Canada Research Data Centres, 7*(1).

Saavedra, P. J., & Harding, R. L. (2009). Reproducing nonresponse adjustments in replicate weights. In *Proceedings of the survey research methods section* (pp. 3826–3829). American Statistical Association.

Salvatore, C., Biffignandi, S., Sakshaug, J. W., Wiśniowski, A., & Struminskaya, B. (2024). Bayesian integration of probability and nonprobability samples for logistic regression. *Journal of Survey Statistics and Methodology, 12*(2), 458–492.

SAS Institute Inc. (2023a). *SAS/GRAPH® 9.4: Statistical graphics procedures guide*. Author.

SAS Institute Inc. (2023b). *SAS® 9.4, SAS/STAT® 15.3. User guide*. SAS Institute, Inc.

Satterthwaite, F. E. (1946). An approximate distribution of estimates of variance components. *Biometrics Bulletin, 2*(6), 110–114.

Savitsky, T. D., & Toth, D. (2016). Bayesian estimation under informative sampling. *Electronic Journal of Statistics, 10*(1), 1677–1708.

Savitsky, T. D., & Williams, M. R. (2022). Pseudo Bayesian mixed models under informative sampling. *Journal of Official Statistics, 38*(3), 901–928.

Schafer, J. L. (1996). *MIX: Multiple imputation for mixed continuous and categorical data*. Software Library for S-PLUS. Written in S-PLUS and Fortran-77.

Schafer, J. L. (1997). *Analysis of incomplete multivariate data*. Chapman & Hall.

Schafer, J. L. (1999). *NORM: Multiple imputation of incomplete multivariate data under a normal model, version 2*. Software for Windows 95/98/NT.

Schafer, J. L., Ezatti-Rice, T. M., Johnson, W., Khare, M., Little, R. J. A., & Rubin, D. B. (1996). The NHANES III multiple imputation project. In *Proceedings of the survey research methods section* (pp. 696–701). American Statistical Association.

Schafer, J. L., & Yucel, R. M. (2002). Computational strategies for multivariate linear mixed-effects models with missing values. *Journal of Computational and Graphical Statistics, 11*, 437–457.

Scharfstein, D., Rotzinsky, A., & Robins, J. M. (1999). Adjusting for non-ignorable dropout in the analysis of voting intention data. *Journal of the Royal Statistical Society, Series C, 4*, 563–577.

Schlesselman, J. J. (1982). *Case control studies: Design, conduct, analysis*. Oxford University Press.

Schmidt, S. C. E., & Woll, A. (2017). Longitudinal drop-out and weighting against its bias. *BMC Medical Research Methodology, 17*(164), 1–11.

Schoenfeld, D. (1982). Partial residuals for the proportional hazards regression model. *Biometrika, 69*(1), 239–241.

Schouten, B., Peytchev, A., & Wagner, J. (2017). *Adaptive survey design*. CRC Press.

Schroeder, H. M., & West, B. T. (2024). Analyzing potential non-ignorable selection bias in an off-wave mail survey implemented in a long-standing panel study. *Journal of Survey Statistics and Methodology*. DOI: https://doi.org/10.1093/jssam/smae039.

Schulz, K. F., & Grimes, D. A. (2002). Case-control studies: Research in reverse. *The Lancet, 359*(9304), 431–434.

Schumacker, R. E., & Lomax, R. G. (2004). *A beginner's guide to structural equation modeling* (2nd ed.). Lawrence Erlbaum.

Seidenberg, A. B., Moser, R. P., & West, B. T. (2023). Preferred reporting items for complex sample survey analysis (PRICSSA). *Journal of Survey Statistics and Methodology, 11*(4), 743–757.

Shah, B. V., Holt, M. M., & Folsom, R. F. (1977). Inference about regression models from sample survey data. *Bulletin of the International Statistical Institute, 41*(3), 43–57.

Shao, J., & Tu, D. (1995). *The jackknife and bootstrap*. Springer-Verlag.

Shao, J., & Wu, C. F. J. (1989). A general theory for jackknife variance estimation. *Annals of Statistics, 17*, 1176–1197.

Shlomo, N. (2023). Statistical disclosure control and developments in formal privacy: In memoriam to Chris Skinner. *Survey Methodology, 49*(1), 5–25.

Shlomo, N., & Skinner, C. (2022). Measuring risk of re-identification in microdata: State-of-the art and new directions. *Journal of the Royal Statistical Society, Series A: Statistics in Society, 185*(4), 1644–1662.

Si, Y., Pillai, N. S., & Gelman, A. (2015). Bayesian nonparametric weighted sampling inference. *Bayesian Analysis, 10*(3), 605–625.

Si, Y., West, B. T., Veliz, P., Patrick, M. E., Schulenberg, J. E., Kloska, D. D., Terry-McElrath, Y. M., & McCabe, S. E. (2022). An empirical evaluation of alternative approaches to adjusting for attrition when analyzing longitudinal survey data on adolescents' substance use trajectories. *International Journal of Methods in Psychiatric Research, 31*(3), e1916. https://doi.org/10.1002/mpr.1916

Signorino, C. S., & Kirchner, A. (2018). Using LASSO to model interactions and nonlinearities in survey data. *Survey Practice, 11*(1).

Sikkel, D., Hox, J., & de Leeuw, E. (2009). Chapter 9: Using auxiliary data for adjustment in longitudinal research. In P. Lynn (Ed.), *Methodology of longitudinal surveys* (pp. 141–155). Wiley.

Silva, P. L. D. N., & Moura, F. A. D. S. (2022). Fitting multivariate multilevel models under informative sampling. *Journal of the Royal Statistical Society, Series A: Statistics in Society, 185*(4), 1663–1678.

Singer, J. D., & Willett, J. B. (1993). It's about time: Using discrete-time survival analysis to study duration and the timing of events. *Journal of Educational and Behavioral Statistics, 18*, 155–195.

Skinner, C., & de Toledo Vieira, M. (2007). Variance estimation in the analysis of clustered longitudinal survey data. *Survey Methodology, 33*(1), 3–12.

Skinner, C. J., & Holmes, D. J. (2003). Chapter 14: Random effects models for longitudinal survey data. In R. L. Chambers & C. J. Skinner (Eds.), *Analysis of survey data*. John Wiley & Sons.

Skinner, C. J., Holt, D., & Smith, T. M. F. (1989). *Analysis of complex surveys*. John Wiley & Sons.

Skrondal, A., & Rabe-Hesketh, S. (2004). *Generalized latent variable modeling: Multilevel, longitudinal, and structural equation models*. Chapman & Hall/CRC Press.

Smith, T. W. (2011). Refining the total survey error perspective. *International Journal of Public Opinion Research, 23*(4), 464–484.

Spencer, B. D. (2000). An approximate design effect for unequal weighting when measurements may correlate with selection probabilities. *Survey Methodology, 26*(2), 137–138.

Sribney, W. M. (1998). Two-way contingency tables for survey or clustered data. *Stata Technical Bulletin, 45*, 33–49.

Stapleton, L. M. (2008). Variance estimation using replication methods in structural equation modeling with complex sample data. *Structural Equation Modeling: A Multidisciplinary Journal, 15*(2), 183–210.

Stapleton, L. M., & Kang, Y. (2018). Design effects of multilevel estimates from national probability samples. *Sociological Methods & Research, 47*(3), 430–457.

StataCorp. (2023). *Stata 18 multiple-imputation reference manual*. Stata Press.

Stefan, M., & Hidiroglou, M. A. (2020). Local polynomial estimation for a small area mean under informative sampling. *Survey Methodology, 46*(1), 29–50.

Stokes, M. E., Davis, C. S., & Koch, G. G. (2002). *Categorical data analysis using the SAS system (2nd ed.)*. SAS Institute Inc.

Striegel-Moore, R. H., Franko, D. L., Thompson, D., Affenito, S., & May, A. (2008). Exploring the typology of night eating syndrome. *International Journal of Eating Disorders*, *41*(5), 411–418.

Strobl, C., Malley, J., & Tutz, G. (2009). An introduction to recursive partitioning: Rationale, application, and characteristics of classification and regression trees, bagging, and random forests. *Psychological Methods*, *14*(4), 323–348.

Sukatme, P. V. (1954). *Sampling theory of surveys, with applications*. Iowa State College Press.

Sullivan, D., & Andridge, R. (2015). A hot deck imputation procedure for multiply imputing non-ignorable missing data: The proxy pattern-mixture hot deck. *Computational Statistics & Data Analysis*, *82*, 173–185.

Sutradhar, B., & Kovacevic, M. (2000). Analyzing ordinal longitudinal survey data: Generalized estimating equations approach. *Biometrika*, *87*, 837–848.

Sverchkov, M., & Pfeffermann, D. (2018). Small area estimation under informative sampling and not missing at random non-response. *Journal of the Royal Statistical Society, Series A: Statistics in Society*, *181*(4), 981–1008.

Tang, J., Hillygus, D. S., & Reiter, J. P. (2024). Using auxiliary marginal distributions in imputations for nonresponse while accounting for survey weights, with application to estimating voter turnout. *Journal of Survey Statistics and Methodology*, *12*(1), 155–182.

Tanner, M., & Wong, W. (1997). The calculation of posterior distributions by data augmentation. *Journal of the American Statistical Association*, *82*, 528–550.

Therneau, T. M., Grambsch, P. M., & Fleming, T. R. (1990). Martingale-based residuals for survival models. *Biometrika*, *77*(1), 147–160.

Thomas, D. R., & Rao, J. N. K. (1987). Small-sample comparisons of level and power for simple goodness-of-fit statistics under cluster sampling. *Journal of the American Statistical Association*, *82*, 630–636.

Thomas, N., Raghunathan, T. E., Schenker, N., Katzoff, M. J., & Johnson, C. L. (2006). An evaluation of matrix sampling methods using data from the national health and nutrition examination survey. *Survey Methodology*, *32*(2), 217–231.

Thompson, K. J. (2005). An empirical investigation into the effects of replicate reweighting on variance estimates for the annual capital expenditures survey. In *Proceedings of the federal committee on statistical methodology*. Office of Management and Budget.

Thompson, M. E. (2015). Using longitudinal complex survey data. *Annual Review of Statistics and Its Applications*, *2*, 305–320.

Thompson, S. K., & Seber, G. A. F. (1996). *Adaptive sampling*. John Wiley & Sons.

Toth, D. (2020). A permutation test on complex sample data. *Journal of Survey Statistics and Methodology*, *8*(4), 772–791.

Toth, D. (2021). *rpms: Recursive partitioning for modeling survey data* (R package version 0.5.1). CRAN. https://CRAN.R-project.org/package=rpms.

Toth, D., & Eltinge, J. L. (2011). Building consistent regression trees from complex sample data. *Journal of the American Statistical Association*, *106*(496), 1626–1636.

Toth, D., & McConville, K. S. (2024). Design consistent random forest models for data collected from a complex sample. *Survey Methodology*, *50*(2), 185–207.

Tufte, E. R. (1983). *The visual display of information*. Graphics Press.

Tukey, J. W. (1977). *Exploratory data analysis*. Addison-Wesley.

University of Michigan (Computer Support Group). (1982). *OSIRIS VI: Statistical analysis and data management software system*. Survey Research Center, Institute for Social Research.

Valliant, R. (1990). Comparisons of variance estimators in stratified random and systematic sampling. Journal of Official Statistics, *6*(2), 115–131.

Valliant, R. (2004). The effect of multiple weighting steps on variance estimation. *Journal of Official Statistics*, *20*(1), 1–18.

Valliant, R. (2020). Comparing alternatives for estimation from nonprobability samples. *Journal of Survey Statistics and Methodology*, *8*(2), 231–263.

Valliant, R., & Dever, J. A. (2011). Estimating propensity adjustments for volunteer web surveys. *Sociological Methods & Research*, *40*(1), 105–137.

Valliant, R., Dever, J. A., & Kreuter, F. (2018). *Practical tools for designing and weighting survey samples* (2nd ed.). Springer.
Valliant, R., Dorfman, A. H., & Royall, R. M. (2000). *Finite population sampling and inference: A prediction approach*. John Wiley & Sons.
Valliant, R., & Rust, K. F. (2010). Degrees of freedom approximations and rules-of-thumb. *Journal of Official Statistics, 26*(4), 585–602.
Van Buuren, S. (2011). Multiple imputation of multi-level data. In J. J. Hox & J. K. Roberts (Eds.), *The handbook of advanced multilevel analysis*. Routledge.
Van Buuren, S. (2012). *Flexible imputation of missing data*. Chapman & Hall.
Van Buuren, S., Boshuizen, H. C., & Knook, D. L. (1999). Multiple imputation of missing blood pressure covariates in survival analysis. *Statistics in Medicine, 18*(6), 681–694.
Van den Brakel, J. A. (2019). Design-based analysis of experiments embedded in probability samples. In P. Lavrakas, M. Traugott, C. Kennedy, A. Holbrook, E. de Leeuw, & B. West (Eds.), *Experimental methods in survey research: Techniques that combine random sampling with random assignment* (pp. 457–479). Wiley.
Vandendijck, Y., Faes, C., Kirby, R. S., Lawson, A., & Hens, N. (2016). Model-based inference for small area estimation with sampling weights. *Spatial Statistics, 18*, 455–473.
Veiga, A., Smith, P. W. F., & Brown, J. J. (2014). The use of sample weights in multivariate multilevel models with an application to income data collected by using a rotating panel survey. *Journal of the Royal Statistical Society, Series C (Applied Statistics), 63*(1), 65–84.
Verbeke, G., & Molenberghs, G. (2000). *Linear mixed models for longitudinal data*. Springer.
Verret, F., Rao, J. N. K., & Hidiroglou, M. A. (2015). Model-based small area estimation under informative sampling. *Survey Methodology, 41*(2), 333–348.
Vieira, M. D. T., & Skinner, C. J. (2008). Estimating models for panel survey data under complex sampling. *Journal of Official Statistics, 24*, 343–364.
von Hippel, P. T. (2020). How many imputations do you need? A two-stage calculation using a quadratic rule. *Sociological Methods & Research, 49*(3), 699–718.
Wang, J. (2021). The pseudo maximum likelihood estimator for quantiles of survey variables. *Journal of Survey Statistics and Methodology, 9*(1), 185–201.
Wang, Z., & Bellhouse, D. (2009). Semiparametric regression model for complex survey data. *Survey Methodology, 35*(2), 247.
Ward, B. W. (2018). Analytic errors in analysis of public health survey data are avoidable. *Preventing Chronic Disease, 15*(E43). https://doi.org/10.5888/pcd15.170426
West, B. T. (2009). A simulation study of alternative weighting class adjustments for nonresponse when estimating a population mean from complex sample survey data. In *Proceedings of the survey research methods section* (pp. 4920–4933). American Statistical Association.
West, B. T., & Andridge, R. R. (2023). Evaluating pre-election polling estimates using a new measure of non-ignorable selection bias. *Public Opinion Quarterly, 87*(S1), 575–601.
West, B. T., Beer, L., Gremel, W., Weiser, J., Johnson, C., Garg, S., & Skarbinski, J. (2015). Weighted multilevel models: A case study. *American Journal of Public Health, 105*(11), 2214–2215.
West, B. T., Berglund, P., & Heeringa, S. G. (2008). A closer examination of subpopulation analysis of complex-sample survey data. *The Stata Journal, 8*(4), 520–531.
West, B. T., & Blom, A. G. (2017). Explaining interviewer effects: A research synthesis. *Journal of Survey Statistics and Methodology*, 5(2), 175–211.
West, B. T., Little, R. J., Andridge, R. R., Boonstra, P. S., Ware, E. B., Pandit, A., & Alvarado-Leiton, F. (2021). Assessing selection bias in regression coefficients estimated from nonprobability samples with applications to genetics and demographic surveys. *The Annals of Applied Statistics, 15*(3), 1556.
West, B. T., & Little, R. J. A. (2013). Nonresponse adjustment of survey estimates based on auxiliary variables subject to error. *Journal of the Royal Statistical Society, Series C, 62*(2), 213–231.
West, B. T., & McCabe, S. E. (2017). Alternative approaches to assessing nonresponse bias in longitudinal survey estimates: An application to substance-use outcomes among young adults in the United States. *American Journal of Epidemiology, 185*(7), 591–600.

West, B. T., & Sakshaug, J. W. (2018). The need to account for complex sampling features when analyzing establishment survey data: An illustration using the 2013 Business Research and Development Innovation Survey (BRDIS). *Survey Methods: Insights from the Field*. https://doi.org/10.13094/SMIF-2018-00001

West, B. T., Sakshaug, J. W., & Aurelien, G. A. S. (2016). How big of a problem is analytic error in secondary analyses of survey data? *PLoS One*, *11*(6), e0158120.

West, B. T., Sakshaug, J. W., & Kim, Y. (2017). Analytic error as an important component of total survey error: Results from a meta-analysis. In P. Biemer, E. de Leeuw, S. Eckman, B. Edwards, F. Kreuter, L. E. Lyberg, N. C. Tucker, & B. West (Eds.), *Total survey error in practice* (pp. 489–508). John Wiley and Sons.

West, B. T., Si, Y., Hu, Y., McCabe, S. E., & Veliz, P. (2024). The role of weighting adjustment for attrition in longitudinal trajectory modeling: A simulation study. *Communications in Statistics: Simulation and Computation*. https://doi.org/10.1080/03610918.2024.2362923

West, B. T., Welch, K. B., & Galecki, A. T. (2022). *Linear mixed models: A practical guide using statistical software* (3rd ed.). Chapman Hall/CRC Press.

Whittaker, G., & Scott, D. W. (1999). Nonparametric regression for analysis of complex surveys and geographic visualization. *Sankhya: The Indian Journal of Statistics*, *61*(Series B, Part 1), 202–227.

Wieczorek, J., Guerin, C., & McMahon, T. (2022). K-fold cross-validation for complex sample surveys. *Stat*, *11*(1), e454.

Williams, M. R., & Savitsky, T. D. (2020). Bayesian estimation under informative sampling with unattenuated dependence. *Bayesian Analysis*, *15*(1), 57–77.

Williams, M. R., & Savitsky, T. D. (2021). Uncertainty estimation for pseudo-Bayesian inference under complex sampling. *International Statistical Review*, *89*(1), 72–107.

Wiśniowski, A., Sakshaug, J. W., Perez Ruiz, D. A., & Blom, A. G. (2020). Integrating probability and nonprobability samples for survey inference. *Journal of Survey Statistics and Methodology*, *8*(1), 120–147.

Wolter, K. M. (2007). *Introduction to variance estimation* (2nd ed.). Springer-Verlag.

Woodruff, R. S. (1971). A simple method for approximating the variance of a complicated estimate. *Journal of the American Statistical Association*, *66*, 411–414.

Wu, C. (2005). Algorithms and R codes for the pseudo empirical likelihood method in survey sampling. *Survey Methodology*, *31*(2), 239–243.

Wu, C. (2016, March 22–24). Empirical likelihood confidence intervals for finite population proportions. In *Presented at statistics Canada's international symposium on methodological Issues*.

Wu, C., & Rao, J. N. K. (2006). Pseudo-empirical likelihood ratio confidence intervals for complex surveys. *The Canadian Journal of Statistics*, *34*(3), 359–375.

Wun, L.-M., Ezzati-Rice, T. M., Diaz-Tena, N., & Greenblatt, J. (2007). On modeling response propensity for dwelling unit (DU) level non-response adjustment in the Medical Expenditure Panel Survey (MEPS). *Statistics in Medicine*, *26*(8), 1875–1884.

Yamaguchi, K. (1991). *Event history analysis*. Sage Publications.

Yang, L., Nandram, B., & Choi, J. W. (2023). Bayesian predictive inference under nine methods for incorporating survey weights. *International Journal of Statistics and Probability*, *12*(1).

Yang, S., Kim, J. K., & Hwang, Y. (2021). Integration of data from probability surveys and big found data for finite population inference using mass imputation. *Survey Methodology*, *47*(1), 29–58.

Yang, S., Kim, J. K., & Song, R. (2020). Doubly robust inference when combining probability and non-probability samples with high dimensional data. *Journal of the Royal Statistical Society, Series B: Statistical Methodology*, *82*(2), 445–465.

Yates, F. (1949). *Sampling methods for censuses and surveys* (1st ed.). Griffin.

You, Y., & Rao, J. N. K. (2002). A pseudo-empirical best linear unbiased prediction approach to small area estimation using survey weights. *The Canadian Journal of Statistics*, *30*(3), 431–439.

You, Y., & Rao, J. N. K. (2003). Pseudo hierarchical Bayes small area estimation combining unit level models and survey weights. *Journal of Statistical Planning and Inference*, *111*(1–2), 197–208.

Yucel, R. M. (2011). State of the multiple imputation software. *Journal of Statistical Software*, *45*(1), 1–5.

Zajacova, A., Dowd, J. B., & Aiello, A. E. (2009). Socioeconomic and race/ethnic patterns in persistent infection burden among U.S. adults. *Journal of Gerontology A: Biological Sciences and Medical Sciences, 64A*(2), 272–279.

Zhang, G., Christensen, F., & Zheng, W. (2015). Nonparametric regression estimators in complex surveys. *Journal of Statistical Computation and Simulation, 85*(5), 1026–1034.

Zhang, L. (2019). On valid descriptive inference from non-probability sample. *Statistical Theory and Related Fields, 3*(2), 103–113.

Zheng, H., & Little, R. J. A. (2005). Inference for the population total from probability-proportional-to-size samples based on predictions from a penalized spline nonparametric model. *Journal of Official Statistics, 21*(1), 1–20.

Zheng, X., & Yang, J. S. (2016). Using sample weights in item response data analysis under complex sample designs. In L. van der Ark, D. Bolt, W. C. Wang, J. Douglas, & M. Wiberg (Eds.), *Quantitative psychology research*. Springer.

Zhou, H., Elliott, M. R., & Raghunathan, T. E. (2016a). Multiple imputation in two-stage cluster samples using the weighted finite population Bayesian bootstrap. *Journal of Survey Statistics and Methodology, 4*, 139–170.

Zhou, H., Elliott, M. R., & Raghunathan, T. E. (2016b). Synthetic multiple-imputation procedure for multistage complex samples. *Journal of Official Statistics, 32*(1), 231–256.

Zhou, H., Elliott, M. R., & Raghunathan, T. E. (2016c). A two-step semiparametric method to accommodate sampling weights in multiple imputation. *Biometrics, 72*(1), 242–252.

Zimmermann, T., & Münnich, R. T. (2018). Small area estimation with a lognormal mixed model under informative sampling. *Journal of Official Statistics, 34*(2), 523–542.

Index

Note: Page numbers in *italics* indicate a figure and page numbers in **bold** indicate a table on the corresponding page.

D

M